PSA 1992

VOLUME ONE

PSA 1992

PROCEEDINGS OF THE 1992
BIENNIAL MEETING
OF THE
PHILOSOPHY OF SCIENCE
ASSOCIATION

volume one

Contributed Papers

edited by

DAVID HULL,
MICKY FORBES
&
KATHLEEN OKRUHLIK

1992
Philosophy of Science Association
East Lansing, Michigan

Library of Congress Catalog Card Number 72-624169

Cloth Edition: ISBN 917586-33-6
Paper Edition: ISBN 917586-32-8

ISSN: 0270-8647

Manufactured in the United States of America

CONTENTS

Preface ix
PSA 1992 Program xi
Synopsis xxi

Part I. Methodology and Explanation

1. *Explanatory Unification and Scientific Understanding* 3
Eric Barnes, Denison University

2. *On Values in Science: Is the Epistemic/Non-Epistemic Distinction Useful?* 13
Phyllis Rooney, Oakland University

3. *Darcy's Law and Structural Explanation in Hydrology* 23
James R. Hofmann, California State University–Fullerton and Paul A. Hofmann, New Mexico Institute of Mining and Technology

4. *Philosophy of Science and Its Rational Reconstructions: Remarks on the VPI Program for Testing Philosophies of Science* 36
Alan W. Richardson, University of Keele

Part II. Experimentation

5. *Some Complexities of Experimental Evidence* 49
Margaret Morrison, University of Toronto

6. *Experimental Reproducibility and the Experimenters' Regress* 63
Hans Radder, Vrije Universiteit, Amsterdam

7. *How Do You Falsify a Question?: Crucial Tests versus Crucial Demonstrations* 74
Douglas Allchin, University of Minnesota

Part III. Quantum Theory I

8. *Value-definiteness and Contextualism: Cut and Paste with Hilbert Space* 91
Allen Stairs, University of Maryland

9. *The Objectivity and Invariance of Quantum Predictions* 104
Gordon N. Fleming, Pennsylvania State University

10. *Relativity, Quantum Mechanics and EPR* 114
Robert Clifton, Cambridge University
Constantine Pagonis, Cambridge University
Itamar Pitowsky, The Hebrew University of Jerusalem

Part IV. Issues in Methodology

11. *Another Day for an Old Dogma* 131
Robert J. Levy, Wittenberg University

12. *The Importance of Models in Theorizing: A Deflationary Semantic View* 142
Stephen M. Downes, University of Utah

13. *Towards an Expanded Epistemology for Approximations* 154
Jeffry L. Ramsey, Rice University

Part V. Problems in Special Sciences

14. *Sociology and Hacking's Trousers* 167
Warren Schmaus, Illinois Institute of Technology

15. *Neo-classical Economics and Evolutionary Theory: Strange Bedfellows?* 174
Alex Rosenberg, University of California–Riverside

16. *Community Ecology, Scale, and the Instability of the Stability Concept* 184
E.D. McCoy and Kristin Shrader-Frechette, University of South Florida

Part VI. Decision Theory

17. *Kings and Prisoners (and Aces)* 203
Jordan Howard Sobel, University of Toronto

18. *Dutch Strategies for Diachronic Rules: When Believers See the Sure Loss Coming* 217
Brad Armendt, Arizona State University

19. *The Collapse of Collective Defeat: Lessons from the Lottery Paradox* 230
Kevin B. Korb, Indiana University

Part VII. Realism: Causes, Capacities and Mathematics

20. *Cartwright, Capacities, and Probabilities* 239
Gürol Irzik, Boğaziçi University

21. *Objects and Structures in the Formal Sciences* 251
Emily Grosholz, Pennsylvania State University

22. *Adding Potential to a Physical Theory of Causation* 261
Mark Zangari, La Trobe University

Part VIII. Kuhnian Themes: SSR at Thirty

23. *Theory-ladenness of Observations as a Test Case of Kuhn's Approach to Scientific Inquiry* 277
Jaakko Hintikka, Boston University

24. *Theory-ladenness of Perception Arguments* 287
Michael A. Bishop, Iowa State University

25. *The* Structure *Thirty Years Later: Refashioning a Constructivist Metaphysical Program* 300
Sergio Sismondo, Cornell University

Part IX. Topics in the Philosophy of Biology

26. *Additivity and the Units of Selection* 315
Peter Godfrey-Smith, Stanford University

27. *Jacques Monod's Scientific Analysis and Its Reductionistic Interpretation* 329
Spas Spassov, University of Montreal

28. *A Kernel of Truth? On the Reality of the Genetic Program* 335
Lenny Moss, Northwestern University

29. *A Defense of Propensity Interpretations of Fitness* 349
Robert C. Richardson, University of Cincinnati and
Richard M. Burian, Virginia Polytechnic Institute and State University

Part X. Quantum Theory II

30. *Locality, Complex Numbers, and Relativistic Quantum Theory* 365
Simon W. Saunders, Harvard University

31. *Reversibility and the Interpretation of Mixtures in Quantum Mechanics* 381
Osvaldo Pessoa, Jr., Universidade Estadual de Campinas–Brazil

32. *Renormalization and the Effective Field Theory Programme* 393
Don Robinson, University of Toronto

33. *Bell's Inequality, Information Transmission, and Prism Models* 404
Tim Maudlin, Rutgers University

Part XI. Realism, Methodology and Underdetermination

34. *Convergent Realism and Approximate Truth* 421
David B. Resnik, University of Wyoming

35. *Realism and Methodological Change* 435
Jarett Leplin, University of North Carolina–Greensboro

36. *Historical Contingency and Theory Selection in Science* 446
James T. Cushing, University of Notre Dame

Part XII. Issues in the Philosophy of Psychology

37. *What Price Neurophilosophy?* 461
Eric Saidel, University of Wisconsin-Madison

38. *Darwin and Disjunction: Foraging Theory and Univocal Assignments of Content* 469
Lawrence A. Shapiro, University of Pennsylvania

39. *Thought and Syntax* 481
William Seager, University of Toronto

40. *Color Perception and Neural Encoding: Does Metameric Matching Entail a Loss of Information?* 492
Gary Hatfield, University of Pennsylvania

Part XIII. Spacetime

41. *When is a Physical Theory Relativistic?* 507
Roland Sypel and Harvey R. Brown, Oxford University

42. *Space-Time and Isomorphism* 515
Brent Mundy, Syracuse University

43. *The Relativity Principle and the Isotropy of Boosts* 528
Tim Budden, Oxford University

44. *A New Look at Simultaneity* 542
Kent A. Peacock, University of Western Ontario

PREFACE

This volume contains the contributed papers for the 1992 Biennial Meeting of the Philosophy of Science Association held in Chicago, Illinois. It is published before the meeting, with a second volume, containing the symposia papers, to be published afterward. The organization of the papers records that of the program. Northwestern University provides generous support for the preparation of these volumes.

The papers were chosen on the basis of individual merit from over 120 individual submissions. Because of very narrow constraints imposed by the time available, some extremely good papers could not be included in the program. Refereeing and selection were done by the Program Committee, consisting of Kathleen Okruhlik (Chair), Cristina Bicchieri, James Robert Brown, Jeremy Butterfield, Daniel Garber, and Elisabeth Lloyd. Kathleen Okruhlik would like to thank Penelope Lister for secretarial assistance and the Department of Philosophy and the Faculty of Arts of the University of Western Ontario for their financial contributions. Micky Forbes, Assistant Editor of the Proceedings, supervised the editing and processing of the papers to produce uniform, camera ready copy. The PSA Business Office saw the copy through to publication.

We wish to thank the Program Committee for the considerable time and energy they contributed and our authors for their cooperation over a tight publication schedule. We are grateful to Northwestern University for financial support and to its Academic Computing Center for the use of its microcomputing facilities. Special thanks are due Wendy Ward whose expertise and imagination in utilizing these facilities shows up on nearly every page of the volume.

Kathleen Okruhlik
Department of Philosophy
University of Western Ontario
London, Ontario

David L. Hull and
Micky Forbes
Department of Philosophy
Northwestern University

Program

Friday, October 30, morning

Methodology and Explanation
Contributed Papers

Chair:	Jane Duran, University of California–Santa Barbara
Papers:	*Explanatory Unification and Scientific Understanding* Eric Barnes, Denison University
	On Values in Science: Is the Epistemic/Non-Epistemic Distinction Useful? Phyllis Rooney, Oakland University
	Darcy's Law and Structural Explanation in Hydrology James R. Hofmann, California State University–Fullerton and Paul A. Hofmann, New Mexico Institute of Mining and Technology
	Philosophy of Science and Its Rational Reconstructions: Remarks on the VPI Program for Testing Philosophies of Science Alan W. Richardson, University of Keele

Chaos Theory
Symposium

Chair:	Roger Jones, University of Kentucky
Speakers:	Stephen Kellert, Indiana University Mark Stone, University of California–Berkeley Robert Batterman, Ohio State University

Altruism: Evolutionary and Psychological
Symposium

Chair:	Elliott Sober, University of Wisconsin–Madison

Speakers: Daniel Batson, University of Kansas
David Sloan Wilson, State University of New York–Binghamton
Gerald Marwell, University of Wisconsin–Madison

Experimentation
Contributed Papers

Chair: Deborah Mayo, Virginia Polytechnic Institute and State University

Papers: *Some Complexities of Experimental Evidence*
Margaret Morrison, University of Toronto

Experimental Reproducibility and the Experimenters' Regress
Hans Radder, Vrije Universiteit, Amsterdam

How Do You Falsify a Question?: Crucial Tests versus Crucial Demonstrations
Douglas Allchin, University of Minnesota

Friday, October 30, afternoon

Cognitive Ethology
Symposium

Chair: Daniel Gilman, Pennsylvania State University

Speakers: Colin Allen, Texas A & M University and Mark Hauser, University of California–Davis
Dorothy L. Cheney and Robert M. Seyfarth, University of Pennsylvania
Dale Jamieson and Marc Bekoff, University of Colorado
Carolyn A. Ristau, Rockefeller University

Bayesian Philosophy of Science
Symposium

Chair: Paul Horwich, Massachusetts Institute of Technology

Speakers: Henry Kyburg, University of Rochester
Patrick Maher, University of Illinois
Colin Howson, London School of Economics

Quantum Theory I
Contributed Papers

Chair: Jeremy Butterfield, Cambridge University

Papers:

Value-definiteness and Contextualism: Cut and Paste with Hilbert Space
Allen Stairs, University of Maryland

The Objectivity and Invariance of Quantum Predictions
Gordon N. Fleming, Pennsylvania State University

Relativity, Quantum Mechanics and EPR
Robert Clifton and Constantine Pagonis, Cambridge University, Itamar Pitowsky, The Hebrew University of Jerusalem

Issues in Methodology
Contributed Papers

Chair: Bradley Wilson, University of Pittsburgh

Papers:

Another Day for an Old Dogma
Robert J. Levy, Wittenberg University

The Importance of Models in Theorizing: A Deflationary Semantic View
Stephen M. Downes, University of Utah

Towards an Expanded Epistemology for Approximations
Jeffry L. Ramsey, Rice University

Problems in Special Sciences
Contributed Papers

Chair: Margaret Schabas, York University

Papers:

Sociology and Hacking's Trousers
Warren Schmaus, Illinois Institute of Technology

Neo-classical Economics and Evolutionary Theory: Strange Bedfellows?
Alex Rosenberg, University of California–Riverside

Community Ecology, Scale, and the Instability of the Stability Concept
Kristin Shrader-Frechette and E.D. McCoy, University of South Florida

Friday, October 30, late afternoon

Special Session on the Occasion of Karl Popper's 90th Birthday
Invited Lecture

Probability and Deduction
David Miller, University of Warwick

Cosmic Censorship
Symposium

Chair: David Malament, University of Chicago

Speakers: John Earman, University of Pittsburgh
Robert Geroch, University of Chicago
Robert Wald, University of Chicago

Decision Theory
Contributed Papers

Chair: James Fetzer, University of Minnesota–Duluth

Papers:

Kings and Prisoners (and Aces)
Jordan Howard Sobel, University of Toronto

Dutch Strategies for Diachronic Rules: When Believers See the Sure Loss Coming
Brad Armendt, Arizona State University

The Collapse of Collective Defeat: Lessons from the Lottery Paradox
Kevin B. Korb, Indiana University

Mill, Whewell, and the Wave-Particle Debate
Symposium

Chair: Ernan McMullin, Notre Dame University

Speakers: Peter Achinstein, Johns Hopkins University
Jed Z. Buchwald, University of Toronto
Larry Laudan, University of Hawaii–Manoa

PSA Business Meeting

Saturday, October 31, morning

Statistical Asymmetries and the Direction of Causation
Symposium

Chair: Steven Savitt, University of British Columbia

Speakers: Frank Arntzenius, University of Southern California
David Papineau, King's College, London
Huw Price, University of Sydney

Thought Experiments: The Theoretician's Laboratory
Symposium

Chair: John Norton, University of Pittsburgh

Speakers: James R. Brown, University of Toronto
David Gooding, University of Bath
Nancy Nersessian, Princeton University

Commentator: Ian Hacking, University of Toronto

Realism: Causes, Capacities and Mathematics
Contributed Papers

Chair: Mark Wilson, Ohio State University

Papers: *Cartwright, Capacities, and Probabilities*
Gürol Irzik, Boğaziçi University

Objects and Structures in the Formal Sciences
Emily Grosholz, Pennsylvania State University

Adding Potential to a Physical Theory of Causation
Mark Zangari, La Trobe University

Kuhnian Themes: SSR at Thirty
Contributed Papers

Chair: Richard Boyd, Cornell University

Papers: *Theory-ladenness of Observations as a Test Case of Kuhn's Approach to Scientific Inquiry*
Jaakko Hintikka, Boston University

Theory-ladenness of Perception Arguments
Michael A. Bishop, Iowa State University

The Structure *Thirty Years Later: Refashioning a Constructivist Metaphysical Program*
Sergio Sismondo, Cornell University

Topics in the Philosophy of Biology
Contributed Papers

Chair: Marc Ereshefsky, University of Calgary

Papers:

Additivity and the Units of Selection
Peter Godfrey-Smith, Stanford University

Jacques Monod's Scientific Analysis and Its Reductionistic Interpretation
Spas Spassov, University of Montreal

A Kernel of Truth? On the Reality of the Genetic Program
Lenny Moss, Northwestern University

A Defense of Propensity Interpretations of Fitness
Robert C. Richardson, University of Cincinnati and Richard M. Burian, Virginia Polytechnic Institute and State University

Saturday, October 31, afternoon

New Directions in the Philosophy of Science: Issues of Gender and Race
Symposium

Chair: Alison Wylie, University of Western Ontario

Speakers:

Sandra Harding, University of Delaware
Janet Kourany, University of Notre Dame
Helen E. Longino, Rice University
Gonzalo Munévar, Evergreen State College

The Dynamics of Rational Deliberation
Symposium

Chair: Cristina Bicchieri, Carnegie-Mellon University

Speakers: William Harper, University of Western Ontario
Richard Jeffrey, Princeton University
Brian Skyrms, University of California–Irvine

Logical Inconsistency in Scientific Theories
Symposium

Chair: Jamie Tappenden, University of Pittsburgh

Speakers: Bryson Brown, University of Lethbridge
John D. Norton, University of Pittsburgh
Peter Schotch, Dalhousie University

Commentator: Ronald E. Laymon, Ohio State University

Quantum Theory II
Contributed Papers

Chair: Edward MacKinnon, California State University–Hayward

Papers: *Locality, Complex Numbers, and Relativistic Quantum Theory*
Simon W. Saunders, Harvard University

Reversibility and the Interpretation of Mixtures in Quantum Mechanics
Osvaldo Pessoa, Jr., Universidade Estadual de Campinas, Brazil

Renormalization and the Effective Field Theory Programme
Don Robinson, University of Toronto

Bell's Inequality, Information Transmission, and Prism Models
Tim Maudlin, Rutgers University

Realism, Methodology and Underdetermination
Contributed Papers

Chair: Malcolm Forster, University of Wisconsin–Madison

Papers: *Convergent Realism and Approximate Truth*
David B. Resnik, University of Wyoming

Realism and Methodological Change
Jarett Leplin, University of North Carolina–Greensboro

Historical Contingency and Theory Selection in Science
James T. Cushing, University of Notre Dame

Saturday, October 31, late afternoon

Presidential Address

From Vicious Circle to Infinite Regress, and Back Again
Bas van Fraassen, Princeton University

Introduction: Thomas S. Kuhn, MIT

Sunday, November 1, morning

Issues in the Philosophy of Psychology
Contributed Papers

Chair: William Demopoulos, University of Western Ontario

Papers:

What Price Neurophilosophy?
Eric Saidel, University of Wisconsin-Madison

Darwin and Disjunction: Foraging Theory and Univocal Assignments of Content
Lawrence A. Shapiro, University of Pennsylvania

Thought and Syntax
William Seager, University of Toronto

Color Perception and Neural Encoding: Does Metameric Matching Entail a Loss of Information?
Gary Hatfield, University of Pennsylvania

Is Foundational Work in Mathematics Relevant to the Philosophy of Science?
Symposium

Chair: W.W. Tate, University of Chicago

Speakers: John Burgess, Princeton University

Solomon Feferman, Stanford University
Harvey M. Friedman, Ohio State University
Geoffrey Hellman, University of Minnesota

What Has the History of Science to Say to the Philosophy of Science?
Symposium

Chair: Michael Ruse, University of Guelph

Speakers: David Hull, Northwestern University
Rachel Laudan, University of Hawaii–Manoa
Robert J. Richards, University of Chicago
Marga Vicedo, Arizona State University West

Spacetime
Contributed Papers

Chair: Roberto Torretti, Puerto Rico

Papers: *When is a Physical Theory Relativistic?*
Roland Sypel and Harvey R. Brown, Oxford University

Space-Time and Isomorphism
Brent Mundy, Syracuse University

The Relativity Principle and the Isotropy of Boosts
Tim Budden, Oxford University

A New Look at Simultaneity
Kent A. Peacock, University of Western Ontario

Synopsis

The following brief summaries, arranged here alphabetically by author, provide an introduction to each of the papers in this volume.

1. *How Do You Falsify a Question?: Crucial Tests versus Crucial Demonstrations.* **Douglas Allchin.** I highlight a category of experiment—what I am calling 'demonstrations'—that differs in justificatory mode and argumentative role from the more familiar 'crucial tests'. 'Tests' are constructed such that alternative results are equally and symmetrically informative; they help discriminate between alternative solutions within a problem-field, where questions are shared. 'Demonstrations' are notably asymmetrical (for example, "failures" are often not telling), yet they are effective, if not "crucial," in inter-paradigm dispute, to legitimate questions themselves. The Ox-Phos Controversy in bioenergetics serves as an integral case study.

2. *Dutch Strategies for Diachronic Rules: When Believers See the Sure Loss Coming.* **Brad Armendt.** Two criticisms of Dutch strategy arguments are discussed: One says that the arguments fail because agents who know the arguments can use that knowledge to avoid Dutch strategy vulnerability, even though they violate the norm in question. The second consists of cases alleged to be counterexamples to the norms that Dutch strategy arguments defend. The principle of Reflection and its Dutch strategy argument are discussed, but most attention is given to the rule of Conditionalization and to Jeffrey's rule for fallible learning. I argue that the first criticism should be rejected, and that the second presents no counterexamples to the rationality of commitment to the rules.

3. *Explanatory Unification and Scientific Understanding.* **Eric Barnes.** The theory of explanatory unification was first proposed by Friedman (1974) and developed by Kitcher (1981, 1989). The primary motivation for this theory, it seems to me, is the argument that this account of explanation is the only account that correctly describes the genesis of scientific understanding. Despite the apparent plausibility of Friedman's argument to this effect, however, I argue here that the unificationist thesis of understanding is false. The theory of explanatory unification as articulated by Friedman and Kitcher thus emerges as fundamentally misconceived.

4. *Theory-ladenness of Perception Arguments.* **Michael A. Bishop.** The first aim of this paper is to adduce a framework for understanding theory-ladenness of perception arguments. The second aim is to begin to assess an important cluster of theory-ladenness arguments—those that begin with some psychological phenomenon and conclude that scientific controversies are resolved without appeal to theory-neutral observations. Three of the arguments (from expectation effects, ambiguous figures, and inverting lenses) turn out to be either irrelevant to or subversive of theory-ladenness. And even if we grant the premises of the fourth argument (from the penetrability of the visual system), it will support at best a mild version of theory-ladenness.

5. *The Relativity Principle and the Isotropy of Boosts.* **Tim Budden.** A class of theories which satisfy the Relativity Principle has been overlooked. The kinematics for these theories is derived by relaxing the 'boost isotropy' symmetry normally invoked, and the role the dynamical fields play in determining the inertial coordinate systems is emphasised, leading to a criticism of Friedman's (1983) practice of identifying them via the absolute objects of a spacetime theory alone. Some theories complete with 'boost anisotropic' dynamics are given.

6. *Relativity, Quantum Mechanics and EPR.* **Robert Clifton, Constantine Pagonis and Itamar Pitowsky.** The Einstein-Podolsky-Rosen argument for the incompleteness of quantum mechanics involves two assumptions: one about locality and the other about when it is legitimate to infer the existence of an element-of-reality. Using one simple thought experiment, we argue that quantum predictions and the relativity of simultaneity require that both these assumptions fail, whether or not quantum mechanics is complete.

7. *Historical Contingency and Theory Selection in Science.* **James T. Cushing.** I argue that historical contingency, in the sense of the order in which events take place, can be an essential factor in determining which of two equally adequate and fruitful, but observationally indistinguishable, scientific theories is accepted by the scientific community. This type of actual underdetermination poses questions for scientific realism and for rational reconstruction in theory evaluation. To illustrate this, I discuss the complete observational equivalence of two radically different, conceptually incompatible interpretations of quantum mechanics and argue that an entirely plausible reordering of historical factors could reasonably have resulted in the causal program having been chosen over the "Copenhagen" one.

8. *The Importance of Models in Theorizing: A Deflationary Semantic View.* **Stephen M. Downes.** I critically examine the semantic view of theories to reveal the following results. First, models in science are not the same as models in mathematics, as holders of the semantic view claim. Second, when several examples of the semantic approach are examined in detail no common thread is found between them, except their close attention to the details of model building in each particular science. These results lead me to propose a deflationary semantic view, which is simply that model construction is an important component of theorizing in science. This deflationary view is consistent with a naturalized approach to the philosophy of science.

9. *The Objectivity and Invariance of Quantum Predictions.* **Gordon N. Fleming.** A recent argument by Pitowsky (1991), leading to the relativity (as opposed to objectivity) of quantum predictions, is refuted. The refutation proceeds by taking into account the hyperplane dependence of the quantum predictions emerging from the three mutually space-like separated measurements, performed on an entangled state of three spin 1/2 particles, that Pitowsky considers. From this hyperplane dependence one finds that the logical step of conjoining the predictions from distinct measurements is ineffective since those predictions apply either, locally, to sets of points with an intersection that is inaccessible to the particles, or globally, to sets of hyperplanes with an intersection that is empty. We also see how explicit reference to the hyperplane dependence of the predictions gives covariant expression to the invariant content of the predictions, which are thereby shown to be objective.

10. *Additivity and the Units of Selection.* **Peter Godfrey-Smith.** "Additive variance in fitness" is an important concept in the formal apparatus of population genetics. Wimsatt and Lloyd have argued that this concept can also be used to decide the "unit of selection" in an evolutionary process. The paper argues that the proposed criteria of Wimsatt and Lloyd are ambiguous, and several interpretations of their views are presented. It is argued that none of these interpretations provide acceptable criteria for deciding units of selection. The reason is that additive variance in fitness can be both a cause of evolution, but also a byproduct of selection at another level.

11. *Objects and Structures in the Formal Sciences.* **Emily Grosholz.** Mathematics, and mechanics conceived as a formal science, have their own proper subject matters, their own proper unities, which ground the characteristic way of constituting problems and solutions in each domain, the discoveries that expand and integrate domains with each other, and so in particular allow them, in the end, to be connected in a partial way with empirical fact. Criticizing both empiricist and structuralist accounts of mathematics, I argue that only an account of the formal sciences which attributes to them objects as well as structure, proper semantics as well as syntax, can do justice to their intelligibility, heuristic force and explanatory power.

12. *Color Perception and Neural Encoding: Does Metameric Matching Entail a Loss of Information?* **Gary Hatfield.** It seems intuitively obvious that metameric matching of color samples entails a loss of information, for spectrophotometrically diverse materials appear the same. This intuition implicitly relies on a conception of the function of color vision and on a related conception of how color samples should be individuated. It assumes that the function of color vision is to distinguish among spectral energy distributions, and that color samples should be individuated by their physical properties. I challenge these assumptions by articulating a different conception of the function of color vision, according to which color vision serves to partition object surfaces into discrimination classes.

13. *Theory-ladenness of Observations as a Test Case of Kuhn's Approach to Scientific Inquiry.* **Jaakko Hintikka.** Kuhn's contribution should be viewed as posing a number of important problems, not as a full-fledged theory of the structure of science. Kuhn's alleged theory-ladenness of observations is examined as a test case in the light of Hintikka's interrogative model of inquiry. A certain superficial theory-ladenness is built into that model. Moreover, the model provides a deeper analysis of theory-ladenness via the two-levelled character of experimental science. A higher-level and a lower-level inquiry rely on different kinds of initial premises and operate with different kinds of "answers" by nature. The model also throws light on the alleged theory-ladenness of meaning.

14. *Darcy's Law and Structural Explanation in Hydrology.* **James R. Hofmann and Paul A. Hofmann.** Darcy's law is a phenomenological relationship for fluid flow rate that finds one of its principle applications in hydrology. Theoretical hydrologists rely upon a multiplicity of conceptual models to carry out approximate derivations of Darcy's law. These derivations provide structural explanations of the law; they require the application of fundamental principles, such as conservation of momentum, to idealized models of the porous media within which the flow occurs. In practice, recognition of the idealized conditions incorporated into models facilitates the empirical clarification of the domain within which the law remains accurate. Structural explanations also contribute to the physical interpretation of phenomenological parameters.

15. *Cartwright, Capacities, and Probabilities.* **Gürol Irzik.** I argue that Nancy Cartwright's largely methodological arguments for capacities and against Hume's regularity account of causation are only partially successful. They are especially problematic in establishing the primacy of singular causation and the reality of mixed-dual capacities. Therefore, her arguments need to be supported by ontological ones, and I propose the propensity interpretation of causal probabilities as a natural way of doing this.

16. *The Collapse of Collective Defeat: Lessons from the Lottery Paradox.* **Kevin B. Korb.** The Lottery Paradox has been thought to provide a reductio argument against probabilistic accounts of inductive inference. As a result, much work in artificial intelligence has concentrated on qualitative methods of inference, including default logics, which are intended to model some varieties of inductive inference. It has recently been shown that the paradox can be generated within qualitative default logics. However, John Pollock's qualitative system of defeasible inference (named OSCAR), does avoid the Lottery Paradox by incorporating a rule designed specifically for that purpose. I shall argue that Pollock's system instead succumbs to a worse disease: it fails to allow for induction at all (a disease sometimes known as "Conjunctivitis').

17. *Realism and Methodological Change.* **Jarrett Leplin.** Some recent theories in theoretical physics are not subject to epistemic evaluation by empiricist standards of evidential warrant. The advantage of these theories is not pragmatic but explanationist; they fail to yield testable consequences that distinguish them from earlier theories. But this is essentially a technological limitation, rather than a theoretical defect. There is an explanation, itself confirmed by empiricist standards, of the unconfirmability of these theories. This paper considers what epistemic stance is proper in this situation, and explores the prospects for justifiable change to an explanationist methodology capable of warranting theories that transcend the range of our experience.

18. *Another Day for an Old Dogma.* **Robert J. Levy.** I propose a modest Bayesian reductionism as an alternative to Quine's moderate confirmational holism. Employing only first-order predicate logic with identity and elementary probability theory, I present two models of confirmation for individual hypotheses within blocks of theoretical sentences. Testing these models, I consider: (1) the old evidence problem; (2) the raven paradox; (3) a version of the the thesis of underdetermination which says that the evidence never provides adequate epistemic grounds for deciding between rival theories; and (4) the confirmation of modal conditionals.

19. *Bell's Inequality, Information Transmission, and Prism Models.* **Tim Maudlin.** Violations of Bell's Inequality can only be reliably produced if some information about the apparatus setting on one wing is available on the other, requiring superluminal information transmission. In this paper I inquire into the minimum amount of information needed to generate quantum statistics for correlated photons. Reflection on informational constraints clarifies the significance of Fine's Prism models, and allows the construction of several models more powerful than Fine's. These models are more efficient than Fine claims to be possible and work for the full range of possible analyzer settings. It also demonstrates that the division of theories into those that violate parameter independence and those that violate outcome independence sheds no light on the question of superluminal information transmission.

20. *Some Complexities of Experimental Evidence.* **Margaret Morrison.** This paper is intended as an extension to some of the recent discussion in the philosophical literature on the nature of experimental evidence. In particular I examine the role of empirical evidence attained through the use of deductions from phenomena. This approach to theory construction has been widely used throughout the history of science both by Newton and Einstein as well as Clerk Maxwell. I discuss a particular formulation of maxwell's electrodynamics, one he claims was deduced from experimental facts. However, the deduction is problematic in that it is not immediately clear that one of the crucial parameters of the theory, the displacement current, can be given an empirical foundation. In outlining Maxwell's argument and his attempts to arrive on an empirically based account of the electromagnetic field equations I draw attention to

the philosophical implications of the constraints on theory that arise in this particular case of deduction from phenomena.

21. *A Kernel of Truth? On the Reality of the Genetic Program.* **Lenny Moss.** The existence claim of a "genetic program" encoded in the DNA molecule which controls biological processes such as development has been examined. Sources of belief in such an entity are found in the rhetoric of Mendelian genetics, in the informationist speculations of Schrödinger and Delbrück, and in the instrumental efficacy found in the use of certain viral, and molecular genetic techniques. In examining specific research models, it is found that attempts at tracking the source of biological control always leads back to the complex state of a cell as a whole, the organizational structure of which is not itself encoded in the DNA. It is argued that the "genetic program" is not an existing biological entity and its philosophical and heuristic status is challenged.

22. *Space-Time and Isomorphism.* **Brent Mundy.** Earman and Norton argue that manifold realism leads to inequivalence of Leibniz-shifted space-time models, with undesirable consequences such as indeterminism. I respond that intrinsic axiomatization of space-time geometry shows the variant models to be isomorphic with respect to the physically meaningful geometric predicates, and therefore certainly physically equivalent because no theory can characterize its models more closely than this. The contrary philosophical arguments involve confusions about identity and representation of space-time points, fostered by extrinsic coordinate formulations and irrelevant modal metaphysics. I conclude that neither the revived Einstein hole argument nor the original Leibniz indiscernibility argument have any force against manifold realism.

23. *A new Look at Simultaneity.* **Kent A. Peacock.** It is generally believed that an invariant notion of a global present or "Now" cannot be defined in special relativity, because of the relativity of optical simultaneity. I argue that this may be a non sequitur since it is not necessarily the case that the psychological "Now" should be thought of as associated with constant time slices in spacetime. By considering a science fictional version of the Twin Paradox due to Robert A. Heinlein, I argue that it is psychologically plausible to associate the common specious present of several observers in relative motion with certain hypersurfaces of proper time of those observers corrected for acceleration history and relative motion in an obvious way. If this is correct then the relativity of optical simultaneity may be simply irrelevant to the question of the relativity of a globally distinguished "present".

24. *Reversibility and the Interpretation of Mixtures in Quantum Mechanics.* **Osvaldo Pessoa, Jr.** This paper examines the problem of the interpretation of mixtures in quantum mechanics, presenting a survey of the philosophical debate between the ignorance interpretation (IgI) and the instrumentalist approach. By defining specific procedures for preparing and analyzing mixed beams of polarized light, we show that an important argument in defense of the IgI is not valid: differently prepared but equivalent mixtures cannot be distinguished by measuring particle fluctuations. We present an alternative argument, based on an experiment to test whether the process of mixing is reversible or not. The expected result of this thought-experiment favors a weak version of the IgI.

25. *Experimental Reproducibility and the Experimenters' Regress.* **Hans Radder.** In his influential book, "Changing Order", H.M. Collins puts forward the following three claims concerning experimental replication. (i) Replication is rarely practiced by experimentalists; (ii) replication cannot be used as an objective test of scientific knowledge claims, because of the occurrence of the so-called experimenters' regress;

and (iii) stopping this regress at some point depends upon the enculturation in a local community of practitioners, who tacitly learn the relevant skills. In my paper I discuss and assess these claims on the basis of a more comprehensive analysis of experimentation and experimental reproducibility. The main point is that Collins' claims are not, strictly speaking, wrong, but rather too one-sided and therefore inadequate. This point also calls for a reconsideration of the radical (social constructivist) conclusions that Collins has drawn from his studies of scientific experimentation.

26. *Towards an Expanded Epistemology for Approximations.* **Jeffry L. Ramsey.** By stressing the act rather than the relation of approximation, I argue that the magnitude of the error introduced should not be used as the sole criterion for judging the worth of the approximation. Magnitude is a necessary but not sufficient condition for such a judgement. Controllability, the absence of cancelling errors, and the approximation's justification are also important criteria to consider when praising or blaming an approximation. Boltzmann's discussion of the types of approximations used in the kinetic theory of gases at the turn of the century illustrates the use of these criteria.

27. *Convergent Realism and Approximate Truth.* **David B. Resnik.** I examine the role that approximate truth plays in arguments for convergent realism and diagnose some difficulties that face attempts to defend realism by employing this slippery concept. Approximate truth plays two important roles in convergent realism: it functions as a truth surrogate and it helps explain the success of science. I argue that approximate truth cannot perform both of these roles. If it adequately fulfills its role as a truth surrogate, then it cannot explain the success of science. If it adequately explains the success of science, then it cannot function as a truth surrogate.

28. *Philosophy of Science and Its Rational Reconstructions: Remarks on the VPI Program for Testing Philosophies of Science.* **Alan W. Richardson.** In this paper I argue that the program of L. Laudan et al for empirically testing historiographical philosophies of science ("the VPI program") does not succeed in providing a consistent naturalist program in philosophy of science. In particular, the VPI program endorses a nonnaturalist metamethodology that insists on a hypothetico-deductive structure to scientific testing. But hypothetico-deductivism seems to be both inadequate as an account of scientific theory testing in general and fundamentally at odds with most of the historiographic philosophies under test. I sketch an account of testing historiographic philosophies of science more consistent with the views about scientific testing of those philosophies and argue that such a program is neither viciously circular nor necessarily self-refuting.

29. *A Defense of Propensity Interpretations of Fitness.* **Robert C. Richardson and Richard M. Burian.** We offer a systematic examination of propensity interpretations of fitness, which emphasizes the role that fitness plays in evolutionary theory and takes seriously the probabilistic character of evolutionary change. We distinguish questions of the probabilistic character of fitness from the particular interpretations of probability which could be incorporated. The roles of selection and drift in evolutionary models support the view that fitness must be understood within a probabilistic framework, and the specific character of organism/environment interactions supports the conclusion that fitness must be understood as a propensity rather than as a limiting frequency.

30. *Renormalization and the Effective Field Theory Programme.* **Don Robinson.** Since 1980 effective field theories (EFT's) have been the focus of much research by quantum field theorists but their philosophical implications have gone mostly unnoticed. Some authors claim EFT's are approximations to some fundamental theory.

Others claim EFT's are ends in themselves, not approximations to some fundamental theory, and that we can use them to bypass the problem of renormalization. In the present work I argue that the EFT programme can bypass the problem if ontological commitments only come from theoretical predictions. Since the history of QFT suggests some form of entity realism, the EFT programme does not allow us to bypass the problem of renormalization.

31. *On Values in Science: Is the Epistemic/Non-Epistemic Distinction Useful?*. **Phyllis Rooney.** The debate about the rational and the social in science has sometimes been developed in the context of a distinction between epistemic and non-epistemic values. Paying particular attention to two important discussion in the last decade, by Longino and by McMullin, I argue that a fuller understanding of values in science ultimately requires abandoning the distinction itself. This is argued directly in terms of an analysis of the lack of clarity concerning what epistemic values are. I also argue that the philosophical import of much of the feminist work in philosophy of science is restricted by any kind of strict adherence to the distinction.

32. *Neo-classical Economics and Evolutionary Theory: Strange Bedfellows?* **Alex Rosenberg.** Microeconomic theory and the theory of natural selection share salient features. This has encouraged economics to appeal to the character of evolutionary theory in defending the adequacy of microeconomics, despite its evident weaknesses as an explanatory or predictive theory. This paper explores the differences and similarities between these two theories and the phenomena they treat in order to assess the force of the economist's appeal to evolutionary theory as a model for how economic theory should proceed.

33. *What Price Neurophilosophy?* **Eric Saidel.** A premise in the recent eliminativist arguments of Paul and Patricia Churchland is the power of connectionist-type models to solve problems facing cognitive science. I argue that their demonstrations of this power do not challenge folk psychology. Implicit in the Churchlands' arguments is the premise that folk psychology will fail to reduce to neuroscience. In the remainder of the paper I argue that just as the failure of classical genetics to reduce to molecular genetics does not suggest the elimination of classical genetics, so the possible future failure of folk psychology to reduce to neuroscience would not of itself argue for the elimination of folk psychology.

34. *Locality, Complex Numbers, and Relativistic Quantum Theory.* **Simon W. Saunders.** A heuristic comparison is made of relativistic and non-relativistic quantum theory. To this end the Segal approach is described for the non-specialist. The significance of antimatter to the local and microcausal properties of the fields is laid bare. The fundamental difference between relativistic and non-relativistic (complex) fields is traced to the existence of two kinds of complex numbers in the relativistic case. Their relation to covariant and Newton-Wigner locality is formulated.

35. *Sociology and Hacking's Trousers.* **Warren Schmaus.** For Hacking, the word "real", like the sexist expression "wear the trousers", takes its meaning from its negative uses. In this essay, I criticize Hacking's reasons for believing that the objects of study of the social sciences are not real. First I argue that the realism issue in the social sciences concerns not unobservable entities but systems of social classification. I then argue that Hacking's social science nominalism derives from his considering social groups in isolation from the entire social system. I conclude that the objects of study of the social sciences do not relegate them to an inferior status.

36. *Thought and Syntax.* **William Seager.** It has been argued that Psychological Externalism is irrelevant to psychology. The grounds for this are that PE fails to individuate intentional states in accord with causal power, and that psychology is primarily interested in the causal roles of psychological states. It is also claimed that one can individuate psychological states via their syntactic structure in some internal "language of thought". This syntactic structure is an internal feature of psychological states and thus provides a key to their causal powers. I argue that in fact any syntactic structure deserving the name will require an external individuation no less than the semantic features of psychological states.

37. *Darwin and Disjunction: Foraging Theory and Univocal Assignments of Content.* **Lawrence A. Shapiro.** Fodor (1990) argues that the theory of evolution by natural selection will not help to save naturalistic accounts of representation from the disjunction problem. This is because, he claims, the context 'was selected for representing things as F' is transparent to the substitution of predicates coextensive with F. But, I respond, from an evolutionary perspective representational contexts cannot be transparent: only under particular descriptions will a representational state appear as a "solution" to a selection "problem" and so be adaptive. Only when we construe representational states as opaque in this manner are the generalizations of branches of evolutionary theory, like foraging theory, possible.

38. *Community Ecology, Scale, and the Instability of the Stability Concept.* **Kristin Shrader-Frechette and E.D. McCoy.** We examine the evolution of the concept of stability in community ecology, arguing that biologists have moved from an emphasis on biotic communities characterized by static balance, to one of dynamic balance (returning to equilibrium after perturbation), to the current concept of stability as persistence. Using Wimsatt's (1987) analysis of how false models can often lead to better ones, we argue that failed attempts to link complexity with stability have significant heuristic value for community ecologists. Nevertheless, we argue that, (A) because there is no common characteristic that stability terms presuppose, community ecology might be better served by abandoning the concept of stability and by employing instead specific terms such as 'persistence', 'resistance', and 'variability'. (B) The current emphasis (of stability terms) on persistence of species provides little basis for explaining possible mechanisms that might account for persistence.

39. *The* Structure *Thirty Years Later: Refashioning a Constructivist Metaphysical Program.* **Sergio Sismondo.** The Thomas Kuhn of "The Structure of Scientific Revolutions" is often seen as an idealist or Neo-Kantian, as holding a constructivist as opposed to realist position. A close reading of the texts in question, keeping in mind Kuhn's interests as a historian, doesn't support this position, though it uncovers other interesting metaphysical commitments. In particular, Kuhn sees a degree of complexity in the world that entails that there will often be some conventionality in our theories. Some reasons for the readings of Kuhn's as a "constructivist" are explored.

40. *Kings and Prisoners (and Aces).* **Jordan Howard Sobel.** What we make of information we come to have should take into account that we have come to have it, and how we think we have come to have it. I relate this homily to several puzzles. One has to do with three cards one of which is a king, and another with three prisoners one of whom will be released. Then I take up the question, What is it that makes that homily remarkable for these cases, and makes them puzzles? An appendix takes up second-ace problems.

41. *Jacques Monod's Scientific Analysis and Its Reductionistic Interpretation.* **Spas Spassov.** Numerous critiques of Monod's biophilosophical ideas have created a picture of an eminent scientist whose philosophical considerations are founded on the old-fashioned cartesian world-view, and he is still cited as one of the strongest and most radical reductionists in contemporary biology. Yet in his scientific papers, published in the early sixties, Monod defends an idea which suggests a quite different interpretation of the cellular control systems. This idea persists in his later book "Chance and Necessity", contradicting his own mechanistic and reductionistic inferences. In his scientific analysis Monod reveals not only the molecular mechanisms of the teleonomical performances of the cell, but also their biological specificity, which does not allow for their reduction to merely chemical processes, although they are grounded on such processes and can be explained by them. However, in his philosophical interpretation of teleonomy he remains at the level of the chemical basis of these processes and does not take into account their biological specificity. If one abstracts Monod's scientific analysis from his philosophical interpretation, one can find in the former some ideas suggesting a different, non-mechanistic interpretation of biological phenomena.

42. *Value-definiteness and Contextualism: Cut and Paste with Hilbert Space.* **Allen Stairs.** I begin with an appeal to the GHZ/Mermin state to illustrate the allure of contextualism and value-definiteness. I then point out that standard contextualism, with its special status for non-degenerate operators, faces some embarrassing questions. Further, there is an alternative that apparently does not have the same problems. A modest re-pasting of Hilbert space makes the honors almost even between these two varieties. The paper closes with some reflections on the peculiarities of contextualism.

43. *When is a Physical Theory Relativistic?* **Roland Sypel and Harvey R. Brown.** Considerable work within the modern 'space-time theory' approach to relativity physics has been devoted to clarifying the role and meaning of the principle of relativity. Two recent discussions of the principle within this approach, due to Arntzenius (1990) and Friedman (1983), are found to contain difficulties.

44. *Adding Potential to a Physical Theory of Causation.* **Mark Zangari.** Several authors have recently attempted to provide a physicalist analysis of causation by appealing to terms from physics that characterise causal processes. Accounts based on forces, energy/momentum transfer and fundamental interactions have been suggested in the literature. In this paper, I wish to show that the former two are untenable when the effect of enclosed electromagnetic fluxes in quantum theory is considered (i.e. the Aharonov-Bohm effect). Furthermore, I suggest that even in the classical and non-relativistic limits, a theory of fundamental interactions should not be reduced to either a theory of forces or of energy/momentum transfer, but should be understood as a classical account of mutual interactions. Causal links are therefore correctly characterised by generalised potentials. This leads to some speculation regarding the fundamental ontology of interactions and, in particular, the role of the quantum mechanical phase.

Part I

METHODOLOGY AND EXPLANATION

Explanatory Unification and Scientific Understanding[1]

Eric Barnes

Denison University

1. Introduction

The theory of explanatory unification was first proposed by Michael Friedman (1974) and developed in more detail by Philip Kitcher (1981, 1989). The primary motivation for this theory is the case that the theory adequately accounts for the genesis of scientific understanding. Standard models of explanation are, moreover, allegedly constructed on the basis of a misconception of the nature of understanding. My thesis is that the unificationist account of understanding proposed by Friedman and Kitcher is false, and hence, that the theory of explanatory unification is fundamentally misconceived.

2. Standard Models of Explanation and Understanding

How does an explanation of some explanandum render the explanandum as 'understood'? Carl Hempel wrote of the D-N covering law model that it shows that "given the particular circumstances and the laws in question, the occurrence of the phenomenon *was to be expected*; and it is in this sense that the explanation allows us to *understand why* the phenomenon occurred." (Hempel 1965, p.327) But it is obvious that the notion of rational expectability hardly provides us an adequate account of the genesis of understanding. A storm may be expected on the basis of a barometer reading and the relevant law, though this basis for storm expectation gives no glimmer of understanding of why the storm occurred.

Friedman cites the view of P.W. Bridgman (1968, p.37) and William Dray (1964, p.79-80) that explanations produce understanding by reducing unfamiliar phenomena to familiar phenomena. But the history of science is replete with examples of explanations which, quite to the contrary, reduced the familiar to the unfamiliar (e.g., the observable reflection of light to the behaviour of electromagnetic waves).[2] Friedman also considers the 'intellectual fashion' view which he attributes to Stephen Toulmin (1963) and N.R. Hanson (1963). According to this view, explanations produce understanding of some phenomenon by relating it to some 'ideal of natural order', a phenomenon that is, during the period of history in question, considered by scientists to be not in need of further explanation. Friedman acknowledges the historical support underwriting this view, but he notes that the proffered account of understanding is unpleasantly subjec-

---

PSA 1992, Volume 1, pp. 3-12

tive, in that this notion of 'understanding' will allow the criteria for a phenomenon's being understood to vary from one historical epoch to another.

Friedman points out that all these accounts of the genesis of understanding assume that an explanation renders its explanandum as understood because of some special epistemological property possessed by the explanans. The explanans is 'familiar', 'already understood', or 'an ideal of natural order.' Let us gloss these descriptions and refer to the special property attributed to the explanans by these accounts as simply 'intelligibility.' On the proposed account, the intelligible quality of the explanans flows down the argument, as it were, across the line dividing explanans and explanandum, and into the explanandum, displacing the latter's mysterious quality. Let us deem this account of the genesis of understanding the 'intelligibility transfer' account of understanding.

Friedman considers as an example the reduction of the Boyle-Charles law of gases to the kinetic theory of gases. (1974, p.14) The latter theory, it is usually maintained, does in fact explain why this law holds true; it seems to do so by describing the underlying molecular mechanisms that result in the observed relations between pressure, volume, temperature, and mass in gas behaviour described by the Boyle-Charles law. Now Friedman wants to pose the following question: on the assumption that we do not understand why the kinetic theory of gases itself holds true, why do we take the reduction of the Boyle-Charles law to the kinetic theory to render the behaviour of gases as more fully understood—as less mysterious—at all? Such a reduction does, of course, supply more information about gases than we possessed prior to the reduction, but the mere gain in information seems not to amount to a gain in understanding for Friedman. The reduction of the Boyle-Charles law to the kinetic theory of gases "merely replaces one brute fact with another." (Friedman 1974, p.14) It is at precisely this moment in Friedman's analysis, I suggest, that we must attend most carefully to what he would try to teach us about understanding.

To argue that we understand why the Boyle-Charles law holds true because we can show it to be derivable from the kinetic theory of gases is about as reasonable, Friedman seems to be suggesting, as to argue that we understand why the earth does not fall down in space by claiming that it rests on the back of a large elephant. The latter claim would not increase our understanding of why the earth does not fall at all (even if it were true!), because the question why the elephant itself does not fall is no less pressing than the original question about the earth's failure to fall. Similarly, the former claim ignores the fact that the question why the kinetic theory holds true is no less pressing than the original question about the Boyle-Charles law.

In general, to derive a fact from any premises whatsoever, Friedman is suggesting, is never *by itself* sufficient to render the fact as less mysterious than it was before, because the premises are inevitably ultimately mysterious. If the premises are themselves derivable from some deeper theory T, this theory will either be mysterious (in the sense that its truth is unexplained) or derivable from some still deeper theory T' which... Ultimately, all explanations will ideally trace back to the ultimate laws of physics. But Friedman argues that these ultimate laws will themselves be essentially and forever mysterious, simply because they are brute facts which admit of no explanation.[3] But in this event, the very bedrock on which all explanations will ultimately be constructed (at some vastly distant point in the future of science, when all such fundamental processes are 'understood' and all phenomena reduced to them) is itself mysterious; hence, from Friedman's viewpoint, it is entirely unclear how understanding of any explanandum is (or will ever be) possible at all. The punchline is that if there is to be such a thing as scientific understanding, it cannot be produced merely by the derivation of an explanandum from some explanans. I shall refer to this conclusion as Friedman's Thesis of Derivational Skepticism.[4]

3. Explanatory Unification and Global Understanding

The only way out of the quandary in which we find ourselves, Friedman concludes, is to radically re-formulate our operative notion of understanding. Friedman does not want to deny, e.g., that the kinetic theory of gases did in fact substantially increase our understanding of gas behaviour. But the reason for this gain in understanding was not merely that it entailed the Boyle-Charles law, for had it done only this, as we have seen, Friedman believes it would not have increased our understanding at all. He reasons as follows:

> But (the entailing of the Boyle-Charles law) is not all the kinetic theory does—it also permits us to derive other phenomena involving the behavior of gases, such as the fact that they obey Graham's law of diffusion and (within certain limits) that they have the specific heat capacities that they do have, from the same laws of mechanics. The kinetic theory offers a significant unification in what we have to accept. Where we once had three independent brute facts...we now have only one—that molecules obey the laws of mechanics. (1974, p.14-5)

This, Friedman argues, is the reason the kinetic theory increased our understanding: this theory effected a reduction in the number of independent phenomena that scientists were aware of but unable to derive from more fundamental phenomena. This unificationist account of understanding suggests a new account of scientific explanation. He proceeds to write that "...this is the essence of scientific explanation—science increases our understanding of the world by reducing the total number of independent phenomena that we have to accept as ultimate or given. A world with fewer independent phenomena is, other things equal, more comprehensible than one with more." (1974, p.15) Friedman's Unificationist Thesis of Understanding, as I shall henceforth refer to it, asserts that our understanding of the world is directly proportional to the degree of unification in our theoretical picture of the world.

Friedman thus identifies the explanatory force of a scientific theory with the propensity of the theory to effect a reduction in the number of independent laws scientists are forced to accept in their overall picture of the world. To say that scientists increase their understanding of nature by the acceptance of some theory T is thus to say, for Friedman, simply that the number of ultimately independent laws (laws not derivable from other statements in the totality of belief) is reduced as a result of adding T to the totality. This account has the virtues of, first, accommodating the insight that understanding is not generated by any sort of intelligibility transfer and, second, providing an answer to the question of how it helps our understanding to derive one puzzling phenomenon from another (and thus provides an account not susceptible to the Thesis of Derivational Skepticism).[5] The answer is that "we don't simply replace one phenomenon with another. We replace one phenomenon with a more comprehensive phenomenon, and thereby effect a reduction in the total number of accepted phenomena. We thus genuinely increase our understanding of the world." (1974, p.19) The conception of understanding advocated here Friedman describes as 'global', for understanding thus construed constitutes a cognitive relation that holds between a subject and the totality of belief. Friedman's analysis of the standard models prompts his rejection of the notion of 'local understanding', the thesis that understanding is a relation that holds between a subject and a particular fact.

Philip Kitcher has developed the theory of explanatory unification in considerable detail. While there are many important differences between Kitcher's and Friedman's accounts, the spirit of both approaches is sufficiently similar for all present purposes that I shall not discuss these differences. Kitcher seems to me by and large to maintain that the fundamental motivations for this theory are those identified by Friedman. It is to these fundamental motivations that I now wish to turn.

4. A Question

I would like to point out what seems to me to be a striking omission from Friedman's and Kitcher's writings on explanation. Neither philosopher, it seems to me, provides any clearly identifiable argument for the Unificationist Thesis of Understanding, i.e., for the claim that our understanding of the world is correctly measured in terms of the degree of theoretical unification inherent in our world picture. Friedman's (1974) is a case in point. After mounting his intriguing and at least *prima facie* compelling critique of the standard models of explanation vis-à-vis their inability to account for the genesis of understanding, Friedman seems to me to simply assert that the reason the kinetic theory of gases increases our understanding is its unifying power. The quotations given above suggest a strong confidence on Friedman's part that this assertion will strike his readers as utterly plausible, but, as noted, I find no explicit argument for this thesis in this article. Neither have I been able to locate a point in Kitcher's writings where he attempts to provide the missing argument. For example, when Kitcher writes that "As Friedman has pointed out, we can easily connect the notion of unification with that of understanding" (1981, p.509) we are left with the impression that Kitcher senses that the nature of the connection will have simply been obvious to any casual reader of Friedman's (1974) article. Kitcher's (1989) seems to reveal the same attitude.

Why do Friedman and Kitcher take the connection between unification and understanding to be such a straightforward one? What is the putatively obvious reason for regarding the Unificationist Thesis of Understanding to be even *prima facie* plausible? The answer might indeed seem to be straightforward. We understand the world better when we countenance fewer independent phenomena because the presence of fewer independent phenomena in our picture of nature amounts to the presence of fewer fundamental mysteries. The fewer fundamental mysteries we are forced to accept, the less mystified we are: the better we understand the world. QED. I do suspect that this sort of answer is the one we are supposed to accept. But is it not rather obvious that there is a serious tension between this answer and the Thesis of Derivational Skepticism articulated at some length by Friedman and apparently accepted by Kitcher? For the very point of this thesis was that the standard models are based on the erroneous view that we come to eliminate the mysterious quality of explananda, and thus understand them, by deriving them from some suitable explanans or other. But the apparent plausibility of the Unificationist Thesis of Understanding seems to depend entirely on the intuition that an underived, independent phenomenon represents a mystery not represented by a derived, 'dependent' phenomenon. If this latter intuition is not held by Friedman and Kitcher, why in heaven's name does it matter vis-à-vis our global understanding how many underived, independent phenomena we countenance? It seems to me, in other words, that the Unificationist Thesis of Understanding ought to be considered as plausible only if we antecedently assume some account of understanding according to which individual explananda are rendered understood, in some sense, by means of their standing under a suitable explanans. But the falsehood of any such account was the punchline of the Thesis of Derivational Skepticism, which is in turn supposed to be one of the most crucial motivations for adopting the Unificationist Thesis of Understanding. If we are to reject, with Friedman and Kitcher, the local nature of explanation and understanding on the grounds of the Thesis of Derivational Skepticism, the question why we should take understanding to be proportional to the unified status of our global theoretical picture emerges as a very pressing one. In what follows, I consider on what other grounds one might defend the Unificationist Thesis of Understanding.

(a) Simplicity. The degree of unification intrinsic to the global scientific picture is a measure of the 'simplicity', in one straightforward sense, of this picture, and the claim that simplicity is a desideratum of scientific theories is certainly a popular one. Does this point assist the Unificationist Thesis of Understanding? At least two difficulties face such a proposal. The first is that it thrusts upon us the notoriously difficult challenge of saying in objective terms what the simplicity of a theory consists of. The second, still more worrisome, difficulty is to explain why one would think that one understands the world better the simpler one's theory of the world happens to be. I am inclined to say that the simplicity of a theory determines how easy or difficult it is for a subject to grasp the theory, and thus to understand the world as described by the theory. But while 'simplicity' thus construed is certainly to be desired, it is no measure of the degree of understanding conferred by the theory on the subject at all, but simply a measure of the degree of cognitive effort required of the subject in order to understand the world. But surely it ought not to matter at all, vis á vis an objective notion of understanding, how much sheer mental force a subject may have to exert to grasp the content of a theory. What matters for us is not how hard or easy it is to acquire understanding, but what understanding itself consists of.

(b) Historical Evidence. I suspect that among the more impressive considerations on behalf of the Unificationist Thesis of Understanding for some is the fact that, in the history of science, there are many examples of theories which have been both powerful unifiers and indisputably rich sources of scientific understanding. In motivating the Unificationist Thesis Kitcher writes that "the acceptance of some major programs of scientific research—such as, the Newtonian program of eighteenth century physics and chemistry, and the Darwinian program of nineteenth century biology—depended on recognizing promises for unifying, and thereby explaining, the phenomena." (1981, p.509) Friedman's analysis of the kinetic theory of gases also provides this sort of historical evidence for the unificationist account of understanding; it is an example of a theory that undeniably both unified and rendered understood the behaviour of gases. But of course, merely from the fact that a theory possesses two properties (such as 'having great unifying power' and 'generating great understanding'), it does not follow that the one property accounts for the presence of the other, and once more I find little in Friedman's or Kitcher's analysis that specifically describes the mechanism of the unification/understanding connection in these historical examples.

I shall confess that, given the *prima facie* inconsistency between Derivational Skepticism and the most obvious argument for the Unificationist Thesis of Understanding, the considerations of simplicity and the historical evidence are the only possible arguments on behalf of the Unificationist Thesis of Understanding I have been able to imagine. Having set down why I find these arguments invalid, I am left with the task of explaining what I take understanding to be.

5. Local Understanding Revisited

I am rather startled at Friedman's failure to consider in his analysis of standard models of explanation what I would have thought to be the most obvious and plausible candidate for the mechanism by which a particular fact is understood: a particular fact (event or law) is understood when its causal basis is known (hereafter, the Causal

Thesis of Understanding). The causal basis of a particular fact will consist of either the relevant causal ancestry which accounts for the fact (in the case of, say, a particular event) or the underlying causal mechanism which accounts for the fact (as in the case of a scientific law like the Boyle-Charles law).[6] Perhaps the present obviousness of this suggestion, and Friedman's failure to acknowledge it, can be explained by the fact that causal models of explanation were not so prevalent in 1974 as they are today. But the views of Wesley Salmon (1984), Paul Humphreys (1989), and others on the essentially causal nature of explanation are so well known today as to hardly require citation. If we were to agree with such authors that to explain an empirical fact F, at least in some cases, is to describe its causal basis, ought there to be any concern or controversy over the derivative claim that we understand F just in case we apprehend a correct explanation (in this causal sense) of F? Such an analysis of understanding, would, of course, be entirely local.

What might Friedman say about the ability of causal theories of explanation to generate an account of understanding? I conjecture that Friedman would press at least two questions upon this proposal: (1) why do we understand an event in virtue of knowing its causal basis? and (2) how do causal theories escape the critique based on Derivational Skepticism (e.g. the elephant analogy)? Both questions deserve careful consideration.

The intuition that we understand an event in virtue of knowing its causal basis is no less sound that the intuition that we explain the fact by citing this causal basis. For almost any empirical fact F, the causal basis for F contains the answer to the question 'Why is F the case?' We understand why the Boyle-Charles law holds true of gas behaviour precisely because the kinetic theory of gases describes the underlying causal mechanism of molecular interaction of matter in a gas phase that accounts for this law. We understand why a storm occurred precisely to the extent that we apprehend the prior meteorological state of affairs that induced the storm. To seek to understand almost any empirical F is just to seek the knowledge of its causal basis.

How does such a causal account of understanding escape Friedman's defense of Derivational Skepticism? The first point to make is one with which Friedman will agree: explanations are often successful despite the fact that their explanans are not understood, familiar, or otherwise intelligible. It is a mistake to insist that the explanans of an adequate explanation be antecedently intelligible. But the unanswered question at this point is how the understanding of the explanandum is to be gleaned if not by way of a transfer of explanans intelligibility to the explanandum—unless we are to go the way of the unificationist.

Let us recall how Friedman puts his defense of Derivational Skepticism. The defense depends in no small part on his claim that, e.g., the reduction of the Boyle-Charles law to the kinetic theory of gases would not have increased our understanding of gas behaviour at all if this were all the kinetic theory did. For in such an event, the reduction would merely have "replaced one brute fact with another," on the assumption that the kinetic theory itself is regarded as a brute fact, i.e., on the assumption that it is not understood why this theory is true. It is at this point, apparently, that the scales are supposed to fall from our eyes and the inevitable failure of any local account is supposedly rendered obvious. Now Friedman could, if he wanted to, simply apply this same point against the Causal Thesis of Understanding, and argue that a cause C cannot make its effect E understood if C is itself merely yet another 'brute fact.' If it is not understood why C is true, then why regard ourselves as any better off with respect to understanding after the causal explanation than before?

It is my view that the skeptical intuition Friedman tries to inspire in the guise of Derivational Skepticism is simply wrong. I claim that it most certainly would have increased our understanding of gas behaviour to reduce the Boyle-Charles law to the kinetic theory of gases, even if this were the only empirical phenomenon the kinetic theory entailed. For such a reduction would have described the causal basis for the Boyle-Charles law; we would thus have come to understand why the Boyle-Charles law is true. I find it unwise to deny that our understanding of gas behaviour would have consequently increased, no matter whether the kinetic theory could itself be explained, and no matter what other purposes the kinetic theory might also have served.

If this reply is to satisfy, it must explain why the elephant story is not analogous to a causal explanation of some explanandum. Why, e.g., is the kinetic theory-based explanation of the Boyle-Charles law better, more explanatory, or less question begging than the elephant-based explanation of why the earth does not fall? The question is not a trivial one, and it surely could be answered in more than one way. I prefer to put the matter as follows: we must first heed the recommendation of van Fraasen (1980), Hattiangadi (1978), and others that we look carefully at the context in which why-questions are posed before deciding what a proper reply will require. I would argue that the question "Why does the earth not fall?" likely arises out of what Hattiangadi regards to be the essence of a scientific problem: an inconsistency in the total fabric of belief. The inconsistency here, it might seem, is between the following two statements: (1) it is the nature of all objects to fall and (2) there are some objects (including the earth and everything it supports) that do not fall. The elephant-based explanation does no work toward the production of understanding of the topic of the question precisely because, it seems to me, it does no work toward eliminating or even reducing the inconsistency arising out of the conjunction of (1) and (2). For the elephant works to support the earth only because it also fails to fall; adding the claim that the earth rests on the back of the elephant to the totality of belief leaves (1) and (2) fully intact and no less inconsistent, hence our intuition that the elephant-based explanation leaves us right where it found us, with one and the same inconsistency. Now the question "Why is the Boyle-Charles law true?", I would think, may or may not arise out of any perceived inconsistency between the truth of that law and other statements accepted by the physicists of the day. If it does so arise, then I would argue that the kinetic theory of gases, if it is to supply the desired understanding of the Boyle-Charles law, must as a necessary condition eliminate or at least substantially reduce this inconsistency in the belief fabric. But, on the other hand, it seems somewhat more probable that the question "Why is the Boyle-Charles law true?" does not arise out of any appearance of inconsistency between it and other accepted statements but merely out of a generic curiosity about the causal basis of this interesting law. In this case the kinetic theory works to fill the missing gap in our causal knowledge and thus answer the why-question. Now, of course, it is true that at this point we are left with the additional question "Why is the kinetic theory true?", which on the reading I advocate amounts to "What is the causal basis of the kinetic theory?" But the crucial point is that this question is by no means the same question we began with; it is an entirely separate question. In short, the reduction of the Boyle-Charles law to the kinetic theory represents a gain in understanding because, in this episode, we begin with two gaps in our causal knowledge (we know neither the causal basis for the Boyle-Charles law nor that of the kinetic theory) but we end up with only one gap (that pertaining to the kinetic theory). This contrasts sharply with the elephant-based explanation of the earth's failure to fall, which on the present analysis begins and ends with one and the same inconsistency.

6. Conclusion

I have attempted to argue that the fundamental motivation for the theory of explanatory unification is the Unificationist Thesis of Understanding, and that this latter account is misconceived. For it ultimately makes sense to claim that scientists understand the world better the fewer independent phenomena they must countenance only if it is already agreed that an independent phenomenon represents a mystery not represented by a 'dependent' (i.e. suitably derived) phenomenon, but this agreement is impossible in the face of Derivational Skepticism; the latter thesis, however, cannot be renounced without embracing a local conception of understanding incompatible (according to Friedman and Kitcher) with the unificationist approach. I have proposed the renunciation of the Thesis of Derivational Skepticism and have opted to describe this local account in causal terms.

The Causal Thesis of Understanding is hardly an innovative one—it is far less innovative than the Unificationist Thesis, for example. But despite the Causal Thesis's mundane appearance, I would maintain that it, unlike the Unificationist Thesis, has the advantage of being a true account, and it is for many the very essence of common sense. For example, the causal account of understanding has no trouble explaining why theories with a lot of unifying power have been such great sources of understanding in the history of science. For theories that unify a group of phenomena also typically do so by describing the causal bases for the various phenomena in the group (as Newtonian mechanics both unifies and describes the causal bases for gas behaviour, the movements of the planets, the motion of falling objects, etc.). The more greatly unifying such a theory is, then, naturally the more understanding it provides. But what is doing the work of manufacturing understanding is the multiple descriptions of the causal bases of the various phenomena offered by the theory, not simply its unifying power *per se*.

In adopting a causal account of the genesis of understanding I will of course alienate a variety of philosophers. Kitcher, for example, would claim for the advantages of his theory of explanatory unification that it, like Hempel's D-N model, presupposes no antecedently available notion of causation; in fact, Kitcher argues for an anti-realist view of causation according to which our causal beliefs are derived from the explanatory stories we are taught by our culture—his proposal is motivated by his concern that one cannot, post-Hume, combine causal realism with the belief in the possibility of causal knowledge. Now if it should turn out to be possible that an adequate theory of explanation could be pieced together without requiring a realist notion of causation this would certainly be a good thing. But my point here is that the theory of explanation proposed is itself entirely unmotivated qua account of explanation; to cite its ancillary advantages thus strikes me as a non-sequitur.

Let it be clear that I do not regard the Causal Thesis of Understanding to be a universally applicable one; its irrelevance to fields like mathematics, e.g., is entirely obvious. I do not make the claim that all empirical explanations are causal; but it does seem to me that this account fits a great many cases of the genesis of understanding. Nor do I regard the work of Friedman and Kitcher on this subject as at all insignificant, for these philosophers have offered strikingly insightful analyses of a metascientific concept whose historical and philosophical significance is undeniable: that of 'unification' itself. Friedman and Kitcher have, in my judgement, discovered the philosophical problem of describing the nature of 'unification' and have made major inroads toward an adequate description. I claim here merely that it is a mistake to think that this investigation will illuminate the nature of scientific explanation.

Notes

[1]I am grateful to the National Endowment for the Humanities for a summer seminar stipend which supported research culminating in this paper. I am indebted to Paul Humphreys both for comments on this paper and for teaching an excellent seminar.

[2] A similar objection applies to Michael Scriven's claim (1970, p.202) that scientific theories produce understanding of phenomena that are not antecedently understood by reducing them to those that are antecedently understood.

[3] Friedman (1974, p.11).

[4]By the use of the term 'derivational' I do not mean to limit the applicability of Friedman's skeptical thesis to models of explanation, like the D-N model, which emphasize a purely logical relation between explanans and explanandum. Any relation between explanans and explanandum (be it causal, probabilistic, pragmatic, etc.) in which the explanans compels acceptance of the explanandum (and thus 'entails' the latter in some sense) will be susceptible to the same point: where the explanans is not antecedently understood, it cannot be deployed to render the explanandum as understood simply by insuring that the latter will be true.

[5]There is another (quite impressive) consideration that would have recommended Friedman's account if the account were to have proved otherwise adequate, namely that it provides a solution to the old Hempelian puzzle (described in what Salmon (1989 pp.9-10) calls the 'notorious footnote 33' of the Hempel and Oppenheim (1948) paper) of why the conjunction of two laws does not count as an adequate explanation of either of the conjuncts. "K & B" does not explain "B", for Friedman, precisely because adding the former to the totality of belief (which is assumed to antecedently include "K" and "B") does not reduce the number of independent laws in this totality. (1974,p.17)

[6]Some may object to my claim that the derivation of a general law from more fundamental theoretical principles amounts to a causal explanation of the general law (cf., e.g., Hempel (1965, p.352)). But it seems to me quite straightforward to claim that the cause of, e.g., the truth of the Boyle-Charles law is that the events at the level of gas microstructure are such as are described by the kinetic theory. In any case, if the label 'causal explanation' offends here, I would be willing to modify my account accordingly and supply a different label—all I require is that the microstructure description offered by the kinetic theory does explain the Boyle-Charles law—regardless of what else the kinetic theory might also explain (more below).

References

Bridgman, P.W. (1968), *The Logic of Modern Physics*. New York: MacMillan.

Dray, W. (1964), *Laws and Explanation in History*. New York: Oxford.

Friedman, M. (1974), "Explanation and Scientific Understanding", *Journal of Philosophy* 71: 5-19.

Hanson, N.R. (1963), *The Concept of the Positron*. New York: Cambridge.

Hattiangadi, J.N. (1978), "The Structure of Problems, Part I", *Philosophy of Social Science* 8: 345-365.

Hempel, C.G. and Oppeneheim, P. (1948), "Studies in the Logic of Explanation," *Philosophy of Science* 15: 135-75.

Hempel, C.G. (1965), *Aspects of Scientific Explanation.* New York: Free Press.

Humphreys, P. (1989), *The Chances of Explanation.* Princeton: Princeton University Press.

Kitcher, P. (1976), "Explanation, Conjunction and Unification", *Journal of Philosophy* 73: 207-12.

Kitcher, P. (1981), "Explanatory Unification", *Philosophy of Science* 33: 337-59.

______. (1989), "Explanatory Unification and the Causal Structure of the World", in *Scientific Explanation, Minnesota Studies in the Philosophy of Science,* P. Kitcher and W. Salmon (eds.). *Volume XIII*, Minneapolis: University of Minnesota Press, pp. 410-505.

Salmon, W. (1984), *Scientific Explanation and the Causal Structure of the World,* Princeton: Princeton University Press.

_______. (1989), *Four Decades of Scientific Explanation.* Minneapolis: University of Minnesota Press.

Scriven, M. (1970), "Explanations, Predictions, and Laws", in *Minnesota Studies in the Philosophy of Science*, H. Feigl and G. Maxwell (eds.). Volume III. Minneapolis: University of Minnesota Press.

Toulmin, S. (1963), *Foresight and Understanding.* New York: Harper & Row.

van Fraasen, B. (1980), *The Scientific Image.* Oxford: Clarendon Press.

On Values in Science: Is the Epistemic/Non-Epistemic Distinction Useful?

Phyllis Rooney

Oakland University

1. Introduction

The debate about values *in* science in the last decade or so reflects an important shift in what many acknowledge as the "post-positivist" era in philosophy of science. It reflects the erosion of the fact-value distinction in at least one of its more simplistic forms (facts belong in science, values outside), and it marks the path to a more enhanced understanding of the roles of both facts and values in scientific inquiry. This discussion has, above all, contributed to what we might call a revaluation of value: from the point of view of the epistemologist or philosopher of science values are neither uniform nor uniformly "bad". Values are acknowledged in their variety of form and function, and some are seen to play a necessary role in the rational and cognitive development of scientific knowledge: these have been called alternatively *epistemic* values (McMullin 1983), *cognitive* values (Laudan 1984), *constitutive* values (Longino 1990).

However, as these terms suggest, a relatively firm distinction is still endorsed between epistemic and non-epistemic values (or constitutive and contextual values—Longino), with social, cultural, and personal values grouped together on the non-epistemic side. A central focus in the recent debate is whether, and if so how, non-epistemic values influence the "internal" development of science. I will argue that the development of an understanding of science in social context is impeded by this distinction as it is often deployed, that a more effective analysis must include a more explicit examination of the usefulness of the distinction itself. My approach is twofold. My more direct criticism includes an analysis of the lack of clarity or agreement about how the set of epistemic values is constituted, even among those who make substantial use of the distinction. A more indirect approach is motivated by an effort to understand how the feminist work in philosophy of science might best be developed within an account of the role of values in science. Here it is generally understood that a central concern is with the role of "androcentric values" in the development of science. While this focus has been effective in places, it should not be seen to circumscribe the extent of the feminist contribution to the debate. I will argue that much of the feminist work also encourages a more direct reevaluation of the epistemic/non-epistemic distinction itself.

PSA 1992, Volume 1, pp. 13-22

2. Epistemic Values: Fuzzy Sets of Fuzzy Criteria

Epistemic values are those that are usually taken as constitutive of the knowledge- and truth-seeking goals of the enterprise of science. Accuracy and consistency, for example, might be taken as the most obvious and least contestable of these values. However, when attempts are made to make these more precise (and especially epistemic values like fruitfulness and simplicity), various problems arise, not least among which is the issue about what constitutes scientific truth and knowledge in the first place.

One is immediately alerted to a problem here when one notices that among those who grant epistemological or philosophical weight to such a set of epistemic criteria there isn't anything resembling convergent agreement on what exactly is in that set. Thomas Kuhn lists his "standard criteria for evaluating the adequacy of a theory" as: accuracy, consistency, scope, simplicity, and fruitfulness (1977, p.322). Ernan McMullin "reworks [Kuhn's] list just a little" and lists his criteria in order: predictive accuracy, internal coherence, external consistency, unifying power, fertility, and then lists simplicity as "one other more problematic candidate" (1983, p.15-16). Helen Longino lists the "governing values and constraints [that] are generated from an understanding of what counts as a good explanation" as: truth, accuracy, simplicity, predictability, and breadth (1990, p.4). One notices immediately that the lists are not exactly identical. Kuhn's "scope" is presumably something like Longino's "breadth" and perhaps something like McMullin's "unifying power" though their subsequent discussions do not shed much light on the matter. It is interesting to note that Longino counts "truth" as a constitutive value whereas McMullin does not, and elsewhere he states that he calls his characteristic values "epistemic" because they are presumed to promote the truth-like character of science, "its character as the most secure knowledge available to us of the world we seek to understand" (p.18). In other words, McMullin takes truth to be something like a second-order value ("a sort of horizon-concept or ideal") determining in part what is to count as a first-order epistemic value. This slight difference touches upon another significant debate about whether to take truth as a primitive or derivative notion in developing our understanding of science. What this suggests is that we cannot naively assume that in positing the seemingly straightforward epistemic/non-epistemic distinction we automatically sidestep some of these difficult problems that already have a long history.

The fact that there is no clear consensus about what is included among the epistemic or constitutive values does not overly concern many of those who make the distinction. Kuhn readily admits that he selects five characteristics "not because they are exhaustive, but because they are individually important and, collectively, sufficiently varied to indicate what is at stake" (p. 321). (What exactly is it that is at stake?) Though McMullin makes a point about reworking Kuhn's list "just a little" he doesn't explain why that was necessary. Perhaps he thought that his terms were more descriptive of the values or characteristics in question, yet even then the lists do not match up. He does admit that the decision about whether a value is epistemic or non-epistemic in a particular context can sometimes be a difficult one to make, though he thereby implies that the decision can in theory, in time, be made. Longino prefaces her list with a "for example", indicating presumably that she is not presenting it as a comprehensive or exhaustive list.

One of the main reasons for the lack of strictness here is that these theorists use the distinction in large part to offset or demarcate the non-epistemic values, the personal, social and cultural values that have garnered special interest in debates about the social dimensions of science. A central question is whether or to what extent these latter values can be seen to operate in science. If it can be shown that such values play some role in the cognitive development of science, in the development of scientific rationality itself, then we need to reconceptualize the objectivity and rationality of science in a way that reflects this and thus differs from traditional articulations in terms of "value-freedom". I argue in this paper that an analysis of the operation of the social within science warrants closer examination of the "epistemic" values themselves, and that this examination is to proceed in a way that undermines the usefulness of the epistemic/non-epistemic distinction itself. The fact that there is no consensus about what exactly the epistemic values are surely provides our first clue here. We haven't seen anything resembling a clear demarcation of epistemic values because there is none to be had.

For a start, it is not at all clear that epistemic or cognitive values exert the kinds of epistemic or consensus-forming force among scientists that philosophers sometimes project. An argument by Larry Laudan has a bearing on this point. He argues that there need be no agreement about basic cognitive values for scientists to agree about methodological matters and a broad range of factual claims, that "axiological differences can coexist with factual-level and methodological agreement" (1984, p.45).[1] On the other hand, McMullin stresses that the skills of (epistemic) evaluation are typically learned through practice in the scientist's training and experience and this has a lot to do with consensus-formation regarding proper methodology around data acquisition and evaluation. One must be careful then with the kinds of claims one makes using a methodology/axiology distinction. It is with this emphasis on the communal-practice acquisition of skills in the development of scientists' *value-judgments* that McMullin's explicit use of the term "value" (in citing epistemic criteria like accuracy and simplicity) becomes apparent. He notes particularly that the operation or "application" of these criteria is much more like a value-judgment than the application of an algorithm. (In general, as we know, they cannot function as algorithms—in only very limited contexts might there be an algorithmic test for consistency, for example. This is, of course, one of the many things we have learned from positivism.) Value-judgment enters here in two ways, evaluation and valuing. It is possible for different scientists to *evaluate* the fertility of a particular theory differently, for example, and different scientists might weigh or *value* these different epistemic criteria differently. We can now cite many examples from social and historical studies of science where a specific theory scored well in terms of predictive accuracy, for instance, yet a competing one scored better in terms of simplicity and fruitfulness, and where their respective proponents differed in how they weighed or valued the relevant epistemic criteria. Though, as will soon become apparent, we would generally do well not to lay the differences to rest there.

The two "openings" here that encourage naming these criteria "values" leave a lot of room for the operation of what are termed "non-epistemic values", and this works in a way that surely fundamentally challenges the distinction itself. Given that epistemic values do not—more exactly cannot—function as algorithms (even among those who might be said to agree in valuing a particular criterion highly) there is no set procedure that dictates when and how such a value is applied. The crucial factor here is not that a particular criterion operates but *when* and *how* it does and, I argue, non-epistemic factors are encoded within the *when* and the *how*, not, as is regularly supposed, in some cognitive gap that is left over after accepted epistemic criteria and rules of inference are applied to the evidentiary base. That there is such a cognitive "gap" somewhere is generally taken to be established by the underdetermination thesis. Where and how this gap is filled in is what is centrally at issue here, and it has non-trivial implications for the way in which we articulate the rationality and objectivity of science and for the role that the epistemic/non-epistemic distinction is seen to play in those articulations.

This point can be illustrated with the well-known example concerning the disagreement between Bohr and Einstein about the acceptability of quantum theory. McMullin notes that the predictive successes of the theory counted much more heavily with Bohr than with Einstein who was in turn concerned that the theory did not rate well in coherence and in consistency with the rest of physics. They thus had, in McMullin's estimation, very different views as to what constituted a "good" theory. He adds, however, that disagreement in "substantive metaphysical beliefs" about the nature of the world "also played a part" (1983, p.17). But surely this is the crucial link that needs to be filled out. How did differences in metaphysical belief inform the way each understood the simplicity or coherence of the theory or even the significance of predictive success? Simplicity is a complex multi-faceted characteristic and all sorts of "non-epistemic" factors can determine how a particular theory is perceived as simple for a particular scientist in a particular context. It is surely not simply a matter of their both apprehending the same simplicity or coherence in the theory but valuing it differently... and adding that *in addition* there were other non-epistemic factors to take into account.

Metaphysical beliefs can, McMullin claims, function as "non-standard epistemic values" in some contexts. This is perhaps the most interesting of McMullin's axiological groupings and one that substantially undermines his epistemic/non-epistemic division. These non-standard epistemic values are typically embedded in a philosophical or theological worldview: he concedes, for example, that theology functioned for Newton as an epistemic factor, as "a set of reasons that Newton thought were truth-bearing" (pp.19-20). Elsewhere he argues that in order to understand Descartes' science it is necessary to grasp the importance for him of his epistemological principles concerning "clear and distinct ideas". The Bohr-Einstein disagreement was rooted in a "deep metaphysical divergence" about whether the universe displayed a fundamental coherence and order (1984, pp.130-31). The sense of "metaphysics" deployed here is of course broader than that relating to the fundamental nature of physical matter—it is a term that acquires wider philosophical and theological import. However, McMullin stresses the *epistemic* role of these factors and seeks to distance himself from the sociologists of science at this point, at least insofar as he sees them as simply providing a socio-psychological analysis of such factors. The debate here in many cases doesn't revolve around when and where these "non-standard" factors operate, but around the different epistemological / sociological / psychological theories one constructs around their operation. We do well to remember these "theories" are just that, and are thus themselves the site of contestation concerning explanatory power. One consequence of this debate is the realization that the epistemological-sociological border is continually being redefined (though McMullin might want to say "refined").

With this set of "non-standard epistemic values" hasn't McMullin undermined the force of his epistemic/non-epistemic distinction? Here are values that have a specific cultural-historical location yet are providing clear cognitive value in the form of metaphysical or epistemological background principles. He thinks that in time, however, the progressive march toward conformity between theory and world will "sift out" any non-epistemic factors that may have been instrumental in original theory-formation. He thus invokes something of a context of discovery/ context of justification distinction in supposing that as science progresses, in the "slow process of shaping our thought to the world" in the continued application of epistemic value-judgment, the non-epistemic (which may have served a necessary epistemic function at one time) will gradually sift out from the epistemic. Yet, if some social and cultural factors get grafted into epistemic value-judgment—for example, get normalized into the language of science itself or get embedded into the structures and models therein—how can the continued application of epistemic value-judgment be expected to yield such a sifting? Given McMullin's earlier admission that to the extent that scientific observation is theory-dependent it is also indirectly value-impregnated (p. 14), it is not clear that one can readily appeal to some neutral value-free court of future observation and experiment that will in time provide the "right" distillation. As he well admits, the "first" epistemic value of predictive accuracy itself involves value-judgment. This is not to deny that all sorts of things are continually being sifted in and out of science as it progresses in time; it is simply to say that it is not at all clear that we can tell the full story of this "progress" (not even the full "rational" story) in terms of this progression of the "epistemic," the "non-standard epistemic," and the "non-epistemic."

Longino also sets out to understand the rational and the social in science by examining the role of "constitutive" and "contextual" values in the development of science.[2] She motivates a central thesis in her book with a rhetorical question: "The traditional interpretation of the value freedom of modern natural science amounts to a claim that its constitutive and contextual features are clearly distinct from and independent of one another. Can this distinction, as commonly conceived, be maintained?" (1990, p.4). Longino's answer is clearly "no" but it is not always clear what exactly this "no" entails. It means at least claim (1): it is wrong to suppose that science is value-free in that we can accommodate its cognitive development and import solely within an account of constitutive values, while granting that contextual values are of interest solely within an historical or sociological account of scientific development. Claim (1) essentially vitiates the constitutive/contextual distinction *only insofar* as it is readily aligned with the rational/social (or the now maligned internal/external) distinction. However, for the purposes of my argument in this paper I am especially concerned with whether or to what extent Longino is suggesting a stronger claim (2): a better or fuller account of the cognitive development of science requires abandoning the constitutive/contextual distinction altogether (though it may be useful for preliminary discussion). There are places where she comes close to (2). At one point she discusses how various ways of conceptualizing the objects of (scientific) knowledge can "help to show how contextual values are transformed into constitutive values" (p. 100). However, the contextual/constitutive distinction remains central in her book and with specific examples she is careful to note when she takes a particular value to be constitutive or contextual. Yet, I would argue that some of those same examples taken one step further entail (2), and thus there is some tension in the book around where exactly she stands between (1) and (2).

Longino is not simply interested in science as theoretical *product* and, in particular, in what happens during times of revolutionary change, of grand theory contestation (as in the oft-cited Ptolemy-Copernicus or Bohr-Einstein examples). Some of the epistemic issues have been skewed, she thinks, by the overemphasis on these kinds of examples. She argues that by drawing attention to the practices of normal science, to science as a process, we can gain valuable insight into the influence of contextual values on these practices, on the description of data, the determination of questions as worthwhile or not, and on the specific and global assumptions that motivate background assumptions facilitating inference from evidence to (otherwise underdetermined) hypothesis. This stress on science as *process*, (like McMullin's stress on the role of community practice and skills in the formation of epistemic value-judgment), is surely something that is very valuable in this discussion. Yet, this stress also shifts the focus away from an understanding of epistemic values as criteria that are somehow applied from the outside, as epistemic measurements used to evaluate competing theories during times of theory contestation. The shift to science in process parallels the shift in our appreciation of epistemic values that I am arguing in this paper—the shift to an appreciation of the development of epistemic factors *within* the context of theory development, as encoded within the many complex elements that go into the making of a sophisticated scientific theory. With their adherence to their respective distinctions, it is this extra shift that both Longino and McMullin don't seem to want to make, and this leaves something missing in the analysis of their examples they draw upon.

Longino pays significant attention to the linear-hormonal model as that is used in the explanation of physical and cognitive human behavior, and specifically as it used to explain purported sex differences. She compares this model with an alternative explanatory model, the selectionist or social-cognitive model. She compares them first with respect to their constitutive-methodological differences. She argues that neither theoretical perspective can muster constitutively based arguments sufficient to exclude the other, thus suggesting that it is contextual interests and values that motivate adherence to one or the other theoretical framework. In the following chapter (ch. 8) Longino outlines some of the political and social interests served by the adoption of one model over the other, and, in particular, the implications for reinforcing gender dimorphism that the linear-hormonal model supports. One is thus left with the impression (which Longino may ultimately not want to leave us with) that while both theories can be maintained on some constitutive-evidentiary grounds, theory choice comes down to one's anticipation or dislike of the social-contextual implications of the two theories. What seems to be missing here is an analysis of the way in which these background interests or values get worked into the theories right from the start, get embedded into the forms of language, data description, and question formation as these develop along with the development of the theory in question... and ultimately get embedded in the ways in which constitutive values like simplicity and fruitfulness get understood within the context of the particular theory.[3]

In comparing the constitutive bases of the two models Longino argues that the linear-hormonal model scores high in theoretical unification and simplicity, both because it supports unification across mammalian species, and also because on that model certain aspects of social behavior can be treated on a continuum with other physiological effects of prenatal gonadal hormonal exposure. But surely part of the constitutive force of "simplicity" in this model is due to the fact that in it gender dimorphism is motivating in part the very understanding of biological determinism itself, rather than the other way around. Increasingly, feminist work in this area is suggesting that we simply cannot suppose that a clear theoretical understanding of biological determinism was developed and then theorists "happened to notice" that this was somewhat different for women and men! What is suggested instead is that gender dimorphism became constitutive of the understanding of biological functioning, and thus in effect became constitutive also of the "constitutive" value of the simplicity or fruitfulness of the linear-hormonal model itself insofar as it was constructed to "explain" biological determinism. Yet, an understanding of the impact or value of gender dimorphism clearly has contextual social and cultural underpinnings in a society that supports gender order and hierarchy, and political power issues loom on the horizon when it becomes apparent that those who got to develop the theory represented overwhelmingly only one of the dimorphic forms in question. A fuller understanding of the complex cognitive development in this and other such examples requires careful historical scrutiny clearly, and as I indicate in this example, ultimately involves adopting position (2) above and not simply (1). In discussing and comparing the constitutive and contextual features of these models separately (even while she argues that both are *necessary* for a full understanding of the cognitive adoption of either theory) Longino clearly implies that they work somewhat separately in the overall picture.

In examining these two important accounts of the role of values in science I have drawn attention to the significance of the distinction between epistemic-constitutive and nonepistemic-contextual values in the two accounts, yet I have also shown that some of the most interesting arguments that emerge from those same accounts go a long way toward undermining that same distinction. McMullin does this in large part with his introduction of non-standard epistemic values, and an exploration of the workings of Longino's contextual values requires an understanding of the way in which these become constitutive of particular scientific research programs. Thus, in a sense, these and other theorists who present us with these sets of epistemic values are correct in presenting them as fuzzy sets of fuzzy criteria, but perhaps not for the reasons they think. Values like simplicity and fruitfulness, even accuracy and coherence, are multi-faceted. We might think of simplicity as a pragmatic or aesthetic quality, simplicity with respect to language, or with respect to cognitive or perceptive models used, simplicity with respect to ease in understanding...where all of these may overlap in interesting ways. As we have seen, the "application" of a "constitutive" criterion in a scientific context requires a complex background of languages, practices, and skills within which all sorts of constitutive-contextual features are already encoded.

3. Feminism, Science, and Values

I was also drawn into a closer examination of this issue through an exploration of feminist work in philosophy of science. Here the role of "patriarchal" or "androcentric" values is taken as a central concern. The project is then seen to involve locating these prime examples of contextual social and cultural values, and showing how they operate in the scientific contexts in question. However, this approach does not encompass the full impact or possibility of the feminist work. In addition, this is often the way that those unfamiliar with the full range of feminist work in this area seek to project it. This can have the effect of marginalizing it as at most a "subarea" in social studies of science, thereby constricting its philosophical import as these theorists construe such import.

This is not to say, however, that the ongoing work of locating the workings of androcentric values is not a very important and necessary part of the feminist project, and significant insights about the development of the biological and anthropological sciences have been achieved with this focus (Hubbard 1982, Bleier 1984, Fausto-Sterling 1985). Numerous examples are given where androcentrism operates in the description of behavior, and especially in the way automatic valuings are assigned to purported gender differences. Ruth Hubbard provides examples of the operation of androcentric and Victorian values in Darwin's evolutionary theory, in the predictable roles accorded "active" males and relatively "passive" females. However, as I argued in connection with the linear-hormonal model earlier, the discovery of separable articulable androcentric values does not exhaust the full impact of this work. Anne Fausto-Sterling details assumptions about male-female differences in the different biological theories of "higher cognitive functioning" in this past century (pp.13-60). She doesn't simply locate specific points where androcentric bias sneaked in. She details in effect how these assumptions became constitutive of the scientific articulation of higher cognitive functioning itself (in the X-linked hypothesis, for example), something which also presents a potentially serious problem for naturalistic circumscriptions of rationality. In evolutionary theory, on the other hand, attention has been given to the importance of variability as a mechanism of survival: a feminist analysis must also include an examination of the way in which the assumption that the male was the "more variable" sex helped circumscribe the theoretical significance and placement of the concept itself within the overall theoretical framework. Donna Haraway's work in primatology (1989) is not adequately described in terms of the articulation of androcentric and ethnocentric values. She provides insights into the inextricable link between cultural valuations and the development of the rhetoric and discourse of primatology studies, insights that can be exploited in our analysis of the rhetorical dimensions of other areas of scientific inquiry. Similarly, Evelyn Fox Keller's work on gender and science details insights into the ways in which culturally-infused language has had an impact both in the development of such fields as molecular biology and quantum mechanics, and in the theoretical interpretations of the "objectivity" of the knowledge that emerged in such fields (1985, pt.3; 1990). Arguments to the effect that all of these influences did not significantly affect the "internal" cognitive development of the sciences in question become especially tricky when we note that the clarity of the demarcation of the "internal" is itself called into question with feminist work on the historical cultural gendering of such divisions as objective/subjective, rational/social, and science/nonscience (Rouse 1991). In other words, it is not at all evident how one can readily make such arguments without thereby making assumptions that are essentially question-begging.

There are various reasons why these more far-reaching implications of the feminist work are not generally acknowledged. First, while some feminists might describe their work simply in terms of locating and uprooting androcentric values, this is much more likely to come from non-feminists describing that work. Michael Ruse's discussion provides a good example here (1984). He frequently refers to the values that concern or interest "the feminists". Working from McMullin's epistemic/non-epistemic framework he takes the values that are "of particular interest to our feminist writers" to be non-epistemic and wants to allow room for the values that feminists *do* endorse by questioning whether non-epistemic values are to be gradually sifted out as McMullin wants. Ruse oversimplifies the feminist work here, and to a certain extent misrepresents McMullin's thesis also. In particular, he doesn't include a discussion of McMullin's non-standard epistemic values which may well be the most interesting ones from the point of view of incorporating all of the feminist work, especially when the insights that emerge in that work (like the non-standard epistemic values) go a long way toward undermining the force of the epistemic/non-epistemic distinction itself.

While, contra Ruse, feminist work in science cannot simply be identified with sets of values that "the feminists" uniformly reject or endorse, that work does shed some light on the workings of various values in science. It shows, first of all, that the influence of "androcentric" and other cultural values in science is much more complex than we might initially suppose. Though these values have what McMullin would probably agree are "non-epistemic" origins, it is not at all clear that the automatic prognosis for them over time pointed to their eventually being sifted out rather than grafted into the various sciences, informing the constitutive formations of those sciences. Why, we might ask, have some of the fairly obvious examples of gender and race bias become visible only relatively recently, despite philosophers' and scientists' long-held proclamations about the value-freedom of science? In addition, discussions about androcentrism regularly meet with resistance or hostility from many of those within science, so one cannot say that science as we have known it to date necessarily welcomes calls to examine and sift out certain kinds of non-epistemic factors, calls that according to its stated truth-seeking, value-free mission it surely ought to welcome. The gradual separation of the non-epistemic from the epistemic does not seem to be as *constitutively* guaranteed as McMullin would like. It sometimes seems to require specific "external" historical political shifts, such as the admission of women into science, or the development of second-wave feminism. These clearly did not come from constitutive "truth-seeking" impulses internal to the institutions and practices of science itself.

Such a truth-promoting impulse is at the heart of McMullin's epistemic/non-epistemic distinction, and my argument here provides additional reasons for us to question that distinction. In addition to the more direct problems outlined earlier in the lack of clarity of the distinction, I have also maintained that the distinction is largely undermined by insights that have resulted from the development of feminist work in science. With continuing development of this work we start to get new glimpses of the ways in which values we *now* recognize as "non-epistemic" became encoded into constitutive features of specific theories—in effect became part of what constituted the truth or simplicity of those theories—and thus could remain for so long impervious to the "progress" of science as that was facilitated by those same constitutive features.

The way forward then is not to seek to regroup the epistemic/non-epistemic division but to adopt different attitudes toward it. First, it may still be quite fruitful to use the epistemic/non-epistemic as something like a continuum scale in given scientific contexts. It may well be important to identify specific factors (in a given context) as non-epistemic with respect to the agreed-upon truth-seeking practices of a given scientific community at a particular time. One might even add that that is precisely what constitutes particular research programs as "scientific". Second, now that the epistemic/non-epistemic "divide" emerges as a continuum site of active renegotiation in the continual reconstitution of specific scientific inquiries, we are surely invited to become active humanistic philosophers of science again. We are no longer passive observers charting out the philosophical landscape of a divide antecedently or transcendentally mapped out by something called Truth, or the Progress of Science. Once we understand how cultural and social values can in time, with systematic theoretical cognitive rearticulation, become encoded into constitutive features of the rationality and objectivity of particular scientific endeavors, into features that are genuinely epistemically compelling for given scientific communities, we are invited to gain greater insight into how that occurs, and we are invited to develop that insight within the context of specific philosophical concerns and questions. We get to show how the undermining of an epistemic/non-epistemic divide need not involve dismissing heretofore "epistemic" criteria as hopelessly reductive or relative. We do not have to deny the importance of a relatively stable epistemic background for a given scientific community over an extended time. We get to ask questions about how particular ways of valuing inform particular ways of knowing, of belief formation, of rational compulsion, without thereby dismissing the significance of our various ways of knowing, or without suggesting that epistemology disappears into value theory. We are then in a better position to more effectively address questions about whether our various ways of knowing in science are reflecting the kinds of values we seek to promote into a viable future, and to understand how to bring about change when we do not like what we see emerging in our future. Above all, we need to continually monitor how our forms of scientific inquiry are answering to the overarching interconnecting values of environmental integrity and universal human dignity. For it is these surely that rest at the heart of the impulse toward science.

Notes

[1]It should be noted here, however, that in this context Laudan seems to take "cognitive values" to be much the same thing as "cognitive goals" or "aims" and this detracts somewhat from the relevance of his argument to the issue at hand. Kuhn, Longino, and McMullin have somewhat similar conceptions of epistemic "value" in these discussions, and would distinguish these values from what might be termed the broader aims or goals of science such as knowledge, truth, or good explanation, however contested these latter might be. In this context at least these latter are more the focus of Laudan's discussion.

[2]One cannot assume here that McMullin's epistemic/non-epistemic distinction is exactly the same as Longino's constitutive/contextual one, especially since neither addresses the other's work in this context, and also since they proceed to do somewhat different things with their respective distinctions. (A distinction is as good as the use to which it is put, one might argue!) My argument in this paper does not depend on their being identical, and in fact suggests that they cannot since neither distinction can be precisely circumscribed.

[3]In places Longino seems to indicate that this is precisely the kind of long-term analysis that her thesis invites. In this book she might be seen to be mapping out the philosophical groundwork of an understanding of science and values within which that project is to be carried out. However, what I am concerned about is whether the way in which she lays that out—specifically with respect to the constitutive/contextual distinction—ultimately restricts the full development of that project

References

Bleier, R. (1984), *Science and Gender: A Critique of Biology and its Theories on Women*. New York: Pergamon Press.

Fausto-Sterling, A. (1985), *Myths of Gender: Biological Theories About Women and Men*. New York: Basic Books.

Haraway, D. (1989), *Primate Visions: Gender, Race, and Nature in the World of Modern Science*. New York: Routledge.

Hubbard, R. (1982), "Have Only Men Evolved?", in *Discovering Reality*, S. Harding and M.B. Hintikka (eds.). Dordrecht: Reidel, pp. 45-69.

Kuhn, T. (1977), "Objectivity, Value Judgment, and Theory Choice", in his, *The Essential Tension*. Chicago: University of Chicago Press, pp. 320-339.

Keller, E.F. (1985), *Reflections on Gender and Science*. New Haven: Yale University Press.

_______. (1990), "From Secrets of Life to Secrets of Death", in *Body/Politics: Women and the Discourses of Science*, M. Jacobus, E. F. Keller, and S. Shuttleworth (eds.). New York: Routledge

Laudan, L. (1984), *Science and Values: The Aims of Science and Their Role in Scientific Debate*. Berkeley: University of California Press.

Longino, H.E. (1990), *Science as Social Knowledge: Values and Objectivity in Scientific Inquiry*. Princeton: Princeton University Press.

McMullin, E. (1983), "Values in Science", in PSA 1982, Volume 2, P.D. Asquith and T. Nickles (eds.). East Lansing: Philosophy of Science Association, pp. 3-28.

________. (1984), "The Rational and the Social in the History of Science", in *Scientific Rationality: The Sociological Turn*, J.R. Brown (ed.). Dordrecht: Reidel, pp. 127-163.

Rouse, J. (1991), "The Politics of Postmodern Philosophy of Science", *Philosophy of Science* 58: 4.

Ruse, M. (1984), "Biological Science and Feminist Values", in *PSA 1984*, Volume 2, P.D. Asquith and P. Kitcher (eds.). East Lansing: Philosophy of Science Association, pp. 525-542.

Darcy's Law and Structural Explanation in Hydrology

James R. Hofmann
California State University, Fullerton

Paul A. Hofmann
New Mexico Institute of Mining and Technology

1. Introduction

According to a recent argument, models play two essential roles in the argumentative structure of solid state physics and chemistry (Hofmann 1990). On the one hand, models are the culmination of phenomenological description. That is, models are idealized representations of the molecular structures thought to be causally responsible for the processes experimentally monitored and measured. Secondly, theoretical physicists and chemists require that models ultimately be cast in a mathematical form appropriate for the application of the Schroedinger equation. In this respect models become the means through which the Schroedinger equation gives a theoretical unity to what would otherwise be a disparate set of empirical phenomenological laws and descriptions with limited scope. That is, it is an important theoretical goal to show that experimentally generated phenomenological laws can be approximately derived through an application of the Schroedinger equation to a necessarily idealized and simplified mathematical description of the relevant system. The two functions of models are not incompatible, but they do reflect two distinct theoretical orientations toward the interpretation of data.

The present paper extends these themes into the domain of hydrology. We consider a specific phenomenological relationship known as Darcy's law. Nancy Cartwright's discussion of phenomenological laws and fundamental laws provides some useful analytic vocabulary and initial insight, particularly when amplified by Bogen and Woodward's distinction between data and phenomena (Bogen and Woodward 1988, Woodward 1989, and Cartwright 1983). We thus begin with a summary of these topics and emphasize the potentially ambiguous nature of "explanation" due to the multiple functions of models. After a brief account of the origins and subsequent history of Darcy's law, we turn to a detailed analysis of the status of the law for both theoretical and applied hydrology. Two principal conclusions result. First, "structural explanation", in the sense specified by Ernan McMullin, is an accurate description of the derivations of Darcy's law carried out by theoretical hydrologists (McMullin 1978). Secondly, however, hydrologists themselves often refer to the derivation of phenomenological laws as a particularly extended exercise in the *description* of relevant phenomena rather than as an explanation of them. As a result, McMullin's realist in-

PSA 1992, Volume 1, pp. 23-35

terpretation of the implications of structural explanation must be modified in the case of hydrology. This is primarily due to the fact that, in many cases, the emphasis in hydrology is not on the hypothetical affirmation of new theoretical entities such as electrons or moving tectonic plates; hydrologists more often have reason to rely upon a multiplicity of conceptual models with idealized structures that are known to only imperfectly approximate actual materials. Finally, Darcy's law, when combined with the equation of continuity, also becomes the basis for a derivation of Laplace's equation applicable to hydraulic head. Laplace's equation and the relevant boundary conditions then serve as a mathematical model that is solved and applied in an instrumentalist fashion to predict and influence groundwater phenomena. A taxonomy of how models function with respect to Darcy's law thus offers insight into both the argumentative structure and the explanatory and descriptive goals of hydrology.

2. Models, Phenomena, and Structural Explanation

Discussion of theoretical explanation among philosophers of science has benefitted considerably from Bogen and Woodward's insistence that specific data are not the target of scientific explanation (Bogen and Woodward 1988, Woodward 1989, and Hofmann 1990). Data are too idiosyncratically dependent upon unique characteristics of specimens and instrumentation to warrant explicit explanatory attention. Rather, data provide evidence for the phenomenological relationships, conditions, or laws that are the potential subject matter for explanation. Theorists traditionally pay particularly close attention to those phenomena that are described by phenomenological laws.

In *How the Laws of Physics Lie*, Nancy Cartwright argued that the explanation of phenomenological laws in physics typically requires the application of fundamental laws such as Schroedinger's equation to an appropriately "prepared" model of the domain (Cartwright 1983). Starting from this combination of a fundamental law and an idealized model, physical and mathematical approximations generate the derivations that Cartwright originally referred to as theoretical explanations.

The role of models in this account has been misunderstood in a manner that requires some clarification. Kroes and Sarlemijn claim that the distinction between phenomenological laws and fundamental laws is not sufficiently precise (Kroes and Sarlemijn 1989). Using the example of the Van der Waals equation, they point out that the law is phenomenological in the sense that its two parameters must be specified experimentally. On the other hand, they also claim that the law could be considered to be fundamental because it is "derived from first principles" (Kroes and Sarlemijn 1989, p.324). But derivation from fundamental laws does not necessarily generate additional fundamental laws. More typically, the fundamental laws are applied to idealized models together with mathematical approximations. The results of these operations are precisely the phenomenological laws that Cartwright offers as our most reasonable candidates for laws of nature. Kroes and Sarlemijn also mistakenly disagree with Cartwright's provocative thesis that the fundamental laws of physics are false. Her point was that these laws say nothing specific about the real world until they are applied to suitably prepared models. Kroes and Sarlemijn claim that "Cartwright seems to confuse the validity of the boundary conditions with the validity of the fundamental laws" (Kroes and Sarlemijn 1989, p.326). A more correct rendering would emphasize that the approximate derivation of phenomenological laws by means of idealized models does not transform those phenomenological laws into fundamental laws; nor does an acknowledgment that models are highly idealized make fundamental laws, by contrast, "true" laws of nature.

J. Hofmann has clarified the functions of models in derivations of phenomenological laws similar to the Van der Waals equation (Hofmann 1990). On the one hand, models are a culminating stage in the description of phenomena; that is, models stipulate and emphasize selected structural aspects of the domain. Secondly, however, models provide this description in a mathematical form amenable to the application of a fundamental law. In this sense, models are a necessary requirement for what Cartwright called theoretical explanation in *How the Laws of Physics Lie*. In the terminology of that book, the two functions of models thus contribute to both phenomenological description and theoretical explanation.

There are good reasons to clarify these conclusions by emphasizing the more specific concept of structural explanation. Cartwright herself no longer considers the derivations she described in *How the Laws of Physics Lie* to constitute explanations. In a 1989 paper she decided to "reserve the word 'explanation' for scientific treatments that tell why phenomena occur" (Cartwright 1989, p.282). She attributes her revision to the fact that influential physicists such as Edwin Kemble hold that "the function of theoretical physics is to describe rather than to explain" (Cartwright 1989, p.275). From this point of view, the derivation of phenomenological laws from fundamental laws is a demonstration that there is an economical way to classify these laws as various applications of a few fundamental principles; explanatory causes are not addressed.

On the other hand, one of the points that emerged from a study of transition metal oxide models is that the wide variety of modeling techniques and motivations means that it is virtually impossible to give a general characterization that summarizes the relationship between phenomena and models (Hofmann 1990). But in some cases, surely, the structure stipulated by a model can be acknowledged to be part of an explanation of why the associated phenomena take place. This is in fact the interpretation Ernan McMullin refers to as structural explanation. His characterization is worth quoting (McMullin 1978, p.139):

> When the properties or behavior of a complex entity are explained by alluding to the structure of that entity, the resultant explanation may be called a structural one. The term "structure" here refers to a set of constituent entities or processes and the relationships between them. Such explanations are causal, since the structure invoked to explain can also be called the cause of the feature being explained.

Two points are in order here. First, the model alone seldom provides structural explanation. That is, phenomena typically are explained by applying fundamental laws to the structure stipulated by the model. In this sense, the combination of model and fundamental principles constitutes a theory of the phenomena. Secondly, McMullin takes the success of structural explanations to warrant an interpretation of the relevant model as a more or less accurate description of reality. In particular, when revisions of a model provide increasingly accurate derivations of phenomenological laws, McMullin argues that it is justifiable to conclude that the real structure responsible for the phenomena is approximately known. Before considering how these ideas apply to theoretical hydrology, note the following example McMullin cites (McMullin 1978, p.147):

> Geologists assume that a successful macrostructural explanation of such surface phenomena as sonar pulses can give reason to believe in the existence of sub-surface structures like pockets of water or oil. These structures play a role in the explanation of the phenomena similar to that played by molecular structures in the explanation of chemical phenomena. But in the geological case, the existence of the water or the oil can be directly ascertained. And the geologists' belief in the ontological reliability of the retroductive form has turned out to be amply justified.

Although the structure cited in this example is a macroscopic one, theoretical hydrology also makes use of conceptual models for the detailed and unobservable structure of soils and fluids. The resulting complications are best considered through a specific example.

3. Darcy's Law

In 1856 Henry-Philibert-Gaspard Darcy published a lengthy assessment of a proposed upgrading of the public water system for the French city of Dijon (Darcy 1856). His investigation of fountains called for information concerning the flow of water through sand filters; in an appendix to his report he included a description of his experimental work on this subject. His data analysis resulted in a relationship that has since come to be known as Darcy's law; the law is well known to hydrologists and Darcy's appendix has been partially translated into English (Freeze and Back 1983, pp.14-20).

Darcy included a diagram and a description of his apparatus. The sand representing the filter was contained in a vertical tube said by Darcy to be .35 meters in diameter and 2.5 meters in length (although the diagram labels the length at 3.5 meters). Darcy performed a sieve analysis on his sand and estimated the porosity at 38%. Care was taken in packing the columns to minimize entrained air. The height of sand could be varied above a screen and grillwork located .2 meters above the bottom of the column. Water entered the sand column from an adjacent hospital through a pipe near the top of the column and exited through a faucet in the chamber below the column. Both entry and output rates could be regulated. To record pressures, two mercury manometers were installed in the column above and below the sand. For the purposes of these measurements, the bottom of the sand was taken as the datum plane with elevation zero. This elevation coincided roughly with the bottom of the lower manometer arm.

In modern terminology, total hydraulic head is the sum of elevation head and pressure head. Elevation head is the distance between datum plane and the point of interest. Pressure head is the length of fluid registered by a manometer installed at the point of interest. Although Darcy did not explicitly explain all of his terminological conventions, his usage was consistent given the specific circumstances of his apparatus. In particular, since Darcy chose the bottom of the sand as his datum plane, the elevation head at the bottom of the filter was zero. The bottom of the column was open to the atmosphere; thus, using gage pressure conventions, the pressure head at the lower manometer position was zero. As a result, Darcy's calculations of hydraulic head amounted to adding the length of his sand column to the height of fluid in the upper manometer arm.

Darcy initially tabulated four series of data for sand columns with heights of .58, 1.14, 1.71 and 1.70 meters respectively. Water-hammer in the hospital plumbing forced him to use a mean value for the level of mercury in the upper manometer arm. This upper manometer value was reported as the mean pressure for each experiment after he converted his mercury pressure readings to equivalent heights of water.

Darcy performed as few as three and as many as ten different measurements of flow rates for four specific heights and types of sand; in each case he gradually increased the height of water in the upper manometer arm (the mean pressure) by adjusting his inflow and outflow faucets. In his first four sets of measurements, the lower end of the column was open to atmospheric pressure. Darcy observed that, for any given elevation head, the outflow volume invariably increased proportionally with the pressure head.

He then averaged the ratios of hydraulic head (Darcy's *charge*) to flow rate for each set of measurements, obtaining four proportionality constants. Darcy attributed the variation among the constants to differences in grain size and purity between the sands in different columns. He also claimed without argument that the data showed that the flow rates varied in inverse proportion to length of sand column. This conclusion was not obvious since the data provided did not include multiple measurements at fixed heads for different column lengths; however, it is substantiated by comparison of his data for differing column lengths with roughly equal mean pressure values. Darcy then performed a similar set of experiments differing mainly in that the pressure at the bottom of the column could be varied widely above or below atmospheric. He was satisfied that his earlier conclusions held in these cases as well.

One point that stands out in this analysis is the clarity of the distinction between data and phenomena. Darcy did describe the texture of his sand samples in some detail. However, aware of the unique nature of his apparatus and the variable effect of heterogeneities, he attributed little importance to the specific magnitudes of individual data readings. Rather, he emphasized two aspects of the general phenomenon he claimed his data supported: the proportionality of flow rate to total hydraulic head, and the inverse proportionality of flow rate to column length (Darcy 1856, p.593).

> Il parait donc que, pour un sable de même nature, on peut admettre que le volume débité est proportionnel à la charge et en raison inverse de l'épaisseur de la couche traversée.

Darcy assembled his conclusions in the following equation:

$$q = k(s/e)(h + e \pm h^*)$$

where:

q = rate of water flow (volume per time)

k = a coefficient dependent on the "permeability" of the sand

s = cross sectional area of the sand filter

e = length of sand filter

h = reading of the upper manometer arm

h^* = reading of the lower manometer arm

At this point, Darcy made use of the implications of his datum plane convention. Only under this convention, Darcy's law reduces to:

$$q = k\ (s/e)\ (h + e)$$

In modern format, using a particular sign convention, Darcy's law is usually written as:

$$Q = -KA\ dh/dl$$

where:

Q = rate of water flow (volume per time)

K = hydraulic conductivity

$A =$ column cross sectional area

$dh/dl =$ hydraulic gradient, that is, the change in head over the length of interest.

The law is often transformed by dividing through by the cross-sectional area and is then restated as:

$$q = Q/A = -K\, dh/dl$$

where q now has the dimensions of a velocity, and is referred to as the Darcy, or superficial, velocity.

Perhaps due to the ambiguous nature of some of Darcy's terminology and conventions, there was some initial confusion over the content of his law. This was cleared up largely due to the efforts of M. King Hubbert, who wrote several influential essays on Darcy's law beginning in 1940 (Hubbert 1969). Hubbert had been quite dismayed to find that widely used and respected texts in hydrology often stated Darcy's law as a proportionality between flow rate and pressure gradient alone, neglecting elevation head. His essays brought about a new consensus concerning the fact that hydraulic head functions as the potential in the law, and that total head is the sum of elevation head and pressure head.

On the other hand, there seems never to have been any doubt concerning the empirical basis of Darcy's law. For example, to quote an authoritative textbook by Freeze and Cherry, "Darcy's law is an empirical law. It rests only on experimental evidence" (Freeze and Cherry 1979, p.17). Darcy's law fully satisfies the requisite criteria to be considered a phenomenological law; it thus is a potential candidate for structural explanation.

In this respect it is somewhat surprising that in a recent discussion of Darcy's law, Shrader-Frechette repeatedly refers to it as either a "theoretical" or "fundamental" law (Shrader-Frechette 1989). She emphasizes the "idealized" nature of the law by claiming that it is "experimentally" verified only by applying the Bernoulli equation to an idealized model. She concludes that "this fundamental or theoretical 'verification' is highly idealized, since actual flow velocity is a function of the microstructure of the medium through which the water is flowing" (Shrader-Frechette 1989, p.335). Similarly, she claims that (Shrader-Frechette 1989, p.337):

> apart from the falsity of Darcy's Law on all three levels (micro, molecular, macro) what is significant is that 'corrections' to it do not come from the theory built into the law itself, but from phenomenological or observational factors not deducible from the theoretical or fundamental law.

There are several misleading aspects to these comments. First, Darcy's law is a phenomenological law rather than a fundamental law. Its limited accuracy and scope were recognized by Darcy in 1856 and have remained apparent to hydrologists ever since. The experimental discovery of the law, as carried out by Darcy himself, is a matter of quantitative measurements with specific soil samples and has no reliance on fundamental laws or theory. Furthermore, although hydrologists do distinguish velocities on several different orders of magnitude, Darcy himself was interested only in macroscopic discharge and its proportionality to directly measurable properties of his filters. The Darcy velocity is a macroscopic parameter influenced by quite possibly unmeasurable microscopic factors.

Although a full discussion of the historical development of Darcy's law is beyond the scope of this paper, one aspect that bears directly on present concerns is the recognition of both lower and upper bounds for the dependable use of the law's stated relationships. Briefly, some authors consider a lower limit below which there is no flow, positing the existence of a minimum threshold hydraulic gradient to motivate flow. The upper limit is of more practical significance; the law has been found, once again experimentally, to be inappropriate when the flow regime is not both laminar and dominated by viscous forces. In laminar flow the molecular velocity vectors are uniformly parallel to macroscopic flow. The determination of laminar flow is in turn dependent upon the magnitude of the Reynold's number, a dimensionless ratio of inertial forces to viscous forces. At low Reynold's numbers, viscous forces dominate, and Darcy's law is valid. There follows a transition zone in which inertial forces become more important; Darcy's law cannot be accurately applied to the nonlinear laminar flow in this zone. Flow in the turbulent zone is both nonlinear and nonlaminar, and deviations from Darcy's law can become very large. In many aquifer materials, the assumption of laminar water flow may not cause any inaccuracies. However, flow analysis predictions based upon Darcy's law in the presence of such matrix material as karstic limestones or highly fractured crystalline rocks can lead to large errors of great significance in problems such as contaminant transport. Flow in such cases cannot be described adequately by a linear relationship such as Darcy's law.

A purely practical difficulty is that virtually any matrix considered, even on the laboratory scale, will be heterogeneous to some extent. Relationships have been developed for estimating the composite conductivity for laminar flow through heterogeneous systems. However, field-scale investigations can fail to detect strata of significantly different conductivities. The presence of such heterogeneities does not theoretically prohibit the use of Darcy's law, but it lends uncertainty to generalizations on a large scale. It is for this reason that statistical models employing probabilistic parameter distributions are sometimes used to model the spatial variability of hydraulic conductivity. Of course, Darcy's own samples were heterogeneous to varying degrees. In this context, it can be seen that Darcy's law remains tied to its empirical roots, and that modern questions of its proper scope can be settled only through a similarly experimental approach.

Hydrologists also use Darcy's law to derive Laplace's equation as a governing equation for spatial variation of hydraulic head. Laplace's equation with its accompanying initial and boundary conditions then becomes a mathematical model applicable to a wide variety of specific sites. The solution of Laplace's equation relies upon either analytic methods or computer approximations, methods referred to as analytic and numerical, respectively (Wang and Anderson 1982). In this process, Darcy's law is taken as an experimentally verified relationship; the empirical foundation and limited scope of the law should always be borne in mind.

The difficulty and importance of delineating the valid application of Darcy's law at field scale were major issues in Shrader-Frechette's analysis of the Maxey Flats radioactive waste dump in Kentucky (Shrader-Frechette 1988). Nevertheless, it appears that in her subsequent more specific discussion of Darcy's law she overlooked the empirical basis of the law and then mistakenly interpreted as an "experimental verification" what was actually an example of structural explanation based upon a conceptual model. To clarify this distinction we must look more closely at structural explanation in hydrology.

4. Structural Explanations of Darcy's Law

Theoretical hydrologists draw several distinctions with respect to the models employed in their discipline. Physical models such as sandboxes with particular packing patterns are sometimes constructed in an attempt to replicate specific aspects of conditions encountered in the field. Electric analogue models have been wired to investigate conductivities and flowlines, using the parallel structures of Ohm's Law and Darcy's Law. The utility of these physical models is limited in that they tend to be highly site specific and therefore often do not generate results amenable to generalization. Theoreticians thus emphasize the use of tractable mathematical idealizations.

Mathematical models in their turn can be either deterministic or statistical. As noted earlier, statistical models include parameters that have probabilistic distributions rather than single values. In the case of deterministic models, a common source of motivation is a conceptual model that approximates the structure of the soil and fluids under study. In a survey of the subject, Faust and Mercer include the following description of how conceptual models are chosen to generate deterministic models (Mercer and Faust 1981, p.2).

> The first step is to understand the physical behavior of the system. Cause-effect relationships are determined and a conceptual model of how the system operates is formulated. For ground-water flow, these relationships are generally well known, and are expressed using concepts such as hydraulic gradient to indicate flow direction.

There is a clear resonance here with McMullin's insistence that descriptions of causal relationships can act as a starting point for what ultimately become structural explanations.

But before turning to more specific discussion of conceptual models, another set of distinctions should be noted. Hydrologists follow the conventions of physics and thermodynamics, generally classifying stipulations of structure as falling within one of three possible viewpoints: molecular, microscopic, or macroscopic. The molecular approach is the most detailed in that it stipulates the path of individual molecules within the fluid in motion. Great physical and mathematical precision is required by models on this scale in order to allow application of statistical techniques. The ultimate interests of hydrology are invariably on a larger scale, as in the case of Darcy himself. Therefore, hydrologists often move to a more coarse-grained approach in which the fluid within any particular pore of material is treated as a continuum rather than a collection of localized particles. The resulting microscopic models then represent various constraints placed upon the idealized continuous fluid in the porous medium.

For example, theoretical hydrologist Jacob Bear has collaborated with Bachmat in the invention and analysis of an elaborate microscopic model to represent fluid flow in a porous medium (Bear 1972, p.92). Bear describes the initial stage in this process as follows (Bear 1972, p.24).

> In the present text we shall adopt the continuum approach. Accordingly, the actual multiphase porous medium is replaced by a fictitious continuum: a structureless substance, to any point of which we can assign kinematic and dynamic variables and parameters that are continuous functions of the spatial coordinates of the point and of the time.

Similarly, Freeze and Cherry emphasize the far-reaching ramifications of structural stipulation, as they extend the discussion to macroscopic flow considerations (Freeze and Cherry 1980, p.17):

> This... may appear innocuous, but it announces a decision of fundamental importance. When we decide to analyze groundwater flow with the Darcian approach, it means, in effect, that we are going to replace the actual ensemble of sand

grains (or clay particles or rock fragments) that make up the porous medium by a representative continuum for which we can define macroscopic parameters, such as the hydraulic conductivity, and utilize macroscopic laws, such as Darcy's law, to provide macroscopically averaged descriptions of the microscopic behavior.

Bear's work provides an excellent illustrative example of how a conceptual model becomes the basis for both a structural explanation of Darcy's law and also for the construction of a much more general mathematical model for fluid flow. Before considering Bear's discussion of Darcy's law, his general conception of his own reasoning process should be noted. The following passage is important enough to be quoted at length; this is Bear's description of the two stages of his procedure that follow upon the introduction of a simplified conceptual model (Bear 1972, p.91).

> Once the model is chosen, the second step is to analyze the model by available theoretical tools, and to derive mathematical relationships that describe the investigated phenomenon. These relationships show how the various active variables (fluxes, forces, etc.) depend on each other. They also show which factors have, according to the chosen model, no influence on the investigated problem. The only way to test the validity of laws derived in this way is to perform controlled experiments in the laboratory (or to observe phenomena in nature). Such controlled experiments, which comprise the third step of this approach, will test the validity of the derived relationships among the variables. No theory developed by this approach can be accepted without first being verified by experiments.

Notice that Bear mentions that the goal of this procedure is to "describe the investigated phenomenon", a passage that calls to mind Cartwright's references to Kemble's conclusions about theoretical physics. We will see that Bear does not entirely avoid the term "explain", but he is reticent to use it because of the way models enter into his reasoning. Let us consider his derivations of Darcy's law.

In his most thorough treatment, he begins with the Bear-Bachmat conceptual model (Bear 1972, p.92). The fluid is idealized as an incompressible continuum and the medium is imagined to be a network of interconnected passages and junctions within a solid that is rigid and does not interact with the fluid. Additional idealizations include the assumption that "the fluid loses energy only during passage through the narrow channels and not while passing from one channel to the next through a junction" (Bear 1972, p.93). Bear then applies a complex averaging procedure in order to be able to assign values to dynamic variables within each representative elementary volume of the idealized continuous fluid. Finally, he applies an equation stating the conservation of linear momentum for a fluid system. The result is a general equation of motion which, when simplified for a homogeneous, incompressible fluid with small inertial forces, is an extension of Darcy's law to three dimensional flow in an anisotropic medium (Bear 1972, pp.104-106). Consequently, it is not surprising that Bear sometimes says that the law simply expresses conservation of momentum during fluid flow through a porous medium.

Although this derivation is Bear's most sophisticated analysis of Darcy's law, he also provides a review of several other derivations in which the mathematics is simplified by assuming at the outset that the fluid is homogeneous. He thus provides derivations from a wide variety of different conceptual models: capillary tube models analyzed by means of the Hagen-Poisseuille law, fissure models, hydraulic radius models, and resistance to flow models. Finally, one of his best known derivations uses a statistical model to take into account the disorder of actual porous media prior to averaging the Navier-Stokes equations over a representative elementary volume. In each case a conceptual model stipulates an idealized structure for the porous medium. Principles of conservation of energy or momentum then are applied to the model and Bear arrives at a version of Darcy's law through a series of approximations and idealizations.

It should be clear that each of these derivations provides an example of what McMullin calls structural explanations. Darcy's law is a phenomenological law generated by experimental data. That is, Darcy argued that the rate of water flow through samples similar to those he employed is proportional to hydraulic head gradient. To explain why this relationship holds, and to explore its limitations, idealized structures are postulated in order to carry out mathematical applications of fundamental physical principles such as conservation of linear momentum. Darcy's law follows only through a series of approximations that may include statistical analysis; it thus remains as much a phenomenological law as it was originally. Nevertheless, it has been brought under the explanatory umbrella of fundamental physical principles via analysis of structures depicted by conceptual models.

At this point we might ask what value these approximate derivations of Darcy's law hold for hydrologists. We have already seen that Bear sometimes writes that such a procedure provides an extended "description" of the phenomenon in question. However, in other passages he uses explanatory language. For example, in referring to his set of derivations of Darcy's law, he makes this comment (Bear 1972, p.92, emphasis added):

> In all these cases, the model is presented as an attempt to simulate, and *thus to explain*, phenomena observed in nature or in the laboratory. Sometimes several models are equally successful in explaining the relationship between observed excitations and responses. However, we must emphasize again that the proof of the validity of a model, and the only way to determine coefficients, is always the experiment.

Bear's ambivalence concerning description *versus* explanation is apparent in his account of the ultimate value of derivations based upon conceptual models (Bear 1972, p.92).

> With these thoughts in mind, a question sometimes arises as to why we bother with the model in the first place, since in any case we must eventually go back to the laboratory to determine the required coefficients. The answer is that in applying the conceptual model approach we gain an understanding of the investigated phenomenon and the role of the various factors that affect it. We also gain an insight into the internal structure of the various coefficients appearing in the equations that describe the investigated phenomenon. All this information is needed for planning the laboratory experiments.

These comments suggest a reassessment of McMullin's position that a sequence of increasingly accurate structural explanations provides an "approximately true" description of the causal components of the structure responsible for the phenomena explained (McMullin 1987, pp.59-60). In Bear's case, the major contrast to the scenarios emphasized by McMullin is that there is not necessarily a progressive modification of a single model with increasingly accurate results. Rather, a multitude of different models may be employed simultaneously or sequentially to explicate various aspects of the coefficients found in phenomenological laws. In the case of Darcy's law, for example, the hydraulic conductivity ultimately is not a "constant", but varies with both fluid and soil type. How the value of this coefficient varies with physical conditions is explored through a variety of models without claiming that any one of them provides a full account of the actual conduction process, even with future modifications in mind.

Analysis of the components of the conductivity "constant" in Darcy's law is an interesting example of this procedure. Darcy described the conductivity constant as primarily a function of grain size and sand purity. Modern versions of this coefficient include additional properties of both the fluid and the soil matrix, such as viscosity and tortuosity. These modern expressions do not always neglect interaction between the fluid and matrix. Conceptual models are chosen to reflect the significance assigned to various parameters, forces and relationships. Mathematical analysis of specific models generates experimental tests of these hypothetically dominant parameters.

For example, one of the structural explanations of Darcy's law provided by Jacob Bear is based on the capillary tube conceptual model. In this model, the void space within the solid matrix of the porous medium is imagined as a collection of uniform, parallel cylindrical tubes of diameter δ and length dimension s. The areal porosity, n, is the percentage of void space in a cross sectional area taken normal to the tubes. The fluid density is ρ, and the dynamic viscosity is μ.

The fundamental law to be applied to this model is the relevant version of the conservation of momentum principle, namely, the Hagen-Poisseuille law. Given a hydraulic head of ϕ and steady laminar flow of an incompressible fluid in a single, long cylindrical tube, this law states that:

$$Q = (\pi\delta^4\rho g/128\mu)\partial\phi/\partial s$$

where Q is the volume flow rate through the tube. Applying this equation to the capillary tube model and dividing through by the model's cross-sectional area gives:

$$q = (n\delta^2\rho g/32\mu)\partial\phi/\partial s = (k\rho g/\mu)\partial\phi/\partial s = K\partial\phi/\partial s$$

where $k = n\delta^2/32$.

This relationship is in fact Darcy's law where K is the hydraulic conductivity, and k is the intrinsic permeability but stipulated in terms of porosity and tube diameter. The capillary tube model thus provides insight into the dependency of hydraulic conductivity on two specific properties of the medium. Subsequent experimental measurements of permeability or conductivity provide information about the corresponding properties of the system. At the same time, the relevance of this insight is limited to media that can be approximated fairly accurately by the capillary tube model. The choice of an appropriate model thus is guided in part by decisions about what aspects of the medium are expected to have a major impact on permeability or conductivity. To cite Bear once again (Bear and Verruijt 1987, p.12):

> The real system is very complicated and there is no need to elaborate on the need to simplify it.... Because the model is a simplified version of the real system, *there exists no unique model* for a given groundwater system. Different sets of simplifying assumptions will result in different models, each approximating the investigated groundwater system in a different way.

5. Conclusion

In conclusion we should repeat that structural explanations not only provide an analysis of the factors relevant to the value of the coefficient in phenomenological laws, but also help specify the limitations within which the laws remain accurate. Fundamental laws, typically conservation of momentum, are applied to a wide variety of models to provide approximate derivations of phenomenological laws such as that of Darcy. However, in contrast to the examples emphasized by McMullin, hydrological models do not necessarily constitute a temporal sequence in which accuracy consistently increases in all respects. Furthermore, depending upon their objectives, hydrologists in the field are forced to consider a multitude of models as simultaneously applicable to a given system.

Analysis of the argumentative form of structural explanations also calls to our attention the mutually supportive roles of fundamental physical laws and models in an applied science such as hydrology. Fundamental laws can only be brought to bear upon models that are in an appropriate mathematical form. The idealized conditions incorporated into a model represent assumptions that permit the explanatory derivation to be carried out. Consequently, a statement of these conditions facilitates the empirical clarification of the domain in which phenomenological laws are applicable. Since hydrology has as its domain such a multitude of disparate individual groundwater systems, models function as an important scheme to classify these systems. Unless a system can be accurately represented by at least one model that functions in a structural explanation of Darcy's law, there is good reason to doubt that the law can be successfully applied to that system. As Wilfred Sellars pointed out long ago, an important characteristic of scientific explanation is the understanding it provides concerning why the phenomenological laws to be explained are in fact only approximately correct under limited circumstances.

References

Bear, J. (1972), *Dynamics of Fluids in Porous Media*. New York: American Elsevier Publishing Company.

Bear, J. and Veruijt, A. (1987), *Modeling Groundwater Flow and Pollution*. Dordrecht and Boston: Dordrecht Reidel.

Bogen, J. and Woodward, J. (1988), "Saving the Phenomena", *The Philosophical Review 97*: 303-352.

Cartwright, N. (1983), *How the Laws of Physics Lie*. New York: Oxford University Press.

_________. (1989), "The Born-Einstein Debate: Where Application and Explanation Separate", *Synthese 81*: 271-282.

Darcy, H. (1856), *Les Fontaines Publiques de la Ville de Dijon*. Paris: Victor Dalmont.

Freeze, R.A. and Back, W. (eds.) (1983), *Physical Hydrogeology*. Stroudsburg: Hutchinson Ross.

Freeze, R.A. and Cherry, J.A. (1979), *Groundwater*. Englewood Cliffs: Prentice Hall.

Hofmann, J. (1990), "How the Models of Chemistry Vie", in *PSA 1990*, volume 1, A. Fine, M. Forbes and L. Wessels (eds.). East Lansing: Philosophy of Science Association, pp.405-419.

Hubbert, M.K. (1969), *The Theory of Groundwater Motion and Related Papers.* New York: Hafner Publishing Company.

Kroes, P.A. and Sarlemijn, A. (1989), "Fundamental Laws and Physical Reality", in *Physics in the Making,* A. Sarlemijn and M.J. Sparnaay (eds.). Amsterdam: Elsevier Science Publishers, pp.303-328.

McMullin, E. (1978), "Structural Explanation", *American Philosophical Quarterly* 15: 139-147.

McMullin, E. (1987), "Explanatory Success and the Truth of Theory", in *Scientific Inquiry in Philosophical Perspective,* N. Rescher (ed.). Lanham: University Press of America, pp.51-73.

Mercer, J.W. and Faust, C.R. (1981), *Ground-Water Modeling.* Reston: National Water Well Association.

Shrader-Frechette, K.S. (1988), "Values and Hydrogeological Method: How Not to Site the World's Largest Nuclear Dump", in *Planning for Changing Energy Conditions,* J. Byrne and D. Rich (eds.). New Brunswick: Transaction Books, pp.101-137.

_____________. (1989), "Idealized Laws, Antirealism, and Applied Science: A Case in Hydrogeology", *Synthese 81*: 329-352.

Wang, H.F. and Anderson, M. (1982), *Introduction to Groundwater Modeling: Finite Difference and Finite Element Methods.* San Francisco: W.H. Freeman and Company.

Woodward J. (1989), "Data and Phenomena", *Synthese 79*: 393-472.

Philosophy of Science and Its Rational Reconstructions: Remarks on the VPI Program for Testing Philosophies of Science[1]

Alan W. Richardson

University of Keele

1. Introduction

For a number of years now, we, as philosophers of science, have been enjoined by more and more of our colleagues to understand the task of developing a philosophy of science to be itself a scientific task. We are told that if we want to understand science we have no better (and perhaps indeed no other) path to such an understanding than the path of science itself. We should view ourselves as ultimately attempting to arrive at a relatively complete theoretical understanding of how science proceeds. This is a call to *naturalize* philosophy of science.

Somewhat more recently some philosophers of science have become impatient even with those who have taken up the naturalist banner. It isn't enough, they tell us, to arrive at an understanding of the nature of philosophy of science that proclaims it to be one science among many. This metaphilosophical view is just the beginning; now we have to move on to the hard work of actually coming up with reasonably sufficient theories within the discipline. Thus, in proclaiming their program for testing theories of theory change in science (henceforth, the VPI program), L. Laudan *et al.* (1986, pp. 142f) tell us that:

> Pieties about the importance of empirical testing must give way to the particularities of the testing process itself... Sloganeering on behalf of naturalism in epistemology must now give way to the real thing...

This impatience is occasioned not only by the inherent inertia of philosophers of science. More pressing is the rather embarrassing fact that certain scientists themselves have both espoused and recommended to their disciplines philosophies of science—such as Kuhn's—despite the fact that philosophers of science know that those philosophies are defective in quite important ways. Also, even allegedly naturalized philosophies of science have not been tested against the historical record of theory change in any compelling way (cf. R. Laudan *et al.* 1988, pp. 6f). Thus, at least some philosophers of science are now gearing up to become scientists of science out of a conviction about the nature of the discipline and out of suspicion that philosophy of

PSA 1992, Volume 1, pp. 36-46

science might actually be doing more harm than good in its current form by presenting us as knowing more about the general nature of science than we actually do.

Perhaps the most ambitious attempt to date to do some empirical spadework in pursuit of better theories in philosophy of science is the above mentioned VPI program. It is with this program and some of its peculiarities alone that I wish to deal in this paper. Since I am not developing general objections to programs that consider themselves to be "naturalized", other versions of naturalism such as those of Giere (1985, 1988), Hull (1989), or even Larry Laudan's (1987, 1990) normative naturalism are deliberately left out of account.

2. The VPI Program

What is the purpose of the VPI program? L. Laudan *et al.* (1986, p. 142) claim that the philosophies of science of the historiographical school were alleged to be based on and to illuminate the actual history of the sciences. In this sense they are naturalized philosophies of science that can and ought to be tested against this historical record. Now these philosophies, unlike their positivist predecessors, are meant to capture the dynamics of theory change in science and, hence, must be tested primarily against the record of the history of theory change. Through such a testing procedure we will see if any of these philosophies is substantially correct or, failing that, at least whether any of the theses associated with these theories is substantially correct. This then will be a first step in the direction of arriving at some more adequate general theory of scientific theory change.

Thus, the VPI program is a testing procedure. Specific claims about theory change from the historiographical philosophies of (principally) Kuhn, Lakatos, Laudan, and Feyerabend are to be tested against the history of theory change in the sciences. This is done because it is the "only way to find out" if any have "captured some significant part of the story of scientific change" (L. Laudan *et al.* 1986, p. 143).

But, before actual tests can be constructed, the VPI theorists note two very important problems in testing these philosophies of science. First, the formulation of the theories is such that testable consequences are very difficult to extract. Despite their naturalist predilections, these philosophers have not presented their philosophies as scientific theories and, hence, have not formulated their theories in ways that "emphasize their empirical implications and ... demonstrate their testable, as opposed to speculative, character..." (L. Laudan *et al.* 1986, p. 143). Second, these philosophies are all presented in vocabulary idiosyncratic to their authors and this makes comparison of their claims unnecessarily difficult. For example, in their discussions of high level families of theories sharing important methodological and (perhaps) metaphysical presuppositions, Kuhn employs the term "paradigm" whereas Lakatos uses "research programme", and Laudan "research tradition". This terminological difference masks a clear convergence in the philosophical views put forward by these authors.

Thus, in their original monograph, the VPI theorists did not address the testing of the philosophies themselves but undertook two preliminary tasks that they felt must be resolved or else "the business of testing and empirical evidence cannot begin" (L. Laudan *et al.* 1986, p. 144). These tasks are, first, the delimitation of specific theses that clearly have empirical import and, second, the paraphrase of these theses into a neutral vocabulary for the purpose of comparison and mutual test. This vocabulary is "neutral" in the sense that it "does not presuppose the machinery of any of the models under review" (L. Laudan *et al.* 1986, p. 145). Thus, for example, they substitute "guiding assumptions" for all of the idiosyncratic terms from Kuhn, Lakatos, and

Laudan canvassed above. Where this quest for neutral terminology required the invention of new terms of their own, the VPI theorists offered definitions of them (L. Laudan *et al.* 1986, pp. 161f). In this way we have empirical hypotheses that are formulated in an understood language and are, therefore, testable across the boundaries of the theories of theoretical change in question.

3. Philosophy of Science and Its Rational Reconstructions

The picture of the empirical status of theories of scientific theory change that we get from the VPI program is this: We have a variety of theories which appear to make substantive empirical claims about how scientific change is in fact effected, but which are presented in vocabulary specific to each theory and vague within any particular theory. To see whether any of these theories is substantially correct we must test them against the history of scientific theory change, but this is not possible until we (a) extract their testable claims and (b) translate those claims into neutral vocabulary for the purpose of comparative test.

This picture is somewhat startling and highly ironic. For we have the following elements:

(1) The theories of theory change are hypothetico-deductive theories that, despite their authors' protestations to the contrary, "were [all] conceived either *a priori* to solve specific philosophical conundrums or *post hoc* to fit a small number of preselected examples" (L. Laudan *et al.* 1986, p. 143).

(2) These extant theories must prove their adequacy by virtue of being confirmed or disconfirmed by testing against the historical record.

(3) This testing requires explicit theses deduced from the theories which are presented in a language that is neutral between the theories.

These theses might, indeed ought to, remind you of the most fundamental theses of hypothetico-deductivism in confirmation as developed in the logical empiricist era. Theories are arrived at hypothetico-deductively, are confirmed or disconfirmed by substantive testing, and when alternatives are in the field these alternatives must be in a position to recognize when a test confirms one and disconfirms the other or, alternatively, must be able to see that they are merely reformulations of one another in different linguistic garb. Therefore, in the hands of the VPI theorists the methodology of testing the naturalized philosophies of science of the historiographical schools crucially involves precisely what the logical empiricists claimed that philosophy of science must do: It presents clearly the linguistic frameworks of scientific theories in order to show how putative disagreements either resolve themselves into merely pragmatically motivated disputes over linguistic framework or are substantive disputes over empirical facts. Furthermore, the theories regimented into their frameworks will be in a position to locate exactly where the substantive disputes lie, and, with sufficient ingenuity on the part of scientists in creating test situations, one will be confirmed and the other disconfirmed.

The VPI program begins then with a rational reconstruction of the naturalist philosophies of science that are the object of their testing procedure. In this way and seemingly unnoticed by the VPI authors themselves, they exhibit in their work the power of the nonnaturalist view of philosophy of science that guided the logical empiricists and kindred spirits such as Popper. Standing at a level above the historiographic philosophies of science it wishes to test, the VPI program seeks to overlay a framework on those theories that first exhibits their empirical consequences. The whole tone of the VPI monograph is one of rational reconstruction in order to get naturalist philosophy of science off the ground.

The irony of employing a nonnaturalist metamethodology in the testing of historiographical philosophy of science seems altogether lost on the VPI authors. Indeed, from a nonnaturalistic point of view, it is highly instructive to see how precisely the concerns of the logical empiricists can be reproduced in methodological thinking about how to test empirical theories of scientific change in the 1980s. And surely at some level the logical empiricists and the VPI program are right, theories so vague that they resist all effort to specify any empirical content are scientifically useless. Moreover, if comparative testing is to be possible it is imperative that the various theories can be brought to bear on the same empirical evidence. Regardless of whether there is an observation language that is neutral (in the sense of the earliest logical empiricist thought) among the theories in question, we must know whether a given experiment confirms or disconfirms each theory under consideration.[2]

The VPI program actually imposes greater strictures on rational reconstruction for testing purposes than that imposed by pre-Kuhnian philosophy of science. In their quest for a neutral language into which all the theories they want to test can be cast, the VPI theorists go beyond what the mature Carnapian logical empiricist would require. Carnap wanted neither a neutral observation language nor a neutral theoretical language, but only a non-question begging metalanguage for discussing logical frameworks. Once all our theorists agree to this metalanguage they can state the analytic sentences that constitute their respective linguistic frameworks, the inductive logic for each framework, and then go about seeing if their theories disagree in their empirical content and degree of confirmation. The linguistic trappings of both the theoretical and empirical terms of their theories will typically differ in different frameworks due to different formation and transformation rules, but each framework will specify the form of the observation sentence for each potential experimental outcome. Thus, Carnap hoped to achieve transtheoretical discourse not by translating all the theories into a new, neutral language, but by specifying a metalogical language into which all disputants may ascend and come to understand each other's languages. If feasible such a procedure would be preferable to the VPI approach; for however closely Kuhnian paradigms may approach Lakatosian research programmes there is an inevitable loss of texture in simply covering both of them with the term "guiding assumptions"—a loss of texture that may in fact make Lakatos and Kuhn seem to agree or disagree more than they do, since they now perforce are speaking of precisely the same things (cf. the remarks of R. Laudan *et al.* 1988, pp. 8f).

It might be argued that I am making too much of the nonnaturalism I associate with the H-D methodology of the VPI program. It might be claimed that such methodology is a general minimal necessary condition for scientific research. This objection does not indicate that my claim that it is nonnaturalist is wrong, however. Indeed, the typical way of justifying this insistence on the necessity of H-D testing is presumably that it is an *a priori* condition on appropriate testing. This takes us back into nonnaturalism—not by insisting on the *a priori* but because such a claim itself requires, it would seem, a transcendental or analytic and, hence, purely *a priori* justification. The only way to insist on the H-D method in a way consistent with naturalism would be to claim that the empirical evidence about the methodology of science shows that the H-D method is how in fact evidence always bears on theory. This claim seems mistaken; the best recent work on confirmation in both the Bayesian and bootstrapping schools indicates that the H-D method is fundamentally flawed.

More than this, however, the H-D method is not endorsed by most of the very historiographic philosophies of science under test by the VPI program.[3] If these are our best choices for naturalized philosophies of science, we ought not beg some of the questions among them that we want to test through a nonnaturalist commitment to H-D methodology.

4. Testing in and of Historiographic Philosophy of Science

The last point is crucial and in need of elaboration: If the historiographic philosophies of science are the ones that are subjected to testing, it should be because they are the ones that not only seem to have been arrived at in a quasi-naturalized way but because they are more likely to be correct than others in the field, such as logical empiricism or Popperianism. But the method of testing employed by the VPI program is seriously at odds with the views about testing endorsed by many of the philosophies under test. The insistence on the availability and use of a neutral language potentially (at least) begs the question against the incommensurability theses of Kuhn and Feyerabend. Moreover, the VPI program's radical atomism stands in opposition to the holism implicit or explicit in the views of Lakatos, Kuhn, and Feyerabend (at least) on theory testing.

In particular, their testing procedure commits them in advance to the truth of at least one of the theses allegedly under test and to the falsity of others. Consider theses (21.4) and (21.6) from the original monograph. These theses state (L. Laudan *et al.* 1986, p. 172) that theory appraisal:

> (21.4) is based entirely on phenomena gathered for the express purpose of testing the theory and which would be unrecognized but for that theory. <Lakatos 1978[a], p. 38>
>
> (21.6) is based on phenomena which can be detected and measured without using assumptions drawn from the theory under evaluation. <Laudan 1977, p. 143>

(21.6) is explicitly presupposed in the testing procedure that the VPI program claims to be necessary in order to truly naturalize philosophy of science. They claim, for example, that the language they construct is (R. Laudan *et al.* 1988, p. 9) "neutral with respect to the theories under scrutiny" and that this allows the theories to be tested by historical case studies that do not import question begging theoretical interpretation. Despite their putting this down as a point of agreement of all the theories under test (cf. L. Laudan *et al.* 1986, p. 155, thesis (7)), the most natural way to understand the second half of (21.4) within Lakatos's general view, is that the phenomena used to test any particular theory are recognized as bearing on that theory by virtue of what the theory itself says about the evidence, i.e., that the theory itself tells us what the evidence is that bears on the theory.[4]

By 1988, the VPI authors seem to have realized that (21.4) and (21.6) are at odds and that a commitment to the neutrality of the evidential base required in their testing procedure is very much at odds with most of the historiographic philosophies of science they seek to test. They write (R. Laudan *et al.* 1988, pp. 36f):

> Thesis [21.6] is, in a sense, the mirror image of [Lakatos's 21.4]... More importantly, [21.6] is directed against the claim of writers like Kuhn and Feyerabend, who sometimes seem to insist that observations are not only 'theory-laden' but laden with the very theory which those observations are designed to test... Thesis [21.6] denies that strong sort of parasitism.

But by basing the tests of the historiographic philosophies of science on a prior commitment to (21.6) the VPI program rules out some of the central claims of Lakatosian, Kuhnian, and Feyerabendian philosophy of science by means of an *a priori* commitment to a thesis that they deny.

This problem with the VPI methodology is best evidenced in their discussion entitled "Empiricism and the Philosophy of Science" (§2 of L. Laudan *et al.* 1986, pp. 146-149). Building on the criticism already cited that all the historiographic philosophies of science were "conceived either *a priori* to solve specific philosophical conundrums or *post hoc* to fit a small number of preselected examples" (L. Laudan *et al.* 1986, p. 143), the VPI authors argue that the use of history of science in historiographic philosophy of science is typically illustrative rather than probative. No attempt is made to devise or extract from history stringent empirical tests that would decide among the theories in question. After giving a list of the "epistemic defects" of historiographic philosophy of science—including holistic testing, reliance on secondary history, constant use of standard examples, lack of comparative testing, etc. —the VPI authors scold (L. Laudan *et al.* 1986, p. 148):

> [I]t is clear enough that the very philosophers who practice such halfhearted empiricism in their methodologies for studying science would tolerate nothing comparable in their reconstructions of methodology within science.

This claim strikes me as extremely dubious at best. The very importance of distinctly non-empiricist aspects of science as codified in the paradigms of Kuhn, the "metaphysical hardcore and protective belt" of Lakatosian research programmes, etc. were what the historiographic philosophies of science wanted to account for. The historiographic turn was a reaction to the perceived naive empiricism of the logical empiricists and traditional Popperians. The quest was for an account of the empirical and, hence, progressive nature of science despite the methodological importance of non-empiricist aspects of science.

The very objection the VPI authors raise to the historiographic philosophies of science due to their being formulated to solve a very limited class of problems is fundamentally at odds with their thesis (24), which they ascribe to Feyerabend (L. Laudan *et al.* 1986, p. 174):

> (24) Almost all theories derive their empirical import from a few successful tests and have to be tinkered with, or distorted, in order to cope with the rest of the evidence.

If this is what Feyerabend believes about scientific theories generally and if he has questions about the H-D methodology of the VPI program (as he surely does), then he clearly shouldn't find it objectionable if his philosophical view is based on a small number of illustrative cases rather than on the more stringent empirical tests demanded by the VPI authors. That's the rule rather than the exception in Feyerabend's account of science and there is no reason to be *more* of an empiricist at the metalevel then at the object level.

We have seen that the tone of the commitment of the VPI authors to H-D testing imports a nonnaturalist commitment to methodological principles at odds with the philosophies of science under test. This introduces a large problem of begging methodological questions that one might have thought would be empirical questions in naturalized philosophy of science. There is a second problem here also: the claims about testing made by Kuhn, Feyerabend, and Lakatos seem to be claims that better capture scientific testing than does the H-D methodology of the VPI program itself.

I am not an expert on physics (whence comes most of the evidence that the VPI program claims supports (21.6)) but I have labored a little in the jungles of linguistic pragmatics and semantics. In this labor I have seen little commitment to the type of comparative, atomistic testing methodology endorsed by the VPI program. Accounts do seem to be offered on the basis of a few examples; no particular attempt is made to see whether those examples are also understandable from the perspective of other theories in the field. Moreover, whether a given linguistic phenomenon such as presupposition is considered semantic or pragmatic and thus how and whether a given theory need account for it is dependent on one's commitment to a theoretical framework. There seems to be nothing like a way of describing the phenomena in question that is relatively neutral among the theories in the field and which imports enough structure to indicate how the theories bear on the phenomena at all. Thus, I see a rather strong form of parasitism in linguistic semantics and pragmatics, but one that does not make impossible the comparative assessment of theories both in terms of their internal coherence and their empirical record.

It is open to respond to this (admittedly) rather vague story by saying that to whatever extent linguistics shows these tendencies it is not a mature or true science. Such a response again, however, seems sensible only on view that philosophy of science gives us *a priori* criteria for true science and this seems opposed to the VPI desire to naturalize philosophy of science. In any case, if I were tempted to give methodological advice to linguists on testing their theories, I wouldn't offer a VPI style H-D methodology with a theoretically neutral demarcation between semantic phenomena and pragmatic phenomena, but would rather urge linguists to articulate their theoretical perspectives more sharply so that each theory would say more clearly what it took to be the demarcation between the semantic and the pragmatic. This greater articulation would then lend to a measure of theoretical control over the empirical phenomena and would point out problematic phenomena within each theory. This seems more Lakatosian than H-D. But this type of parasitism still allows cross theoretical judgements of better and worse empirical grounding of the theories since the theories will be understandable by all linguists by virtue of their explicit articulation and there can be generally agreed judgements about their relative empirical strength and internal coherence.[5]

5. Circularity and Pluralism in Naturalized Philosophy of Science

It may be argued that adopting a testing procedure more in line with Kuhn's, Lakatos's, or Feyerabend's views on theory testing would beg just as many questions against opposing views of theory testing as adopting the VPI methodology. This is almost right, but not quite. The nonnaturalist, *a priorist* tone of the pronouncements of the VPI authors that this is how the testing *must* proceed can and should be left out. In its place we could put a self-consciously reflexive position along the lines of Lakatosian metamethodology that recognizes certain methodological commitments to be metaphysical presuppositions of particular research programmes without there being any absolute sense in which they are necessary truths about scientific testing.[6] But clearly, beyond this more consistent naturalism, which is available to but not utilized by the VPI program, any and all such programs will make certain prior commitments about the appropriate form of testing that will influence their methodologies for testing the philosophies of science.

We should note that this notion of prior commitment to certain forms of testing isn't guaranteed to be self-supporting. Just as the VPI program might have committed itself to (21.6) in designing its tests but have resulted in the rejection of (21.6) in science, all other programs based on the historiographic philosophies themselves might make prior commitments to testing procedures that lead to the rejection of precisely those commitments in accord with the results of their scientific tests. Of course, there is a complication. Just as the VPI program's prior commitment to H-D methodology might have lead them to interpret historical episodes as exhibiting H-D methodology in science (since after all few, if any, episodes of the history of science are univocally examples of one method to the exclusion of others), so too, for example, a commitment to a Lakatosian metamethodology may lead to a high likelihood of reconstructing history in accord with Lakatosian methodology.

Nevertheless, it seems that testing of philosophy of science can proceed in accord with the views of historiographic philosophy of science. Let us, as an example, consider for a moment a generally Kuhnian model for testing the historiographic philosophies of science. Someone within the Kuhnian school of thought would think that these theories aren't to be tested by translating them all into a neutral language and testing the theses individually. Rather these theories are to be compared against one another holistically by establishing their own puzzle solving traditions of normal (philosophy of) science wherein each theory attempts to show how various episodes in scientific history can be understood in light of the theory being elaborated. Indeed, Kuhn's own historical work in the history of astronomy and physics might be viewed as exemplars of how historical episodes are to be cast in light of his theory of theory change. The various theories can then be tested by seeing whether any of these puzzle solving traditions has significantly more success in illuminating the history of theory change. This program has obvious affinities to Lakatos's and indeed Laudan's metamethodology.

It may seem that I am endorsing a naturalized philosophy of science that amounts to no more than a plurality of self-refuting or circular research programmes. And indeed the methodological circularity of naturalized philosophy of science has been urged from many quarters and this "circle argument", as Giere (1985, p. 333) has dubbed it, has been taken by many to decisively refute naturalism in philosophy of science. But like Giere (1985) we can endorse what is essentially a skeptical solution to the skeptical question of circularity.[7] Giere (1985, pp. 334-339) argues that no attempt by nonnaturalists, be they logical empiricists, normative metamethodologists, or what have you, has succeeded in giving a nonarbitrary *a priori* response to the problem of circularity. Since nonnaturalism doesn't succeed in solving the very problems that nonnaturalists find with naturalism, naturalists should acknowledge and be sensitive to but not paralyzed by the methodological circularity of their accounts.

Nor need the circularity be vicious and here is where Lakatosian rational reconstruction and methodological pluralism helps. Lakatos insists that his views don't imply that anything goes in science because of two principal factors. First, each research programme is intellectually honest enough to admit to both its successes and whatever failures it has. This is the intellectual scorecard that Lakatos requires for science; "[t]he scores of the rival sides, however, must be recorded and publicly displayed at all times" (Lakatos 1978b, p. 113). Second, the programmes all must fulfill the methodological advice that they be articulated in a sufficiently perspicuous fashion for these failings and successes to be generally agreed on by the community of scholars. Add to this (the perhaps too pious hope that it is) the desire among philosophers of science generally to understand the business of science rather than vindicate some particular naturalist programme or other and we have a naturalist notion of philosophy of science very much in accord with the rather oblique role that Kuhn, Lakatos, and (perhaps) Feyerabend tell us empirical evidence plays in science as a whole.

The pluralism I am envisioning cannot rule the VPI program out of the game, but it insists that they retreat from the nonnaturalism entailed by their insistence that the H-D method is how evidence must bear on naturalized philosophies of science. They must replace this with an explicit commitment to the methodological drive to design H-D tests of philosophies of science as part of the methodological identity conditions (or metaphysical hardcore) of the program. Other programs must make their methodological decisions also and then a Lakatosian scoreboard can be kept to see how things are progressing. I have argued, beyond this, that we have reason to doubt that the H-D methodology will in fact prove the most plausible candidate for scientific testing whether as a finding or a methodological presupposition of naturalized philosophy of science.

6. The Persistence of Logical Empiricism

I end with one final irony. I have argued that the VPI program offers a nonnaturalized metamethodology based on an endorsement of a particularly strong version of one thesis of logical empiricism—the H-D method of scientific testing. Thus, one could say that the VPI program offers us logical empiricism once removed. On the other hand, the type of naturalism that I have sketched as an alternative places great weight on methodological decisions which provide part of the conditions of identity of research programmes within naturalized philosophy of science and that provide the conditions under which episodes of scientific history are theoretically interpreted. This means that precisely the methodological questions raised by conventionalism and the revolutions in mathematics, logic, and physics that the logical empiricists were concerned with and which were codified as analytic sentences of linguistic frameworks by Carnap will not and could not go by the boards simply by endorsing naturalism in philosophy of science. The commitment to the idea that these methodological problems can be resolved purely *a priori* by logical means does disappear, but the importance of conventional decisions, the articulation of frameworks, the bearing of empirical evidence on theories within frameworks, etc., which were the primary concerns of the logical empiricists, reassert themselves as the primary questions of philosophy of science. The persistence of the importance of the problems that exercised the logical empiricists and the suspicion that the trajectory of the logical empiricist program has much light to shed on these issues both positively and negatively should lead us to endorse at least one claim of the VPI authors: "that the philosophy of science [should be] rooted in and responsible to its history" (L. Laudan *et al.* 1986, p. 142).[8]

Notes

[1]A prior version of this paper was read at the 1990 Minnesota Philosophical Association Meetings. I would like to thank my audience there and especially my commentator, James D. Fetzer, for helpful comments. I would like to thank also the University of Minnesota Naturalized Philosophers of Science: Ron Giere, Charles Wallis, Geoffrey Gorham, and Jerry Smerchansky, for inspiration, criticism, comments, and research materials. None of these philosophers should be held responsible for my errors in this paper.

[2]The necessity of the neutrality of the observation language was argued against by Carnap as early as 1932 in his exchange with Neurath on protocol sentences. See Carnap (1932, 1987).

[3]With this claim I part company with Nickles (1986), which I find to be otherwise an extremely trenchant critique of the VPI methodology.

[4]This view has obvious affinities with views about the relation of theory and evidence in van Fraassen's (1981) constructive empiricism, Glymour's (1980) bootstrapping, and certain unificationist views in scientific explanation and their bearing on confirmation and realism (cf. Friedman 1981).

[5]I assume, of course, no incommensurability, which even believers admit is more the exception than the rule in science.

[6]Lakatos seems to rely only on a relativized notion of the *a priori* very reminiscent of early logical empiricist conceptions of conventionalism, especially Reichenbach's (1920). Recent works (Friedman 1991, Reisch 1991; compare also Kuhn 1991) have noted similarities between Carnap's linguistic frameworks and Kuhn's paradigms, if anything the connection between Carnap and Lakatos is more striking.

[7]Giere (1985, pp. 337f) also seems to think of Lakatos's metamethodology as non-naturalist. On the contrary I find typically naturalist self-reflexivity in Lakatosian metamethodology. Lakatos certainly isn't a Quinean naturalist given his endorsement of something like a relativized *a priori* (see endnote 6 above), but naturalists needn't deny the *a priori* but only deny that it plays a different role in philosophy of science than it does in science generally. Lakatosian naturalism is more a completely immanent Kantianism then a radical empiricism.

[8]I admit that I exploit an ambiguity in this phrase from the VPI manifesto. It is clear in context that they mean the 'its' to refer anaphorically to 'science' and not to 'philosophy of science'.

References

Carnap, R. (1932), "Über Protokolsätze", *Erkenntnis* 3:215-228. (Translated as Carnap 1987.)

______. (1987), "On Protocol Sentences", *Nous* 21: 457-470.

Friedman, M. (1981), "Theoretical Explanation", in *Reduction, Time, and Reality,* R. Healey (ed.). Cambridge: Cambridge University Press, pp. 1-16.

________. (1991), "The Re-evaluation of Logical Positivism", *Journal of Philosophy* 88: 505-519.

Giere, R. (1985), "Philosophy of Science Naturalized", *Philosophy of Science* 52: 331-356.

_____. (1988), *Explaining Science*. Chicago: University of Chicago Press.

Glymour, C. (1980), *Theory and Evidence*. Princeton: Princeton University Press.

Hull, D. (1989), *Science as Process*. Chicago: University of Chicago Press.

Kuhn, T. (1991), "The Road Since Structure", in *PSA 1990*, volume 2, A. Fine, M. Forbes, and L. Wessels (eds.). East Lansing, MI: Philosophy of Science Association, pp. 3-13.

Lakatos, I. (1978a), "Falsification and the Methodology of Scientific Research Programmes", in his *Methodology of Scientific Research Programmes*. Cambridge: Cambridge University Press, pp. 8-101.

______. (1978b), "History of Science and Its Rational Reconstructions", in his *Methodology of Scientific Research Programmes*. Cambridge: Cambridge University Press, pp. 102-138.

Laudan, L. (1987), "Progress or Rationality? The Prospects for Normative Naturalism", *American Philosophical Quarterly* 24: 19-31.

______. (1990), "Normative Naturalism", *Philosophy of Science* 57: 44-59.

______. A. Donovan, R. Laudan, P. Barker, H. Brown, J. Leplin, P. Thagard, and S. Wykstra (1986), "Scientific Change: Philosophical Models and Historical Research", *Synthese* 69: 141-223.

Laudan, R, L. Laudan, and A. Donovan (1988), "Testing Theories of Scientific Change", in *Scrutinizing Science*, A. Donovan, L. Laudan and R. Laudan (eds.). Dordrecht: Kluwer, pp. 3-44.

Nickles, T. (1986), "Remarks on the Use of History as Evidence", *Synthese* 69: 253-266.

Reichenbach, H. (1920), *Relativitätstheorie und Erkenntnis A Priori*. Berlin: Springer.

Reisch, G. (1991), "Did Kuhn Kill Logical Empiricism?", *Philosophy of Science* 58: 264-277.

van Fraassen, B. (1981), *The Scientific Image*. Oxford: Oxford University Press.

Part II

EXPERIMENTATION

Some Complexities of Experimental Evidence

Margaret Morrison

University of Toronto

1. Introduction

Traditional accounts of the relationship between experimental evidence and hypotheses have been challenged in the recent work of Hacking (1983), Galison (1987), and Franklin (1986, 1990). These authors have pointed to the independent aspects of the experimental tradition and forced us to rethink the view that experiment is simply the handmaid of theory. As a result we have come to recognize the tremendous intricacies involved in theory testing and experimentation and have come to appreciate the various ways experiment shapes scientific practice.

In what follows I want to extend this discussion by taking a closer look at the way in which theories are supported by empirical evidence. In addition to the traditional testing model the method of deduction from phenomena, made famous by Newton, is also thought to furnish experimental evidence insofar as the theory itself is supposedly *deduced* from empirical data. Although some attention has been paid to this method in the literature, particularly as it functions in the work of Newton and Einstein[1]; I want to investigate a lesser known example of deduction from phenomena, specifically Maxwell's 1865 formulation of his electrodynamics. Maxwell's paper, entitled "A Dynamical Theory of the Electromagnetic Field" (hereafter DT), is especially important since in it he claims that its conclusions are derived from experimental facts; yet the displacement current, one of the key features that enabled him to derive the velocity for the propagation of electromagnetic waves, was interpreted by Maxwell's colleagues as completely lacking experimental evidence and support.

The importance of this case is that it clarifies, in a novel way, some peculiarities in the nature and extent of evidential support that is achieved through the method of deduction from phenomena. Here I am referring not only to the role the empirical evidence plays in formulating and establishing theories but also whether a method like deduction from phenomena can be truly seen as providing an experimental foundation for a theory. Part of the answer to this question will depend on how we view the nature of theories, their function as explanatory vehicles, and whether the phenomena used to "deduce" the theory should be seen as also supporting explanations offered by the concepts and mechanisms that form the core of the theoretical machinery.

PSA 1992, Volume 1, pp. 49-62

Although deduction from phenomena may be an important first step in isolating the experimental basis for particular theories, I argue that in the case of electrodynamics it is unable to sustain necessary features of the theoretical structure, thereby undermining its ability to function as a method for providing a secure empirical foundation.

2. Analogies and Models: Maxwell's Introduction of Displacement

Before discussing the evolution of electrodynamics and the factors leading up to the introduction of displacement a few points need to be stressed. First, not all accounts of electromagnetism required the displacement current; action at a distance theories, especially Weber's, accounted for the phenomena equally well without it. And, even those like Kelvin who were committed to field theory, thought displacement involved an illegitimate appeal to hypotheses. Secondly, because displacement played such a key role in the unification of electromagnetism and optics (without it Maxwell could not have derived the velocity for propagation of electromagnetic waves) it could have been easily justified within a hypothetico-deductive framework. However, Maxwell was extremely critical of such methods preferring to use analogy, deduction from phenomena and what he called "appropriate" mechanical ideas and concepts in the construction of theories. However, as we shall see this approach could not deliver an experimental foundation for displacement.

Maxwell's first formulation of electromagnetism completed in 1861 was not a deduction from phenomena and instead can be traced to work he did in 1856 involving a geometrical illustration of Faraday's lines of force. The 1856 paper, "On Faraday's Lines of Force" (FL) was based on a hydrodynamical analogy between the motion of an incompressible fluid flowing in tubes and a stationary field. The formal equivalence between the equations of heat flow and action at a distance enabled Maxwell to substitute the flow of the ideal fluid for the distant action. Although the pressure in the tubes varied inversely as the distance from the source the crucial difference was that the energy of the system was in the tubes rather than being transmitted at a distance. The direction of the tubes indicated the direction of the fluid in the way that the lines of force indicated the direction and intensity of a current. Both the tubes and the lines of force satisfied the same partial differential equations. Maxwell extended the hydrodynamic analogy to include electrostatics, current electricity and magnetism. Its purpose was to illustrate the mathematical similarity of the laws governing these phenomena and although the fluid was a purely fictional entity it provided a visual representation of this new field theoretic approach to electromagnetism.

What Maxwell's analogy did was furnish a physical "conception" for Faraday's lines of force; a conception that involved a fictional representation yet provided a mathematical account of electromagnetic phenomena.[2] This method of physical analogy, as Maxwell referred to it, marked the beginning of what he saw as progressive stages of development in theory construction. Physical analogy was intended as a middle ground between a purely mathematical formula and a physical hypothesis. The former causes us to loose sight of the phenomena to be explained while the latter clouds our perception by imposing theoretical assumptions that restrict our ability to evaluate alternatives. By contrast, the method of physical analogy allows us to grasp a clear physical conception without full blown commitment to a particular physical theory and prevents us from being drawn away from the subject under investigation by the pursuit of "analytical subtleties" (1890, 1 p.156).[3] So, in a physical analogy we have a partial similarity between the laws of one science and those of another.[4]

Although the analogy did provide a model in some sense, it was merely a descriptive account of the distribution of the lines in space with no mechanism for understanding the forces of attraction and repulsion between magnetic poles. A physical treatment was developed in a paper written by Maxwell in 1861-2 entitled "On Physical Lines of Force" (PL). The goal was to find an account of the physical behaviour of magnetic lines that could give rise to magnetic forces. Prior to this Kelvin had explained the Faraday effect (the rotation of the plane of polarized light by magnets) as the result of the rotation of molecular vortices in a fluid aether. Maxwell used this idea to develop an account of the magnetic field that involved the rotation of the aether around lines of force. The paper also offered an account of the forces that caused the medium (or aether) to move and of the occurrence of electric currents. This required an explanation of how the vortices could rotate in the same direction; the problem which led Maxwell to develop his famous mechanical aether model. The model involved a vortex motion that resulted from a layer of rolling particles called idle wheels that were interspersed between the vortices. Electromotive force was then explained in terms of the forces exerted by the vortices on the particles between them. Although Maxwell was successful in developing the mathematics required for his model he was insistent that the representation be considered provisional and temporary.

> The conception of a particle having its motion connected with that of a vortex by perfect rolling contact may appear somewhat awkward. I do not bring it forward as a connexion existing in nature, or even as that which I would willingly assent to as an electrical hypothesis. It is, however, a mode of connexion which is mechanically conceivable, and easily investigated... . I would venture to say that anyone who understands the provisional and temporary character of this hypothesis, will find himself rather helped than hindered by it in his search after the true interpretation of the phenomena (1890, 1 p.486).

The difficulty with the model was that Maxwell was unable to extend it to electrostatics, a problem that led him to propose a rather different account in part three of the paper. Instead of the hydrodynamic model consisting of a fluid cellular structure he developed an elastic solid model made up of spherical cells endowed with elasticity. In October of 1861 Maxwell wrote to Faraday describing his "conception" of electric forces.[5]

> ...They are called into play in insulating media by slight electric displacements, which put certain small portions of the medium into a state of distortion, which, being resisted by the elasticity of the medium, produces an electromotive force. A spherical cell would, by such a displacement, be distorted.
> I suppose the elasticity of the sphere to react on the electrical matter surrounding it, and press it downwards. (Campbell and Garnett 1884, p.244)

In order to account for the transmission of rotation from the exterior to the interior parts of each cell Maxwell assumed that the substance in the cells possessed elasticity similar to that observed in solids. This assumption of elasticity was further supported by analogy with the luminiferous medium which was assumed to be an elastic solid in order to account to the transverse vibrations required by the the wave theory of light.

On this new account the particles forming the partitions between the cells constituted the matter of electricity while their motion was thought to constitute an electric current. The tangential force with which the particles were pressed by the matter of the cells was termed the electromotive force and the pressure of the particles on each other corresponded to electric potential. What Maxwell wanted to do was explain "the condition of a body with respect to the surrounding medium when it is said to be charged and account for the forces acting between electrified bodies" (SP, 1 p.490). To accomplish this he first examined the relationship between the transmission of electricity in conductors and insulators (dielectrics). Although the capacity of a dielectric was a different quantity than the resistance of a conductor the electric potential was the same regardless of whether it was the potential difference between two insulated conductors (and hence related to the properties of the dielectric) or between two parts of the same conductor. Consequently one could describe the behaviour of a dielectric as being in a state of polarization with every particle having its poles in "opposite conditions"; a description completely analogous to the behaviour of particles of iron under the influence of a magnet. It was at this point that Maxwell first introduced displacement:

> In a dielectric under induction, we may conceive that the electricity in each molecule is so displaced that one side is rendered positively, and the other negatively electrical, but that the electricity remains entirely connected with the molecule, and does not pass from one molecule to another.
> The effect of this action on the whole of the dielectric mass is to produce a general displacement of the electricity in a certain direction... .
> The amount of displacement depends on the nature of the body and on the electromotive force... . (SP, 1 p.491)

Although displacement was not understood as a current (it remained constant at a certain point with the force producing it balanced by the elastic reaction of the medium) the change in displacement did constitute a current **r** which was represented by the displacement **h** in the **z** direction, where **r=dh/dt**. This displacement was related to the electromotive force **R** by Hooke's law where $\mathbf{R}=-4\pi\mathbf{E}^2\mathbf{h}$ and E is a coefficient of elasticity. It is important to point out here that Maxwell understood these relations as "independent of any theory about the internal mechanism of dielectrics" (Ibid.). In other words, they were equations that described the behaviour of elastic bodies and were not derived from particular assumptions about the nature of the aether. They were simply phenomenological laws based on observations of properties of dielectrics subjected to electric forces. The conceptualization of displacement was, however, directly associated with the mechanical aether model.

The introduction of displacement required that Maxwell modify Ampère's law; however, what is particularly interesting from the point of view of "deductions from phenomena" is the lack of any experimental data that necessitated such a change.[6] Maxwell's extension of his theory to electrostatics required a mechanism to account for electric charge which supposedly resulted from an accumulation of electric particles. The motion of the idle wheels in part two of PL were governed by Ampère's law which related electric flux and magnetic intensity (curl $\mathbf{H}=4\pi\mathbf{J}$ where **H** is the magnetic field and **J** is the electric current density). But, because this law applied only to closed currents there was no way of representing the build up of electrical particles that produced charge. As a result Maxwell added a term to Ampère's law to ensure that the current was no longer circuital. With this component $\partial D/\partial t$ (where $\mathbf{D} = (1/c^2)\mathbf{E} + 4\pi\mathbf{P}$ (the polarization vector)) added to the current its force was proportional to the rate of increase of the electric force in the dielectric and therefore produced the same magnetic effects as a true current.[7]

Displacement allowed Maxwell to calculate the coefficient of rigidity for the aether and arrive at a value for the propagation of electromagnetic waves, the crucial step that ultimately led to the unification of electromagnetism and optics.[8] However, this result didn't provide any additional empirical support for displacement or for Maxwell's aether model. Clearly the convergence of values for electromagnetic and light waves would have furnished increased inductive support on an hypothetico-deductive model of theory acceptance; but Maxwell as well as his opponents had little faith in the details of his aether model and did not consider it to be even an approximate representation of reality. He explicitly remarked in PL that he had shown "in what way electromagnetic phenomena may be imitated by an *imaginary* system of molecular vortices"(SP, 1, p.488, emphasis added). Although the law describing displacement was essentially an empirical relation, its conceptualization was intimately linked with the aether model and in that sense could not be thought of as a deduction from phenomena.

Later formulations of the electromagnetic theory did not rely on the aether model but they did include displacement as a crucial parameter. The view presented in DT is especially interesting in that it is an explicit attempt to secure an empirical foundation for the electromagnetic theory. But, unless displacement could also be given an experimental justification its presence could only be justified by appeal to the very methodology Maxwell had explicitly criticized.

> In forming dynamical theories of the physical sciences it has been a too frequent practice to invent a particular dynamical hypothesis and then by means of the equations of motion to deduce certain results. The agreement of these results with real phenomena has been supposed to furnish a certain amount of evidence in favour of the hypothesis. (SP, 2 p.309)

Maxwell's solution to the problem was to adopt what he called the "true method of physical reasoning" which was to "begin with the phenomena and deduce the forces from them by a direct application of the equations of motion"(Ibid.). This was the approach he used in DT.

3. The Experimental Foundation of Electromagnetism

In October 1864 Maxwell wrote to Stokes claiming that he finally had

> ...the materials for *calculating* the velocity of transmission of a magnetic disturbance through air founded on *experimental evidence* without any hypotheses about the structure of the medium or any mechanical explanation of electricity or magnetism. (Larmour 1907, 2 p.26, emphasis added)

Maxwell's stated goal for the dynamical theory was to

> clear the electromagnetic theory of light from any unwarrantable assumptions, so that we may safely determine the velocity of light by measuring the attraction between bodies kept at a given difference of potential, the value of which is unknown in electromagnetic measure.[9]

In addition to the awkwardness of Maxwell's mechanical model the idea that a field theoretic account of electricity involved particles was highly undesirable to Maxwell and to his other supporters, including Faraday (SP,1 p.486). Although the alternative action at a distance theories accounted for the known phenomena there was a long history of difficulty associated with the possibility of providing a mechanical explanation for distant forces. The significance of DT was that Maxwell was able to derive the wave equation, not from an abstract model, but from results that were themselves established by electrical experiments.

The fact that Maxwell chose to establish his theory without the aid of hypotheses is particularly important for understanding his substantive commitments in DT. He began the paper by enumerating the reasons for supposing the existence of a luminiferous medium capable of transmitting motion from one part to another. One in particular was the Faraday effect which showed that this medium could be altered by magnetic force causing the rotation of the plane of polarized light. This phenomenon suggested that there was "warrantable grounds for inquiring" whether there was a motion of the aetherial medium wherever magnetic effects were observed, and whether this motion was rotational with the direction of the magnetic force as its axis. Electrical phenomena also supported the possibility of such a medium. When a body was moved across the lines of magnetic force it experienced what was termed electromotive force causing the body's extremities to become oppositely electrified, thereby producing a current. The force was identified empirically with the observed motion of a body across a field, resulting in the production of currents. If the force did not heat the body, produce a current or decompose it, it put it in a state of electric polarization which disappeared when the force was removed. The result of this polarization was a general displacement of electricity or charge in a certain direction. From the effects observed in electrical and magnetic phenomena Maxwell concluded that one is led to the "conception" of a complicated mechanism capable of a variety of motion and subject to the general laws of dynamics.

This idea of a "conception" is crucial for Maxwell's understanding of dynamical theory but before discussing it further let me first focus on the formulation of the field equations and the experimental facts that Maxwell cites. He began with a discovery of Oersted's—when an electric current is established in a conducting circuit the neighbouring part of the field (or surrounding space) is characterized by magnetic properties. Similarly if two circuits are in the field the magnetic properties of the field due to the two currents are combined. Hence each part of the field is in connection with both currents and the two currents are connected with each other in virtue of their connection with the magnetization of the field. When magnetic forces are exerted in the field their direction and magnitude depends on the form of the conductor carrying the current and when the strength of the current is increased the magnetic effects are increased proportionally. If the magnetic state of the field depends on the motions of a medium then a certain force must be exerted in order to increase or diminish these motions. Hence, the effect of the connection between the current and the electromagnetic field is to endow the current with a kind of momentum in a manner similar to the way in which the connection between the driving point of a machine and a fly wheel endows the driving point with additional momentum. In the case of electric currents the resistance to sudden increase or decrease of strength also produces effects like those of momentum which depend on the shape of the conductor and the relative position of its parts. To explain the connection Maxwell used a dynamical illustration with the medium represented as a body with a specific mass ***C*** and the current represented as independent driving points **A** and **B**. Using Lagrange's elementary laws of dynamics Maxwell was able to calculate the forces acting at *A* and *B* which included the momentum of **C** referred to **A** and **B**. In keeping with the generality of the Lagrangian approach Maxwell stressed that the analogy was used simply to illustrate the notion of a reduced momentum in mechanics (SP, 1, pp.537-538). Although the facts about the induction of currents (which depended on variations of the quantity termed electromagnetic momentum) were based on Faraday's experiments, the use of the words electromagnetic momentum to designate that quantity was intended merely as a guide to understanding the electrical phenomena on analogy with the mechanical ones. No account of the structure of the medium was provided.

The first result of the connection between currents and the field that Maxwell examined was the induction of one current by another and by the motion of conductors in the field. The second result, also deduced from the connection between electric and magnetic phenomena, concerned the mechanical action between conductors carrying currents. The magnetic field due to each current is a vector quantity, hence the fields add vectorially and any decrease or increase of one current will produce a force with or contrary to the other. This phenomenon of induction had been deduced from the mechanical action of currents by Helmholtz and Thomson. Maxwell followed a reverse order, deducing the mechanical action from the laws of induction and describing experimental methods for determining the coefficients of induction, the values of which depended on the geometrical relationship of the conductors. He then applied the phenomena of induction and attraction of currents to the exploration of the electromagnetic field. Next Maxwell specified a system of magnetic lines of force which indicated the magnetic properties of the field and using a uniformly magnetized bar showed the distribution of equipotential surfaces which cut the lines of force at right angles.

Maxwell claimed that it was these results, expressed symbolically, that formed the general equations of the electromagnetic field. The equations all expressed relations between various quantities like Electromotive Force, Magnetic Intensity, Electric Displacement, Electric Potential, etc., and from these quantities he was able to express the intrinsic energy of the field as a function of its magnetic and electric polarization at every point. He stressed that the conclusions were independent of hypotheses about the medium and were deduced from three kinds of experimental facts: (1) the induction of electric currents by the increase or diminution of neighbouring currents according to the changes in the lines of force passing through the circuit; (2) the distribution of magnetic intensity according to the variations of magnetic potential and (3) the induction of statical electricity through dielectrics. From these principles Maxwell demonstrated the existence and laws of mechanical forces acting on a movable conductor carrying an electric current, on a magnetic pole and on an electrified body. The last result gave rise to an independent method of electrical measurement based on electrostatic effects. This work had been done by Maxwell in connection with a committee set up to determine a set of electrical standards. The first experiment was on a standard of resistance and a new system based on energy principles was set up with Maxwell formulating definitions of electric and magnetic quantities related to measure of mass, length and time. The analysis disclosed five different classes of experiments from which **C** might be determined. One was a direct comparison of electromagnetic and electrostatic forces carried out by Maxwell and his assistant Hockin. So, even prior to the writing of DT, Maxwell had found a phenomenological link between electromagnetic quantities and the velocity of light.

Next he calculated the electrostatic capacity of a condenser and the specific inductive capacity of a dielectric. The equations were then applied to the case of a magnetic disturbance propagated through a nonconducting field and Maxwell showed that if these disturbances were to be propagated as plane waves they would have to be transverse vibrations with a velocity very nearly that of light. In order to further demonstrate the use of his equations Maxwell applied them to various other phenomena including the self induction of the coil used in the experiments on electric resistance.

It is important to keep in mind that at the time Maxwell was formulating his account there was no experimental evidence whatsoever for field theoretic phenomena; that is, it had not been shown that any kind of electromagnetic activity was in fact propagated through the field. What Maxwell could do however, and this is where his initial remark to Stokes (above) is crucial, is *calculate* these field theoretic results using the behaviour of electric and magnetic phenomena as the empirical foundation.[10] No qualitative conclusions about field theory had been established but the calculations provided by the equations could, supposedly, be seen to rest on experimental facts.

The relation between the units employed in the equations and the experimental work on resistance depended on what Maxwell referred to as the "electric elasticity" of the medium, a quantity associated with displacement. The interesting twist in the argument however is that electric elasticity, like electromagnetic momentum, was considered to have an illustrative function only (SP, 1, p.564). When Maxwell introduced the qualitative discussion of displacement in DT he claimed that in a dielectric under the action of electromotive force we could "conceive that the electricity in each molecule is so displaced that one side is rendered positively and the other negatively electrical... . The effect of this action on the whole dielectric mass is to produce a general displacement of electricity in a certain direction" (SP, 1 p.531). This displacement was a result of the elasticity of each part.

In section III where Maxwell discussed the general equations of the electromagnetic field he again introduced displacement as dielectric polarization on analogy with magnetic polarization. He claimed that the "variations of the electrical displacement must be added to the currents p, q, r to get the total motion of electricity"(SP, 1, p.554). He then substituted this total motion for the conduction current in Ampère's law and used it to refer to all dielectrics including the aether. Displacement was no longer associated with the change in position of rolling particles; instead Maxwell defined it as the motion of electricity, that is, in terms of a quantity of charge crossing a designated area. The effect of an electromotive force on a dielectric was manifested as a macroscopic polarization or displacement of electric charge, and although it could be described as the result of microscopic polarization this was not necessary for the result established in DT. The crucial methodological point in Maxwell's argument is that displacement in the aether is not *inferred analogically*.

Mary Hesse (1973) criticizes Maxwell for his rather cavalier attitude in passing from "real" electric current and dielectric polarization to displacement current and polarization in aether and hence for regarding the analogy as needing no further support. She claims that his lack of a specific interpretation of the causal nature of displacement prevented him from appealing to the kind of inductive evidence that results from a properly constructed analogical argument. In addition, the ambiguity surrounding displacement carried over into Maxwell's interpretation of charge, which was sometimes described as an incompressible substance satisfying a fluid continuity equation and sometimes as an epiphenomenon of the aetherial medium or field. Hesse claims that in order to be successful Maxwell's analogies require a commitment to one of these physical interpretations.

It seems clear however from Maxwell's discussion in DT that he intended to establish only a *formal* analogy between displacement and magnetic polarization. Although he provided a physical interpretation of displacement using the electromagnetic medium and the mechanisms necessary for describing how polarization takes place, he claimed that these terms are to have an illustrative function only. In contrast to the causal account in PL, DT is devoid of any *hypotheses* concerning the motion and strain of the medium (SP, 1, p.563). In fact, Maxwell explicitly stated that one can speak about the energy of the field, without invoking hypotheses, as simply "magnetic and electric polarization" or with the aid of hypothesis as "the motion and strain of one and the same medium". The experimental facts that he cited (Ibid.) facilitated the introduction of laws describing mechanical forces, laws that were applicable to the medium but did not require the medium for their legitimacy. The dynamical account of energy transformations provided the field representation, but this needn't be equated with an endorsement of the aether hypothesis. Elsewhere in a Royal Institution lecture he remarked that:

> ...those who introduce aetherial, or other media, to account for these actions, without any direct evidence of the existence of such media, or any clear understanding of how the media do their work,... why the less these men talk about their philosophical scruples about admitting action at a distance the better. (SP, 2 p.315)

Because the general equations which represented the intrinsic energy of the field were not directly linked with the mechanical connections between moving bodies it was possible for the fields to exist in space independent of material bodies. Hence fields were thought to be continuous in nature in contrast with the Newtonian version which defined fields in terms of the action of bodies. Despite the mathematical success of the argument Maxwell was unable to reconcile the ways in which the two different forms of energy, potential and kinetic, could manifest themselves without appeal to a material, mechanical representation. Consequently the notion of a medium together with various mechanical concepts reemerge in an illustrative role. The dynamical method allowed Maxwell to focus on formal similarities while supplementing these formal relations with constructive images that provided a "conception" of the phenomena.[11] The generality of the Lagrangian dynamics used in DT allowed Maxwell to neglect the specific mechanisms responsible for electromagnetic effects thereby attaining a degree of certainty unavailable in more concrete formulations of physical theory. In fact, he remarked (SP, 1, p.531) that electromotive force did not correspond to an ordinary mechanical force that would move a body of electricity in a conductor. Instead the analogy was with a complex system whose connections were left unspecified.[12]

At the end of the discussion of the field equations he contrasted the theory presented in DT with this work in "Physical Lines". In the latter context he attempted to describe

> a particular kind of motion and a particular kind of strain, so arranged as to account for the phenomena. In the present paper I avoid any hypothesis of this kind; and in using such words as electric momentum and electric elasticity in reference to the known phenomena of the induction of currents and the polarization of dielectrics, I wish merely to direct the mind of the reader to mechanical phenomena which will assist him in understanding the electrical ones. All such phrases is the present paper are to be considered as illustrative, not as explanatory (SP, 1, p.564).

For Maxwell a truly scientific illustration consisted of a method that enabled the mind to grasp some conception or law in one branch of science by "placing before it a conception or law in a different branch of science, and directing the mind to lay hold of that mathematical form which is common to the corresponding ideas in the two sciences, leaving out...the difference between the physical nature of the real phenomena"(SP, 2, p.219). The vortex aether had been replaced by an analysis of the *relations* between electromagnetic and optical phenomena. In other words, displacement was not to be understood in a substantive way, as providing a mechanical explanation of what takes place in the process of dielectric polarization. Instead it is merely an empirical relation describing the behaviour of macro processes.

But this doesn't really solve the problem. Equations describing total currents, electric elasticity, electromotive force all involved displacement as a parameter, in one form or another. And, although the empirical manifestation of displacement was polarization there needed to be some evidence that displacement was an actual process that took place in the molecules of material dielectrics as well as in the field. In other words, the terms which represented the components of displacement needed to have an empirical foundation if the theory was to be truly derived from experimental facts. This was necessary if displacement was to have a *causal* role rather than simply being identified with an observable *effect* like macro-polarization. As Maxwell defined it, "if a quantity of electricity which would appear on the faces **dy, dz** of an element **dx, dy, dz** cut from a body be **f, dy, dz** then **f** is the component of displacement parallel to **x**" (SP, 1, p.554). Variations of displacement are added to currents *p*, *q*, *r* to get the total motion of electricity **p', q', r'** so that:

$$p' = p + df/dt;\; q' = q + dg/dt;\; r' = r + dh/dt$$

Moreover, displacement had nothing to do with the experimental evidence that Maxwell cited as the basis for his derivation of the conclusions presented in the paper. The question that arises then, is what exactly could Maxwell have meant when he claimed that his theory was derivable from experimental facts?

Interestingly enough Maxwell himself acknowledged the problem in a note written in 1868 where he discussed the relationship between electric displacement and electric current.

> The current produced in discharging a condenser is a complete circuit and might be traced within the dielectric itself by a galvanometer properly constructed. I am not aware that his has been done, so that this part of the theory, though apparently a natural consequence of the former, has not been verified by direct experiment. The experiment would certainly be a very delicate and difficult one. (SP, 2 p.139)[13]

A close look at Maxwell's methodology reveals the structure of his argument. In his earlier work Maxwell claimed that the method of physical analogy did not involve a physical identification of properties or phenomena in the two systems. Instead it was simply taken to mean that two branches of science had the same mathematical form, as in the case of heat flow and electrostatics. A subspecies of physical analogy is what Maxwell called "dynamical analogy" where again both analogues have the same mathematical form but one is concerned with the motion and configuration of material systems.[14] The extension of these dynamical analogies to a more substantive interpretation constituted a "dynamical explanation", where the properties of one system were literally identified with the properties of the other. When one is able to provide this kind of identification then the account of the material system can be understood as a "physical hypothesis". The hypothesis must meet other conditions such as consistency with general laws of dynamics as well as independent experimental verification that the entities it specifies actually exist, if it is to be considered legitimate. Although Maxwell claimed to have provided a dynamical "theory" of the field there is nothing in this work that would qualify as an explanation since he explicitly avoided any hypotheses about the nature of the medium. There were no assumptions about hidden mechanisms or causes that could be used to explain the behaviour of material systems; a methodology facilitated through the use of the Lagrangian formalism.

With this in mind, recall that displacement was introduced as a formal analogue for magnetic polarization. This analogy not only provided the form of the argument for displacement but Maxwell's notion of physical analogy facilitated a qualitative way of thinking about the physical process of displacement. Each of the equations described relationships between quantities that were manifested by observable effects. Hence, the electromagnetic phenomena which gave rise to these relations could be seen as providing the empirical foundation for the equations but not for every parameter. Although one could claim that quantities like displacement benefitted from this indirect evidence, Maxwell was more cautious.

Part of his refusal to sanction a causal inference in the case of the displacement current was his commitment to standards of evidence that required direct verification of causes, independently of their role in an explanatory theory or framework. The other, equally important consideration stems from his vision of what constituted the proper domain of a dynamical theory and what could be legitimately asserted using the method of "deductions from phenomena". Maxwell noted that the difficulty associated with this method is that in the early stages of investigation one is only able to arrive at results that are indefinite. Because there are no terms sufficiently general to express these results one is often forced to introduce ideas that are not strictly deducible from experiments. The solution lay in the method used in DT; a "method of statement by which ideas, precise as far as they go, may be conveyed to the mind, and yet sufficiently general to avoid the introduction of any unwarrantable details (SP, 2, p.309). The goal of dynamical theory was to establish relations or connections between phenomena; in fact, from a scientific point of view, relations were the "most important thing to know" since it is in this way that we are able to reason from the known to the unknown. This process defines what Maxwell terms the "proper question of science" which

> is not - what phenomena will result from the hypothesis that the system is of a certain specified kind? But - what is the most general specification of a material system consistent with the condition that the motions of those parts of the system which we can observe are what we find them to be. (SP, 2 p.781)

Maxwell was right to be worried about displacement since it really could not be legitimated using deduction from phenomena. However, it is somewhat ironic that it was his own methodology that prevented him from arguing for a more justifiable foundation for displacement (using something like H-D) which could then be used to facilitate a broader empirical justification for his field-theoretic account of electromagnetism. This is not to say that Maxwell should have done this, for I think he was right not to. I simply want to point out that he was in fact well aware of what could be achieved using deduction from phenomena and although he clearly wanted to provide an experimental foundation for electromagnetism he was unwilling to mask the difficulties using what he took to be illegitimate methods.

4. Conclusion

The experimental basis for electromagnetism is interesting for a variety of reasons, not the least of which is how the relationship between the phenomena and the theory determined, to a large extent, how Maxwell viewed the overall structure and conclusions of the theory itself. Maxwell's "dynamical theory" which he takes to be a deduction from phenomena yields little more than a phenomenological theory of the electromagnetic field. The qualitative understanding of field-theoretic processes that he provides are furnished by what he takes to be illustrative concepts. However, these "conceptions" have no explanatory power insofar as they occupy a different place than the underlying theoretical structure. They are unable to yield the kinds of causal explanations that we demand from scientific theories and, at least in this case, deduction from phenomena fails to supply the mechanism for fully establishing the empirical basis for the equations. Displacement which was initially introduced by means of a formal analogy could not be given a substantive role nor could it be justified as a formal parameter within the electromagnetic equations. Hence deduction from phenomena provided indirect evidence at best for one of the crucial parameters of Maxwell's electrodynamics.

Notes

[1]See for example the work of William Harper (1991) on Newton and Jon Dorling (1971) on Einstein.

[2]The notion of a 'conception' can be traced to the work of William Whewell (1847). For Whewell a conception bound together particular observations in a coherent fashion, in the way that the notion of an ellipse unified Kepler's data for the orbit of Mars. The conception was something supplied by the mind and in keeping with this formulation Maxwell frequently spoke of a conception that functioned as a way of thinking about or picturing the phenomena to be explained. Maxwell's use of 'conception' was more liberal than Whewell's since it is unlikely that Whewell would have sanctioned the use of fictional representations, something Maxwell frequently did.

[3]All future references to Maxwell's 1890 edition of collected papers will be referred to as SP.

[4]As Maxwell remarked in FL "thus all the mathematical sciences are founded on relations between physical laws and laws of numbers, so that the aim of exact science is to reduce the problem of nature to the determination of quantities by operations with numbers" (SP 1, p.156).

[5]Faraday's account of electrostatic induction involved polarization of the particles of a dielectric; a view which had also been developed by Mossotti who attributed polarization in the aether to a displacement of aetherial particles. It was Mossotti's work that influenced Maxwell in the incorporation of electrostatic induction into the vortex aether model. One of the main difficulties with the vortex model was the lack of any mechanism that could account for interaction between vortices in adjacent molecules. In addition the vortices were not conducive to motion in a fluid aether. The motion was transmitted instantaneously throughout the vortex with the entire cell rotating with the same angular velocity. It was this notion of an elastic solid that actually inspired Maxwell in his account of electrostatic induction.

[6]In FL p.195 Maxwell himself remarked that there was very little known about data regarding the relation between magnetic fields and electric currents in open circuits.

[7]For the case of a charged capacitor with a dielectric material between the plates, the dielectric was seen as the seat of the inductive state with the plates functioning as bounding surfaces where the chain of polarized particles in the dielectric terminated. For the case of closed circuits the added term had a zero value and hence Ampère's law which gives the relationship between the electric current and the corresponding field was sufficient. In open currents where the term took a non-zero value it gave definite predictions for magnetic effects; however, there was no available data against which to test the results. Some of the secondary literature suggests that this modification resulted from an explicit desire to extend Ampère's law to open circuits, and not as a consequence of Maxwell's mechanical model. The lack of experimental data would suggest that this is not the case; moreover, once the qualitative account of displacement was introduced into the mechanics of the model it then became necessary to revise Ampère's law. Although Maxwell claimed that the mathematical laws describing displacement were independent of hypotheses about the aether, it is important to stress the way in which displacement emerged from the vortex model rather than from empirical phenomena that necessitated its postulation.

[8]This began with an assumption by Maxwell about the identity of the electromagnetic and luminiferous media. In a letter to Faraday dated Oct.19, 1861, Maxwell wrote that there was a "strong reason to believe", whether his theory was a fact or not, that the luminiferous and electromagnetic aethers were one.

[9]See Campbell and Garnett (1844) p.340. Maxwell's theory was entitled a dynamical theory because it assumed that in space there is matter in motion which produces the observed electromagnetic phenomena. It is a theory of the electromagnetic field because it concerns that part of space which contains and surrounds bodies in "electric or magnetic conditions" (SP, 1, p.527).

[10]Indeed it was his stated goal as early as 1861 to provide "an exact mathematical expression" for all that was known about electromagnetism without the aid of hypotheses. He was also interested in investigating possible variations of Ampère's law as well as determining whether evidence could be gathered from mathematical expressions that would show the superiority of field theory over direct action theories. See Campbell and Garnett (1884), pp.246-247.

[11]Maxwell made several references to "conceptions" of the phenomena, visual images that enabled the reader to keep in mind mechanical ideas. Conceptions often seemed to take on the role of hypotheses that could not be legitimately introduced into the context of a theory, instead they were usually introduced through the method of physical analogy. In DT Maxwell claimed that from the Faraday effect, the polarization of dielectrics, and the phenomena of optics we are led to the conception of a complicated mechanism capable of a vast variety of motion (SP, 1, p.533). Also in A Treatise on Electricity and Magnetism p.200 Maxwell remarks that the method of Thomson and Tait allows us to "conceive of a moving system connected by means of an imaginary mechanics used merely to assist the imagination in ascribing position, velocity and momentum to purely algebraic quantities". The method of physical analogy allowed "the mind at every step to lay hold of a clear physical conception, without being committed to any theory founded on the physical science from which that conception is borrowed, so that it is neither drawn aside from the subject in pursuit of analytical subtleties, nor carried beyond the truth by a favourite hypothesis (SP, 1, p.156).

[12]See SP,1, p.538 for a discussion of what Maxwell refers to as a dynamical illustration. Francis Everitt (1975) also gives a nice discussion of this issue.

[13]Similarly in A Treatise on Electricity and Magnetism, Volume 2: Section607 Maxwell remarks that:

> We have very little experimental evidence relating to the direct electromagnetic action of currents due to the variation of electric displacement in dielectrics, but the extreme difficulty of reconciling the laws of electromagnetism with the existence of electric currents that are not closed is one reason among many why we must admit the existence of transient currents due to displacement.

[14]An example is Maxwell's analogy between electrostatics and the motion of an incompressible fluid, where the latter is concerned with fluid flow from sources to sinks.

References

Campbell, L. and Garnett, W. (1882), *The Life of James Clerk Maxwell*. London: MacMillan. Reprinted 1969, New York: Johnson Reprint. Second edition printed in 1884.

Dorling, J. (1971), "Einstein's Introduction of Photons: Argument by Analogy or Deduction from Phenomena?" *British Journal for the Philosophy of Science* 22: 1-8.

Everitt, C.W.F. (1975), *James Clerk Maxwell: Physicist and Natural Philosopher*. New York: Scribners.

Franklin, A. (1986), *The Neglect of Experiment*. Cambridge: Cambridge University Press.

_ _ _ _ _ _ _. (1990), *Experiment, Right or Wrong*. Cambridge: Cambridge University Press.

Galison, P. (1987), *How Experiments End*. Chicago: University of Chicago Press.

Hacking, I. (1983), *Representing and Intervening*. Cambridge: Cambridge University Press.

Harper, W. (1991), "Newton's Classic Deductions from Phenomena", in *PSA 1990* volume 2, A. Fine, M. Forbes and L. Wessels (eds.). East Lansing: Philosophy of Science Association, pp.

Hesse, M. (1973), "Logic of Discovery in Maxwell's Electromagnetic Theory" in *Theories of Scientific Method*, R. Giere and R. Westfall (eds.). Bloomington: University of Indiana Press.

Larmour, Sir J. (1907), *Memoirs and Scientific Correspondence of the Late Sir George Gabriel Stokes*, 2 Volumes. Cambridge: Cambridge University Press.

Maxwell, J.C. (1890), *The Scientific Papers of James Clerk Maxwell*, 2 Volumes, W.D. Niven (ed.). Cambridge: Cambridge University Press. Reprinted in one volume in 1965, New York: Dover.

_ _ _ _ _ _ _ _. (1891), *A Treatise on Electricity and Magnetism*, 2 Volumes. Oxford: Clarendon Press. Reprinted in 1954, New York: Dover.

Whewell, W. (1847), *The Philosophy of the Inductive Sciences*, 2 Volumes. London: John W. Parker. Reprinted in 1967, New York: Johnson Reprint.

Experimental Reproducibility and the Experimenters' Regress

Hans Radder

Vrije Universiteit, Amsterdam

Many philosophers of science tend to take for granted the proposition that, as a rule, scientific experiments are reproducible (or, in other words, repeatable, replicable). Very few of them, however, have examined the issue in any detail (a recent exception is Hones 1990). Yet, by now an important body of literature exists, mainly written by sociologists and historians of science, in which the above proposition is discussed and analyzed on the basis of elaborate studies of experimental practice. These studies claim to offer not only a descriptively adequate account of experimentation, but also a number of fundamental philosophical conclusions concerning (experimental) science as a whole.

In the present paper I analyze and evaluate the views of H.M. Collins, an early and influential thinker concerning the issue in question. In particular I discuss what he has called the 'experimenters' regress'. In his 1985 book *Changing Order*, Collins summarizes, systematizes and develops much of his earlier studies. Concerning experimental replication and replicability he makes the following three claims:

(i) Replication is rarely practiced by experimenters.

(ii) In an important class of cases replication turns out to be circular, because of the so-called experimenters' regress. As a consequence, in these cases replication cannot serve as an objective test of the truth of scientific knowledge claims.

(iii) In practice the experimenters' regress is stopped at some stage. This, however, is not a matter of applying explicit, universal criteria but rather of having learned the relevant skills and being enculturated in a local community of experimenters. More in general, the empirical studies of experimental practice confirm the view that science is best interpreted as a loose collection of social practices.

Such claims, or comparable ones, have been rather influential in recent science studies (cf., for instance, Latour and Woolgar 1979; Rouse 1987; Schaffer 1989; Draaisma 1989; Gooding 1990). Given this and given their far-reaching philosophical pretensions, it is worthwhile to make an accurate analysis and evaluation of the above claims.

PSA 1992, Volume 1, pp. 63-73

Below I first describe an alternative account of scientific experimentation and, in particular, of the notion of experimental reproducibility. On this basis I then critically assess Collins' analysis of replication and the experimenters' regress. The emphasis will be on the second and third of the above claims, the first being dealt with only briefly. Generally speaking, I argue not so much that Collins' view of replication is false, but that it is incomplete and hence inadequate. This implies that a number of his philosophical conclusions are also inadequate.

1. Experimentation and Experimental Reproducibility

I shall start with a brief, summary account of my earlier systematic analyses of experimentation in the natural sciences (see Radder 1988, ch. 3; and Radder forthcoming). In this account two basic aspects of experimentation are distinguished. The first is the *theoretical description or interpretation* of an experiment, noted as $p \rightarrow q$. The description q denotes the (intended) experimental result. The composite description p comprises all kinds of premises that are necessary for drawing the conclusion that q is the result of the overall experimental process. For example, in a Stern-Gerlach experiment q may be the claim that the measured spin of electrons is ½, while p will include descriptions of the electron source, the focussing of the electron beam, the magnet and the interaction of the electrons with the inhomogeneous magnetic field, and the instruments (for example, photographic plates) for the final detection of the split-up beam.

A second basic aspect of experimentation is the *material realization*, or performance, of an experiment. The notion of material realization refers to the features of action and production. That is to say, it reflects the fact that science concerns not only thinking, reasoning and theorizing but also doing, manipulating and producing. Materially realizing a Stern-Gerlach experiment, for example, requires performing a number of manipulations with the objects under study and with the experimental devices and, in doing so, producing a number of material results.

The material realization of an experiment can be *described*, and the distinction between theoretical description and material realization *operationalized*, by means of a process of division of labor between a theoretically informed experimenter and a 'lay person'. In this process the experimenter instructs the lay person to actually perform all the required experimental actions. Then, the description of the material realization is just the sum total of all these instructions, phrased in the non-theoretical, daily language in which experimenter and lay person communicate with each other. Therefore, the theoretical description and the description of the material realization are essentially different, even if—in some sense—they describe the same experiment.

Next, this analysis of experimentation makes it possible to distinguish between the following *three types of reproducibility*:

(i) Reproducibility of an experiment under a fixed theoretical interpretation $p \rightarrow q$. In this case we have to do with exact repeatability of the experiment from the point of view of the theoretical interpretation in question.

(ii) Reproducibility of the result of an experiment. We may, for example, have $p \rightarrow q$ and $p' \rightarrow q$, where p and p' are two (possibly radically) different descriptions that—mostly, but not necessarily—will refer to different material procedures for obtaining q. As a matter of terminology, in this case I will speak of *replicability* of the experimental result. A successful reproduction under a fixed interpretation or a successful replication of a result entails that the people involved agree on the correctness or the truth of the interpretation or result in question.

(iii) Reproducibility of the material realization of an experiment. This type of reproducibility requires that, guided by some theoretical interpretation, the same actions are performed and the same experimental situations produced from the point of view of the daily language description of the material realization of the experiment. The point is that, in principle and in practice, reproductions of this type may be obtained on the basis of different members from a class of theoretical interpretations. This implies that reproducing the material realization of an experiment does not require a shared belief in the correctness of one specific interpretation from that class.[1]

We see that the question 'reproducibility *of what*?' may receive different answers, *viz.* reproducibility of the material realization, or of the experimental result, or of the overall experimental process referred to in the theoretical description. Moreover, we can further differentiate these types of reproducibility by adding the question 'reproducibility *by whom*'? In answering this question it is helpful to distinguish between four ranges: reproducibility by any scientist or any human being, in past, present or future; reproducibility by contemporary scientists; reproducibility by the original experimenter; and reproducibility by the lay performers of the experiment.

So much for this account of experimentation and experimental reproducibility. It is still rather schematic, but it will take on more body in the following sections, where I will employ it in a discussion of Collins' views on replication and the experimenters' regress.

2. Reproduction in Experimental Practice

Let me first briefly consider the claim that in actual practice experimental replications are exceptional. Collins (1985, p. 19) writes:

> replication of others' findings and results is an activity that is rarely practiced! Only in exceptional circumstances is there any reward to be gained from repeating another's work. ... A confirmation, if it is to be worth anything in its own right, must be done in an elegant new way or in a manner that will noticeably advance the state of the art.

I do not think that these statements, formulated in this manner, are strictly speaking false. My main objection though is that they are inadequate, because they cover only a part of the issues surrounding experimental reproduction. By distinguishing between reproduction 'of what' and reproduction 'by whom' we obtain a richer notion of experimental reproduction, which enables us to draw a more finely grained map of its role in experimental practice (see Table I).

On the basis of this broader notion of reproducibility it is possible to maintain the claim that reproduction does play a significant part in actual experimental practice, even if some forms will not occur as frequently as others. As I have shown (Radder forthcoming), the boxes 2, 3, 4, 6, 7, 10 and 11 can be exemplified by cases from the recent historical and sociological literature on experimentation. Boxes 5 and 9 are most probably empty. After all, the claimed range of these types of reproducibility would presuppose an unrealistic continuity of belief in the correctness of the theoretical interpretation and result, as well as an unrealistic stability of the material and social conditions required for the material realization of the experiment. For the same reason, box 1 is also probably empty. (Perhaps a few very crude material realizations—for example, producing static electricity by combing one's hair—might approximate the required range.) Boxes 8 and 12 are empty by definition, since the lay persons in question are presumed not to be theoretically informed.

Table I: Types and ranges of reproducibility			
of what? / by whom?	reproducibility of the material realization	reproducibility of the theoretical interpretation	reproducibility of the result of the experiment
by any scientist or any human being, in past, present or future	1	5	9
by contemporary scientists	2	6	10
by the original experimenter	3	7	11
by the lay performers of the experiment	4	8	12

As the above quotation demonstrates, Collins' discussion focuses on reproduction by other scientists, by means of new and better, and thus different, experimental procedures. In other words, he mainly deals with boxes 9 and 10, the activities for which I, in conformity with his usage, have retained the term 'replication'. However, since also boxes 2, 3, 4, 6, 7 and 11 are exemplified by various, concrete experimental episodes, we may conclude that the scope and role of experimental reproduction in actual practice is much larger that Collins assumes it to be.

3. The Experimenters' Regress

According to Collins himself, the experimenters' regress is at the heart of his (social constructivist) interpretation of science (Collins 1985, p. 84). This regress concerns the replication and replicability of experimental results. Suppose that an experimental result q is, for some reason or other, controversial and suppose that we try to resolve the controversy by replicating the experiment. Then, we may be confronted with the following circle or regress:

- whether or not q is believed to be true depends on the result of a correct q-measurement. But:
- whether or not a q-measurement is correct depends on whether or not q is believed to be true.
- etcetera.[2]

Collins concludes from this that in cases where there is no consensus about the correctness of a q-measurement, replicating the experiment cannot bring about the consensus either. In these cases, the similarity of the second experiment and the competence of the second experimenter are particularly likely to be questioned. This means that replication cannot function as a test (of the truth) of a claimed experimental result. In support of his claims Collins offers, among other things, an extensive empirical study of the history of the replications of experiments concerning the existence or non-existence of gravity waves (Collins 1985, ch. 5).

For purposes of further discussion it is helpful to reformulate the experimenters' regress on the basis of the analysis of experimentation presented in section 1. Suppose, the original claim q results from the overall experimental process theoretically interpreted as $p \rightarrow q$; and suppose that this claim is criticized on the basis of a replication: $p' \rightarrow$ not-q. Then, two different ways are open to challenge this replication. The original experimenter may argue that the experimental arrangement described by p' is not an adequate q-meter; that is, not-($p' \rightarrow q$). Alternatively, he or she may claim that p' would have resulted in q, had the experiment been competently performed. In fact, so this claim continues, it is not p' but, say, p'' that has been materially realized, where here indeed: $p'' \rightarrow$ not-q.

These two ways to challenge the (dis)confirming power of replications lead to two different formulations of the experimenters' regress, which both occur in Collins' book, although they are not explicitly distinguished. The first can be found in the gravity radiation case:

> What the correct outcome is depends upon whether there are gravity waves hitting the Earth in detectable fluxes. To find this out we must build a good gravity wave detector and have a look. But we won't know if we have built a good detector until we have tried it and obtained the correct outcome! (Collins 1985, p. 84)

The second formulation reads as follows:

> the *experimenters' regress* ... is a paradox which arises for those who want to use replication as a test of the truth of scientific knowledge claims. The problem is that, since experimentation is a matter of skilful practice, it can never be clear whether a second experiment has been done sufficiently well to count as a check on the results of the first. Some further test is needed to test the quality of the experiment — and so forth. (Collins 1985, p. 2)

For instance, in replications it may be that p and p' are not totally different but have one or more procedures (described by, say, p_1) in common. Therefore, when this is the case, the original experimenter may claim that in the second experiment p_1 has not been skillfully performed, because some tacitly assumed but crucially important aspects have been overlooked.

4. The Experimenters' Regress Reconsidered

How to evaluate the experimenters' regress and its implications? In the present section I will submit six points of critical comment. It will become apparent that reconsidering the regress cannot mean refuting it. Nevertheless, it will also turn out that scientific practice does include a number of 'stabilizing procedures' by which the effects of the regress are significantly *mitigated.*

1. Let me first note that an analogous 'knowers' regress' is a characteristic of *all* methods of human knowledge acquisition. Since the only access we have to reality is through our methods of knowledge acquisition, claims about reality or truth and claims about the adequacy and/or the competent operation of our methods are, in one way or another, intrinsically dependent on each other. There is a certain holism here, which in another tradition has been described as a 'hermeneutical circle' (cf. Hesse 1986). For logical reasons, this general circle cannot be broken and as a consequence some kind of circularity is unavoidable.

This argument implies that, remarkably enough, the experimenters' regress bears no *specific* relation to experimentation. This becomes especially apparent from the first formulation of the regress, which says simply that a correct result presupposes a correctly followed method. The correctness of a particular method may be questioned independently of whether or not non-formalized skills are employed in applying it. After all, one may simply argue that the method in question is inappropriate to the study of certain domains of reality. In this sense some feminists, for instance, claim that the experimental approach to nature is inadequate because it is inherently masculine and oppressive.

However this may be, it is the second formulation of the regress which is the most basic in the view of Collins. Because experimental skills are 'invisible' and cannot be 'fully explicated, or absolutely established' (Collins 1985, p. 129), in replications the problems of reaching an 'objective' consensus about the correctness of the method employed are, according to him, aggravated to the point of becoming insoluble in principle.

2. Although some degree of hermeneutical circularity is unavoidable, the flexibility in experimental practice is reduced by the non-trivial requirement of the reproducibility of the material realization. Now, it is true that reproducing experiments in this sense is possible even in the case of controversy about the theoretical interpretation. After all, one and the same reproducible material realization can be compatible with a class of possibly radically different interpretations. A striking illustration can be found in the controversy between Boyle and Hobbes about a number of Boyle's air-pump experiments. Boyle's theoretically uninformed assistants, who placed a burning candle in the receiver of the air-pump and repeatedly noticed that the candle went out after pumping for some time, performed a reproducible material realization. Theoretically, the result q of the experiments concerned what happened to the air in the receiver. Boyle described the ceasing of the flame as due to the vanishing air, while Hobbes interpreted it as due to the violent winds produced by the pumping (cf. Shapin and Schaffer 1985, chs. II and IV). The point is that the reproducibility of the above material realization is not affected by the existence of this controversy about its theoretical interpretation. Thus, this type of reproducibility entails a form of material stability of experiments by circumventing the experimenters' regress.

Collins, in contrast, focuses on replication as a test of theoretical hypotheses, such as the claimed existence of gravity waves. In doing so, he underexposes the significance of the processes of experimental action and production. This point is nicely illustrated in a section where he discusses a study of experiments to detect solar neutrinos. What is eventually produced in these experiments is 'a wiggly line' on a sheet of paper, but Collins' claim is that the meaning of the experiments lies *exclusively* in the theoretical interpretations of such a curve:

> The point is this: there is little future in the scientist's reporting that $100,000 has been spend on sinking an instrumented tank of perchlorethylene in a gold mine to produce a wiggly line. No one is interested in a wiggly line. But, by virtue of its complete vacuousness, the claim that a wiggly line has been seen is unlikely to be contested. It will change no one's life; it will alter no networks of relationships. (Collins 1985, p. 137)

In general I cannot agree with this. In fact, the fast and controlled material realization of wiggly lines of various forms on a monitor or on paper is what I am doing right now, by means of my word processor. Many people are extremely interested in such wiggly lines and the large-scale introduction of these apparatus has already changed a lot of lives. Of course this does not imply that the meaning of experimental science can be *reduced* to such material realizations and that theoretical interpretations and debate are of no significance. The point is, however, that the stable and reproducible material realization of experiments is a significant achievement of experimental practice. This achievement is, moreover, crucial to the potential utilization of experimental results within technological systems (cf. Radder 1988, ch. 3; and Radder forthcoming).

The latter point is also relevant with respect to one of Collins' own case studies. For, in contradiction to the statements quoted above, in his laser study he readily accepts 'the capacity to vaporize concrete' as a legitimate criterion to determine wether a laser works (see Collins 1985, ch. 3). In Collins' line of thought, someone then might ask: who is interested in vaporizing concrete? Again, *my* answer would be: many people are and especially the military, whose role is not discussed by Collins but who played and continue to play such a decisive part in the whole development of laser theory and laser building.

3. Despite the fundamental nature of the experimenters' regress, apparently, in practice it *is* stopped at some point by the acceptance of some view of what the result q is and what q-measurements have been correctly performed:

> there was never any doubt that the laser could be replicated and never any doubt when it had been replicated. The fact remains that our experience of nearly all natural phenomena is like the experience of laser building; we know that the familiar objects of science are replicable. (Collins 1985, p. 127)

According to Collins, the reasons for this acceptance are non-objective and entirely social in character, since successful replications have a consensual rather than an empirical basis. In particular, he emphasizes the role played by experimental competence and tacit skill-knowledge, which is acquired in a process of enculturation in an 'experimental form of life'.

I think that Collins is right in that non-formalized skills play an important and non-eliminable part in science, especially experimental science. Consider again the operationalization of the notion of material realization through division of labor between a lay person and an experimenter. Also in this case, success in reproducing the material realization requires the possession or learning of certain manipulative skills on the part of the lay person. This learning process may be guided by explicit rules and knowledge, but it is certainly not determined by them. Yet, in spite of my agreement so far, I want to add a number of important qualifications to the role of tacit knowledge in experimental science.

4. In general Collins' view on the role of the inarticulable in science is too one-sided to be true. He states as a central conclusion that:

> Proper working of the apparatus, parts of the apparatus and the experimenter are defined by their ability to take part in producing the proper experimental outcome. Other indicators cannot be found. (Collins 1985, p. 129, italics omitted)

Consequently, explicit, prospective design of the experimental set-up does not play a role in the definition of its eventual success. The proper working of an experiment can only be established retrospectively. For instance, a TEA-laser (or Transversely Excited Atmospheric gas laser) works because it is able to produce the proper outcome of vaporizing concrete.

Yet, the success of a TEA-laser in meeting this criterion is not the only 'explanation' of its working. A satisfactory explanation will refer *both* to the skilful experimental action and production *and* to the prospective design characteristics of the device. Such explicit, theoretical indicators of success *can* be found, and that even in Collins' own account of the case. For example, it was known in advance that building a high-power laser requires a relatively high gas pressure, a high voltage and an accurate construction of the anode and cathode; also, the gas should have a definite and well-known composition and it may not contain any air or other contamination (Collins 1985, pp. 51-54). Characteristics such as these cannot be excluded from the definition of a working TEA-laser. Without this explicit knowledge no amount of skill would have been able to produce a working device.

This objection can be related to the two formulations of the experimenters' regress discussed in the preceding section. As I observed, only the second formulation refers directly to skills. The first formulation concerns the question of what is considered to be a good *q*-meter. As I argued above, the answer to this question is, in part, dependent on explicit, theoretical knowledge.

5. Next, remarkably enough, it is the very procedure of replication of a result *q* by means of different experimental processes that restricts its dependence on specific, local skills. A well-known example is the replication of experiments to test Avogadro's hypothesis, the claim that a grammolecule of a substance always contains the same number of molecules N. This claim was tested in the early decades of this century by means of a large number of very different experimental replications, *viz.* through Brownian motion, alpha decay, X-ray diffraction, blackbody radiation and electrochemical processes, among others (cf. Salmon 1984, pp. 214-20).

If a replication is successful, *q* becomes less dependent on any one of the specific experimental processes *p*, *p*', *p*" etc., and therefore also less dependent on the specific skills that are necessary to produce these particular processes. In replicating experiments the scientists *abstract* from (a part of) the local situations in which *q* has been produced. Therefore, the meaning of a replicated experimental result, in some way, transcends the separate contexts in which it has been produced. The successfully replicated result *q* attains a more systematic position, mostly within the context of a body of theoretical knowledge. This is very clearly illustrated in the replications of Avogadro's hypothesis.

Collins does not deny the occurrence of this kind of 'delocalization', but he does not fully account for its meaning and role in science either.[3] As mentioned above, it is the experimenters' regress upon which he bases his interpretation of science. His discussion of the network character of scientific concepts (see Collins 1985, ch. 6) is valuable, but he does not feed it back into his earlier conclusions concerning the more local experimental episodes (cf. also Hesse 1986). If we take full account of the non-local aspects of experimental and theoretical science, some of Collins' views have to be revised. This applies especially to the claim that the proper working of a particular experiment and the proper expertise of a particular experimenter can be exclusively derived from their ability to produce the correct outcome of this one particular experiment.[4]

6. A final point that is relevant to the issue of the experimenters' regress and the role of skills is that experimental practice itself shows a clear tendency towards standardization. An important stage in experimenting is making experiments more robust and therefore less dependent on the skills of specific individuals. Faraday's experiments on magnetic rotation in the early 1820s offer a striking illustration of the significance of standardization. Having successfully performed a first experiment for himself, he immediately went on to build a more standardized, 'pocket' version of his rotation device. Gooding (1985, p. 121) discusses this experimental episode and concludes:

He made it easy for others to reproduce the effect by sending them the device. This also shows how aware he was of the possibility of failure. It reduced the risk by making it unnecessary for others to acquire all of the practical skills and tacit knowledge that Faraday had so laboriously built up.

Just like replication, standardization is an important procedure by means of which experimenters attempt to delocalize their results. It is true that experimenting essentially involves particular, skilful action and production. But it is equally true that these activities do not exhaust experimental practice, a substantial aspect of which is trying to make the results reproducible by others. The continual reconstruction of knowledge for the purpose of making it more stable is an essential feature of experimental (and indeed of all scientific) practice (cf. Nickles 1988).

5. Conclusions

In the above discussion I have pointed out that experimental reproducibility is a complex notion and that as such it does play a significant role in scientific practice. I have also shown that the reality and even the unavoidablity of the experimenters' regress does not imply that experimental results are entirely constituted by tacit knowledge or that their validity is necessarily restricted to their local production context. The reason for this is the existence of a number of stabilizing procedures, to wit: stabilization through reproducing the material realization; stabilization through explicit, theoretical design; stabilization through replication; and stabilization through standardization.

The arguments of the paper also have a more general philosophical relevance. The experimenters' regress is one of the corner-stones of social constructivism. This is the view that scientific facts are no more than social constructions, that they are merely the result of local negotiations between scientists. Discussing this strongly anti-realist claim would clearly go beyond the framework of the present paper. However, as I have shown elsewhere (Radder 1988), the more comprehensive account of experimentation and experimental reproducibility discussed above can function as one of the central elements of a different, a referentially realist, interpretation of scientific knowledge claims.

Notes

[1]For a much more detailed account, including the mutual relations between these three types of reproducibility, see Radder forthcoming.

[2]Other authors have proposed similar views. For instance, Latour and Woolgar (1979, p. 64) state: 'the spectrum produced by a nuclear magnetic resonance ... spectrometer ... would not exist but for the spectrometer. It is not simply that phenomena *depend on* certain material instrumentation; rather, the phenomena *are thoroughly constituted by* the material setting of the laboratory.' Hacking 1988 and Brown 1989, pp. 76-86, offer two (different) philosophical assessments of these statements.

[3]From the discussion of the types and ranges of reproducibility in section 2 it will be clear that delocalization procedures such as replication do not simply lead to *universally* valid results, but rather to what I call *non-local* results. In general, the range of these non-localities cannot be assumed *a priori*, but should be investigated with the help of historical research.

[4]Brown 1989, pp. 86-93, makes a number of clarifying comments on the issues of tacit knowledge and the experimenters' regress, that are to a certain extent in agreement with the point made above. Yet, in his discussion he wrongly jumps from the relativist extreme to its contrary, the rationalist extreme. Brown does acknowledge the significance of skilful experimental action (p. 87), but he appears to shrink away from the consequences of this fact. He first separates the issue of tacit knowledge from the issue of the experimenters' regress by focussing on what I have called the 'first formulation' of the regress. On this premise, he then claims that background theories 'can be employed to block the regress' (p. 91). This claim, however, straightforwardly contradicts the plausible part of Collins' views: explicit, theoretical knowledge is in general *not sufficient* for competently materially realizing experiments. Brown seems to sense the difficulty but, instead of argumentatively substantiating his views, in the end he relapses into the familiar but unconvincing 'in principle' strategy: 'We have a large number of background beliefs which will tell us what gravity waves are and how they can, *in principle*, be detected' (p. 90, my italics).

References

Brown, J.R. (1989), *The Rational and the Social*. London: Routledge.

Collins, H.M. (1985), *Changing Order*. London: Sage Publications.

Draaisma, D. (1989), "Voorbij het getal van Avogadro: de Benveniste-affaire", *Kennis en Methode* 13: 84-107.

Gooding, D. (1985), "'In Nature's School': Faraday as an Experimentalist", in *Faraday Rediscovered*, D. Gooding and F.A.L.J. James (eds.). New York: Stockton Press, pp. 105-35.

_______. (1990), *Experiment and the Making of Meaning*. Dordrecht: Kluwer Academic Publishers.

Hacking, I. (1988), "The Participant Irrealist at Large in the Laboratory", *British Journal for the Philosophy of Science* 39: 277-94.

Hesse, M. (1986), "Changing Concepts and Stable Order", *Social Studies of Science* 16: 714-26.

Hones, M.J. (1990), "Reproducibility as a Methodological Imperative in Experimental Research", in *PSA 1990*, Volume I, A. Fine, M. Forbes and L. Wessels (eds.). East Lansing: Philosophy of Science Association, pp. 585-99.

Latour, B. and Woolgar, S. (1979), *Laboratory Life*. Beverly Hills: Sage Publications.

Nickles, T. (1988), "Reconstructing Science: Discovery and Experiment", in *Theory and Experiment*, D. Batens and J.P. van Bendegem (eds.). Dordrecht: Reidel Publishing Company, pp. 33-53.

Radder, H. (1988), *The Material Realization of Science*. Assen: Van Gorcum.

______. (forthcoming), "Experimenting in the Natural Sciences. A Philosophical Approach", in *The Autonomy of Experiment/The Sovereignty of Practice*, J.Z. Buchwald (ed.).

Rouse, J. (1987), *Knowledge and Power*. Ithaca: Cornell University Press.

Salmon, W. (1984), *Scientific Explanation and the Causal Structure of the World.* Princeton: Princeton University Press.

Schaffer, S. (1989), "Glass Works: Newton's Prisms and the Uses of Experiment", in *The Uses of Experiment*, D. Gooding, T. Pinch and S. Schaffer (eds.). Cambridge: Cambridge University Press, pp. 67-104.

Shapin, S. and Schaffer, S. (1985), *Leviathan and the Air-Pump*. Princeton: Princeton University Press.

How Do You Falsify a Question?: Crucial Tests versus Crucial Demonstrations

Douglas Allchin

University of Minnesota

1. Introduction

How is the deep disagreement between what Kuhn characterized as paradigms ultimately resolved and how do we interpret such debates epistemically? A close analysis of the Ox-Phos Controversy in bioenergetics from the 1960s and 70s (§§2-3 below) suggests that one justifies a set of questions through an ensemble of empirical demonstrations. This contrasts to decisions between theoretical alternatives through 'crucial experiments'. When viewed along with other historical episodes, this case suggests a philosophical category of 'demonstrations', distinguished from crucial tests and complementary in justificatory status to falsifying instances. The distinction also suggests specific strategies for scientists (§4).

The Ox-Phos Controversy is an especially valuable case for studying theory development and scientific change, and for investigating the problems of disagreement, originally highlighted by Kuhn, where two incompatible conceptual or experimental gestalts converge on the same empirical domain (see also Gilbert and Mulkay 1984a; 1984b; Robinson 1984; 1986; Rowen 1986; Weber 1986; 1991; Allchin 1990; 1991). The debate in this case centered on perhaps the most significant stage of energy-processing in the cell, oxidative phosphorylation, or ox-phos (pronounced as an assonant, nearly rhyming 'OX-FOSS'). The basic problem was how energy was transferred from the stage where we ultimately use the oxygen we breathe to the stage where we produce ATP, the molecule that provides energy for virtually all our cellular functions. Though originating in a relatively specialized area of biochemistry, the controversy soon surfaced in introductory biology texts (e.g., Keeton 1972; Dyson 1975; Becker 1977; Curtis 1979) where normally only consensual knowledge is presented. The reconceptualization that emerged was the occasion for the 1978 Nobel Prize in Chemistry, awarded to Peter Mitchell for his 'chemiosmotic theory' (Ernster 1979).

Originally, in the 1950s, energy was regarded as passing from chemical bond to chemical bond through a series of enzymatic reactions—much like buckets of water along a fire brigade. For chemists, the experimental challenge was crudely to tear apart the cell, isolate its essential components—especially a set of high-energy intermediate compounds—combine them all together again in a test tube and thus reconstitute the re-

PSA 1992, Volume 1, pp. 74-88

action system *in vitro*. Peter Mitchell, on the other hand, conceived the intermediate energy state as an electrochemical gradient of ions across the two sides of the membrane in which the system was embedded. Under this 'chemiosmotic' scheme, the directional orientation of components in the membrane and the structural integrity of the membrane were critical. They set the tasks of mapping the vectorial structure, and measuring the gradients and membrane permeability. Two overlapping, but incompatible causal networks were thus proposed. And they emerged from two divergent sets of questions, research goals and even domains of supposedly relevant phenomena: no single discriminating or "crucial" experiment was possible (see also reviews by Greville 1969; Racker 1970). Experimental findings were, nonetheless, collectively effective in resolving the dispute—and it is critical epistemically to understand how.

2. Falsification and Anomaly-Localization Revisited

First, however, one needs to understand how empirical results failed to be effective. In particular, one needs to appreciate how anomalous results, or potentially 'falsifying' instances, were variously interpreted. Researchers differed in their theoretical backgrounds or cognitive resources and, therefore, in the questions they asked. The variable response suggests that the role of experiment merely in reducing theoretical alternatives depends on those questions, and that we may look for other ways that they shape the development of theory or research programs or help in arbitrating disputes.

Among philosophers, the notion of simple or 'naive' falsification has been heavily criticized and largely abandoned (e.g., Lakatos 1970). Among scientists, however, the concept may function as a heuristic and rhetorical device. At least this was the case in the ox-phos episode. Peter Mitchell, for example, self-consciously framed his novel chemiosmotic hypothesis according to Popperian principles (Boyer et al. 1977, p.996; Mitchell 1980, pp.184-190; 1981a, p.17; 1981b, p.611); others, as well, appealed to the scientific authority of falsification (Azzone 1972; Huszagh and Infante 1989). Still others argued, without explicit philosophical reference, that single experiments were decisive against the opposing theory (e.g., Chance and Mela 1966; Chance, Lee and Mela 1967; Slater 1967; Tupper and Tedeschi 1969).

But it is also clear that the researchers often did not follow their own advice or adhere rigidly to their own rhetorical claims. Elements of Mitchell's original hypothesis, for example, failed repeatedly in the early development of the theory to match actual observations. Data about the direction of the energy gradient (Mitchell 1961b), the magnitude of the gradient (Mitchell 1966a) and an important intermediate ion ratio (the H^+/O ratio; Mitchell 1966a) all challenged Mitchell's initial proposals—and the discrepancies certainly did not escape the notice of critics (Chance, Lee and Schoener 1966; Slater 1966; 1967; 1971). Yet Mitchell persisted in his broader, more general program. One particularly recalcitrant problem was the arrangement of molecular components (including the cytochrome b pair) that would provide the necessary directional orientation (redox loop) that Mitchell postulated. Mitchell admitted later that the data "had always been regarded by [Britton] Chance and other people as anomalous." At one point, in fact, he was "feeling more and more that this might be a point where we could succeed in falsifying the chemiosmotic hypothesis" (1980 interview, quoted in Weber 1991). Yet Mitchell did not capitulate. On the verge of crisis (perhaps), Mitchell dramatically revised the theory and introduced an arguably *ad hoc* concept (the "Q cycle"), deemed later by some as the most elegant achievement of Mitchell's theorizing (Weber 1991; Slater 1981). In brief—and perhaps to no philosopher's surprise—this episode exhibited no falsifications in large-scale intertheoretic debate.

The most informative lessons one can draw here, however, are not about naive falsification, but about how the anomalous results were interpreted. Indeed, many results did not match theoretical expectations. In Lakatos' more 'sophisticated' version of falsification (1970), anomalous results should lead to progressive theory development. But Lakatos did not consider the dynamics of such a process, and he notably failed to address the common criticism that mere falsification does not identify where in a network of concepts and methods, items need to be amended. By contrast, Glymour (1975), Wimsatt (1980; 1981) and Darden (1991), for example, have each offered a rich repertoire of methods by which we might indeed isolate or localize the weaknesses in a theory. The special virtue of such analyses is that they avoid abstract and often vague scales of 'progressiveness' or 'problem-solving power' and aim to identify specifically where justification is or is not warranted. These procedures thus carry the bulk of the epistemic work, at least within a research programme or tradition. The ox-phos case suggests, however, that even the application of such methods is heavily context-dependent, or based on a researcher's orientation, and thus may be ineffective in resolving the deepest forms of intertheoretic disagreement. That is, the interpretation of anomalies and the localization of problems may themselves depend on the questions one asks.

When Mitchell originally introduced his chemiosmotic alternative, for example, he noted six "facts" that he claimed were anomalous or epistemically threatening to the conventional chemical hypothesis. Chemists acknowledged the same "facts," but interpreted them quite differently. Mitchell had observed that "it is not clear why phosphorylation [the last step in the energy-transfer process] should be so closely associated with membranous structures" (1961b, p.145). For Mitchell, the membrane separated inside from outside and was essential in preserving an energized gradient. The chemists certainly recognized the close association of the process with membranes and at first they considered it a "nuisance" in their efforts to reconstruct the enzyme system *in vitro* (Cooper and Lehninger 1957). Albert Lehninger later suggested, however, that the severe technical problems might reveal "a biological necessity for structural organization of these catalysts in a moderately rigid, geometrically organized constellation" (1960, p.952): the "problem" thus dissolved into a promising research enterprise. Where Mitchell saw a defeating anomaly or falsifying instance, chemists saw instead the exciting potential for exploring a whole new dimension of biological organization.

Most dramatically, the chemical hypothesis postulated a series of high-energy intermediate compounds, none of which had been isolated or identified. For Mitchell they were not found—nor would they ever be found—because, simply, they did not exist. He claimed one had to stop asking how energy was passed only from chemical bond to chemical bond, and ask instead how it could be channeled through the movement of protons or ions and, say, create a membrane gradient. Chemists, however, advanced numerous reasons—mostly technical—why the energy-rich intermediates were, as Mitchell had phrased it, "elusive to identification" (1961b, p.144): they would only need to exist in small concentrations; they were unstable and short-lived; and/or they were tightly bound to other molecules in the membrane (Griffiths 1963; 1965; Chance, Lee and Mela 1967; Greville 1969). David Griffiths characterized the experimental task of isolation as "formidable" (1963, p.1064)—though Efraim Racker only went so far as to call it "rather formidable" (1970, p.137). Chemists had localized the "anomaly" of the high-energy intermediates in yet unsolved technical puzzles. Mitchell, on the other hand, had "localized" it globally, in the whole way investigations were conceived. There was accord on the experimental record, but not on how to interpret the problems it generated.

Finally, from the reciprocal view, one may note that anomalies for the chemiosmotic approach were also not interpreted uniformly. E.C. Slater, one of the most prominent critics of Mitchell's framework, described how the sequence of components that he studied and knew best (the cytochrome b complex) simply could not be made to fit any version of the chemiosmotic hypothesis (1971, pp.44-45). For him, there was no reason to consider the alternative any further: this single mismatch was enough to reject even the plausibility of the chemiosmotic enterprise. As noted above, Mitchell took the conceptual deficit quite seriously—but decidedly not as a reason for abandoning his problem frame. Instead, he used it as an occasion to focus on several anomalous results and to reconstruct a more acceptable answer to the same set of questions about vectorial chemistry. In this and the other cases cited above, Mitchell preserved his orientation to the problem in the face of anomalous findings and worked on details, while others saw the failures as justification to disregard it entirely.

Close analysis of the ox-phos case suggests that observations were not so "theory-laden" that anomalous data could not be recognized. Researchers did repeatedly reassess (and even revise) their beliefs based on the evidence. Rarely, however, did researchers abandon the questions that motivated their research. One finds, in fact, that these central questions, more than mere theoretical commitment, guided their varying responses to the same data. This suggests that we might best characterize each research enterprise *interrogatively*, by its problem-field, problem-frame or set of *questions* (see also Laudan 1977; Nickles 1980, pp.33-38; Allchin 1990).[1] By regarding the questions as primary, one can understand exactly how the response to anomalous results was shaped in each case. The philosophical lesson is perhaps best expressed in my title, through a query at once rhetorical and self-referential: namely, "how do you falsify a question?"

3. The 'Crucial' Role of Demonstrations

The epistemic challenge in profound scientific debate, then, may sometimes become articulating how one unasks a question—or comes to ask a different one (the "replacement problem" of Nickles 1981, pp.95-96). Philosophically, one wants to understand how one justifies a question or set of related questions, particularly in relation to others that may be similar.

One may examine, for instance, how scientists legitimate individual research projects in the opening section of their papers. Research studies are generally formally embedded in a context of experimental practice and extant theory (Griesemer and Wimsatt 1989), so that their results are positioned at important junctures for channeling reasoning or resources (Knorr-Cetina 1981, Chap. 6; Latour 1987, pp.108-121). An analysis of the ox-phos case (below) allows one to see further how such arguments are received, not merely presented—that is, how researchers can construe experiments as warranting certain questions and, indirectly, their pursuit.

The most dramatic feature of even sophisticated versions of falsification is their use of editing, selection, or other eliminative procedures (Hacking 1983, pp.3-5). Justification is always unfinished, and a burden of further proof thus perpetually remains. While effective within a research enterprise, such methods (as shown above) may be ineffective where questions themselves are "in question". One may also see experiments, however, in their more positive or productive role (e.g., Franklin 1986). How do empirical studies generate or create justification, rather than merely limit or qualify it?

In the ox-phos episode, controversy was resolved through numerous studies, which are now entering a canon of "classic experiments" in the field of bioenergetics. One of the earliest and now most renowned set of experiments was originally done as part of another, intersecting research enterprise. André Jagendorf and his coworkers at Cornell merely measured (or "observed") the energized gradient earlier only hypothesized by Mitchell. Soon thereafter, however, they induced an artificial gradient by plunging the membrane-bound vesicles into an "acid-bath" (Jagendorf and Uribe 1966). The sudden pH differential across the membrane generated the final energy product, ATP, where there was no natural source of energy—and under conditions that were not at the time considered to exist in nature. As a result of these acid-bath experiments, researchers—in this case, even Mitchell himself—began to address the chemiosmotic framework in ox-phos more seriously: why?

The acid-bath experiments were dramatic, in part, because they followed an interventive strategy (Hacking 1983). More importantly, though, they revealed a hitherto unknown phenomenon that could well be related to the central question of energy transfer in the cell. By showing how the membrane gradient was causally connected to ox-phos, the novel results potentially altered the range of relevant phenomena or 'domain' *(sensu* Shapere 1974) included in any complete theory or explanation of ox-phos. In so doing, they also thrust chemiosmotic questions, which addressed precisely this reconfigured domain, into explicit consideration. The acid-bath experiments did not limit or qualify the chemiosmotic theory or problem-field; rather, they legitimated them. They played an epistemic role complementary to that of falsification: namely, *demonstration.*

In many conceptions of scientific explanation or justification, confirmation of predictions (or specifically novel or "risky" predictions) plays a central role (e.g., Hempel 1966). In this capacity, a 'demonstration' may likewise function as an empirical "benchmark" for a conceptual or theoretical "map." But the role of the acid-bath studies went much deeper: they began to redraw the very boundary of phenomena then considered relevant to ox-phos. They did so by showing empirically how a conceptually new class or type of phenomena was causally connected to those already known. Before the acid-bath experiments, the range of relevant phenomena according to the chemiosmotic approach was merely plausible theoretically; afterwards—through a largely ostensive exercise—it was also plausible experimentally. Jagendorf's results 'demonstrated', at least within a local domain, that the framework for posing chemiosmotic questions was empirically well framed.[2]

The major role of the acid-bath demonstrations, however, was not merely along a single theory-evidence axis. Jagendorf's results were also key in moving intertheoretic debate "downstream" and, ultimately, in resolving the controversy. Still, the acid-bath studies were not structured as a crucial test between the two available alternatives. The conditions of the experiments did not even strictly address the standard chemical interpretation, and certainly did not directly falsify or challenge any element of its approach. That is, they were not designed to evaluate the two problem-fields symmetrically, or in parallel. But the demonstrations did challenge the chemical approach indirectly. As Robinson (1984) has noted, the acid-bath experiments forced chemists to retreat from their assumptions or claims about the irrelevance of chemiosmotic concerns. Originally, they conceived energy transfer in terms of a chemical intermediate with a high-energy bond (scheme A, following page), while the chemiosmotic framework used a membrane gradient (scheme B):

(A) oxidation ⟷ high-energy intermediate ⟷ phosphorylation (ATP)

(B) oxidation ⟷ membrane gradient ⟷ phosphorylation (ATP)

(C) oxidation ⟷ high-energy intermediate ⟷ phosphorylation (ATP)
↕
membrane gradient

Jagendorf's findings obliged proponents of the chemical hypothesis to acknowledge the phenomenon as causally relevant, though initially they regarded it as only a peripheral or side reaction (scheme C; see Chance, Lee and Mela 1967, pp. 1341-42; Slater 1967, pp. 321-22). Disagreement persisted, but on a substantially different issue. The former issue was whether gradients were part of the domain of ox-phos; the new issue was whether the central mechanism of ox-phos could occur without gradients. That is, the epistemic task of debate had shifted— "crucially" so—from establishing warrant for the chemiosmotic approach, to finding ways to dismiss it (if possible). The demonstrations, though not designed as discriminating two-way tests, nevertheless carried the horizon of debate forward and, notably, also shifted the burden of proof.

While Jagendorf's demonstrations were crucial within a local domain, they did not justify the entire chemiosmotic enterprise across its whole domain or in all areas of application. Data on artificial gradients could not shift empirical contexts and be applied to claims about, say, the specific directional arrangement of components in the membrane—despite the prior construction of a hypothesis which linked them. Further evidence was necessary to legitimate Mitchell's claims that other local domains were also simultaneously relevant to the domains already well understood. There were thus many demonstrations, each crucial in establishing an empirical benchmark in a different local domain. These included: showing that the membrane was relatively impermeable and could thus preserve a gradient, once formed (Mitchell 1961a; Mitchell and Moyle 1965a; 1967); finding that the intermediate gradient was crudely quantitatively positioned (H^+/O and H^+/ATP ratios; Mitchell and Moyle 1965a); measuring the membrane gradient more definitively using the movement of synthetic ions (Skulachev 1970); and demonstrating that elements from systems that had evolved in divergent organisms, when recombined in a chimeric vesicle, could function as an ensemble (Oesterfelt and Stoeckenius 1973; Racker and Stoeckenius 1974). Mitchell (1966c) could also draw on earlier studies, not previously considered relevant: Lee and Ernster (1966) had in a different context noted that the membrane was "sided" or had different features on either side; and further results indicated that vesicles that were inside-out with respect to each other behaved differently (by generating reversed gradients). Each of these demonstrations served to anchor the chemiosmotic "map" to the empirical landscape or, perhaps more appropriately expressed, served as crucial knots in tying a new causal network together. One should note, additionally, that this process of shifting questions or problem-fields was not one of gradually increased support according to some single abstract scale of justification or a set of increasingly rigorous evaluative standards (Laudan and Laudan 1989). Rather, the task of justification was distributed across several local domains, with each initial demonstration playing a crucial role in warranting the questions for further investigation in the respective range of phenomena.[3]

The existence of 'crucial demonstrations' did not exclude, of course, the possibility for counter demonstrations. In fact, there were several experimental findings that had the potential, at least initially, to reclaim for the conventional chemists some of their threatened domain. Among these were several claims to have successfully isolated or identified the high-energy intermediate molecule which, according to the chemiosmotic formulation, did not exist. In each case, however, the results—though "repeatable" even now—did not fit into their proposed locus in the larger causal nexus. That is, the context of the claims could not be substantiated, and the findings were attributed to either different or highly circumscribed domain (Allchin 1991, pp.180-195). Even failing in their broader implications, though, the claims represented critical turns in the development of the field and provided momentary warrant or plausibility for further investigation.

Another domain of opportunity for defenders of the rear guard was in demonstrating ox-phos without closed membranes. Mitchell's questions implied that ox-phos could not (or would not?) take place in open, ruptured vesicles or with only fragments of membrane. Such a universal prohibitive claim was difficult to defend. One could only appeal, as Mitchell (1961b; 1966a) did, to the prolonged absence to the contrary. By contrast, the chemiosmotic hypothesis would be "immediately and irrevocably refuted," according to Greville, if oxidative phosphorylation could be demonstrated in a solution without closed membrane compartments (1969, p.71). Thus, one researcher was able to note the exceptional attention given to one reported finding:

> I remember in [a] meeting in 1972, somebody had written an abstract saying that they had demonstrated oxidative phosphorylation in a membrane-free system derived from a bacterium . . . . Normally these ten-minute papers, not many people attend. But I noticed that the room was filled, and the usual anti-chemiosmotic gang were all there like vultures. But the evidence that there were no membranes there wasn't very satisfactory. You could see them going away a little disappointed (Gilbert and Mulkay 1984b, p.29).

In fact, several claims to have demonstrated membrane-free ox-phos were published (e.g., Painter and Hunter 1970; Wilson et al. 1972; Komai et al. 1976; Tedeschi 1980) and each became a focal point of attention. Despite the obvious interest in such findings (due to their potential import epistemically), the results could not be repeated, in some cases "in several laboratories" (Racker and Horstman 1972). A demonstration, even if "crucial" in its implications, still had to survive further investigation or continued development (see also Hacking 1983, pp.249-50). Promising experimental results that suggested a problem or domain was plausible or worth pursuing never guaranteed generating the "right" answers. The demonstrations did, however, fuel further significant research.

Given the dual justificatory-suggestive status of demonstrations, one may be tempted to use them to construct some sort of comprehensive criterion of progressiveness, probable belief, or level of opportunity or novel problems posed, by which the alternative research enterprises in the ox-phos episode were (or should have been) assessed. But this would blatantly disregard the situatedness of each demonstration in validating certain questions only locally and in warranting only specific prospective work. Again, the domain or scope of justification for each demonstration was local, or limited to phenomena that (with given experience) could be classified as similar (recalling, perhaps, problems posed by Goodman, 1978, about exemplification and "fair samples" and, 1965, about induction classes). Still, the multiple demonstrations cited above were collectively effective at discriminating between the two hypotheses, or research programs. Because the two traditions in ox-phos were conceptually and experimentally incompatible—that is, incapable of recombining their parts (see also Allchin 1990; 1991, pp.168-241)—they interacted as wholes. Yet warrant for each was nevertheless established piecemeal. It was the domain, or substrate of study, however, that was "reduced" or decomposed, not the problems or conceptual framework (as in Simon 1969; Wimsatt 1980). There were no crucial tests between the major theories or even between individual corresponding concepts. But there were demonstrations that, as research successfully deepened, established for each research lineage authority over certain relatively distinct local domains.

Attention to specific local domains is critical because it allows one to understand fully how the controversy was resolved. When debate finally subsided, marked by an exceptional joint six-author review article (Boyer et al. 1977), both research traditions remained: how? The demonstrations detailed above had largely vindicated the novel chemiosmotic problem-field, which now formed the central framework for interpreting the transfer of energy from oxidation to phosphorylation. The original chemical lineage, its concepts and problems, however, also persisted— though in a substantially more limited domain associated with certain finer-scale problems. Indeed, even a third major hypothesis (Boyer's conformational hypothesis), representing yet another lineage, contributed to the final interpretation. In this case, there was no single solution to the problem(s) of ox-phos; there was only resolution among originally competing, overlapping explanations. One must jettison the either-or, winner-take-all terms of most models of theory-choice (e.g., Kuhn 1962; Lakatos 1970; Laudan 1977; Howson and Urbach 1989) and instead characterize the outcome as a differentiation of domain or the distribution of authority among several theories (see also Whitt 1990, pp.474-476, for similar conclusions in the case of the debate between Dalton and Berzelius over the proper problem domains in chemistry). The experiments had established by exemplification or ostension the constellation of local domains appropriate to each set of questions. In this sense, the demonstrations were "crucial" not only in guiding research, but also in resolving disagreement. That is, they not only justified each theory, especially in the face of criticism, but they allowed one to partition the domain into separate contexts or domains, and thereby resolve the deep interparadigm conflict.

4. From Historical Case to Philosophical Principle and Scientific Strategy

In detailing the acid-bath experiments, etc., from the ox-phos episode, I hope that other historical examples that resonate with them will be highlighted. Runcorn's polar wandering curves, in the context of criticisms of continental drift; Lavoisier's measurements of the conservation of mass during combustion specifically when oxygen was included, in the context of interpretations using phlogiston; and Young's documentation and measurement of light interference, in the context of wave-corpuscular questions—as three immediate cases—were all critical to intertheoretic debate. All filled the role of concrete, domain-claiming demonstrations that promoted certain conceptions of the problem while not specifically refuting alternatives. If I do not analyze these or other examples here in the same depth as the ox-phos episode, they can nevertheless indicate that the case I have presented is hardly unique; and I trust that the details of the ox-phos case "demonstrate" their particular significance.

But further, one may hope to generalize from these cases, using historical clues as an occasion to develop more formal philosophical principles (e.g., Darden 1991; Wimsatt 1987; 1992). Discussion of the ox-phos episode above has been oriented, in fact, specifically to highlight the significant epistemic features of what I am calling 'demonstrations'. A demonstration may be seen as a uniquely significant sample confirmation, and (as noted above) complementary in its role to a (model) falsifying instance or anomaly. That is, while falsifications function negatively, selectively, or eliminatively to reduce theory (as noted above), demonstrations function positively or constructively to expand extant theory. Their primary feature is to legitimate a conceptual frame with respect to a certain local domain, perhaps construed as a variable or a dimension or range of measurement. In this respect, demonstrations in science may be compared with precedents in the development of common law. Largely through ostension or exemplification, the demonstration also makes a question or problem-frame concretely plausible, thus indirectly giving warrant to continued pursuit of similar or related questions in that area.

Just as one cannot always effectively localize a falsifying instance in a body of theory and method, so, too, one cannot explicitly distribute the "credit" of a demonstration. But when the demonstration is successful (inherently so—see below), there is little immediate need to isolate any unreliable, unnecessary or redundant element, if any exists: the system works, the problem-frame is effective at getting solutions, and there is no explicit error to handicap further research. The productive result validates the ensemble of factors as an integrated ensemble.

A demonstration may be further distinguished from a 'test'. Tests exhibit symmetry, or parallel experimental conditions, based on the alternative answers for a given question at issue; thus (when it approaches the ideal of their unambiguous design) a test's outcome can distinguish clearly between two or more conceptual alternatives. In the ox-phos case, such tests were able to address (within the chemical framework) such questions as whether there were one or more steps in the energy-transfer sequence; how many molecules were involved; at what step phosphate was introduced, etc. (Allchin 1990, pp.55-57; 1991, pp.144-167). (I assume that the notion of a test, or a 'crucial test', is familiar and that one need not enumerate its features further.) One may note, however, that tests may only be effective within paradigms—that is, for cases where questions are shared and background conditions are similar for each alternative solution.

A demonstration, by contrast, is distinctly asymmetrical. That is, the alternative results of the experiment do not have equal import. If "successful," a demonstration is meaningful: it shows, sometimes quite dramatically, how a (the) problem can be solved. If unsuccessful, however, the experiment implies very little other than perhaps a lack of imagination or luck in finding the right combination of experimental procedures and theoretical parameters (note Kuhn on puzzle-solving, 1962, esp. p.37). The acid-bath demonstrations, for example, strongly legitimated the chemiosmotic problem-frame, while the failure to isolate the high-energy intermediates, while clearly frustrating to the chemists, did not directly "falsify" or challenge their questions.

In the context of disagreement, the acid-bath and other demonstrations each represented a sort of "territorial claim" in their respective domains. (This is an especially apt metaphor where one construes theories or their models as conceptual maps). As noted above (§3), the claim essentially shifted the burden of proof. For the opposing chemists to re-claim their territory, they would have to "advance" an even stronger alternative explanation, or show how the original result was merely a residual artifact (by demonstrating how the proposed cause could be screened off by another variable in producing the same result; Salmon 1984). Interparadigm debate (between problem-fields) thus proceeded by escalation. Demonstrations, each with more rigorous experimental demands or forms of completeness, may be stacked on a local domain, much as chess pieces may all be concentrated on a particular board-position—until the demonstrative resources one research tradition can muster are depleted. This would be the case, at least, where domains overlapped and could not be partitioned or more finely "resolved" as they were in the ox-phos controversy.

Lastly, one may expect that demonstrations may also function much like Kuhnian exemplars (1962, pp.viii,10,23,80-81). That is, they may serve as explicit points of departure for further research, notably where the demonstration makes an unexplored domain accessible through an effective problem-frame or set of questions. Identifying certain experiments as having such a role fits comfortably with characterizations of scientists as pursuing "the path of opportunity" or (perhaps more cynically) as "opportunists" (Pickering 1984, p.10; in the ox-phos case, Robinson 1984).

The concept of a 'demonstration', then, when set against the notions of 'tests' and falsifying anomalies, provides a way to interpret interparadigm debate, as exemplified in the ox-phos controversy. But the touchstone of value for this concept, like any philosophical concept of science, may be whether it provides tools or strategies to practice science more effectively. Indeed, the distinction between tests and demonstrations and the role of demonstrations in warranting questions suggests that scientists cannot neglect the context in which they present their arguments: only certain types of experiments will be effective in different modes of disagreement. Where researchers share questions or problem-fields, one expects that crucial tests will be effective. Where they must argue for the questions themselves, however, one must not attempt to construct crucial 'tests'—or worse, try to falsify an opponent's questions: one must *demonstrate*. Of course, scientists must recognize the signals of the two forms of debate—not always obvious. This study thus complements a diagnosis of disagreement that distinguishes between intra- and interparadigm debate (Allchin 1990)—and aids scientists in identifying the specific contexts under which demonstrations versus tests will be crucially effective in resolving disagreement.

Notes

[1]The interrogative orientation relates to Kuhn's sense of a paradigm as a problem-field (1962, pp. 103, 147-48, 155, 157), and contrasts to a 'hard core' *(sensu* Lakatos, 1970) of conceptual commitments. The focus here clearly points to the need for philosophers to reflect more deeply on the nature of 'problems' (how they originate, how they are considered solved, etc.), how problems may be framed differently, and how they may fit in ensembles of problem-fields—work suggested and begun, for example, by Shapere (1974) and Nickles (1981). This orientation may also guide a reading of Kuhnian 'incommensurability' as the clash between questions or problem-fields, not meanings or world views (Allchin 1990).

[2]This may, in turn, have contributed to an image of "promise" for the chemiosmotic problem-field. But my emphasis here is not on the cognitive roles of surprise, prior expectation, or opportunity in posing new questions (though these may have occurred secondarily—see Robinson 1984). Rather, I want to underscore the justificatory role of the demonstrations. By concretely embodying answers when questions were posed in a certain way, they exemplified how further experimental phenomena could be constructed "downstream" and, equally important, they shifted the burden of proof (see below).

[3]Support may also be seen as distributed across a community of researchers, where each member is viewed as having different cognitive resources and commitments to explaining different local domains (Giere 1988; Whitt 1990, pp. 476-479; Allchin 1991, pp. 278-291). In this episode, one may be especially impressed by the case of E.C. Slater, whose trenchant criticism against the chemiosmotic hypothesis persisted until he became satisfied that it solved the "35-year-old paradox" of the cytochrome b complex, the portion of the system that he studied and knew most intimately (Slater 1981a).

References

Allchin, D. (1990), "Paradigms, Populations and Problem Fields: Approaches to Disagreement", in *PSA 1990*, Volume 1, A. Fine, M. Forbes and L. Wessels (eds.). East Lansing: Philosophy of Science Association, pp.53-66.

_______. (1991), *Resolving Disagreement in Science: The Ox-Phos Controversy, 1961-1977*. Ph.D. dissertation, Comm. on the Conceptual Foundations of Science. Chicago: University of Chicago.

Azzone, G.F. (1972), "Oxidative Phosphorylation, A History of Unsuccessful Attempts: Is It Only an Explanatory Problem?", *Journal of Bioenergetics* 3: 95-103.

Becker, W.M. (1977), *Energy and the Living Cell*. Philadelphia: J.B. Lippincott.

Boyer, P.D., Chance, B., Ernster, L., Mitchell, P., Racker, E., and Slater, E.C. (1977), "Oxidative Phosphorylation and Photophosphorylation", *Annual Review of Biochemistry* 46: 955-1026.

Chance, B., Lee, C.P., and Schoener, B. (1966), "High and Low Energy States of Cytochromes. II. In Submitochondrial Particles", *Journal of Biological Chemistry* 241: 4574-76.

_______. and Mela, L. (1966), "A Hydrogen Concentration Gradient in a Mitochondrial Membrane", *Nature* 212: 369-72.

_______., Lee, C.P. and Mela, L. (1967), "Control and Conservation of Energy in the Cytochrome Chain", *Federation Proceedings* 26: 1341-54.

Cooper, C. and Lehninger, A.L. (1957), "Oxidative Phosphorylation by an Enzyme Complex from Extracts of Mitochondria", *Journal of Biological Chemistry* 224: 547-578.

Curtis, H. (1979), *Biology*, 3rd edition. New York: Worth.

Darden, L. (1991), *Strategies for Theory Change: The Case of the Theory of the Gene*. Oxford: Oxford University Press.

Dyson, R.D. (1975), *Essentials of Cell Biology*. Boston: Allyn and Bacon.

Ernster, L. (1979), "The Nobel Prize for Chemistry", *Les Prix Nobels, Nobelstiftelsen*. Stockholm: Almqvist and Wiksell, pp. 24-26.

Giere, R. (1988), *Explaining Science*. Chicago: University of Chicago Press.

Gilbert, G.N. and M. Mulkay. (1984a), "Experiments are the Key: Participants' Histories and Historians' Histories of Science," *Isis* 75: 105-25.

_________________. (1984b), *Opening Pandora's Box*. Cambridge: Cambridge University Press.

Glymour, C. (1975), *Theory and Evidence*. Princeton: Princeton University Press.

Goodman, N. (1965), *Fact, Fiction, and Forecast*, 2nd edition. Indianapolis: Bobbs-Merrill.

________. (1978), *Ways of Worldmaking*. Indianapolis: Hackett.

Greville, G.D. (1969), "A Scrutiny of Mitchell's Chemiosmotic Hypothesis of Respiratory Chain and Photosynthetic Phosphorylation", *Current Topics in Bioenergetics* 3: 1-78.

Griesemer, J.R. and Wimsatt, W.C. (1989), "Picturing Weismannism: A Case Study of Conceptual Evolution", in *What the Philosophy of Biology Is*, Michael Ruse (ed.). Boston: Kluwer Academic, pp.75-137.

Griffiths, D. (1963), "A New Phosphorylated Derivative of NAD, an Intermediate in Oxidative Phosphorylation", *Federation Proceedings* 22: 1064-70.

________. (1965), "Oxidative Phosphorylation", *Essays in Biochemistry* 1: 91-120.

Hacking, I. (1983), *Representing and Intervening*. Cambridge: Cambridge University Press.

Hempel, C.G. (1966), *Philosophy of Natural Science*, Englewood Cliffs: Prentice-Hall.

Howson, C. and Urbach, P. (1989), *Scientific Reasoning: The Bayesian Approach*. La Salle: Open Court.

Huszagh, V.A. and Infante, J.P. (1989), "The Hypothetical Way of Progress", *Nature* 338: 109.

Jagendorf, A. and Uribe, E. (1966), "ATP Formation Caused by Acid-Base Transition of Spinach Chloroplasts", *Proceedings of the National Academy of Sciences*, 55: 170-77.

Keeton, W.T. (1972), *Biological Science*, 2nd edition. New York: W.W. Norton.

Knorr-Cetina, K.D. (1981), *The Manufacture of Knowledge*. Oxford: Pergamon Press.

Komai, H., Hunter, D.R., Southward, J.H., Haworth, R.A., and Green, D.E. (1976), "Energy Coupling in Lysolecithin-treated Submitochondrial Particles", *Biochemical and Biophysical Research Communications* 69: 695-704.

Kuhn, T.S. (1962), *The Structure of Scientific Revolutions*. Chicago: University of Chicago Press.

Lakatos, I. (1970), "Falsification and the Methodology of Scientific Research Programmes", reprinted in J. Worrall and G. Currie (eds.), *The Methodology of Scientific Research Programmes*, Philosophical Papers (Volume 1). Cambridge: Cambridge University Press (1978), pp.8-101.

Latour, B. (1987), *Science in Action*. Cambridge: Harvard University Press.

Laudan, L. (1977), *Progress and Its Problems*. Berkeley: University of California Press.

Laudan, L. and Laudan R. (1989), "Dominance and the Disunity of Method", *Philosophy of Science* 56: 221-37.

Lehninger, A.L. (1960), "Oxidative Phosphorylation in Submitochondrial Systems", *Federation Proceedings* 19: 952-62.

Mitchell, P. (1961a), "Conduction of Protons Through the Membrane of Mitochondria and Bacteria by Uncouplers of Oxidative Phosphorylation", *Biochemical Journal* 81: 24P.

_______. (1961b), "Coupling of Phosphorylation to Electron and Hydrogen Transfer by a Chemi-Osmotic Type of Mechanism", *Nature* 191: 144-48.

_______. (1966), *Chemiosmotic Coupling in Oxidative and Photosynthetic Phosphorylation.* Bodmin, U.K.: Glynn Research Laboratories.

_______. (1980), "The Culture of the Imagination", *Journal of the Royal Institute of Cornwall* 3(3): 173-91.

_______. (1981a), "Bioenergetic Aspects of Unity in Biochemistry: Evolution of the Concept of Ligand Conduction in Chemical, Osmotic, and Chemiosmotic Reaction Mechanisms", in *Of Oxygen, Fuels and Living Matter* (Part 1), G. Semenza (ed.). New York: John Wiley & Sons, pp.1-56.

_______. (1981b), "From Black-Box Energetics to Molecular Mechanics: Vectorial Ligand-Conduction Mechanisms in Biochemistry", in V.P. Skulachev and P.C. Hinkle (eds.) *Chemiosmotic Proton Circuits in Biological Membranes.* London: Addison-Wesley, pp.611-33.

Nickles, T. (1980), "Introductory Essay," *Scientific Discovery, Logic, and Rationality, Boston Studies in the Philosophy of Science*, Volume 56, T. Nickles (ed.). Dordrecht: Reidel, pp.1-59.

______. (1981), "What is a Problem That We May Solve It?" *Synthese* 47:85-118.

Painter, A.A. and Hunter, F.E., Jr. (1970), "Phosphorylation Coupled to Oxidation of Thiol Groups (GSH) by Cytochrome *c* with Disulfide (GSSG) as an Essential Catalyst. I-IV.", *Biochemical and Biophysical Research Communications* 40: 360-95.

Pickering, A. (1984), *Constructing Quarks.* Edinburgh University Press.

Racker, E. (1970), "Function and Structure of the Inner Membrane of Mitochondria and Chloroplasts", in E. Racker (ed.) *Membranes of Mitochondria and Chloroplasts.* New York: Van Nostrand Reinhold, pp.127-171.

Racker, E. and Horstman, L.L. (1972), "Mechanism and Control of Oxidative Phosphorylation" in *Energy Metabolism and the Regulation of Metabolic Proceses in Mitochondria,* M.A. Mehlman and R.W. Hanson (eds.). New York: Academic Press, pp.1-25.

Robinson, J. (1984), "The Chemiosmotic Hypothesis of Energy Coupling and the Path of Scientific Opportunity", *Perspectives in Biology and Medicine* 27: 367-383.

_ _ _ _ _ _ _. (1986), "Appreciating Key Experiments", *British Journal of the History of Science* 19: 51-56.

Rowen, L. (1986), *Normative Epistemology and Scientific Research: Reflections on the "Ox-Phos" Controversy, A Case History in Biochemistry.* Ph.D. dissertation, Nashville: Vanderbilt University.

Salmon, W. (1984), *Scientific Explanation and the Causal Structure of the World.* Princeton: Princeton University Press.

Shapere, D. (1984), "Scientific Theories and Their Domains", in *The Structure of Scientific Theories,* F. Suppe (ed.). Champaign: University of Illinois Press, pp.518-565.

Simon, H. (1969), "The Architecture of Complexity", in *The Sciences of the Artificial.* Cambridge: MIT Press, pp.192-229.

Slater, E.C. (1966), "Oxidative Phosphorylation", in *Comprehensive Biochemistry,* Volume 14, M. Florkin and E.H. Stotz (eds.). Amsterdam: Elsevier, pp.327-96.

_ _ _ _ _ _. (1967), "An Evaluation of the Mitchell Hypothesis of Chemiosmotic Coupling in Oxidative and Photosynthetic Phosphorylation", *European Journal of Biochemistry* 1: 317-26.

_ _ _ _ _ _. (1971), "The Coupling Between Energy-Yielding and Energy-Utilizing Reactions in Mitochondria", *Quarterly Review of Biophysics* 4: 35-71.

_ _ _ _ _ _. (1981), "The Cytochrome *b* Paradox, the BAL-labile Factor and the Q Cycle", in *Chemiosmotic Proton Circuits in Biological Membranes,* V.P. Skulachev and P.C. Hinkle (eds.). London: Addison-Wesley, pp.69-104.

Tedeschi, H. (1980), "The Mitochondrial Membrane Potential", *Biological Reviews* 55: 171-206.

Tupper, J.T. and Tedeschi, H. (1969), "Microelectrode Studies in the Membrane Properties of Isolated Mitochondria", *Proceedings of the National Academy of Sciences* 63: 370-77.

Weber, B. (1986), "The Impact of the Prague Symposium on the Conceptual Development of Bioenergetics: A Retrospective and Prospective View", in *Ion Gradient-Coupled Transport,* INSERM Symposium No. 26, F. Alvardo and C.H. van Os (eds.). Amsterdam: Elsevier.

_ _ _ _ _. (1991), "Glynn and the Conceptual Development of the Chemiosmotic Theory: A Retrospective and Prospective View", in *Bioscience Reports* 11(6).

Whitt, L.A. (1990), "Theory Pursuit: Between Discovery and Acceptance", in *PSA 1990*, Volume 1, A. Fine, M. Forbes and L. Wessels (eds.), Philosophy of Science Association, pp.467-483.

Wilson, D.F., Dutton, P.L., Erecinska, M., Lindsay, J.G. and Sato, N. (1972), "Mitochondrial Electron Transport and Energy Conservation", *Accounts of Chemical Research* 5: 234-41.

Wimsatt, W.C. (1980), "Reductionistic Research Strategies and Their Biases in the Units of Selection Controversy", in *Boston Studies in the Philosophy of Science,*Volume 60, T. Nickles (ed.),. Dordrecht: Reidel, pp.213-59.

________. (1981), "Robustness, Reliability and Overdetermination," in *Scientific Inquiry in the Social Sciences,* M. Brewer and B. Collins (eds.). San Francisco: Jossey-Bass, pp.124-63.

________. (1987), "False Models as a Means to Truer Theories," in *Neutral Models in Biology,* M. Nitecki and A. Hoffman (eds.). Oxford University Press, pp.23-35.

________. (1992), "Golden Generalities and Co-opted Anomalies: Haldane vs. Muller and the Drosophila Group on the Theory and Practice of Linkage Mapping", in *Fisher, Haldane, Muller and Wright: Founders of the Modern Mathematical Theory of Evolution,* S. Sarkar (ed.). Dordrecht: Martinus-Nijhoff.

Part III

QUANTUM THEORY I

Value-definiteness and Contextualism: Cut and Paste with Hilbert Space

Allen Stairs

University of Maryland

My topics are two theses about quantum mechanics that are widely doubted but curiously tempting: value-definiteness and contextualism. Value-definiteness holds that every observable has a definite value; contextualism follows from value-definiteness, and claims that one Hermitian operator can represent many observables. In fact, contextualism is my deeper interest. Although it is not a view with an army of defenders, it has a long history of discussion in the literature. (See, for example, Belinfante 1973, van Fraassen 1973, Shimony 1984, Redhead 1987). The bulk of this paper will be concerned with a puzzle about degenerate operators that arises out of the standard accounts of contextualism. But before we get to the details, let us begin with a lurid example.

We consider three spatially-separated electrons, A, B and C, each with its own z-axis, and in the GHZ/Mermin state

$$\Psi = 1/\sqrt{2}(|z+\rangle|z+\rangle|z+\rangle - |z-\rangle|z-\rangle|z-\rangle).$$

The kets are eigenvectors of z-spin. I denote the A, B and C spin operators by ***Ax***, ***By***, etc., and the corresponding quantities by Ax, By, etc. EPR-style locality considerations suggest that these quantities are unambiguous. Ψ is an eigenstate of each of the following three operators, with eigenvalue +1 (in units 1/2 h-bar) in each case:

$$\boldsymbol{Ax\otimes By\otimes Cy},\ \boldsymbol{Ay\otimes Bx\otimes Cy},\ \boldsymbol{Ay\otimes By\otimes Cx}.$$

Suppose we want to predict the value of Ax without measuring it directly. Then we can measure By and Cy and take the product of the results. That product is the predicted value of Ax. Similar remarks hold for Ay, Bx, By, Cx and Cy. At this point, familiar arguments tempt us to accept value-definiteness, at least for these few quantities.

Of course this is too simple. Ψ is an eigenstate of Ax⊗Bx⊗Cx, with eigenvalue *minus* one. If we treat Ax, Ay, etc. as unique quantities with well-defined numbers [Ax], [Ay], etc. as values, we are led to the following argument:

$$[Ax]\cdot[By]\cdot[Cy] = [Ay]\cdot[Bx]\cdot[Cy]=[Ay]\cdot[By]\cdot[Cx]= +1$$

PSA 1992, Volume 1, pp. 91-103

$\therefore$ [Ax]·[By]·[Cy]·[Ay]·[Bx]·[Cy]·[Ay]·[By]·[Cx]= +1
=[Ax]·[Bx]·[Cx]·$[Ay]^2$·$[By]^2$·$[Cy]^2$
$\therefore$ [Ax]·[Bx]·[Cx]= +1 (since $[Ay]^2 = [By]^2 = [Cy]^2 = +1$)

But the expected value of Ax·Bx·Cx in Ψ is *minus* one.

The situation is wonderfully intriguing. Looking from one point of view, we conclude that either all three or exactly one of Ax, Bx, Cx takes the value +1. Looking from another point of view, we conclude that either none or exactly two of Ax, Bx, Cx take the value +1. If value-definiteness is true, then "Ax", "Bx" and "Cx" cannot denote well-defined quantities, but must be identified *via* "contexts", to which we now turn.

Somewhat surprisingly, the four product operators ***Ax⊗By⊗Cy***, ***Ay⊗Bx⊗Cy***, ***Ay⊗By⊗Cx***, and ***Ax⊗Bx⊗Cx*** are pairwise commuting. Any three of them form a complete commuting set in $H^2 \otimes H^2 \otimes H^2$, and the product of any three is *minus* the remaining fourth operator. Their common eigenbasis is the set of eight vectors

$$1/\sqrt{2}(|z+\rangle|z+\rangle|z+\rangle + |z-\rangle|z-\rangle|z-\rangle),\ 1/\sqrt{2}(|z+\rangle|z+\rangle|z+\rangle - |z-\rangle|z-\rangle|z-\rangle),$$
$$1/\sqrt{2}(|z+\rangle|z+\rangle|z-\rangle + |z-\rangle|z-\rangle|z+\rangle), \ldots\ 1/\sqrt{2}(|z-\rangle|z+\rangle|z+\rangle - |z+\rangle|z-\rangle|z-\rangle).$$

These vectors pick out a non-degenerate operator ***O*** on $H^2 \otimes H^2 \otimes H^2$. Each of the product operators is a function of ***O*** but also of another obvious maximal operator. For example, ***Ax⊗Bx⊗Cx*** can be viewed as a function of ***O*** or of the maximal operator determined by ***Ax***, ***Bx*** and ***Cx***, an operator that we will call ***Ax#Bx#Cx***.[1] However, these perspectives are very different, as is emphasized by considering the associated measurement contexts. The value of Ax#Bx#Cx, and hence of Ax⊗Bx⊗Cx, can be ascertained by purely local operations. But so long as the particles are well-separated, ***no*** purely local operations could determine a value for O. Our choice of Ψ ensures that ***Ax⊗Bx⊗Cx*** will take the same value no matter how measured. Nonetheless, one could be pardoned for wondering whether such radically different measurement contexts should really be thought of as two ways to ascertain a value for one quantity.

To be sure, the dialectical situation hardly points unequivocally to contextualism. EPR reasoning begins with the assumption that local operators pick out unique quantities and then argues for value-definiteness. Unfortunately, the conclusion of the argument conflicts with the major premise. We can retain the conclusion and reject the premise, but this is rather like sawing off the branch on which we sit. In spite of this, our ruminations on the contexts ***Ax#Bx#Cx*** and ***O*** suggest that even the mildest operationalist tendencies give contextualism charms independent of the breakdown of EPR arguments.[2]

As for value-definiteness, it too is attractive independently of EPR arguments: it holds the promise of solving the measurement problem. Thus, the combination seems to be worth exploring further. In what follows, we restrict ourselves to discrete operators. We will also adopt explicitly a principle that we have relied on implicitly up to now, what Michael Redhead calls the *Value Rule:* if ϕ yields probability zero that Q takes value v, then Q does not take the value v. Formally:

$$\text{VR: } \text{prob}_{\phi}([Q]^{\phi} = v) = 0 \rightarrow [Q]^{\phi} \neq v.$$

where $[Q]^{\phi}$ is the value taken by Q in the state ϕ.

1. Value-Definiteness and Contextualism

Kochen and Specker have a particular way of stating the cost of value-definiteness: if every Hermitian operator picks out a unique quantity, the values of the quantities will violate the functional relations among the corresponding operators. This leads to one familiar way of characterizing contextualism. *Standard contextualism* holds that a degenerate operator splits into one quantity for each maximal operator of which it is a function; *non-degenerate* operators pick out the fundamental quantities. Here we will follow Redhead's presentation (1987, pp. 134-8).

Suppose $\boldsymbol{A}$, $\boldsymbol{B}$, and $\boldsymbol{C}$ are Hermitian, that $\boldsymbol{A}$ and $\boldsymbol{B}$ are non-degenerate and non-commuting, that $\boldsymbol{C}$ is degenerate, and that $f(\boldsymbol{A}) = \boldsymbol{C} = g(\boldsymbol{B})$. Each of $\boldsymbol{A}$ and $\boldsymbol{B}$ picks out a unique quantity. Hence, assuming value-definiteness, the numbers $f([A])$ and $g([B])$ exist. However, since $\boldsymbol{AB} \neq \boldsymbol{BA}$, we may have $f([A]) \neq g([B])$. Therefore, we introduce two quantities C_A and C_B, defined by the relations

$$[C_A] = f([A]),\ [C_B] = g([B])$$

The subscripts represent *ontological contexts;* they serve to individuate quantities and permit us to distinguish what Hilbert space would identify. Thus, in one sense contextualism preserves functional relations, but the price is the severing of Hilbert space connections.

There is another way around Gleason's theorem. It is the trivial hidden variable theory of Kochen and Specker (1967). For a given quantum state ψ, associate a classical probability space with *each* self-adjoint operator in the obvious way. Then take the σ-product of all these spaces, with the measure given by the product of the measures on the factor spaces. This device will reproduce all the statistics of individual measurements, but will permit $[C] \neq f[A]$, even if $\boldsymbol{C} = f(\boldsymbol{A})$.

Call this the *product representation.* Superficially, it presents a very different metaphysical picture from contextualism: in the product representation, there is a one-one correspondence between quantum mechanical operators and physical quantities, but functional connections are severed. In standard contextualism, quantities multiply but functional relations are preserved. However, a little reflection makes clear that this is quite misleading. Return once again to our three operators $\boldsymbol{A}$, $\boldsymbol{B}$, and $\boldsymbol{C}$. According to standard contextualism, there are four quantities here: A, B, C_A, and C_B. At first blush, the product representation provides only three quantities—random variables—which we will denote by A, B, and C. However, for standard contextualism, C_A and C_B are in effect just definitional extensions of A and B. Thus, C_A just *is* that quantity such that $[C_A] = f([A])$. The quantities A and B are capable of exactly the same sort of definitional extension, yielding quantities $f(\mathsf{A})$ and $g(\mathsf{B})$. These are distinct from and algebraically independent of C. This gives us *five* quantities: A, B, C, $f(\mathsf{A})$ and $g(\mathsf{B})$. Hence, there is a clear sense in which the product representation posits *more* quantities than standard contextualism.

I suggest that the product representation *is* a form of contextualism, but with more contexts than standard contextualism. We could obviously write C_A instead of $f(\mathsf{A})$. But if $\boldsymbol{D} = h(\boldsymbol{C})$, then in contrast to standard contextualism, we could also write D_C, even though $\boldsymbol{C}$ is degenerate. Of course the product representation has an additional feature: it represents the contexts as random variables over a common probability space. From one point of view this is completely gratuitous, as Arthur Fine has insisted for years. However, the use of a common space does have one virtue. It establishes that the interpretation is compatible with a classical ignorance interpretation of the probabilities. Furthermore, it is obvious that standard contextualism could also use a product space to establish its probabilistic *bona fides*. One need not take seriously the product measure or any other measure on the full product space. What is important is the *existence* of such measures.

We shall use the term *full contextualism* to denote the view that each Hermitian operator picks out (at most) one quantity, and to mark the contrast with *standard contextualism*, according to which only non-degenerate operators can individuate quantities. Is there any point to the extra quantities of full contextualism? On the face of it, the answer is yes. As Shimony notes, (1984, p. 29), virtually all real measurements are degenerate. Indeed, consider our three-particle example. If we perform local measurements on only one or two of the particles, the obvious context, algebraic *or* measurement, is non-maximal. On the usual presentation of contextualism, we would have to claim that any apparent non-maximal measurement—in effect, virtually every measurement—is really a maximal measurement in disguise. But which maximal measurement? And by virtue of what?[3]

This may seem to be a small puzzle. But for contextualism it is not small at all. Contextualism aims to make intellectual sense of quantum theory. If standard contextualism leaves us unable to say what is really going on in the most common of quantum mechanical situations, then it is a woefully inadequate interpretation. On the other hand, full contextualism is no prize either. The idea that degenerate operators should be able float free of the maximal operators of which they are functions may solve our puzzle about measurement, but it leaves with with a serious problem about what these operators really represent. As it turns out, however, standard contextualism can go a long way toward rehabilitating itself.

2. Measurements

Begin with a simple example. Suppose A and B are two electrons in the singlet state. Suppose we "measure" Ax, the x-component of spin on A. (The shudder quotes are because standard contextualism claims we are really measuring something else.) Pretending that the spin space is the whole Hilbert space, Ax is locally maximal but corresponds to the degenerate operator $\boldsymbol{Ax \otimes I}$ on $H^2 \otimes H^2$. Although there are infinitely many candidates for what maximal quantity is really being measured, the correct choice seems obvious. The singlet state allows us to predict the result of a Bx "measurement" from the result of an Ax "measurement". Thus, the direct Ax "measurement" is also an indirect Bx "measurement", and the quantity actually measured is Ax#Bx.

Two principles seem to lie behind this story:

(i) The quantity actually measured in an apparent measurement of O must be represented by an operator $\boldsymbol{M}$ of which $\boldsymbol{O}$ is a function.

(ii) The quantity M actually measured must be one whose value is inferable from the result of the "measurement" of O.

This second principle implies that the state is crucial in picking out the quantity actually measured. In the abstract, this might seem objectionable. In the concrete, it seems exactly right. Thus, suppose that instead of being in the singlet state, our two electrons had been in the state

$$1/\sqrt{2}\,(|x+\rangle|y-\rangle - |x-\rangle|y+\rangle).$$

In that case, an Ax "measurement" would have allowed us to infer the result of a "measurement" of By rather than Bx. Therefore, it seems plausible that for this state, the real context is Ax#By rather than Ax#Ax.

To evaluate (i) and (ii) we need to say something more general about measurement. If ***A*** is maximal, with eigenvectors $\{\alpha_i\}$, then an ideal measurement of A takes the form

$$\Sigma_i c_i \alpha_i \otimes \phi_A \rightarrow \Sigma_i c_i \alpha_i \otimes \xi_i,$$

where ϕ_A is the ready state of the A-apparatus, $\Sigma_i c_i \alpha_i$ is the state of the measured system and the ξ_i are distinct and orthogonal. However if ***A*** is degenerate, not all of the ξ_i will be distinct. Thus, suppose the Hilbert space has three dimensions, and that ***A*** has two eigenvalues, a_1 and a_2, associated with the subspaces determined by $\{\alpha_1, \alpha_2\}$ and $\{\alpha_3\}$ respectively. Then a "measurement" of A will exhibit interference; it will be governed by an interaction of the form

$$(c_1\alpha_1 \otimes \phi_A + c_2\alpha_2 \otimes \phi_A + c_3\alpha_3 \otimes \phi_A) \rightarrow ((c_1\alpha_1 + c_2\alpha_2) \otimes \zeta_1 + c_3\alpha_3 \otimes \zeta_2).$$

There is a tidier, unified way to describe non-degenerate *and* degenerate measurement interactions. Let ***O*** be the operator representing the quantity to be measured (or "measured"), and let $\boldsymbol{O} = \Sigma_i o_i \mathbf{Po}_i$, where all the o_i are distinct and the $\mathbf{Po}_i$ are the projectors onto the associated eigenspaces. If ψ is the state of the system to be measured, an ideal measurement-type interaction will take the form

$$\Sigma_i \mathbf{Po}_i(\psi) \otimes \phi_A \rightarrow \Sigma_i \mathbf{Po}_i(\psi) \otimes \xi_i,$$

where all the ξ_i are distinct. Now consider the set of *non-zero* vectors $\{\mathbf{Po}_j(\psi)\}$. This set picks out a family of *non-degenerate* operators $\{\boldsymbol{M}_O\}$ such that each $\mathbf{Po}_j(\psi)$ is an eigenvector of each $\boldsymbol{M}_O$. These operators have a special distinction: in the state ψ, a measurement of any one of them is formally indistinguishable from a "measurement" of O. Thus, let ***O*** be the operator ***A*** considered above. Here $\mathbf{Pa}_1(\psi)$ is just $c_1\alpha_1 + c_2\alpha_2$. Let β_1 denote the normalized version of this vector. Let $\beta_3 = \alpha_3$, and let β_2 be any of the collinear unit vectors in H^3 that are orthogonal to β_1 and β_3. These three vectors form an eigenbasis for a non-degenerate operator ***B*** on H^3. For an arbitrary state χ, a measurement of B would take the form

$$((\beta_1,\chi)\beta_1 \otimes \phi_B + (\beta_2,\chi)\beta_2 \otimes \phi_B + (\beta_3,\chi)\beta_3 \otimes \phi_B) \rightarrow$$
$$((\beta_1,\chi)\beta_1 \otimes \xi_1 + (\beta_2,\chi)\beta_2 \otimes \xi_2 + (\beta_3,\chi)\beta_3 \otimes \xi_3).$$

However, in the state $\psi = (c_1\alpha_1 + c_2\alpha_2 + c_3\alpha_3)$, this becomes

$$((\beta_1,\psi)\beta_1 \otimes \phi_B + (\beta_3,\psi)\beta_3 \otimes \phi_B) \rightarrow ((\beta_1,\psi)\beta_1 \otimes \xi_1 + (\beta_3,\psi)\beta_3 \otimes \xi_3).$$

What is the difference between this interaction and a "measurement" of A? Formally, there is none. $(\beta_1,\psi)\beta_1$ is just $\mathbf{Pa}_1(\psi)$; $(\beta_3,\psi)\beta_3$ is just $\mathbf{Pa}_2(\psi)$. As for ϕ_A and ϕ_B, there is no doubt that they differ materially. An A-instrument will in general differ from a B-instrument. However, both are simply ready states of a measuring device. Similarly, ζ_1 and ξ_1 will differ, as will ζ_2 and ξ_3. But each is just an indicator state of a measuring device. From a formal point of view, an ideal measurement is constituted by the fact that eigenstates of an operator on the observed system are correlated with indicator states on the observing system. And from that point of view, the "measurement" of A is indistinguishable from a *measurement* of B.

Let us say that two operators are *ψ-equivalent* if ψ is a sum of their common eigenstates. What I am proposing, on grounds that have nothing special to do with contextualism, is that we adopt the following principle:

> *Measurement Equivalence:* If ***O*** and ***M*** are ψ-equivalent, and if ***M*** is non-degenerate, then an apparent measurement of O on a system in state ψ is also an actual measurement of M.

In light if this, return to (i) and (ii). The analysis in terms of measurement equivalence provides a clear rationale for (ii): if ***A*** and ***B*** are *measurement equivalent with respect to* ψ, as we shall say, then I can infer the value of B from a "measurement" of A because the "measurement" of A is at the same time a measurement of B. Indeed, *inference* is not the issue. If I am ignorant of the precise state, I may be unsure which maximal magnitudes are "really" being measured. What the principle of measurement equivalence does (assuming that quantum states are objective) is to ensure that for standard contextualism, there is a fact of the matter about what is really measured when an apparently degenerate measurement is performed.

As for (i), if we accept Measurement Equivalence, we will say that (i) is insufficiently general in two respects. First, it presupposes that there is *one* quantity that is "really" measured in an apparent measurement of O. However, *any* quantity that is ψ-equivalent to O will do, and it would be perverse to insist that just one of the ψ-equivalent quantities is the one actually measured. It is not just that *we* have no way of figuring out which one it is; the theory itself provides no basis for singling out a single quantity.

We can illustrate this by substituting a pair of spin-3/2 particles for the spin-1/2 particles of our earlier example, and replacing the singlet state with

$$\Phi = 1/\sqrt{2}(|x{=}{+}1/2\rangle|x{=}{-}1/2\rangle - |x{=}{-}1/2\rangle|x{=}{+}1/2\rangle).$$

Here a "measurement" of Ax will allow us to infer the value of Bx. However, because correlations depend only on vectors, this same measurement will allow us to infer the value of *any* of the infinitely many locally maximal B-magnitudes $\{B_\beta\}$ of which $|x{=}{+}1/2\rangle$ and $|x{=}{-}1/2\rangle$ are eigenstates. Note: it is not just that we have no way of discerning which of the magnitudes $Ax\#B_\beta$ was actually measured; it is that there is no obvious *sense* to the idea that one of these magnitudes but not another was measured.

Of course, Measurement Equivalence goes further, and here we come to the second respect in which (i) is insufficiently general. In addition to the magnitudes $Ax\#B_\beta$, there are infinitely many other magnitudes that are Φ-equivalent to Ax, represented by operators that do not commute with $\boldsymbol{Ax}\otimes\boldsymbol{I}$. Return to the simpler case of two spin-1/2 systems. Consider the four vectors

$$|x+\rangle|x-\rangle,\ |x-\rangle|x+\rangle,\ (|x+\rangle|x+\rangle + |x-\rangle|x-\rangle),\ (|x+\rangle|x+\rangle - |x-\rangle|x-\rangle).$$

These are eigenstates of a non-degenerate operator ***M*** on $H^2\otimes H^2$. Clearly $\boldsymbol{Ax}\otimes\boldsymbol{I}$ is not a function of ***M***; neither $(|x+\rangle|x+\rangle + |x-\rangle|x-\rangle)$ nor $(|x+\rangle|x+\rangle - |x-\rangle|x-\rangle)$ is an eigenstate of $\boldsymbol{Ax}\otimes\boldsymbol{I}$. However, from the point of view of the singlet state, an ideal "measurement" of Ax is equivalent to a measurement of M. Hence, on the view I am recommending, M is among the quantities actually measured in an apparent measurement of Ax in the singlet state.

Here the contextualist might protest. After all, the theory does distinguish between maximal operators with which ***O*** commutes and all the rest. Indeed, I will not deny that the contextualist can make do with a weaker account than the one I am proposing. But there are other reasons for adopting the principle of measurement equivalence. Contextualism, whether standard or full, pulls Hilbert space apart: if Q and Q' share the eigenray γ, Q might take the value associated with γ and Q' might take some unrelated value. To some, this might have little bearing on the plausibility of contextualism. Others (and I am one of them) find it hard to believe that the way in which Hilbert space knits quantities together has nothing to do with quantum mechanical success. The principle of measurement equivalence provides a basis for a partial but principled restoration of some of these severed connections.

3. Standard Contextualism: A Modest Re-Pasting

Michael Redhead (1987, p.89) considers the following principle for ideal measurements.

> *Faithful Measurement (FM):* The result of a measurement is numerically equal to the value possessed by an observable immediately prior to measurement.

FM is a plausible principle for anyone who holds that all observables have values. Why ascribe values if they cannot even be discovered through ideal measurement? Suppose, however, that we add Measurement Equivalence to FM. The result is a version of standard contextualism that still satisfies Value Definiteness, still obeys the Value Rule, still allows a classical representation of the probabilities, but is less "contextual".

Let us begin with value definiteness. There are many ways to see that all quantities can be assigned values, but one way is particularly useful for present purposes. Define a *frame* as a maximal orthogonal set of rays. The axiom of choice ensures that there is a function f that selects a ray from each frame. As Gleason's theorem teaches us, there are certain things we can't expect from f. If ϕ is a vector, let $[\phi]$ be the ray spanned by ϕ. Suppose $[\phi] \in \Gamma$ and $[\phi] \in \Gamma'$, where Γ and Γ' are two distinct frames. Then if $f(\Gamma) = [\phi]$, there can be no guarantee that $f(\Gamma') = [\phi]$. Nonetheless, every (discrete) maximal magnitude determines a frame. And assigning a value to a maximal magnitude is equivalent to selecting a ray from the corresponding frame. Thus, f assigns a value to every maximal magnitude.

The function f as defined so far does not satisfy VR. ψ may be orthogonal to some element γ of Γ. If f selected γ, and if γ determined the value of G, VR would be violated. Thus, call a ray γ *ψ-permissible* if γ is not orthogonal to ψ. Let Γ_ψ denote the set of ψ-permissible rays of Γ. We will call Γ_ψ the *ψ-reduced frame determined by Γ*. Note that if $\boldsymbol{G} = \Sigma_i g_i \mathbf{P}g_i$, then the associated ψ-reduced frame Γ_ψ is the set $\{[\mathbf{P}g_j(\psi)]\}$ for all non-null $\mathbf{P}g_j(\psi)$. Note also that many distinct operators may share the same ψ-reduced frame Γ_ψ. Now re-define f so that its domain becomes the set of all ψ-reduced frames on H. The axiom of choice guarantees that f, as re-defined, is well-defined: for every Γ_ψ, there is a unique ray $f(\Gamma_\psi)$ such that $f(\Gamma_\psi) \in \Gamma_\psi$. Since $f(\Gamma_\psi)$ is still a member of Γ, f still determines a value for every maximal magnitude. But now we can re-state our earlier proposal. We simply require that any acceptable assignment of values to the maximal magnitudes must be determined by a function f defined on the set of *reduced* frames. Thus,

> *VR_{SC}:* For every state ψ there exists a function f defined on the set of all ψ-reduced frames such that (i) $f(\Gamma_\psi) \in \Gamma_\psi$, and (ii) if $\boldsymbol{G}$ is a maximal operator with ψ-reduced frame Γ_ψ, then G takes the eigenvalue associated with $f(\Gamma_\psi)$.

VR_{SC} incorporates VR and spells out the effect of FM and Measurement Equivalence for value definiteness. It embodies the *Pasting Principle:*

> *Pasting Principle:* If ψ is a sum of shared eigenvectors $\{\alpha_i\}$ of two maximal quantities Q and R, then Q takes the value associated with α_i just in case R does as well.

The economy and elegance of the Pasting Principle seem to me to recommend it quite apart from any justification it may have *via* FM and Measurement Equivalence. Indeed, it may make full Measurement Equivalence more attractive to the nervous contextualist. However, so far we have not discussed probability. We turn to that task next.

It is well known that there can be consistent value assignments that do not mesh with a classical interpretation of the probabilities. The best-known example is the case of the singlet state. Here it is possible to assign values to all the local spin quantities, and even to do so in a way that respects the mirror-image correlations. Nonetheless, it is impossible to do so in a way that allows the probabilities to be represented as mere ignorance of the underlying values. Thus, the fact that VR_{SC} provides values for all quantities does not by itself guarantee that the quantum probabilities can be reconstructed. However, providing this guarantee is simple enough. Pick a state ψ. Consider the set $\{\Gamma_\psi\}$ of all ψ-reduced frames. If γ is a member of some Γ_ψ, then in general γ will represent *many* propositions: one for each maximal quantity of which Γ_ψ is the reduced frame. Thus, suppose two operators ***G*** and ***G'*** share the same ψ-reduced frame Γ_ψ and that γ is an eigenray of ***G*** and of ***G'***, with associated eigenvalues g and g' respectively. Then γ represents the propositions *G takes value g* and *G' takes value g'*. Now for each Γ_ψ, let the elements of Γ_ψ be the atoms of a (σ-)algebra of sets $\wp(\Gamma_\psi)$. Let each of the atoms be assigned the obvious probability determined by ψ. This generates a measure μ_ψ on each $\wp(\Gamma_\psi)$. From here it is easy to show that all these probability spaces can be made to mesh in classical style. One way is simply to treat all the separate spaces $\wp(\Gamma_\psi)$ as independent. This is equivalent to taking the σ-product $\prod\wp(\Gamma_\psi)$ of all the separate spaces $\wp(\Gamma_\psi)$ and letting the measure on this larger space be the product measure μ. This gives us exactly what we need. $\prod\wp(\Gamma_\psi)$ provides a non-zero representative for each proposition corresponding to a permissible ray. When *and only when* VR_{SC} requires that two atomic propositions be identified, $\prod\wp(\Gamma_\psi)$ identifies them. Further, if proposition *P* corresponds to a ray, then $\mu(P) = p_\psi(P)$. Other quantum propositions are taken care of automatically.[4]

Once again, the product space and the associated measure are a mere device for establishing the possibility of an ignorance interpretation of the probabilities. In fact, the individual probability spaces $\wp(\Gamma_\psi)$ are very much like Fine's statistical variables (Fine 1984). The main difference is in the way in which one space may represent many quantities simultaneously.

For those who still prefer full contextualism, a simple extension of what we have said so far provides a re-pasting appropriate for that view. First, we generalize the notion of a reduced frame. If ***O*** is an *arbitrary* (discrete Hermitian) operator, let $\{o_i\}$ be the set of ψ-permissible eigenvalues of ***O***—i.e., those eigenvalues o_i such that $\mathbf{Po_i}(\psi) \neq \varnothing$. We will say that the *$\psi$-reduced frame determined by **O*** is the set $\{[\mathbf{Po_i}(\psi)]\}$. If ***O*** is non-degenerate, this reduces to the notion introduced earlier. If ***O*** is degenerate, then each eigenvalue of ***O*** will be associated with at most *one* ray. The full contextualist will extend VR to

> *VR_{FC}:* For every state ψ and every time *t*, there exists a function *f* defined on the set of all ψ-reduced frames such that (i) $f(\Gamma_\psi) \in \Gamma_\psi$, and (ii) if G is a mag-

nitude (maximal *or* non-maximal) with ψ-reduced frame Γ_ψ, then G takes the eigenvalue associated with $f(\Gamma_\psi)$.

An extension of our earlier remarks about VR_{SC} shows that VR_{FC} provides consistent value assignments to *all* quantities, and does so in a way that permits a classical representation of the probabilities.

4. Choosing Between Standard and Full Contextualism

The last two sections have been concerned with rescuing standard contextualism from a potential embarrassment: having no way to say what is really going on when an apparently degenerate measurement is performed. Since that challenge appears to have been met, it might seem that there is no reason to retain the extra contexts of full contextualism. However, up to now we have restricted our attention to pure states. What happens if we consider mixtures? If the mixtures represent mere ignorance of the pure state, then nothing new is called for. It can simply be claimed that whatever pure state may be, all the previous discussion applies. The interesting case arises if the mixture is not the result of combining ensembles of systems in various pure states. Here the cases divide into two. In the first, the system of interest is part of a larger system that is in some pure state, and the mixture is obtained by taking a partial trace. We have, in effect, already dealt with this case: take the Hilbert space in which the pure state sits to be the appropriate Hilbert space, and use the frames in that space.

The remaining possibility is that some mixtures are neither representations of ignorance nor local shadows cast by states of larger systems. Even here there is some hope for standard contextualism. Suppose that a system S is in a "deep" mixture, represented by a density operator $\boldsymbol{W}$. Let $\boldsymbol{O}$ correspond to the measurement (or "measurement") we want to perform. The reader may have noticed that the appeal to reduced frames thus far takes more than a little inspiration from Lüders' rule. One way to think of Lüders' rule is as the restriction of a state to a subalgebra.[5] Thus, consider the subalgebra **A** consisting of all Hermitian operators that commute with $\boldsymbol{O}$. Among these will be various density operators, and one of these, call it $\boldsymbol{W'}$, will be the restriction of $\boldsymbol{W}$ to **A**. That is, for every $\boldsymbol{O'}$ in **A**, we will have $\mathrm{tr}(\boldsymbol{WO'}) = \mathrm{tr}(\boldsymbol{W'O'})$. This state, it turns out, is the one that we would arrive at by using Lüders' rule to generate a post-measurement state for an ensemble of systems in $\boldsymbol{W}$, all of which were subjected to a "measurement" of O. If we are fortunate, the spectral representation of $\boldsymbol{W'}$ will be a weighted sum of projectors onto *rays*, with no degeneracies among the non-zero components. If so, the standard contextualist can use $\boldsymbol{W'}$ to pick out a unique reduced frame, and hence a unique set of maximal quantities that can be taken to be the ones really measured. However, if $\boldsymbol{W'}$ contains any further degeneracies, no such unique set will be selected. In that case, full contextualism wins by a nose. But it seems prudent simply to leave the issue at this point.

5. Whither Contextualism?

So much for our internal review of contextualism. The obvious question is whether the game is worth the candle. On the one hand, contextualism is a recurring theme in the literature. If for no other reason than that, it would seem to be worth being clear on its internal constitution. On the other hand, contextualism has many vices, and even the streamlined versions presented here have most of them. Notice that even with our re-sewing, full contextualism leaves Hilbert space in tatters. In three-dimensional Hilbert space, the reader can easily verify that *no* maximal quantities will have their values linked by VR_{SC} or VR_{FC} except when ψ is one of their common eigenstates.

But VR alone, which virtually everyone is prepared to accept, already makes this connection. Furthermore, contextualism with VR_{SC} or VR_{FC} will still be highly non-local. Consider, for example, the contexts {***Ax*, *Bx*, *Cx***} and {***Ax*, *By*, *Cy***} from our previous discussion of GHZ/Mermin. These contexts pick out maximal operators, which we will denote by ***Ax#Bx#Cx*** and ***Ax#Bx#Cx***. The question is whether VR_{SC} induces any connections among the corresponding quantities. The answer is straightforwardly no. The frames associated with these contexts are, respectively,

{[|x+>|x+>|x+>], [|x+>|x+>|x->], ... [|x->|x->|x>]},

and

{[|x+>|y+>|y+>], [|x+>|y+>|y->],... [|x->|y->|y->]}.

Clearly they have no members in common. In fact the only way we could even get a hint of locality would be if two locally maximal operators ***X*** and ***Y*** on one subsystem had some of their eigenvectors in common.

Of course, none of this would be troublesome to the committed contextualist, who was already prepared to live with an even more ragged remnant of Hilbert space than we have at this point. It simply shows how far we still are from conventional quantum mechanics. I therefore turn, briefly, to a couple of problems that seem to me a little more serious.

We noted earlier that one motivation for the combination of value-definiteness and contextualism is that it might provide a way of dealing with the measurement problem. In fact, the situation is not so clear. The problem is that the operator representing the apparatus indicator-observable will almost certainly be degenerate. Thus, it will be a function of many distinct maximal operators, some pairs of which have *no* eigenvectors in common. On both standard and full contextualism, each of the corresponding quantities will have a value, and there is no reason, even given our revisions of contextualism, to expect those values to mesh. Thus, it will be as though the apparatus had several "pointer readings" at once.

On the face of it, the full contextualist fares a little better here. S/he can at least say that there really is one observable that is "fundamentally" picked out by the operator in question, and that it has a unique value. However, it is not actually so clear why the other quantities whose statistics are given by this same operator have any less claim to being the "real" pointer reading. After all, by virtue of being less degenerate, they provide more information about the actual physical situation of the apparatus.

It might be replied that multiple pointer-readings (or ersatz pointer-readings) are puzzling but not necessarily serious. We would have real trouble if multiple readings were *observed*—if you and I could be expected regularly to disagree about the result of the measurement when we looked at the same apparatus. However, the contextualist will argue that there is no reason to expect this. Leaving aside the case of ineliminable mixtures, a measurement will correlate a particular basis in the apparatus Hilbert space with a particular basis in the observer's Hilbert space. Furthermore, which basis this is depends only on the state of the apparatus (or the apparatus plus environment). Hence, if you and I look at the same apparatus, we should see the same result. The problem with this is that this is so only if we look simultaneously. Suppose you look first and then I look a moment later. Our experience is that the pointer reading will be stable over such brief intervals. But it is not at all clear that contextualism predicts this. The state of the apparatus (plus environment) will almost certainly have changed during the interval, even if only slightly. The consequence will be that when the state is projected into the eigenspaces of the indicator observable, a different set of vectors will be selected and hence a different family of maximal magnitudes. We have been given no reason to think that their values should link up with the earlier set. Clearly the contextualist owes us some sort of stability theorem.

A deeper issue centers around locality. I am one of those who believe that quantum mechanics involves some form of non-locality, non-separability, holism or whatever. The usual view is that it is a holism of *states*. Contextualism presents a different option: the primary site of holism is the very quantities about which the states provide information. (See Redhead 1987, p. 139.) However intriguing this may sound, what strikes me most about contextualist holism is how unsatisfying it is. In the usual quantum mechanical story, the "holistic" character of such states as the GHZ/Mermin state can be expressed clearly in terms of the mathematics of the theory itself. (For pertinent discussions, see Hughes 1989 and Stairs 1983.) On the other hand, consider a very simple situation as described by contextualism. Let A and B be two well-separated quantum systems. Let ***A1*** be a locally maximal A-operator and let ***B1*** and ***B2*** be locally maximal B-operators. Now consider the two quantities A1#B1 and A1#B2. Explicit measurement of A1#B1 will also yield a value for $A1_{A1\#B1}$; explicit measurement of A1#B2 will also yield a value for $A1_{A1\#B2}$; $A1_{A1\#B1}$ and $A1_{A1\#B2}$ are different quantities. They may take distinct values, and if they do, measurements of the sort just described would yield different results at the A-site. But the algebraic representation simply makes this appear miraculous. Somehow, the A-apparatus must know which aspect of the system to respond to: the $A1_{A1\#B1}$-aspect or the $A1_{A1\#B2}$- aspect. Which one it should respond to depends not just on an intrinsic fact about the system being measured, even if we view the system as constituted indivisibly by the pair. It depends on the facts about the system *and* on the type of apparatus to which the B-component of the system is subjected. But the assignment of distinct values to $A1_{A1\#B1}$ and $A1_{A1\#B2}$ tells us nothing about how the A-apparatus knows what it needs to. Rather, it tells us what the local apparatus will do if only it can figure out how its distant partner is oriented.

To some this objection may not seem very potent. Mathematical elegance aside, the orthodox understanding of an ordinary Bell-type situation is not without controversy. One witness to this is the scheme proposed by David Bohm, in which the sort of non-locality just described is built in at the outset. But even leaving contextual theories aside, the quantum mechanics of the singlet state of two spin-1/2 particles is odd as it stands. *If* the distant spin apparatus is aligned parallel to the local one, and *if* the distant particle registers spin-up, then the local particle must register spin-down. And one might ask how the local particle knows what it needs to know about the distant setting/result combination.

I think there are many things that might be said here. To mention just one, on the usual quantum mechanical story we at least have locality in Jarrett's sense: local outcomes do not depend on the choice of settings made by a distant experimenter. Nonetheless, I will concede that the difference between failures of what Jarrett calls locality and what he calls completeness may not be as great as they are often taken to be, especially in light of the results of Jones and Clifton (unpublished).

Thus we end on an inconclusive note. Contextualism is both somehow attractive and somehow offputting, and both the attraction and the aversion are hard to articulate precisely. Although I don't believe contextualism, I'm not sure I can say why it isn't true. But then if you want to know the truth about quantum mechanics, you will have to stand in line with the rest of us.

Notes

[1]Of course there are many distinct maximal *operators* determined by ***Ax***, ***Bx*** and ***Cx***. However, we will follow Redhead in treating them all as picking out a single *quantity*.

[2]It is also worth remembering that Bell's own objections to using Gleason's theorem as a no-hidden-variables proof had contextualist motivations.

[3]I was inspired to think about this by a paper that Bas van Fraassen read in 1978 at the University of Western Ontario. His solution rested on exploiting a concept of a mixed test developed by Robin Giles. Unfortunately, when I spoke to him about it, van Fraassen no longer had a copy of his talk, and no longer remembered the details of what he presented. My solution, in any case, is quite different.

[4]There are various possible ways of dealing with joint probabilities, but the details are not germane here.

[4]For a somewhat related discussion, see Fine 1987.

References

Belinfante, F.J. (1973), *A Survey of Hidden Variable Theories*, Oxford: Pergamon Press.

Bohm, David (1980), *Wholeness and the Implicate Order*. London: Routledge, Kegan and Paul.

Fine, A. (1974), "On the Completeness of Quantum Theory", *Synthese* 29, 257-89.

Fine, Arthur (1987), "With Complacency or Concern: Solving the Quantum Measurement Problem", in *Kelvin's Baltimore Lectures and Theoretical Physics: Historical and Philosophical Perspectives* Robert Kargon and Peter Achinstein (eds.). Cambridge: The MIT Press, pp. 491-506.

Hughes, R.I.G. (1989), "Bell's Theorem, Ideology and Structural Explanation", in *Philosophical Consequences of Quantum Theory: Reflections on Bell's Theorem* James T. Cushing and Ernan McMullin (eds.). Notre Dame: University of Notre Dame Press, pp. 195-207.

Jarrett, Jon P. (1984), "On the Physical Significance of the Locality Conditions in the Bell Arguments", *Noûs* 18: 569-589.

Jones, Martin and Clifton, Robert K. (unpublished), "Incompleteness and Superluminal Signalling".

Kochen, S. and Specker, E.P. (1967), "The Problem of Hidden Variables in Quantum Mechanics", *Journal of Mathematics and Mechanics* 17: 59-87.

Redhead, Michael (1987), *Incompleteness, Nonlocality and Realism: A Prolegomenon to the Philosophy of Quantum Mechanics*. Oxford: Clarendon Press.

Shimony, Abner (1984), "Contextual Hidden Variable Theories and Bell's Inequalities", *British Journal for the Philosophy of Science* 35: 25-45

Stairs, Allen (1983), "On the Logic of Pairs of Quantum Systems", *Synthese* 56: 437-460

Van Fraassen, B. (1973), "Semantic Analysis of Quantum Logic", in C.A. Hooker (ed.) *Contemporary Research in the Philosophy and Foundations of Quantum Theory*. Dordrecht: Reidel, pp. 80-113.

The Objectivity and Invariance of Quantum Predictions[1]

Gordon N. Fleming

Pennsylvania State University

1. Introduction

In a recent article Itamar Pitowsky (1991)[2] has analyzed some measurement possibilities presented by an entangled state of three spin 1/2 particles. From that analysis he has concluded that the valid *predictions* of results of possible future measurements, made by observers who actually carry out mutually space-like separated measurements, can be incompatible with one another in the sense that their conjunction contradicts the quantum theory. From this result Pitowsky infers the inherently observer specific, or relative, character of quantum *predictions* as opposed to the results of *actual* observations or measurements, which, he claims, are covariant; i.e. (I take it) compatibly translatable from one observer to another.

This conclusion opposes the objective interpretation of the quantum theory[3] according to which the possible outcomes of possible measurements and the probabilities of occurrence of the various possible outcomes are as objective, as covariant (compatibly translatable) as the actual results of observations. Pitowsky claims that his argument, employing only quantum theoretic expectation values, does not, itself, depend on any interpretation of the theory.

I offer a refutation of Pitowsky's argument. The principal contention of my refutation is that the "valid predictions" of Pitowsky's observers are ambiguous, that the ambiguity concerns the "time-like domains" to which they apply, that when the ambiguities are removed the resulting limited applicability of the predictions makes their conjunction not applicable to any "time-like domain" in which contradiction with quantum theory would result. The second contention of my refutation is that the "valid predictions" of Pitowsky's observers, suitably clarified, are as compatibly translatable from one observer to another, *so that all observers can agree on, and accept, the physical content of the predictions of any observer*, and therefore are as covariant, as objective, as are the descriptions of the results of actually performed measurements. Indeed, it is in the terms of a naturally covariant language that one most easily identifies the "time-like domains" in which the "valid predictions" are applicable. In fact, the whole issue of covariance and translatability of the description of predictions or observations could be avoided by the adoption of a frame independent, *invariant* language. It is not the

PSA 1992, Volume 1, pp. 104-113

physical content of either predictions or observation reports that requires translation but just the language in which they are presented so as to gaurantee that the content will be invariant! Accordingly, I will, when appropriate, employ an invariant language to clarify the content of the predictions of Pitowsky's observers.

2. Review of Pitowsky's Argument

The quantum system of interest consists of three spin 1/2 particles in an entangled state, the spin structure of which is given by, (up to normalization)

$$|\Psi\rangle = |+--\rangle + |-+-\rangle + |--+\rangle, \tag{1}$$

where the + and - signs refer to the z-components of the spins of the 1st to the 3rd particle, reading from left to right within any one ket vector (since Pitowsky regards the locations of the particles to be spatially widely separated when the measurements he considers are to be made, we will distinguish the particles spatially and will not concern ourselves with whether or not they are identical particle types). Measurements of the z-component of spin are now assumed to be made on particles 1,2, and 3, by observers O_1, O_2, and O_3 respectively, at times and spatial locations such that the measurements are all mutually space-like with respect to one another. A fourth observer, O_4, is associated with a frame of referance in which the three measurements occur roughly simultaneously. The observers, O_1, O_2, and O_3, have rest frames of referance in which, *for each of them*, their measurement is performed first and, if the spatial separation is large enough, the time intervals between their measurement and the other two may be very long.

If the outcome of the O_1 measurement is $\sigma_z^1 = -1$, then O_1 can infer that a subsequent measurement of $\sigma_x^2\sigma_x^3$ will yield the result, $\sigma_x^2\sigma_x^3 = +1$. Alternatively, if the outcome of the O_1 measurement is $\sigma_z^1 = +1$, then , O_4 will eventually learn, according to $|\Psi\rangle$, that the O_2 and O_3 results were both -1. In that case O_2 can infer that subsequent measurements of $\sigma_x^3\sigma_x^1$ will yield the result $\sigma_x^3\sigma_x^1 = +1$ and O_3 can infer that subsequent measurements of $\sigma_x^1\sigma_x^2$ will yield the result $\sigma_x^1\sigma_x^2 = +1$. But the conjunction of these last two compatible predictions yields,

$$(\sigma_x^3\sigma_x^1)(\sigma_x^1\sigma_x^2) = \sigma_x^2\sigma_x^3 = +1, \tag{2}$$

the same result inferred by O_1 in the original case with $\sigma_z^1 = -1$. Consequently $\sigma_x^2\sigma_x^3 = +1$ regardless of the result of the measurement of σ_z^1 by O_1. But this conclusion contradicts quantum theory since,

$$\langle\Psi|\sigma_z^1|\Psi\rangle \neq \langle\Psi|\sigma_z^1\sigma_x^2\sigma_x^3|\Psi\rangle. \tag{3}$$

3. Refutation of the Argument

As a preliminary step, I want to consider the question of just what quantities are being measured and inferred about in this discussion. The point is that since any two of the considered observer-apparatus complexes are in relative motion (since the time sequence of measurement events is different for each of them) the designation of z or x-components of spin is not unambiguous without further specification. I suggest, as one of many possible options, regarding each term in the given state $|\Psi\rangle$ as a product state with the i-th factor being a z-component spin eigenstate for the i-th particle, *as such a state would be defined by the observer-apparatus complex O_i*. Similarly for any references in the argument to x or y components of spin. This option has the slightly unconventional consequence that all observer-apparatus complexes O_1, O_2, O_3, and O_4 are describing some of their measurements and predictions in a language not referring directly to their own rest frame coordinates. The advantage of this option for the argument, however, is that when we consider conjoining the inferences of O_2 and O_3, we can be sure they are both referring to the same quantity as σ_x^1 and that that quantity is the same quantity called σ_x^1 by O_1. I repeat that there are many other equally acceptable ways to render this matter precise so as to proceed with the argument. In particular, there is nothing special about the status of O_4 which, with the present option, makes no descriptions or predictions referring directly to the O_4 rest frame coordinates.

With that taken care of we now consider the possible results of the O_1 measurement. Suppose it is $\sigma_z^1 = -1$. Then O_1 can infer that any measurement of $\sigma_x^2\sigma_x^3$ *made between the time of the O_1 measurement and the later measurements of O_2 and O_3* would yield $\sigma_x^2\sigma_x^3 = +1$. O_1 would certainly not infer any such result for an hypothetical measurement of $\sigma_x^2\sigma_x^3$ made *after* either of the measurements of O_2 or O_3. For such later measurements quantum theory predicts, and O_1 would infer,

$$<\sigma_x^2\sigma_x^3> = 0. \tag{4}$$

Alternatively suppose $\sigma_z^1 = +1$. Then, not only O_4, but O_1, as well, can infer that the later (for O_1) measurements of O_2 and O_3 will yield $\sigma_z^2 = -1$ and $\sigma_z^3 = -1$. Furthermore, O_1 can also deduce that each of O_2 and O_3 can separately infer that if measurements of $\sigma_x^3\sigma_x^1$, $\sigma_x^1\sigma_x^2$, respectively, are performed *between the O_2, O_3, times, respectively, of their own measurements and the later (for them) measurements of the other two observer-apparatus complexes* then the results will be

$$\sigma_x^3\sigma_x^1 = +1 \quad \text{and} \quad \sigma_x^1\sigma_x^2 = +1, \quad \text{respectively.} \tag{5}$$

But neither of these predictions, *as they stand and without further assumptions*, refer to *any* definite times or time intervals in the rest frames of either of the observer apparatus complexes O_1 or O_4. In particular,*there is no time interval in the O_1 frame of referance to which both of these predictions of O_2 and O_3 apply!* Consequently, their is no time interval in the O_1 frame to which the *conjunction* of the predictions (5.) and, therefore, the *product* of the predicted values, and, therefore, the *inferred* value of $\sigma_x^2\sigma_x^3$ applies. Thus there is no time interval in the O_1 frame in which one can assert the value of $\sigma_x^2\sigma_x^3$ to be unity regardless of the outcome of the measurement of σ_z^1. *Without further assumptions the contradiction with the quantum theory result (3.) can not be derived!*

Are there natural further assumptions that one might make that would yield the contradiction? I will consider some "natural" auxilliary assumptions here and show why they do not succeed in reinstating the contradiction. In the course of the discussion I will suggest and adopt a model for the frame independent spatio-temporal referance of measurements and predictions of the relevant dynamical variables which makes it impossible to derive the contradiction. Elsewhere (Fleming 1966, 1987), I have argued at length that the general structure of Lorentz Invariant Quantum Theory dictates for us the frame independent spatio-temporal referance of measurements and predictions of *all* dynamical variables occurring in the theory.

The physical systems to which these predictions refer are intrinsically spatially extended since they are two particle subsystems of an extended three particle system. Consequently assertions about what will happen with such systems at a definite time, from the standpoint of one inertial frame, can not, *without further assumptions*, be translated into assertions about happenings *at any definite time* from the standpoint of a relatively moving frame of referance.

Suppose one specifies the locations of the constituents of the two particle subsystems at the spin measureing events referred to in the predictions (i.e. suppose one *assumes* that the spin measurements require localizations of the particles). Then spin measurements, actual or hypothetical, would refer to space-time points (or, more realistically, highly localized space-time regions) and such spatio-temporal referance would be associated with definite times or time intervals in all referance frames. It then turns out, however, that the only space-time points at which calculation of σ_x^2 σ_x^3 must comply with *both* the predictions of O_2 (for $\sigma_x^3\sigma_x^1$) and O_3, (for $\sigma_x^1\sigma_x^2$) in the case of $\sigma_z^1 = -1$, are space-time points which are space-like with respect to the original localized measurements of σ_z^2 and σ_z^3. This follows from the considerations that, since O_2's inferance applies only to O_2 times *later* than the measurement of σ_z^2, while O_3's inferances apply only to O_3 times, *earlier* than the measurement of σ_z^2, the calculation of σ_x^2 σ_x^3 must apply *after* the measurement of σ_z^2, as assessed by O_2, but *before* the measurement of σ_z^2, as assessed by O_3. Similarly for the measurement of σ_z^3. But this relativity of the time ordering of the points to which the calculation of σ_x^2 σ_x^3 applies with respect to the σ_z^2 and σ_z^3 measurements implies space-like separation between those points and the measurements. Now let us further suppose that the quantum particles do not have superluminally propagating position representation state functions, i.e. they propagate strictly subluminally. *If so, then the matter is settled, since the predictions for localized spin measurements jointly apply only to restricted regions of space-time which are space-like separated from the actual σ_z measurements and in such regions the particles would never be found.*

Upon relaxing the assumption of localized spin measurements we need an assumption about how the spatio-temporal referance of a non-local measurement or prediction made in one inertial frame is to be construed in another relatively moving inertial frame. The simplest "natural" assumption, it seems to me, is to identify the *set* of space-time points that comprise the spatio-temporal referance in the original frame of referance and to regard that same *set* as defining the spatio-temporal referance in all frames of referance, however much the *description* of the *set* may change from one frame to another. Thus measurements or predictions referring to definite times or time intervals in one referance frame would refer to definite space-like hyperplanes or bounded sets of parallel hyperplanes in any referance frame, since one frames definite time is another frames definite space-like hyperplane. Indeed the space-like hyperplane characterization of the spatio-temporal referance is an invariant characterization since the hyperplane in question does not change from frame to frame, only its parameterization and relationship to the frame changes. For readers not familiar with the concept of space-like hyperplanes the remarks in the opening paragraphs of section 4. may be helpful. *With this assumption for the spatio-temporal referance of non-local spin measurements the drawing of Pitowsky's conclusion is prevented again, this time by the predicted measurement results, for $\sigma_x^3\sigma_x^1$ by O_2, and $\sigma_x^1\sigma_x^2$ by O_3, referring to disjoint sets of space-like hyperplanes, namely, sets of hyperplanes which are instantaneous in O_2 or instantaneous in O_3.*

Even more questionable than the assumption of localized spin measurements is the assumption of subluminal propagation of the position representation state functions for relativistic quantum particles. In fact, as I and others have argued elsewhere (Fleming 1965a,b; Hegerfeldt 1974, 1985; Ruijsenaars 1981; Fleming and Bennett 1989; Fleming and Butterfield 1991), the assumption of subluminal propagation appears to be incompatible with the general principles of Lorentz Invariant *Quantum Theory* (LIQT). Furthermore, compatiblity between superluminal propagation and Lorentz invariance alone is accomplished via the hyperplane dependence of the position probability amplitudes. In other words, all probability amplitudes for finding a particle at a particular point of space-time are qualified as probability amplitudes for finding the particle at the point in question *but regarded as a point belonging to a particular space-like hyperplane*. The probability amplitudes can depend strongly on the hyperplane even when the point of localization belonging to the hyperplane is held fixed! In short, the probability amplitudes are associated with ordered pairs of points and space-like hyperplanes containing those points. Therefore, once again Pitowsky's conclusion is avoided since even though the particles can now reach the space-like separated *points* to which the predictions may jointly apply, *the predictions, in an essential way, now refer to ordered pairs of points and hyperplanes, and the hyperplanes referred to in the different predictions belong to mutually disjoint sets. There are, consequently, no ordered pairs of points and hyperplanes to which the conjunctions of the predictions refer.*

This still leaves an interesting problem of the internal consistency of such predictions. In particular the question of how the predictions change when we pass from considering space-like hyperplanes that are "later" than one of the σ_z measurements and "earlier" than the other two, to space-like hyperplanes that are "later" than two of the σ_z measurements and "earlier" than only one, and then finally to space-like hyperplanes that are "later" than all of the σ_z measurements. We turn to these considerations in the next section.

4. Quantum Predictions on Space-like Hyperplanes

The flat, three dimensional, space-like slices of space-time, which, individually, would be designated as an instantaneous with respect to some Minkowski inertial frame, are commonly referred to as space-like hyperplanes, or hyperplanes for short. The instantaneous hyperplanes of one inertial frame are not instantaneous for any relatively moving inertial frame and may be said to be "tilted" (with respect to the time axis) in that relatively moving frame. There is a widespread tendency to regard as somehow more real and less formal those hyperplanes that are instantaneous in whatever inertial frame is being employed. This undoubtedly springs from our intuitive experience of "time" as being more real than space-time, but in late twentieth century physics it is a wholly unwarranted attitude for which Minkowski's (1908) ringing words should be ample antidote. In any case I am taking the position here that the predictions for definite times, or time intervals, of any of our observer-apparatus complexes, O_i, can be interpreted as predictions for definite tilted hyperplanes, or sets of such hyperplanes, by any other of the complexes. *This is the manner in which predictions or observations, initially expressed in terms of the language of one inertial frame, can be translated, without prejudicial dynamical assumptions, into the language of any other inertial frame.* It is furthermore a move which expands the available language for expressing predictions and observations within one inertial frame since it does not require the expression to refer to any definite time or time interval. The expression may refer to tilted hyperplanes instead.

Every hyperplane divides the points of space-time into three disjoint sets; those that are earlier than the hyperplane in the inertial frame in which the hyperplane is instantaneous, those that are later than the hyperplane in the same inertial frame, and those that lie in the hyperplane. We will say of any hyperplane that it is "later" than the first set of points, "earlier" than the second set of points, and "contains" the third set of points. This characterization is invariant under transformations from any inertial frame to any other.

To analyze the whole family of spin measurement predictions, of the kind in question here, that O_1, O_2, and O_3, can make, it will be convenient and instructive, I think, to adopt the standpoint of the interpretations of quantum theory which take state reductions to be objective real processes. One of the widely held concerns about state reductions is that they can not be made objectively compatible in the presence of space-like separated measurements (Penrose 1989; Shimony 1986). I have argued elsewhere (Fleming 1987, 1989) that it is precisely the hyperplane language that enables one to so make them and our present case is an ideal instance for illustration. Furthermore, all of the statistical predictions that follow from such state reductions can be extracted from the original state vector by calculating expectation values of appropriate products of projection operators or other observables as Pitowsky stressed in his paper. The virtue of the hypothesis of state reduction is that it provides the *explanation* for the probabilities that hold "after" measurements are completed. They hold because the state has changed in a specific way as a consequence of the completed measurement. Accordingly I will employ the state reduction language.

Let h_0 denote the set of all hyperplanes that are earlier than the σ_z measurements of all the O_i. Let h_i denote the set of all hyperplanes that are later than the measurement of $\sigma_z{}^i$ by O_i and earlier than the other two σ_z measurements by the other two complexes. Let h_{ij} denote the set of all hyperplanes that are later than the measurements of both $\sigma_z{}^i$ by O_i and $\sigma_z{}^j$ by O_j and earlier than the remaining σ_z measurement by the remaining complex. Finally, let h_{123} denote the set of all hyperplanes that are later than all three σ_z measurements (see Fig.1.). In the Heisenberg picture of dynamical evolution, in which the state vector represents an entire history of unitary evolution and changes only as a consequence of state reduction, the state vector for our system has the spin structure given by (1.) on all the hyperplanes of the set h_0[4]. The state vectors associated with other hyperplanes will depend on the outcomes of the σ_z measurements. The possibilities are listed below with the corresponding spin structure assignments for the state vectors of each of the sets of hyperplanes we have just defined.

Dynamical Variables	Possible Results of Measurements		
$\sigma_z{}^1$:	+	-	-
$\sigma_z{}^2$:	-	+	-
$\sigma_z{}^3$:	-	-	+

Hyperplane Sets	Reduced State Vectors Assigned to the Hyperplane Sets for the above Possible Results of Measurements		
h_1 :	$|{+}{-}{-}\rangle$,	$|{-}{+}{-}\rangle + |{-}{-}{+}\rangle$,	$|{-}{+}{-}\rangle + |{-}{-}{+}\rangle$
h_2 :	$|{+}{-}{-}\rangle + |{-}{-}{+}\rangle$,	$|{-}{+}{-}\rangle$,	$|{+}{-}{-}\rangle + |{-}{-}{+}\rangle$
h_3 :	$|{+}{-}{-}\rangle + |{-}{+}{-}\rangle$,	$|{+}{-}{-}\rangle + |{-}{+}{-}\rangle$,	$|{-}{-}{+}\rangle$
h_{12} :	$|{+}{-}{-}\rangle$,	$|{-}{+}{-}\rangle$,	$|{-}{-}{+}\rangle$
h_{23} :	$|{+}{-}{-}\rangle$,	$|{-}{+}{-}\rangle$,	$|{-}{-}{+}\rangle$
h_{31} :	$|{+}{-}{-}\rangle$,	$|{-}{+}{-}\rangle$,	$|{-}{-}{+}\rangle$
h_{123} :	$|{+}{-}{-}\rangle$,	$|{-}{+}{-}\rangle$,	$|{-}{-}{+}\rangle$

We see that, as a consequence of the uniqueness of any third σ_z value once two such values are determined, the reduced state assignment to *any* hyperplane later than two or more of the σ_z measurements is the same. For hyperplanes later than only one σ_z measurement, however, the reduced state vector assignment depends on which measurement precedes the hyperplane, and this notwithstanding the fact that any two hyperplanes in distinct sets, h_i, and h_j, $i \neq j$, intersect, i.e. they have points in common. This seems to point out the global nature of these state vector assignments; they are assignments to the hyperplane as a whole and not assignments to the points comprising the hyperplanes. But then we are rendered once again anxious about the possibility of combined position-spin measurements which, even if they can not be used to reach Pitowsky's conclusion, would seem to give rise to conceptual difficulty for some points common to hyperplanes from distinct sets.

For example, consider any point common to a member of h_0 and a member of h_i for $i \neq 0$. In the *absence* of superluminal propagation of particles, particle i can not occur (has zero probability of being detected) at the common point *on* h_i and yet can occur (have non-zero probability of being detected) at the common point *on* h_0*!* In the *presence* of superluminal propagation the particle can occur at the common point on both hyperplanes, but with, in general, very different probabilities and accompanied by very different spin structure states! Risking redundancy, I remind the reader again that it has been known for a long time now that it is very difficult and probably impossible to mount physically interpretable position representations for particles in Lorentz invariant quantum theory without admitting superluminal propagation (Fleming 1965b; Hegerfeldt 1974, 1985; Ruijsenaars 1981). I have also argued elsewhere that it is precisely in the conceptual framework provided by hyperplane dependence that such superluminal propagation can be given a coherent *and covariant* interpretation (Fleming 1966, 1987, 1989; Fleming and Bennett 1989; Fleming and Butterfield 1991). The theory of the hyperplane dependence of position and spin observables yields *incompatibility* of position observables of a particle associated with distinct hyperplanes and compatibility of position and spin of a particle only when each refer to one and the same hyperplane (Fleming 1965a, 1966, 1987). Consequently, the position-spin measurement probabilities predicted by the different O_i can not be conjoined, when referring to the same space-time points, not only because those points are associated with distinct hyperplanes, but also because that association renders them probabilities for incompatible observables. The most dramatic instances of this incompatibility are provided by the position-spin eigenstates, which, for a particular hyperplane confine the particle to a single point with a definite spin eigenstate. At every other point of *that* hyperplane the particle position probability density is strictly zero, yet for any other hyperplane intersecting the original hyperplane at one of those other points, the particle position probability density will not be zero, nor will it, in general, be associated with any definite pure spin state. It appears, then, that unlike the non-relativistic limit, in which all possible hyperplane orientations coellesc, the particles of LIQT may be regarded as possessing a multitude of mutually incompatible position and other observables associated with the various possible orientations of space-like hyperplanes. The hyperplane orientation dependence of these observables is, however, generally smooth and expressible in differential equations very similar in form to the Heisenberg picture equations of motion of non-relativistic quantum theory (NRQT). This suggests regarding such a 'multitude' of related observables instead as a single observable with a more elaborate spatio-temporal dependence than the simple time dependence of NRQT. For different purposes different attitudes are appropriate. *In any case a particle of LIQT is even less of a spatio-temporal localized entity than a particle of NRQT!*

Finally, I must mention the important distinction, in the presence of state reductions and in the Heisenberg picture, between the hyperplane dependence of state vector assignments, due solely to state reduction, and the hyperplane dependence of dynamical variables undergoing unitary evolution, which is independent of state reduction. This distinction gives at least formal meaning to the probability distribution, over the eigenvalue spectrum, of a dynamical variable evaluated at one hyperplane in a Heisenberg picture state which, due to state reduction, is not associated with that hyperplane. There are at least two ways in which such a probability distribution can be given a physical interpretation. First, counterfactually, as the probability distribution the dynamical variable would have had at the given hyperplane in the absence of (some) state reductions. Second, counterfunctionally, by using the unitary evolution of the dynamical variable to reexpress it as a function of dynamical variables evaluated at a hyperplane to which the state vector in question does apply. This second interpretation allows, in principal, for the measurement of the probability distribution of interest in the state in question. The simplest instances of such a counterfunctional interpretation are provided by conserved dynamical variables, such as the four-momentum of a free particle, which have no hyperplane dependence due to unitary evolution. The next more complex instances are provided by dynamical variables which are conserved except for a purely kinematical dependence on the orientation of hyperplanes. There are usefull characterizations of spin angular momentum, which, for a free particle, are of this second type[5].

In the context of our present problem, in which the particles are free except for the spin measurements made upon them, examples of the kind of questions raised in the preceeding paragraph are afforded by the probability distributions, in the state associated with hyperplanes of the set h_i, say, for the momenta, spins or positions of particle j evaluated at a hyperplane belonging to the set h_k. These considerations help to bring home to us the subtle richness of the kinematical, dynamical and statistical relationships that hold in LIQT!

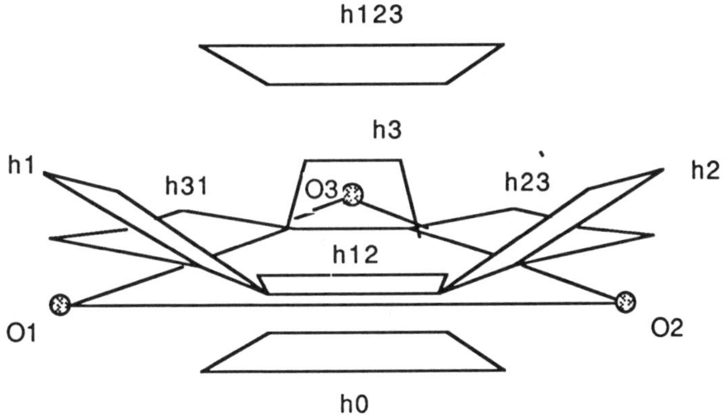

Figure 1. Representative hyperplane segments from the sets discussed in the text.

Notes

[1]I wish to thank Jeremy Butterfield, Robert Clifton and Constantin Pagonis for extensive discussion which led to considerable clarification of an earlier draft of this paper.

[2]I wish to thank Jeremy Butterfield for drawing my attention to this paper A more recent article by R. Clifton, C. Pagonis and I. Pitowsky (1992), discusses the same issues much more carefully and places them in a broader context of interpretational questions with the result that no challenge to the objectivity of quantum predictions is mounted. The relation of the present work to this later article is to indicate further the clarifications and sharpening of concepts that result from a focus on the space-like hyperplane dependence of dynamical variables and state reduction assignments.

[3]Representative examples of the many variants of this interpretation are A. Shimony (1989) and R. Penrose (1989). For comparative analysis see M.L.G. Redhead (1987).

[4]Strictly speaking, the state vectors assigned by O_1, O_2, O_3 and O_4 to any hyperplane are never identical, but rather, are the unitary Lorentz transforms of one another. The physical content of those state vector assignments, to any single hyperplane, is, of course, invariant, but distinct state vectors are required to account for it, from the standpoint of relatively moving inertial frames, since it is the observable relationship of the physical content to the inertial frame that is directly represented by the state vector. Accordingly each O_i assigns a single Heisenberg picture-state vector to every hyperplane in any one of the h sets. The state vectors assigned by each O_i to hyperplanes in distinct h sets are different and the difference depends on the outcome of the measurements that separate the sets. In each h set the four distinct state vectors assigned to every hyperplane in that set by the four O_i are the unitary Lorentz transforms of one another. For the hyperplanes in h_0 all the state vectors correspond to the spin structure indicated in (1.).

[5]The Pauli-Lubyanski 4-vector provides a characterization of spin angular momentum which does not have this kinematical dependence on hyperplane orientation and for a free particle is a dynamical variable of the first type mentioned. It has the disadvantage, however, of not being compatible with any positive energy position operator for the particle on any hyperplane. See A. S. Wightman (1960).

References

Clifton, R., Pagonis, C., and Pitowsky, I. (1992), "Relativity, Quantum Mechanics and EPR". Cambridge: Cambridge University Press.

Fleming, G.N. (1965a), "Covariant Position Operators, Spin and Locality", *Physical Review* 137: B188-97.

_________. (1965b), "Nonlocal Properties of Stable Particles", *Physical Review* 139: B963-68.

_________. (1966), "A Manifestly Covariant Description of Arbitrary Dynamical Variables in Relativistic Quantum Mechanics", *Journal of Mathematical Physics* 7: 1959-81.

_________. (1987), "Towards a Lorentz Invariant Quantum Theory of Measurement", *Proceedings of the First Workshop on Fundamental Physics at the University of Puerto Rico,* A. Rueda (ed.). University of Puerto Rico, pp. 8-114. (revised version available as Penn State University preprint).

_________. (1989), "Lorentz Invariant State Reduction and Localization", *Philosophy of Science Association Proceedings1988* volume 2, A. Fine and M. Forbes (eds.), pp. 112-26.

_________. and Bennett, H. (1989), "Hyperplane Dependence in Relativistic Quantum Mechanics", *Foundations of Physics* 19: 231-67.

_________. and Butterfield, J. (forthcoming), "Is There Superluminal Causation in the Bell Experiment?", in the proceedings of the 1991 international conferance on *Bell's Theorem and the Foundations of Modern Physics.* Cesena, Italy.

Hegerfeldt, G.C. (1974), "Remark on Causality and Particle Localisation", *Physical Review* D10: 3320-21.

__________. (1985), "Violation of Causality in Relativistic Quantum Theory", *Physical Review Letters* 54: 2395-98.

Minkowski, H. (1908), Address to the 80th Assembly of German Natural Scientists and Physicians, Cologne. English translation as "Space and Time." in (1964) *Problems of Space and Time*, J.J. Smart (ed.). New York: MacMillan, pp. 297-312.

Penrose, R. (1989), *The Emporers New Mind.* Oxford: Oxford University Press, ch. 6.

Pitowsky, I. (1991), "The Relativity of Quantum Predictions". *Physics Letters A* 156: 137-39.

Redhead, M.L.G. (1987), *Incompleteness, Nonlocality and Realism.* Oxford: Clarendon Press.

Ruijsenaars, S.N.M. (1981), "On Newton-Wigner Localization and Superluminal Propagation Speeds", *Annals of Physics* 137: 33-43.

Shimony, A. (1986), "Events and Processes in the Quantum World." *Quantum Concepts in Space and Time* R. Penrose and C. Isham (eds.). Oxford: Oxford University Press, pp. 182-203.

_______. (1989), "Search for a Worldview Which can Accomodate our Knowledge of Microphysics." *Philosophical Consequences of Quantum Theory*, J.T Cushing and E. McMullin (eds.). South Bend: University of Notre Dame Press, pp. 25-7.

Wightman, A. S. (1960), "L'Invariance dans la Mecanique Quantique Relativiste", *Dispersion Relations and Elementary Particles*, C. de Wit and R. Omnes (eds.). New York: John Wiley and Sons, pp. 160-226.

Relativity, Quantum Mechanics and EPR[1]

Robert Clifton and Constantine Pagonis
Cambridge University

Itamar Pitowsky
The Hebrew University of Jerusalem

The Einstein-Podolsky-Rosen argument for the incompleteness of quantum mechanics involves two assumptions: one about locality and the other about when it is legitimate to infer the existence of an element-of-reality. Using one simple thought experiment, we argue that quantum predictions and the relativity of simultaneity require that both these assumptions fail, whether or not quantum mechanics is complete.

1. The Einstein-Podolsky-Rosen Argument

EPR's (1935) argument for the incompleteness of QM turns upon the following sufficient condition:

> "If, without in any way disturbing a system, we can predict with certainty (i.e., with probability equal to unity) the value of a physical quantity, then there exists an element of physical reality corresponding to this physical quantity."

EPR apply this condition in a case where knowledge gained from measuring one quantum system allows one to predict with certainty a value for a physical quantity on another. So EPR need to assume that a system is undisturbed by measurements made on other distant systems, i.e., they need to make a 'locality' assumption, as well as to assume the inference from a prediction with certainty to an element-of-reality, i.e., a 'reality' assumption. Let us start by recalling how EPR's conclusion that QM is incomplete follows from these two assumptions. For this, we employ the same thought experiment that we will later use to run our own arguments.

Suppose three spin-1/2 particles 1, 2 and 3, leave a source in different directions towards spin-meters 1, 2 and 3 that can measure their spins at spacelike separation. Let each particle 'carry' its own orthogonal coordinate system, where the Z-axis for each is picked out by its line of flight. So particle 1 carries coordinate system (X_1,Y_1,Z_1), and similarly (X_2,Y_2,Z_2) and (X_3,Y_3,Z_3) for particles 2 and 3. For simplicity, suppose the three Z-axes lie in a plane, as in Figure 1. The spin-meters then either measure the X- or Y-spin component of their particle, with possible results $x_1=\pm1$, $y_2=\pm1$, etc.

PSA 1992, Volume 1, pp. 114-128

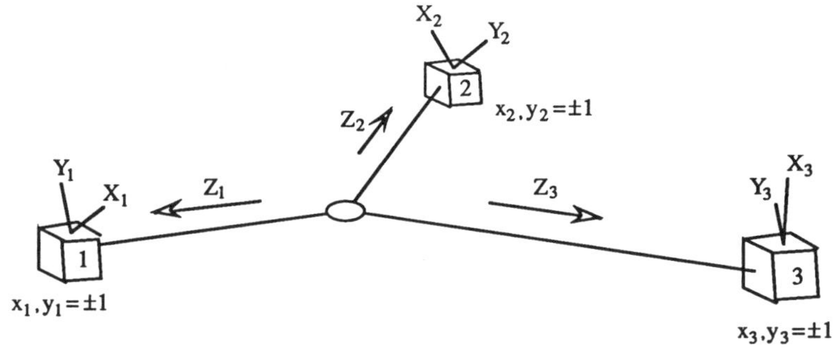

Figure 1. Thought Experiment

Now consider the four operators: $\sigma_{X1}\sigma_{Y2}\sigma_{Y3}$, $\sigma_{Y1}\sigma_{X2}\sigma_{Y3}$, $\sigma_{Y1}\sigma_{Y2}\sigma_{X3}$ and $\sigma_{X1}\sigma_{X2}\sigma_{X3}$. Since $\sigma_{Xi}\sigma_{Yj} = (1-2\delta_{ij})\sigma_{Yj}\sigma_{Xi}$ and $(\sigma_{Xi})^2 = (\sigma_{Yj})^2 = I$, these four operators commute and multiply to -I. Thus they share at least one eigenstate, $|\psi\rangle$, that assigns them values multiplying to -1. This follows from the fact that any simultaneous eigenstate, $|\psi\rangle$, of a set A, B, ... of commuting operators must assign them values $[A]_{|\psi\rangle}$, $[B]_{|\psi\rangle}$, ... that respect the *product rule*: $[AB\cdots]_{|\psi\rangle} = [A]_{|\psi\rangle}[B]_{|\psi\rangle}\cdots$; for $A|\psi\rangle = [A]_{|\psi\rangle}|\psi\rangle$, $B|\psi\rangle = [B]_{|\psi\rangle}|\psi\rangle$, ... entails $AB\cdots|\psi\rangle \equiv [AB\cdots]_{|\psi\rangle}|\psi\rangle = [A]_{|\psi\rangle}[B]_{|\psi\rangle}\cdots|\psi\rangle$. Applying this rule to our four operators gives:

$$(\exists|\psi\rangle)([\sigma_{X1}\sigma_{Y2}\sigma_{Y3}]_{|\psi\rangle}[\sigma_{Y1}\sigma_{X2}\sigma_{Y3}]_{|\psi\rangle}[\sigma_{Y1}\sigma_{Y2}\sigma_{X3}]_{|\psi\rangle}[\sigma_{X1}\sigma_{X2}\sigma_{X3}]_{|\psi\rangle} = -1)\,. \quad (1)$$

Since the square of each operator is I, another application of the product rule implies that each factor in (1) must be ±1. So there are eight possible states, $|\psi\rangle$, that can make (1) true. For our thought experiment, let us arbitarily choose the state |GHZ> (the 'Greenberger-Horne-Zeilinger' state of recent fame—see Mermin (1990)) that assigns the values:

$$\begin{aligned}
&[\sigma_{X1}\sigma_{Y2}\sigma_{Y3}]_{|GHZ\rangle} = +1\,,\\
&[\sigma_{Y1}\sigma_{X2}\sigma_{Y3}]_{|GHZ\rangle} = +1\,,\\
&[\sigma_{Y1}\sigma_{Y2}\sigma_{X3}]_{|GHZ\rangle} = +1\,,\\
&[\sigma_{X1}\sigma_{X2}\sigma_{X3}]_{|GHZ\rangle} = -1\,. \qquad (2)
\end{aligned}$$

Now for the purposes of EPR, suppose neither of the particles in state |GHZ> is actually measured. Nevertheless, the +1 in (2)'s first equation entails that were particle 1's X-spin and particle 2's Y-spin measured and some particular results x_1 and y_2 obtained, then a subsequent measurement of particle 3's Y-spin would, with certainty, yield x_1y_2. Applying the 'reality' part of EPR's sufficient condition, this means there would then exist a Y_3-spin element-of-reality. But surely measuring along X_1 and Y_2 could not bring a spacelike separated element-of-reality into existence! So using the 'locality' part of EPR's condition, a Y_3-spin element-of-reality—corresponding to the observable σ_{Y3} - must also exist in the actual world.

The trouble is QM fails to capture this state of affairs in its ascription of the state |GHZ> to the particles at all times in the actual world (remember: no measurements are actually performed). For since |GHZ> assigns values to the products $\sigma_{Y1}\sigma_{Y2}\sigma_{X3}$ and $\sigma_{X1}\sigma_{X2}\sigma_{X3}$, it is prevented from assigning a value or element-of-reality to any observable, like σ_{Y3}, that fails to commute with this pair of operators. Given their sufficient condition, EPR are thus right to conclude that the quantum description of physical reality is incomplete.

Is the EPR argument sound? We aim to use the very same thought experiment to show that the argument is unsound on, not one, but *two* counts. For we shall show that both the 'reality' and 'locality' parts of EPR's sufficient condition fail when spelled out in a natural counterfactual way. To establish this, our argument for the failure of locality (in section 2) will make no use of EPR's inference to elements-of-reality, nor will our argument for the failure of that inference (in section 3) make any use of locality. Furthermore, neither argument will assume the completeness of QM; only the truth of the select set of predictions in (2) that are inductively well-confirmed. In addition, we assume the well-established relativity of simultaneity for our argument against EPR's inference to elements-of-reality. Although we do not explicitly employ the formalism of relativistic QM in these arguments, all predictions on which they turn carry over to the relativistic theory since, as seen above, they follow directly from the (invariant) algebraic structure of (2)'s spin operators.

2. The Failure of Locality

We shall take locality to be the denial of causal dependence, in Lewis' (1986, p. 166) counterfactual sense, between spacelike separated events. Let A and B denote propositions (or the events they describe: we will not need to distinguish the two) and $A \Box\!\!\rightarrow B$ denote the counterfactual conditional: '*If* A were true, *then* B would be true'. Then Lewis proposes:

> If $c_1, c_2, \ldots$ and $e_1, e_2, \ldots$ are two pairwise distinct families of events (i.e., c_1 is distinct from e_1, c_2 from e_2, etc.) such that no two of the cs and no two of the es are compossible, then the 'e-family' deterministically causally depends upon the 'c-family' at world w *iff* $c_1 \Box\!\!\rightarrow e_1$, $c_2 \Box\!\!\rightarrow e_2$, ... are true at w.

Although Lewis (1986, p. 162) does not concern himself with the case of partial causation, where several events jointly cause another, the above definition can accommodate this if we allow each c_i to be composed of sub-events $c_i^1, c_i^2, \ldots$ that act as partial causes of e_i. In fact, our argument will only aim to show that Lewisian causation exists between an event and some of its partial causes at spacelike separation. Do not be alarmed that our use of Lewis' definition for deterministic causal dependence commits us to determinism at the outset; it does not. Rather, with the brute quantum predictions in (2), we shall prove that there must be spacelike separated events that satisfy Lewis' definition, even if those events fail to be predetermined by any other events in their causal history.

To establish notation, let X_1 be the proposition (event) that particle 1's spin is measured with spin-meter 1 set to 'X_1', and similarly for Y_1, X_3, etc. For conjunctions of settings, we write for simplicity '$X_1Y_2Y_3$' in place of '$X_1\&Y_2\&Y_3$', etc. Let x_1 be a variable whose possible values (±1) are propositions (or events) describing the possible X_1-spin measurement results, and similarly for x_2, y_3, etc. Again, for conjunctions of results we just write '$y_1x_2y_3$', etc. Finally, let λ be the proposition (event) describing the complete state of the particles as they leave their source on a particular occasion. For example, λ is just |GHZ> if QM is complete, but λ consists of |GHZ> plus the particles' positions in Bohm's (1952) theory.

We can now formulate (2)'s predictions using this notation. (2)'s first equation says that were $X_1Y_2Y_3$ true, the product of results obtained from these measurements would be +1, and similarly for (2)'s other three equations. Thus on the particular occasion that λ obtains, the following are true:

$$(X_1Y_2Y_3\lambda) \Box\!\!\rightarrow (x_1y_2y_3 = +1),$$
$$(Y_1X_2Y_3\lambda) \Box\!\!\rightarrow (y_1x_2y_3 = +1),$$

$$(Y_1Y_2X_3\lambda) \;\Box\!\!\rightarrow\; (y_1y_2x_3 = +1),$$
$$(X_1X_2X_3\lambda) \;\Box\!\!\rightarrow\; (x_1x_2x_3 = -1). \tag{3}$$

To prevent any of these counterfactuals from being vacuously true, we need to assume that λ is compatible with all four measurement combinations. This will be so if: (a) the settings are all fixed just before the measurement events occur, so that λ lies in the backwards light-cones of the setting events, and (b) the choice of which settings to fix is determined by (pseudo-)random number generators whose causal history is sufficiently disentangled from λ. We need only assume that conditions (a) and (b) obtain in a nonzero measure of trials.

We shall suppose, for simplicity, that all four counterfactuals in (3) are strictly contrary-to-fact, i.e., no measurements are actually made. Our argument will be that these counterfactuals entail, either that there is partial Lewisian causation between spacelike separated measurement results, or that there is such causation between results and spacelike separated settings. The argument is best broken down into the two parts below, with 'locality' understood as absence of Lewisian causation.

Part I: Determinism of Results & Setting-to-Result Locality & (3) $\Rightarrow$ Contradiction ,
Part II: (3) & Result-to-Result Locality $\Rightarrow$ Determinism of Results .

While determinism is assumed for Part I, it is derived in Part II, so it drops out of the argument leaving us with the conclusion we seek by *reductio*.

Beginning with Part I, determinism of results is just:

$$(\exists x_1,y_2,y_3)(X_1Y_2Y_3\lambda \;\Box\!\!\rightarrow\; x_1y_2y_3),$$
$$(\exists y_1,x_2,y_3)(Y_1X_2Y_3\lambda \;\Box\!\!\rightarrow\; y_1x_2y_3),$$
$$(\exists y_1,y_2,x_3)(Y_1Y_2X_3\lambda \;\Box\!\!\rightarrow\; y_1y_2x_3),$$
$$(\exists x_1,x_2,x_3)(X_1X_2X_3\lambda \;\Box\!\!\rightarrow\; x_1x_2x_3), \tag{4}$$

with '$(\exists x_1,y_2,y_3)$' a short-hand for $(\exists x_1)(\exists y_2)(\exists y_3)$, etc. Now suppose that the value for y_3 that makes the first counterfactual in (4) true is different from that which makes the second counterfactual true. Then we could straightforwardly derive:

$$(\exists y_3)(X_1Y_2Y_3\lambda \;\Box\!\!\rightarrow\; y_3 \;\;\&\;\; Y_1X_2Y_3\lambda \;\Box\!\!\rightarrow\; -y_3). \tag{5}$$

This satisfies Lewis' definition for (partial) deterministic causal dependence between y_3-results and the spacelike separated settings X_1Y_2 and Y_1X_2. So to block this, we must take the y_3 values that make (4)'s first two counterfactuals true to be the same. A similar argument suffices for the first and third counterfactuals being made true by the same y_2 value, the first and fourth counterfactuals being made true by the same x_1 value, etc. (six arguments in all). The upshot is that to block setting-to-result causation, we must make each quantifier in (4) range over all four counterfactuals at once:

$$(\exists x_1,x_2,x_3,y_1,y_2,y_3)\,((X_1Y_2Y_3\lambda \;\Box\!\!\rightarrow\; x_1y_2y_3) \;\&\; (Y_1X_2Y_3\lambda \;\Box\!\!\rightarrow\; y_1x_2y_3)$$
$$\&\; (Y_1Y_2X_3\lambda \;\Box\!\!\rightarrow\; y_1y_2x_3) \;\&\; (X_1X_2X_3\lambda \;\Box\!\!\rightarrow\; x_1x_2x_3)). \tag{6}$$

Now we can bring in (3)'s counterfactual quantum predictions. These assert that whatever results make (6) true, they must be such that the product $x_1y_2y_3$ equals +1, the product $y_1x_2y_3$ equals +1, etc. So (3) and (6) jointly entail that there must exist particular results x_1, x_2, x_3, y_1, y_2 and y_3 satisfying:

$$x_1y_2y_3 = +1\,,$$
$$y_1x_2y_3 = +1\,,$$
$$y_1y_2x_3 = +1\,,$$
$$x_1x_2x_3 = -1\,. \qquad (7)$$

But there can be no such results—(7)'s equations cannot be satisfied! For since each factor that appears is ±1, multiplying the four equations together gives the contradiction: +1 = -1.

Let us move then to Part II of the argument which uses (3) and result-to-result locality to deduce determinism. We focus on deriving the first of (4)'s counterfactuals which asserts determinism of results for the measurement $X_1Y_2Y_3$. Similar arguments suffice for the other three counterfactuals (measurements).

From the first of (3)'s predictions, we see that the product of the three results having to be +1 leaves us with four possibilities:

$$\begin{aligned}(X_1Y_2Y_3\lambda) \;\Box\!\!\rightarrow\; ((&x_1 = +1)(y_2 = +1)(y_3 = +1) \vee \\ &(x_1 = -1)(y_2 = -1)(y_3 = +1) \vee \\ &(x_1 = +1)(y_2 = -1)(y_3 = -1) \vee \\ &(x_1 = -1)(y_2 = +1)(y_3 = -1))\,, \end{aligned} \qquad (8)$$

where '∨' is the exclusive 'or'. It is convenient to express (8) in terms of Lewis' (1973) semantics for counterfactuals:

> A □→ B is true at world w *iff* all the closest A-worlds (i.e., worlds where A is true) to w are B-worlds.

For simplicity we have adopted the 'Limit principle', and conditions (a) and (b) following (3) allow us to omit the vacuous case where the antecedent of the counterfactual is impossible. Thus (8) just says that the closest $X_1Y_2Y_3\lambda$-worlds to the actual world can come in up to four varieties, depending upon which of the four combinations of results in (8) is obtained in each world.

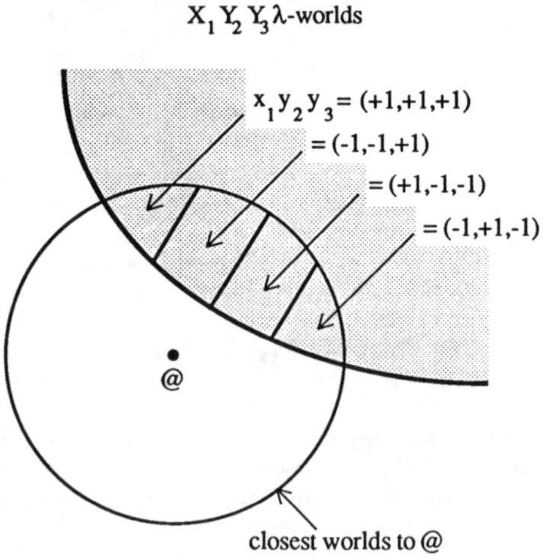

Figure 2. Lewis' Semantics for Part II

Now suppose that worlds of all four varieties obtain in the worlds closest to the actual world, @, as illustrated in Figure 2. One can now argue that to block result-to-result causation, *at most one* variety of world (i.e., combination of results) can appear within the closest worlds to @. For suppose worlds where $(x_1, y_2, y_3) = (+1,+1,+1)$ obtains *and* where $(x_1, y_2, y_3) = (-1,-1,+1)$ obtains are amongst the closest, just as Figure 2 says. Then using Lewis' semantics, the truth of the following pair of counterfactuals can be read straight off Figure 2:

$$(X_1Y_2Y_3\lambda)(y_2 = +1)(y_3 = +1) \;\Box\!\!\rightarrow (x_1 = +1) \;\; \& \\ (X_1Y_2Y_3\lambda)(y_2 = -1)(y_3 = +1) \;\Box\!\!\rightarrow (x_1 = -1)\,. \qquad (9)$$

Since (9) delivers Lewisian (partial) causal dependence between the spacelike separated x_1 and y_2 results, such dependence can only be blocked by abandoning our supposition that *both* pairs of worlds, defined by the result combinations (+1,+1,+1) and (-1,-1,+1), are amongst the closest $X_1Y_2Y_3\lambda$-worlds to @. So, instead, at most one of these two combinations can be. A similar argument goes through for any two varieties of worlds pictured in Figure 2 (six arguments in all). Thus, contrary to Figure 2, *one* combination of results must be closer to @ than any of the other three. Under Lewis' semantics, this automatically renders the first of (4)'s counterfactuals true, as required.

As one might expect, our argument does not specify which of the two types of spacelike causation, result-to-result or setting-to-result, must obtain; nor does it specify which results must be determined if result-to-result causation fails. Both of these 'choices' are left to the particular theory that is attempting to recover quantum predictions.

Thus in Bohm's theory, which denies result-to-result causation, one can show (under simplifying assumptions—cf. Clifton (1991), Ch. 4) that results are deterministically fixed in the following way. Suppose the position parts of the statevector for the particles just before they are measured are narrow wavepackets peaked at the origin of the coordinate system each particle carries. Let Rs represent settings (= X or Y), rs their corresponding results (= x or y) and $\overline{R}$s represent the actual positions of the particles in the R direction just prior to measurement within the aforementioned wavepackets. Then the jth result is given by just:

$$r_j = \mathbf{sgn}(\overline{R}_j) = [\sigma_{Rj}\sigma_{Rk}\sigma_{Rk'}]_{|GHZ>}\mathbf{sgn}(\overline{R}_k)\mathbf{sgn}(\overline{R}_{k'}) \;, \qquad (10)$$

in terms of the pre-measurement positions of the other two particles, k and k', and the sign function, defined by **sgn**(x) = +1, if x>0; -1, if x<0; and 0, if x=0. The product of the other two results is just:

$$r_kr_{k'} = \mathbf{sgn}(\overline{R}_k)\mathbf{sgn}(\overline{R}_{k'}) = [\sigma_{Rj}\sigma_{Rk}\sigma_{Rk'}]_{|GHZ>}\mathbf{sgn}(\overline{R}_j)\,. \qquad (11)$$

By (10), each result explicitly depends upon $[\sigma_{Rj}\sigma_{Rk}\sigma_{Rk'}]_{|GHZ>}$. Thus, by (2), each in general carries a dependence on all three measurement settings. So Bohm's theory explicitly embraces setting-to-result causation in Lewis' sense—as it must, since the above predictions also make the quantum predictions in (2) and (3) true. We shall make use of the predictions of Bohm's theory in the form (11) at the end of the next section to illustrate the implications of our second argument against EPR's inference to elements-of-reality.

As for QM itself (the case λ = |GHZ>), it clearly rejects determinism of results. Thus all four combinations of results in (8) would be genuinely possible outcomes were the measurement $X_1Y_2Y_3$ to occur. This means that we must allow all four varieties of worlds in Figure 2 to be in the set of closest worlds to @, as is in fact shown in that figure. But then we can just read off that figure the conclusion that, whatever three results are obtained upon measuring $X_1Y_2Y_3$, each causally depends upon the others given a fixed value for the third (and similarly for the other three measurement combinations). Surely a causal loop of the worst kind: Does this not tell against Lewis' analysis of causal dependence?

We believe it does not. Causal loops *per se* are no contradiction. Paradox looms only if some external agent is allowed to intervene in the loop and exploit its links to undercut the grounds for the intervention, as with tachyon paradoxes. But QM fails to allow deterministic, or even stochastic, control of results without destroying the very quantum state, $|GHZ\rangle$, upon whose predictions our argument for spacelike causation rests. So no paradox need arise. Thus if one takes relativity to prohibit spacelike causation only insofar as it leads to causal paradox, then there seem little grounds for rejecting the verdict of Lewis' analysis of causation in this case.

Indeed, something similar can be said for Lewisian spacelike causation in Bohm's theory. It is at best unclear whether the positions of the particles could be controlled in that theory, and so setting-to-result signalling achieved, without destroying the nonlocal action of the 'quantum potential'. For that action relies on the QM state remaining nonfactorizable (as is $|GHZ\rangle$) once control of particle positions has been achieved. Thus tachyon paradoxes again need not arise.

Some might wish to interpret relativity more strongly as prohibiting spacelike causation *per se*. They may then be tempted to strengthen Lewis' analysis of causation with requirements like spatio-temporal contiguity (Cartwright 1989, Ch. 6) or robustness (Redhead 1992) in order to evade our argument's conclusion. However, although it is interesting to see how different analyses of causation yield different verdicts on whether quantum predictions entail spacelike causation (see e.g., Skyrms 1984 and Clifton 1991), it is far from clear that relativity prohibits spacelike causation in any other sense than prohibiting causal paradox. And since all that is needed to threaten paradox is the truth of the counterfactuals that Lewis takes as constitutive of causation between spacelike separated events, his analysis is clearly appropriate to assessing the 'peaceful coexistence' between quantum predictions and relativity. More importantly, Lewis' analysis is compatible with keeping that peace, as we have argued.

3. The Failure of the Inference to Elements-of-Reality

For our second argument, we follow Redhead (1987, p. 72) in taking EPR's inference to an element-of-reality, free from any assumption of locality, to be of the form:

> If we can predict with certainty some result q of measuring at time t a physical quantity Q, then at t (or just before) there exists a corresponding element-of-reality $[Q]_t = q$ that fixes the measurement result to be q.

This allows the prediction with certainty to be acquired before t by a 'disturbance of the system', if need be. So our focus is now solely on elements-of-reality after any disturbances have occurred, not before.

The above formulation goes a step beyond EPR's own by stipulating that the 'correspondence' between $[Q]_t$ and Q be such that the fact that $[Q]_t$ equals q fixes the measurement result to be q. But if $[Q]_t$ did not *fix* result q, then it could not explain why result q is bound to occur, leaving little motivation for positing $[Q]_t$'s existence in the first place! And it is hardly too strong to require that $[Q]_t$ *equal* q. Certainly quantum orthodoxy says that when a prediction with certainty is verified, the value $[Q]_t$ equaling a particular eigenvalue is what fixes the predicted result (cf. Dirac1958), p. 46). But so do theories like Bohm's. (10) and (11) make clear that measuring the spin of any spin-1/2 particle in that theory amounts to looking for its position (whether it be brought about nonlocally or not) above or below a plane orthogonal to the spin-meter's orientation. So the element-of-reality that fixes an actually measured result is just the sign of the particle's position in the direction of the spin-meter's orientation (when the origin is taken to lie in the plane orthogonal to that orientation). Since this sign equals ±1 (almost always), as do the possible results of the measurement, the requirement that $[Q]_t = q$ is borne out once again. Henceforth we shall frequently refer to elements-of-reality as just 'values' because they have this property. This is in accord with our earlier use of square brackets to denote values.

So far we have emphasized what can be inferred, whether or not QM is complete, in the case when a prediction with certainty is actually verified. However, EPR's inference to a value is not meant to be restricted to this case: the value should still be 'out there' even if it actually fails to perform its job of fixing the predicted result. This feature of the inference, along with one further *necessary* condition for the existence of an element-of-reality, will be crucial to our argument. For together they will allow us to show that, after all, the inference is incompatible with both orthodox QM and any theory that takes QM to be incomplete, like Bohm's. To formalize the fact that actual verification of the prediction with certainty is unnecessary, let M_Q denote the measurement of Q and the variable q range over possible measurement results. Then the inference to a value says:

$$(\forall q)((M_Q \;\Box\!\!\rightarrow q)_t \;\Rightarrow\; (\exists [Q]_t)([Q]_t = q)) \,, \tag{12}$$

where t is the time at which the prediction with certainty is true.

The necessary condition we shall employ is:

If [Q] exists and equals q within a spacetime region R with respect to one spacelike hyperplane H containing R, then it exists and has the same value in R with respect to any other spacelike hyperplane H' containing R.

More formally:

$$[Q]_R = q \text{ with respect to } H \supseteq R \Rightarrow (\forall H' \neq H)(H' \supseteq R \Rightarrow \\ [Q]_R = q \text{ with respect to } H') \,. \tag{13}$$

(13) follows from $[Q]_R$'s being an element-of-*reality*; for if it is, it surely cannot spontaneously pop into and out of existence within R depending upon how an observer moves! Furthermore, since the elements-of-reality at issue are just scalars, there is no difficulty with the demand that two observers equipped with simultaneity planes H and H' should ascribe the *very same* value $[Q]_R$ within R.

Let us now return to our three spin-1/2 particle thought experiment. This time we adopt a slightly different counterfactual reading of (2)'s predictions than that used in section 1's argument: we read them as 'nested' counterfactuals. For example, (2)'s first equation says that were X_1 true and some result x_1 obtained, then it would subsequently be true that were Y_2Y_3 true, then the product of the measurements results for Y_2Y_3 would be x_1 (and similarly for (2)'s other equations). For simplicity, we shall assume that in the actual world, @, the measurement $X_1X_2X_3$ is performed yielding the particular results x_1, x_2 and x_3 whose product is -1, in accordance with the last of (2)'s predictions. Then the other three predictions, in nested counterfactual form, are just:

$$\begin{aligned}(\forall t,t'{>}t)\{(X_1x_1)_t &\Rightarrow (Y_2Y_3 \;\Box\!\!\rightarrow\; (y_2y_3 = x_1))_{t'}\}, \\ (\forall t,t'{>}t)\{(X_2x_2)_t &\Rightarrow (Y_1Y_3 \;\Box\!\!\rightarrow\; (y_1y_3 = x_2))_{t'}\}, \\ (\forall t,t'{>}t)\{(X_3x_3)_t &\Rightarrow (Y_1Y_2 \;\Box\!\!\rightarrow\; (y_1y_2 = x_3))_{t'}\}. \end{aligned} \qquad (14)$$

Clearly these predictions are only valid if no other measurements, incompatible with the nested counterfactual suppositions in each line above, are performed between t and t'. For our argument, we shall assume that no such intervening measurements take place.

As in section 1, the argument is best broken into two parts:

Part I: (14) & Relativity of Simultaneity & (12) $\Rightarrow$ Certain Elements-of-Reality Exist,
Part II: Part I's Elements-of-Reality & (13) $\Rightarrow$ Contradiction.

Since (13) is necessary for an element-of-reality to exist, by *reductio* Parts I and II entail the conclusion we seek: that the predictions of relativistic QM are incompatible with inference (12).

For Part I we must specify the frames relative to which the times t and t'>t in (14)'s predictions are to be judged before we can exploit them to infer values via (12). To do this, we need a further sketch of our thought experiment in spacetime. Suppose, for the moment, that in @ the position parts of the statevector for the particles are nearly delta functions at all times relative to any state of motion. (We shall shortly see that this idealization is actually incompatible with relativistic QM, but that it is easily relaxed without blocking our argument!) Since all the particles move in the same plane (cf. Figure 1), we can represent their trajectories 1, 2 and 3 as shown in Figure 3's 2+1 spacetime diagram. Suppose, in addition, that the spin-meters and their operators, observers 1, 2 and 3, move alongside their respective particles outwards from the source with the same velocity. Then, at some point, each observer moves their spin-meter into their particle's path to measure its spin. Label the three X-spin measurement *events* that obtain in @ (with results x_1, x_2 and x_3 multiplying to -1) by the points **1**, **2** and **3**, chosen so that they are pairwise spacelike separated.

Now suppose observer 1 (with particle 1) is moving so that she is equipped with the family of tilted hyperplanes shown in Figure 3. So for her, event **1** occurs first and events **2** and **3** occur simultaneously later on. Let the intermediate plane of time $t_{2'3'}$ shown in the figure correspond to some time in the interval strictly between the measurements according to observer 1. Similarly, let observer 2 move (with his particle) so that event **2** occurs first and events **1** and **3** occur simultaneously later on, with the intermediate plane of time $t_{1'3'}$ lying between the measurements, and likewise for observer 3. Clearly there exists a state of motion for each observer that allows their intermediate planes ($t_{2'3'}$ for 1, $t_{1'3'}$ for 2 and $t_{1'2'}$ for 3) to satisfy these conditions. Thus diagrams similar to Figure 3 could be drawn for observers 2 and 3.

Beginning with observer 1, let us now make use of (14)'s first prediction and inference (12). Once observer 1 registers her result x_1, she can well assert that: $(Y_2Y_3 \;\Box\!\!\rightarrow\; (y_2y_3 = x_1))_{t2'3'}$. So by inference (12), she will ascribe the value $[\sigma_{Y2}\sigma_{Y3}]_{t2'3'} = x_1$ (a value which we can allow to have been brought about by her measurement/result!). This gives us the *time* of observer 1's value ascription, but we can also determine the (invariant) *spacetime* region to which that value pertains. It is just the bold line segment **2'3'** of Figure 3 lying in the plane $t_{2'3'}$. The reason is that points **2'** and **3'** are the spacetime locations of the measurement events that would have obtained had particles 2 and 3 been measured at $t_{2'3'}$ (as required by (14)'s first counterfactual supposition). And since $\sigma_{Y2}\sigma_{Y3}$ is a nonlocal observable on the joint system consisting of particles 2 and 3, its value cannot be attributed to either of points **2'** or **3'** separately—only the extended spacetime region **2'3'**. Denote the value observer 1 ascribes at $t_{2'3'}$ along **2'3'** using the invariant notation $[\sigma_{Y2}\sigma_{Y3}]_{2'3'} = x_1$.

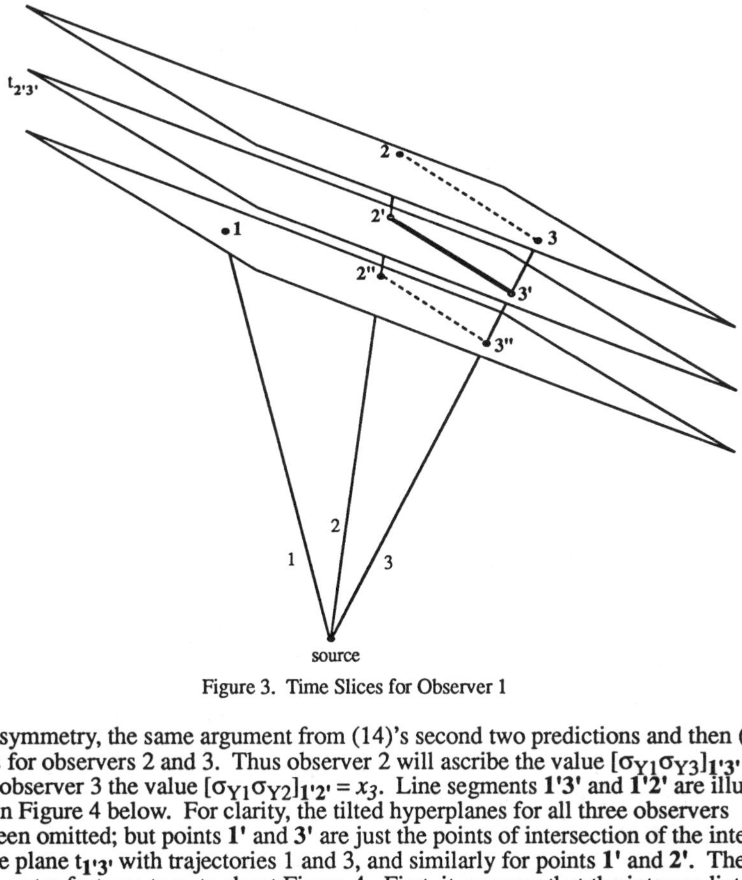

Figure 3. Time Slices for Observer 1

By symmetry, the same argument from (14)'s second two predictions and then (12) applies for observers 2 and 3. Thus observer 2 will ascribe the value $[\sigma_{Y1}\sigma_{Y3}]_{1'3'} = x_2$ and observer 3 the value $[\sigma_{Y1}\sigma_{Y2}]_{1'2'} = x_3$. Line segments **1'3'** and **1'2'** are illustrated in Figure 4 below. For clarity, the tilted hyperplanes for all three observers have been omitted; but points **1'** and **3'** are just the points of intersection of the intermediate plane $t_{1'3'}$ with trajectories 1 and 3, and similarly for points **1'** and **2'**. There are two extra features to note about Figure 4. First, it assumes that the intermediate planes $t_{1'3'}$ and $t_{1'2'}$ intersect trajectory 1 in the *same* point **1'**, and similarly for points **2'** and **3'**. Clearly this is possible for intermediate planes $t_{1'3'}$, $t_{1'2'}$ and $t_{2'3'}$ with appropriately chosen orientation (which amounts to making the appropriate choice for the velocities of the particles alongside which their corresponding observers move). Second, depending upon how these intermediate planes are chosen, triangle **1'2'3'** can lie anywhere in the interval strictly *between* triangles **123** and **1"2"3"** for the purposes of our argument—which is why Figure 4 indicates the latter with dotted lines.

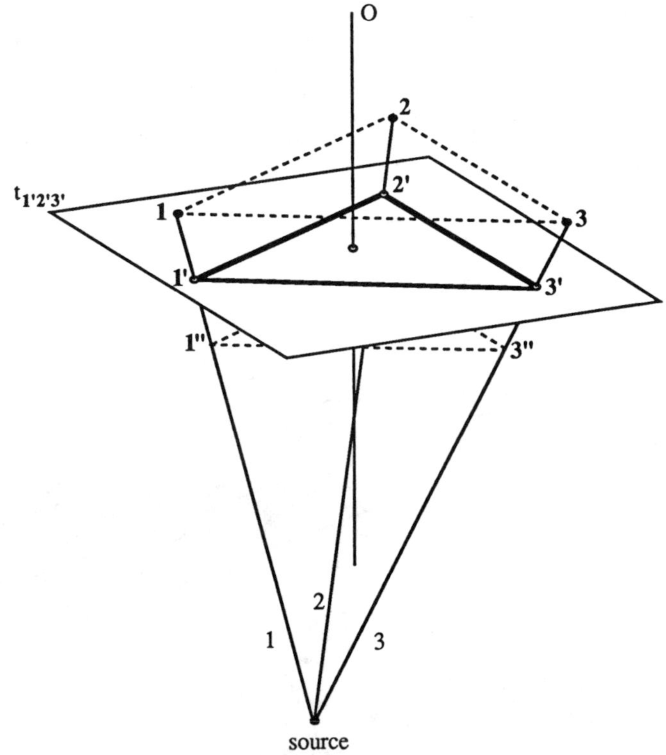

Figure 4. Spacelike Triangle **1'2'3'** to which the Observers' Values Pertain

We now turn to Part II of our argument against inference (12). As indicated in Figure 4, triangle **1'2'3'** lies in a plane, call it $t_{1'2'3'}$. Consider a fourth observer in @, O, at rest relative to the source so that plane $t_{1'2'3'}$ is one of her planes of simultaneity. By our necessary condition, (13), O must agree at $t_{1'2'3'}$ to assign the value that observer 1 ascribes along the line **2'3'**, i.e., O must assign the value $[\sigma_{Y2}\sigma_{Y3}]_{2'3'} = [\sigma_{Y2}\sigma_{Y3}]_{t1'2'3'} = x_1$. Similarly for the values ascribed by 2 and 3, i.e., $[\sigma_{Y1}\sigma_{Y3}]_{t1'2'3'} = x_2$ and $[\sigma_{Y1}\sigma_{Y2}]_{t1'2'3'} = x_3$, since all three sides of the triangle lie in the same plane $t_{1'2'3'}$. The punchline is that, whether or not QM is complete, O cannot ascribe these values at $t_{1'2'3'}$ as required by (13).

Take first the case where QM *is* complete. The values O ascribes at $t_{1'2'3'}$ must satisfy:

$$[\sigma_{Y2}\sigma_{Y3}][\sigma_{Y1}\sigma_{Y3}][\sigma_{Y1}\sigma_{Y2}] = x_1 x_2 x_3 = -1 \, . \tag{15}$$

However, there is no simultaneous QM eigenstate for the operators $\sigma_{Y2}\sigma_{Y3}$, $\sigma_{Y1}\sigma_{Y3}$ and $\sigma_{Y1}\sigma_{Y2}$ that O can use to ascribe them values in accordance with (15). For if there were such an eigenstate then, because these three operators multiply to I, (15) entails that the simultaneous eigenstate would have to violate the product rule, which we know from section 1 is not possible! So the completeness of the quantum description must be abandoned if EPR's inference to values is maintained.

Now suppose we allow O to posit additional values to 'underpin' those QM itself ascribes in order to alleviate the trouble. So suppose she posits the existence of values $[\sigma_{Y2}]$ and $[\sigma_{Y3}] = \pm 1$ at $t_{1'2'3'}$ such that $[\sigma_{Y2}][\sigma_{Y3}] = [\sigma_{Y2}\sigma_{Y3}]$, and similarly for the other product observables: $[\sigma_{Y1}][\sigma_{Y3}] = [\sigma_{Y1}\sigma_{Y3}]$ and $[\sigma_{Y1}][\sigma_{Y2}] = [\sigma_{Y1}\sigma_{Y2}]$. Each of these additional values, attributed now to local observables, can be thought of as attached to the spacetime points **1'**, **2'** and **3'** where the particles are located at $t_{1'2'3'}$. Furthermore, for consistency in O's value assignments at $t_{1'2'3'}$ in @, the value $[\sigma_{Y1}]$ that goes into determining the value for $[\sigma_{Y1}\sigma_{Y3}]$ must be the same as that which goes into determining $[\sigma_{Y1}\sigma_{Y2}]$, etc. Otherwise she would be assigning different values in @ to the same observable at the same spacetime point! Now clearly these extra values cannot avoid the trouble, since when inserted into (15) the Y-spin values for each particle appear twice on the left-hand side yielding the contradiction: +1 = -1. So whether O only ascribes values on the basis of quantum predictions, or whether she adds more (assuming QM incomplete), the contradiction cannot be avoided. And so inference (12) cannot be maintained.

We now wish to relax two idealizations we have made and show that without them both parts of our argument still go through (as promised earlier). One is our assumption that the particles follow well-defined trajectories at all times, which ignores the usual spreading of wavepackets in time. The other is our assumption that at any given point along its trajectory, a particle can be localized at that point for all observers, regardless of their state of motion. This ignores the fact that, in relativistic QM, localization of a particle along one spacelike hyperplane about some point is incompatible with such localization along any 'tilted' hyperplane through the same point (Newton and Wigner 1949). Since the particles' trajectories were used to define triangle **1'2'3'**, relaxing these two 'definite trajectory' assumptions means that it no longer immediately follows that the values attributed by observers 1, 2 and 3 pertain to the three sides of triangle **1'2'3'**.

However, our argument is easily modified as follows. Suppose that the trajectories in Figures 3 and 4 are now just worldlines of the points of entry into each observer's spin-meter, but that *relative to O's frame* the position states of the particles *at* $t_{1'2'3'}$ are virtually delta functions centered on the vertices of triangle **1'2'3'**. (Of course this means observer 1's probability distribution for finding particle 1 will be 'smeared' along plane $t_{12''3''}$. But she will still have a nonzero probability of detecting particle 1 at point **1**, and similarly for observers 2 and 3. So our hypothesis that three X-spin measurements occur at points **1**, **2** and **3** in @ is not undermined.) Using the delta functions at $t_{1'2'3'}$, O can make a prediction with certainty about the particles being found at vertices **1'**, **2'** and **3'**, even though position measurements are not actually performed at $t_{1'2'3'}$. Thus she can use (12) to infer *position* elements-of-reality for the particles at the vertices **1'**, **2'** and **3'**. (13) then entails that observer 1, moving relative to O, must also ascribe position elements-of-reality to the particles at the points **2'** and **3'**, and similarly for observers 2 and 3 with regard to points **1'** and **3'**, and **1'** and **2'** respectively. So when it comes to the observers locating the *spin* values they attribute to the particles, they will locate them, as in Part I, along the sides of the triangle **1'2'3'**. This leads us back (again, via (13)) to O's spin value contradiction in Part II, whether or not QM is taken to be complete.

There are at least two possible responses one could make to our *reductio* on EPR's inference: one for the case where (relativistic) QM is taken to be complete and the other assuming it is not. Both responses (in our view, quite rightly!) retain the realist requirement that the truth of (14)'s counterfactuals for their respective observers needs to be grounded in the existence of values, and that these must be invariantly associated with regions in spacetime. However, the nature of values in these two approaches is quite different.

The first approach, essentially Fleming's (1989,1992), is to regard values as supervening on entire hyperplanes, not just on bounded spacetime regions, like the sides of triangle **1'2'3'**. The rationale for this is as follows. As noted above, relativistic QM entails that if, say, particle 2 is certain to be found at point **2'** on hyperplane $t_{\mathbf{1'2'3'}}$, then there is a nonzero probability of it being found at any point along the tilted hyperplane $t_{\mathbf{2'3'}}$. So the formalism treats 'position along $t_{\mathbf{1'2'3'}}$' and 'position along $t_{\mathbf{2'3'}}$' as distinct noncommuting physical quantities. If this is taken to have ontological significance, then it breaks the above argument from (13) to the need for observer 1 to go along with observer O and locate particle 2's position element-of-reality at **2'**. For that argument (and its analogues for observers 2 and 3) treats two distinct physical quantities of position as being the same. Indeed, if no extra values are assumed beyond those one can infer from QM's predictions alone (i.e., if QM is complete), then 'position along $t_{\mathbf{2'3'}}$' has no value, while 'position along $t_{\mathbf{1'2'3'}}$' has a value at point **2'**.

Spin values also must supervene on entire hyperplanes, i.e., be 'hyperplane-dependent', as a consequence of the hyperplane-dependence of position. For where one locates a particle's spin surely depends upon what position the particle has! Thus one can see, on this view, how the rest of our argument (specifically, Part II) breaks down. First, the value $[\sigma_{Y2}\sigma_{Y3}]_{t2'3'}$ ascribed by observer 1 and the value $[\sigma_{Y2}\sigma_{Y3}]_{t1'2'3'}$ by O need no longer be the same because they correspond to distinct physical quantities. And second, if no extra values are assumed so that QM is complete, $[\sigma_{Y2}\sigma_{Y3}]_{t2'3'}$ will have a value by observer 1's prediction with certainty, while the value $[\sigma_{Y2}\sigma_{Y3}]_{t1'2'3'}$ will fail to exist since |GHZ> fails to supply a prediction with certainty to O for the observable $\sigma_{Y2}\sigma_{Y3}$ at $t_{\mathbf{1'2'3'}}$. None of this requires invariance of values to be abandoned; for all observers can still agree on what value, if any, to ascribe to any given observable *defined with respect to a given hyperplane.*

The second response to our argument is to retain an ontologically more conservative treatment of values as supervening on bounded spacetime regions, but reject EPR's inference by allowing values of observables not measured in @ to be different than they would be were they actually measured! This is precisely how Bohm's (1952) theory can escape our argument. Recall again that the pre-existing value of a measured spin component in that theory is the sign of the particle's position, $\overline{X}$ or $\overline{Y}$, within the wavepacket just prior to measurement. Suppose the actual world is as above, i.e., three X-spin measurements with outcomes satisfying $x_1x_2x_3 = -1$, strongly peaked wavepackets about the vertices of triangle **1'2'3'** relative to O, etc. Let superscript 'cf' denote a value which would underpin a prediction with certainty were it verified (contrary-to-fact) and '@' denote an actual value. Then assuming Bohm's predictions are true for observers 1, 2, and 3, (11) and (2) entail:

$$
\begin{aligned}
&[\sigma_{Y1}\sigma_{Y2}]^{cf}_{\mathbf{1'2'}}\,[\sigma_{Y2}\sigma_{Y3}]^{cf}_{\mathbf{2'3'}}\,[\sigma_{Y1}\sigma_{Y3}]^{cf}_{\mathbf{1'3'}} \\
&\quad = ((\mathbf{sgn}\,\overline{Y}{}^{cf}_{1})_{\mathbf{1'}}(\mathbf{sgn}\,\overline{Y}{}^{cf}_{2})_{\mathbf{2'}})((\mathbf{sgn}\,\overline{Y}{}^{cf}_{2})_{\mathbf{2'}}(\mathbf{sgn}\,\overline{Y}{}^{cf}_{3})_{\mathbf{3'}})((\mathbf{sgn}\,\overline{Y}{}^{cf}_{1})_{\mathbf{1'}}(\mathbf{sgn}\,\overline{Y}{}^{cf}_{3})_{\mathbf{3'}}), \\
&\quad = [\sigma_{Y1}\sigma_{Y2}\sigma_{X3}]_{|GHZ>}(\mathbf{sgn}\,\overline{X}{}^{@}_{3})_{\mathbf{3}}[\sigma_{X1}\sigma_{Y2}\sigma_{Y3}]_{|GHZ>}(\mathbf{sgn}\,\overline{X}{}^{@}_{1})_{\mathbf{1}} \\
&\qquad\qquad \times [\sigma_{Y1}\sigma_{X2}\sigma_{Y3}]_{|GHZ>}(\mathbf{sgn}\,\overline{X}{}^{@}_{2})_{\mathbf{2}}, \\
&\quad = x_1x_2x_3, \\
&\quad = -1, \\
&\quad = -1\times((\mathbf{sgn}\,\overline{Y}{}^{@}_{1})_{\mathbf{1'}}(\mathbf{sgn}\,\overline{Y}{}^{@}_{2})_{\mathbf{2'}})((\mathbf{sgn}\,\overline{Y}{}^{@}_{2})_{\mathbf{2'}}(\mathbf{sgn}\,\overline{Y}{}^{@}_{3})_{\mathbf{3'}})((\mathbf{sgn}\,\overline{Y}{}^{@}_{1})_{\mathbf{1'}}(\mathbf{sgn}\,\overline{Y}{}^{@}_{3})_{\mathbf{3'}}), \\
&\quad = -[\sigma_{Y1}\sigma_{Y2}]^{@}_{\mathbf{1'2'}}\,[\sigma_{Y2}\sigma_{Y3}]^{@}_{\mathbf{2'3'}}\,[\sigma_{Y1}\sigma_{Y3}]^{@}_{\mathbf{1'3'}}\,. \qquad (17)
\end{aligned}
$$

The equality of the first and last expressions in this chain of equivalences entails that at least one of the observables $\sigma_{Y2}\sigma_{Y3}$, $\sigma_{Y1}\sigma_{Y3}$, or $\sigma_{Y1}\sigma_{Y2}$ must have a 'cf'-value that is different from its '@'-value. And that just means that the value would be different from what it actually is were the prediction with certainty to be verified.

4. Conclusion

If EPR's locality and reality assumptions are given natural counterfactual readings then, whether or not QM is complete, each of them is already incompatible with quantum predictions and the relativity of simultaneity as applied within the same simple thought experiment. Not only do these results doubly undercut EPR's argument for incompleteness, but they impose a strong constraint on any interpretation of relativistic QM.

Note

[1]Sections 2 and 3 of this paper clarify and extend the arguments given in Clifton (1991, Ch. 4) and Pitowsky (1991), respectively. We are especially grateful to Jeremy Butterfield and Gordon Fleming for their comments. For their generous support, Rob Clifton thanks Christ's College and the Social Sciences and Humanities Research Council of Canada; Constantine Pagonis thanks the Arnold Gerstenberg Fund, the British Academy and Wolfson College; and Itamar Pitowsky thanks the Sidney Edelstein Center for the History and Philosophy of Science at The Hebrew University. Truly a formidable financial package.

References

Bohm, D. (1952), "A Suggested Interpretation of Quantum Theory in Terms of 'Hidden' Variables, I and II", *Physical Review* 85: 166-193.

Cartwright, N. (1989), *Nature's Capacities and their Measurement.* Oxford: Clarendon Press.

Clifton, R.K. (1991), *Nonlocality in Quantum Mechanics: Signalling, Counterfactuals, Probability and Causation.* Doctoral Dissertation, Cambridge University.

Dirac, P.A.M. (1958), *The Principles of Quantum Mechanics.* Fourth Edition (Revised). Oxford: Clarendon Press.

Einstein, A., Podolsky, B. and Rosen, N. (1935), "Can Quantum-Mechanical Description of Physical Reality be Considered Complete?", *Physical Review* 47: 777-780.

Fleming, G. (1989), "Lorentz Invariant State Reductions and Localization", in *Philosophy of Science Association 1988*, Volume II, A. Fine and M. Forbes (eds.). East Lansing: Philosophy of Science Association, pp. 112-126.

_______. (1992), "The Objectivity and Invariance of Quantum Predictions", this volume.

Lewis, D. (1973), *Counterfactuals*. Oxford: Blackwell Press.

_ _ _ _ _ _ . (1986), *Philosophical Papers*, Volume II. Oxford: Oxford University Press.

Mermin, N.D. (1990), "What's Wrong With These Elements-of-Reality?", *Physics Today* 43: 9-11.

Newton, T.D. and Wigner, E.P. (1949), "Localized States for Elementary Systems", *Reviews of Modern Physics* 21: 400-406.

Pitowsky, I. (1991), "The Relativity of Quantum Predictions", *Physics Letters A* 156: 137-139.

Redhead, M.L.G. (1987), *Incompleteness, Nonlocality and Realism*. Oxford: Oxford University Press.

_ _ _ _ _ _ _ _ _ _ . (1992), "Propensities, Correlations and Metaphysics", *Foundations of Physics* 22: 381-394.

Skyrms, B. (1984), "EPR: Lessons for Metaphysics", *Midwest Studies in Philosophy* 9: 245-255.

Part IV

ISSUES IN METHODOLOGY

Another Day for an Old Dogma

Robert J. Levy

Wittenberg University

1. Introduction

In *Theory and Evidence,* Clark Glymour asserts that many philosophers are convinced that "evidence can only bear on the entire body of our beliefs, and cannot be parceled out here and there"(1980,p.5). Attributing this view to Quine, Glymour says:

> No working scientist acts as though the entire sweep of scientific theory faces the tribunal of experience as a single, undifferentiated whole; nor, I think, does any working person act so with regard to his beliefs. On the contrary, much of the scientist's business is to construct arguments that aim to show that a particular piece of experiment or observation bears on a particular piece of theory, and such arguments are among the most celebrated accomplishments in the history of our sciences (1980,p.3).

Some philosophers might counter that Glymour's criticisms apply properly to a view Quine had significantly modified two decades before the publication of *Theory and Evidence*. In "Two Dogmas," Quine claims that our statements about the external world can only be confirmed or disconfirmed in the context of "the totality of our so-called knowledge or beliefs" or in the context of "total science"(1961,pp.42-3). However, in *Word and Object,* Quine suggests that the holism of "Two Dogmas" is excessive (see 1960,p.13). The holism in his more recent writings is more moderate. In "Epistemology Naturalized", he says "theoretical sentences have their evidence not as single sentences but only as larger blocks of theory" and adds that "most sentences, apart from observation sentences, are theoretical"(1969, pp.80-81). In "Five Milestones of Empiricism", he says that "it is an uninteresting legalism . . . to think of our scientific system of the world as involved *en bloc* in every prediction. More modest chunks suffice . . . (1981,p.71). In *Pursuit of Truth,* he holds that although some single observation statements are testable, "for the most part . . . a testable set or conjunction of sentences has to be pretty big, and such is the burden of holism"(1990,p.17).

The moderate confirmational holism described in *Pursuit of Truth* is proposed as "an obvious but vital correction of the naive conception of scientific sentences as each

---

PSA 1992, Volume 1, pp. 131-141

endowed with its own separable content"(1990,p.160). This repeats the rejection in "Two Dogmas" of the reductionist claim that "each statement, taken in isolation from its fellows, can admit of confirmation or infirmation at all"(1961,p.41).

However, the shift from excessive to moderate confirmational holism fails to articulate the logic of confirmation and disconfirmation of individual hypotheses within these blocks or chunks of theory. Glymour's most interesting criticism remains unanswered.

2. Quine's Confirmational Holism

In *Pursuit of Truth,* Quine says that he is "undertaking to examine the evidential support of science" which he construes as a "relation of stimulations to scientific theory"(1990,p.2). He presents the problem of evidential support as that of specifying certain logical relations between theory which "consists of sentences, or is couched in them" and "observation sentences"(1990,p.2).

Observation sentences, on Quine's view, are occasion sentences which are "directly and firmly associated with our stimulations" and "should command the subject's assent or dissent outright, without further investigation and independently of whatever he may have been engaged in at the time"(1990,p.3). He adds that an observation sentence "must command the same verdict from all linguistically competent witnesses of the occasion"(1990,p.3). Viewing the observation sentence as "the means of verbalizing the prediction that checks a theory"(1990,p.4), he says that "the requirement that it command a verdict outright is what makes it a final checkpoint," but "the requirement of intersubjectivity is what makes science objective"(1990,p.5).

Central to Quine's analysis of confirmation in *Pursuit of Truth* is his account of observation categoricals. Observation categoricals are standing sentences which have the form 'whenever this, that' and are compounds of observation sentences. Claiming that observation categoricals are implied by or explained by scientific theories, Quine views these sentences as solving "the problem of linking theory logically to observation, as well as epitomizing the experimental situation"(1990,p.10).

An observation categorical is "not conclusively verified by observations that are conformable to it, but it is refuted by a pair of observations, one affirmative and one negative"(1990,p.12). For example, 'When the sun rises, the birds sing' "is refuted by observing sunrise among silent birds"(1990,p.12). According to Quine, by refuting an observation categorical, "you have refuted whatever implied it"(1990,p.12). If a refuted observation categorical is deduced not from a given hypothesis alone but only in conjunction with other theoretical sentences, parts of mathematics, and even some "common-sense platitudes," then it refutes "the conjunction of sentences that was needed to imply the observation categorical"(1990,p.13). Hence, Quine says, "we do not have to retract the hypothesis in question; we could retract some other sentence of the conjunction instead"(1990,p.14).

In revising a body of sentences S which imply a false observation categorical, Quine says that we should employ both (1) a holistic "maxim of minimum mutilation"(1990,p.14) to limit the revisions to those members of S required for this implication and (2) a holistic policy of maximizing simplicity. The goal of the revision, says Quine, is to "maximize future success in prediction: future coverage of true observation categoricals"(1990.p.15).

In *Pursuit of Truth,* there is a suggestion of a non-holistic account of the confirmation of hypotheses. Noting that we reason "not only in refutation of hypotheses but in support of them"(1990,p.13), Quine describes this as "a matter of arguing logically or probabilistically from other beliefs already held"(1990,p.13). He does not develop this non-holistic analysis of the logic of confirmation. The upshot is that, in his most recent book, Quine presents a holistic view of confirmation which fails to articulate the logic of confirmation and disconfirmation of particular hypotheses with blocks of theory. On Quine's developed view, only holistic criteria such as simplicity and conservatism are relevant to the context of confirmation and disconfirmation.

While I have no wish to deny the utility of such criteria, I believe that it is possible to go beyond confirmational holism and articulate the logic of the confirmation and disconfirmation of individual hypotheses within blocks of theory. Using the linguistic resources of Quine's less austere language of science, viz., probability theory and first-order predicate logic with identity, I propose a nonsubjective Bayesian analysis of the logic of confirmation of hypotheses by instances. In so doing, I offer an alternative to both the reductionism which Quine rejects and the confirmational holism which he embraces.

3. Bayesian Confirmation

The following nonsubjective Bayesian proposal attempts to clarify the notion that evidence reports of specific forms confirm, disconfirm, or are irrelevant to universal hypotheses of the form '(y)(Ry ⊃ By)', where '⊃' is the symbol for the truth-functional 'if-then'. Incorporating the fiction of a rational inquirer whose beliefs necessarily conform to the requirements of (1) logical and mathematical consistency and (2) probabilistic coherence, this proposal uses the following version of the positive relevance criterion of confirmation.

> Given background knowledge K: (1) evidence report E confirms hypothesis H if and only if, for *all* rational inquirers, $P(H/E\&K)>P(H/K)$; (2) evidence report E disconfirms H if and only if, for *all* rational inquirers, $P(H/K\&E)<P(H/K)$; and (3) evidence report E is irrelevant to H if and only if E neither confirms nor disconfirms H.

Because this proposal regarding confirmation uses the positive relevance criterion of confirmation and Bayes' Theorem in the form $P(H/E\&K) = P(H/K) \times P(E/H\&K)/P(E/K)$, confirmation of H by E requires that $P(E/K)>0$ and $1>P(H/K)>0$. No H such that $P(H/K) = 0$ or $P(H/K)=1$ is confirmable on either model of confirmation presented in this paper.

3.1. Confirmation on the Basis of Certain Knowledge

Discussing the confirmation of the hypothesis 'all ravens are black', Horwich suggests that there is a significant evidential difference between (1) reporting that an object selected at random is observed to be a black raven, (2) reporting that a *known* raven is found to be black, and (3) reporting that a *known* black thing is found to be a raven. Horwich says the report that a *known* raven is found to be black provides the best evidence, for in that case, the hypothesis "is subject to the maximum risk of falsification and has passed the most severe test"(1982,p.58).

Let us refer to the '*' operator which appears in (G1) through (G8) below as the epistemic operator. Like the truth-functional negation symbol '~', the epistemic operator '*' applies only to the sentence to its immediate right. Let us refer to the sentence 'Zu' which is flagged by the epistemic operator in evidence reports of the form '*Zu&...' as the epistemic condition. In '*Zu&Qu' the epistemic operator applies only to 'Zu' in contrast to' *(Zu&Qu)' where the epistemic operator applies to the conjunction 'Zu&Qu'.

I take the evidence report '*Zu&...' as a non-tautological claim which says that an individual **u** is *known with certainty* to be a Z on the basis of background knowledge K alone. '*Zu&Qu' is to be read as the report that an individual **u** which is *known with certainty* to be a Z on the basis of K is found to have the property Q. By contrast, 'Zu&Qu' merely says that an individual **u** has the properties Z and Q.

I interpret an evidence report of the form '*Zu&...' to imply that P(Zu/K)=1. I interpret the formula 'P(*Zu&.../K)' to mean 'P(Zu&.../K)' under the condition that P(Zu/K)=1 and the formula 'P(*Zu&.../H&K)' to mean 'P(Zu&.../H&K)' under the condition P(Zu/K)=1.

Extending a suggestion made by Horwich, I consider a hypothesis H of the form '(y)(Zy ⊃ Qy)' and its relation to the following forms of evidence reports: (G1) '*Zu&Qu'; (G2) '*Zu&~Qu'; (G3) '*~Qu&Zu'; (G4) '*~Qu&~Zu'; (G5) '*Qu&~Zu'; (G6) '*Qu&Zu'; (G7) '*~Zu&Qu'; (G8) '*~Zu&~Qu' and (G9) 'Zu&Qu'. I show below that some of these forms confirm this hypothesis, some disconfirm it, and some are merely conforming instances. A hypothesis is directly confirmable on this model of confirmation if and only if it is confirmable by reports of at least one of the forms (G1) through (G9).

This model of confirmation incorporates a requirement that direct confirmation of hypothesis H occurs only in conditions in which it is possible to falsify H. I formulate this requirement as follows:

> (RC1) If H is of the form '(y)(Zy ⊃ Qy)' and there is an evidence report either (i) of the form '*Zu&...' which says that **u** is *known with certainty* to be a Z on the basis of background knowledge K alone and that is our only relevant background knowledge regarding **u** or (ii) of the form '*~Qu&...' which says that **u** is *known with certainty* to be a ~Q on the basis of K alone and that is our only relevant background knowledge regarding **u**, then 1> P(Zu&~Qu/K)>0.

Before considering any of the forms (G1) through (G8), it is useful to show that if P(Zu/K)=1 and P(K&H)>0, then P(Zu/K&H)=1. Using indirect proof, assume P(Zu/K)=1, P(K&H)>0. and P(Zu/K&H)<1. Hence, P(Zu&K)=P(K), P(Zu&K&H)<P(K&H), and P(Zu&K&H)+P(Zu&K&~H)=P(K&H)+P(K&~H). This entails the impossibility that P(Zu/K&~H)>1.

Suppose H is a directly confirmable hypothesis of the form '(y)(Zy ⊃ Qy)' and there is an evidence report of the form (G1),'*Zu&Qu',which implies that P(Zu/K)=1. From (RC1) and P(Zu/K)=1, we obtain 1>P(Zu&Qu/K)>0. Using Bayes' Theorem, we obtain P(H/Zu&Qu&K)=P(H/K) X P(Zu&Qu&H&K) / [P(H&K) X P(Zu&Qu/K)]. Since P(Zu/H&K)=1, P(H&K)=P(Zu&H&K). Because 'Zu&H' is logically equivalent to 'Zu&Qu&H', P(Zu&H&K)=P(Zu&Qu&H&K). Hence, P(H&K)=P(Zu&Qu&H&K) and P(H/Zu&Qu&K) = P(H/K) X 1/P(Zu&Qu/K). Because 1>P(Zu&Qu/K)>0, P(H/Zu&Qu&K)>P(H/K). The evidence report '*Zu&Qu' confirms '(y)(Zy ⊃ Qy)' according to the positive relevance criterion given earlier. A parallel argument shows that (G4), '*~Qu&~Zu', also confirms this hypothesis.

Both (G2),'*Zu&~Qu', and (G3),'*~Qu&Zu', disconfirm H. In these two cases, P(H/K)>0 and P(H/Zu&~Qu&K)=0 and thus P(H/K)>P(H/E&K).

Evidence reports of the forms (G5) and (G6) imply that P(Qu/K)=1. It follows that P(Qu/K)=P(Qu/K&H). Similarly, reports of the forms (G7) and (G8) allow the derivation of P(~Zu/K)=P(~Zu/K&H)=1. Hence, we may derive :

P(Qu&Zu/K)+P(Qu&~Zu/K)=P(Qu&Zu/K&H)+P(Qu&~Zu/K&H)

and

P(~Zu&Qu/K)+P(~Zu&~Qu/K)=P(~Zu&Qu/K&H)+P(~Zu&~Qu/K&H) .

Because it is possible to assign values in these equations such that P(Qu&~Zu/K&H)=P(Qu&~Zu/K), P(Qu&Zu/K&H)=P(Qu&Zu/K), and P(~Zu&~Qu/K&H)=P(~Zu&~Qu/K), (G5) through (G8) report conforming rather than confirming instances of H. Similarly, (G9), 'Zu&Qu', which lacks the epistemic operator, is compatible with P(Zu&Qu/K)=P(Zu&Qu/K&H) and reports a conforming instance of H.

In the next section, I present a model of confirmation of universal hypotheses by instances which makes no requirement of certainty.

3.2. Confirmation Not Requiring Certainty

Let us refer to the '#' operator which appears in (F1) through (F8) below as the epistemic operator. Like the truth-functional negation operator '~', the epistemic operator '#' applies only to the sentence to its immediate right. Let us refer to the sentence 'Zu' which is flagged by the epistemic operator in evidence reports of the form '#Zu&...' as the epistemic condition. In '#Zu&Qu' the epistemic operator applies only to 'Zu' in contrast to '#(Zu&Qu)' where the epistemic operator applies to the conjunction 'Zu&Qu'.

I take the evidence report '#Zu&...' as a non-tautological claim which says that **u** is known to be a Z on the basis of background knowledge K alone. '#Zu&Qu' is to be read as the report that an individual **u** which is known to be a Z on the basis of K alone is found to have the property Q. By contrast, 'Zu&Qu' merely says that an individual **u** has the properties Z and Q.

I interpret an evidence report of the form '#Zu&...' to imply that P(Zu/K)>0. I interpret 'P(#Zu&.../K)' to mean 'P(Zu&.../K)' under the condition that P(Zu/K)>0 and the formula 'P(#Zu&.../H&K)' to mean 'P(Zu&.../H&K)' under the condition P(Zu/K)>0.

I consider a hypothesis H of the form '(y)(Zy ⊃ Qy)' and its relation to the following forms of evidence reports: (F1) '#Zu&Qu'; (F2) '#Zu&~Qu'; (F3) '#~Qu&Zu'; (F4) '#~Qu&~Zu'; (F5) '#Qu&~Zu'; (F6) '#Qu&Zu'; (F7) '#~Zu&Qu'; (F8) '#~Zu&~Qu'; and (F9) 'Zu&Qu'. I show below that some of these forms confirm the hypothesis H, some disconfirm it, and others are merely reports of conforming instances. A hypothesis is directly confirmable on this model if and only if it is confirmable by reports of at least one of the forms (F1) through (F9).

This model of confirmation incorporates a requirement that direct confirmation of a hypothesis H of the form '(y)(Zy ⊃ Qy)' occurs only under conditions in which it is possible to falsify H. I formulate this requirement as follows:

(NC1): If H is of the form '(y)(Zy ⊃ Qy)' and there is an evidence report either (1) of the form '#Zu&...' which says that **u** is known on the basis of background knowledge K alone to be a Z and that is our only relevant background knowledge regarding **u** or (2) of the form '#~Qu&...' which says that **u** is known on the basis of background knowledge K alone to be a ~Q and that is our only relevant background knowledge regarding **u**, then P(Zu&~Qu/K)>0.

This model also incorporates the following requirement:

(NC2) : Where 'Mu' is the non-tautological epistemic condition of an evidence report of the form '#Mu&...' and H is of the form '(y)(Zy ⊃ Qy)', and where 'Mu' is replaced by 'Zu', '~Zu', 'Qu', or '~Qu', H is not negatively relevant to 'Mu' given K, i.e., P(Mu/H&K) ≥ P(Mu/K).

Suppose P(Zu&Qu/K)=0, given (NC1), (NC2), and the evidence report '#Zu&...'. It follows that P(Zu/K)>0 and P(Zu&K)>0. Hence, P(Qu/Zu&K)=0 and P(~Qu/Zu&K)=P(Zu&~Qu/Zu&K)=1. But 'Zu&~Qu' entails the denial of H and thus P(H/Zu&K)=0. This contradicts P(H/Zu&K)>0 which follows from '#Zu&...', (NC1), (NC2), and Bayes' Theorem. Thus, if there is an evidence report of the form '#Zu&...' which satisfies the conditions (NC1) and (NC2), 1>P(Zu&Qu/K)>0 and 1>P(Zu&~Qu/K)>0. A parallel argument may be constructed to show that, given (NC1), (NC2), and an evidence report of the form '#~Qu&...', 1>P(~Qu&Zu/K)>0 and 1>P(~Qu&~Zu/K)>0.

Given (NC1), (NC2), and the positive relevance criterion stated above, it is possible to show that: (i) evidence reports of the forms (F1), '#Zu&Qu', and (F4), '#~Qu&~Zu', confirm H, i.e., '(y)(Zy ⊃ Qy)' ; (ii) evidence reports of the forms (F2),'#Zu&~Qu' , and (F3),'#~Qu&Zu', disconfirm H; and (iii) the remaining forms (F5) through (F9) neither confirm nor disconfirm H but report conforming instances for this hypothesis.

(i) Suppose that there is an evidence report of the form '#Zu&Qu'. Given (NC1) and (NC2), we derive P(Zu/H&K) ≥ P(Zu/K)>0 which is equivalent to P(Zu&Qu/H&K) + P(Zu&~Qu/H&K) ≥ P(Zu&Qu/K)+P(Zu&~Qu/K)>0. Since P(Zu&~Qu/H&K)=0, P(Zu&Qu/K)>0, and P(Zu&~Qu/K)>0, it follows that P(Zu&Qu/H&K)>P(Zu&Qu/K)>0. By Bayes' Theorem this implies that P(H/Zu&Qu&K) > P(H/K) > 0, i.e., (F1),'#Zu&Qu', confirms H. Replacing 'Zu' by '~Qu' and 'Qu' by '~Zu' in the formulae just above, and using standard rules of logical equivalence, it is possible to show that (F4) ,'#~Qu&~Zu', also confirms H.

(ii) Given an evidence report of the form (F2),'#Zu&~Qu', or of the form (F3),'#~Qu&Zu', we obtain P(Zu&~Qu/K)>0 by (NC1). Since P(Zu&~Qu/H&K)=0, it follows that P(H/Zu&~Qu&K) < P(H/K), i.e., reports of forms (F2) and (F3) disconfirm H.

(iii) Given an evidence report of the form (F7) or (F8), represented generically as '#~Zu&...', we derive P(~Zu/H&K) ≥ P(~Zu/K)>0 which is equivalent to P(~Zu&Qu/H&K)+P(~Zu&~Qu/H&K) ≥ P(~Zu&Qu/K)+P(~Zu&~Qu/K)>0. Since a rational inquirer is free to assign probability values so that P(~Zu&Qu/H&K) = P(~Zu&Qu/K) and P(~Zu&~Qu/H&K) = P(~Zu&~Qu/K), reports of forms (F7) and (F8) neither confirm nor disconfirm H but merely express conforming instances for H. A parallel demonstration showing that (F5) and (F6) also report merely conforming instances is easily constructed. Similarly, (F9), which lacks the epistemic operator '#', reports a conforming instance for H.

'Zu', 'Qu', etc. in the preceding discussion need not be interpreted as observation sentences. Where 'Zu', 'Qu', etc. are theoretical, we can use background knowledge, including, as Quine suggests, "our backlog of accepted theory"(1990,p.9), to obtain the information that some individual **u** has the property Z, Q, etc.. The confirmation of at least some hypotheses by some evidence reports need not concern directly the relation of stimulations to scientific theory.

Thus, it is possible to articulate a notion of the confirmation and disconfirmation of individual hypotheses within a larger body of theory using only first-order predicate logic with identity and elementary probability theory. Remaining within the linguistic resources described in *Pursuit of Truth,* we need not rest with Quine's confirmational holism.

4 Extensions and Applications

The models of confirmation presented above and the Bayesian approach have the resources to avoid some interesting paradoxes and problems, e.g., the problem of old evidence, Hempel's Raven paradox, and a strong form of relativism which utilizes Goodman's grue paradox.

4.1. The "Old Evidence" Problem

The problem of so-called "old evidence" is widely held to constitute a major objection to any theory of confirmation which includes essentially Bayes' Theorem and the positive relevance criterion of confirmation. Glymour (1980,p.86) and Chihara (1987,p.552) contend that if E is old evidence, i.e., any evidence that is known prior to the time of formulating or introducing or confirming H, E cannot confirm H according to the positive relevance criterion. Neither Glymour nor Chihara present the argument in detail, but it might be formulated as follows:

> If E is old evidence relative to K, then K contains and entails E. Hence, if $P(K) > 0$ and $P(H\&K) > 0$, then $P(E/K) = P(E/K\&H) = 1$. So, $P(H/K) = P(H/K\&E)$ and E does not confirm H.

The claim that $P(E/K)=1$, where E is any old evidence relative to K, is presented by Glymour and Chihara without argument. Glymour dismisses the counterclaim that, for some old evidence, $P(E/K)<1$, saying that "the acceptance of old evidence may make the degree of belief in it as close to unity as our degree of belief in some bit of evidence ever is"(1980,p.87). Glymour not only waffles here by using the phrase 'as close to unity', but also ignores the views of Neurath (1959,p.201), Jeffrey (1965,p.169), Lakatos (1970,p.108), and Horwich (1982,pp.78-9) regarding the uncertainty of at least some old evidence reports.

As a historian of science, Glymour recognizes that some old evidence may be uncertain. Noting that the anomaly of the perihelion of Mercury was known more than a half century prior to Einstein's 1915 explanation, Glymour says that no single event "makes the perihelion event virtually certain" and suggests that, in the period from 1890 to 1920, belief in the evidence report of the anomaly "waxed, waned, and waxed again" (1980,p.88). Waxing and waning of belief in this bit of old evidence E suggests that, for some background knowledge K during this period, $P(E/K)<1$.

Contrary to the claims of Glymour and Chihara, the so-called "old evidence" problem is not essentially one of the age of the evidence, but of the application of Bayes' Theorem and the positive relevance criterion of confirmation where $P(E/K)=P(E/H\&K)=1$. The two models of confirmation presented above avoid this problem by requiring that in order for E to confirm H, given K, $1>P(E/K)>0$.

4.2. The Raven Paradox

The raven paradox is stated by Hempel (1965,pp.14-5) in terms of objects and their properties. Translated into talk of evidence reports, the raven paradox makes four claims. (RP1) An evidence report of the form '~Qu&~Zu' confirms H. (RP2) An evidence report of the form '~Zu&...' confirms H. (RP3) An evidence report of the form '~Zu v Qu' confirms '(y)[(Zy v ~Zy) ⊃ (~Zy v Qy)]' which is logically equivalent to H and thus confirms H. (RP4) Some formulae logically equivalent to H, e.g., '(y)[(Zy&~Qy) ⊃ (Zy & ~Zy)]', are not confirmable.

Neither model of confirmation presented above is correctly understood as supporting (RP1), (RP2), or (RP3). None of the evidence reports '~Zu&~Qu', '~Zu&..', or '~Zu v Qu' is taken on either model as confirming H or any hypothesis logically equivalent to H. In each case, the evidence report lacks the epistemic operator '*' or '#' and can report a conforming but not a confirming instance for H.

(RP4) is the most plausible of Hempel's remarks concerning the paradoxes of confirmation. On the models presented earlier, if H is directly confirmable by an evidence report of the form '*Zu&...' or '#Zu&...', 1>P(Zu&~Qu/K)>0 and 1>P(Zu&Qu/K)>0. This excludes cases where 'Zu' is self contradictory or 'Qu' is either self-contradictory or tautological. Similarly, if H is directly confirmable by an evidence report of the form '*~Qu&...' or '#~Qu&...', 1>P(~Qu&Zu/K)>0 and 1>P(~Qu&~Zu/K)>0. This excludes cases where 'Qu' is tautological or 'Zu' is either self-contradictory or tautological. If H is directly confirmable, it is confirmable by reports of the forms '*Zu&Qu', '*~Qu&~Zu', '#Zu&Qu', and '#~Qu&~Zu'. In no hypothesis of the form '(y)(Zy ⊃ Qy)' which is directly confirmable by reports of these forms is either 'Zu' or 'Qu' either tautologous or self-contradictory. Thus, neither (H1) '(y)[(Zy&~Qy) ⊃ (Zy&~Zy)]' nor (H2) '(y)[(Zy v ~Zy) ⊃ (~Zy v Qy)]' is directly confirmable on either model by evidence reports of any of the following forms: (1) '*(Zu v ~Zu) & (~Zu v Qu)'; (2) '*(Zu & ~Qu) &(Zu & ~Zu)'; (3) '#(Zu v ~Zu) & (~Zu v Qu)'; (4) '#(Zu & ~Qu) &(Zu & ~Zu)'. To this extent, the present proposal agrees with Hempel. However, it might be suggested that '(y)[(Zy&~Qy) ⊃ (Zy&~Zy)]' and '(y)[(Zy v ~Zy) ⊃ (~Zy v Qy)]' are *indirectly* confirmable by reports of the forms '*Zu&Qu', '*~Qu&~Zu', '#Zu&Qu', and '#~Qu&~Zu' because each of these forms directly confirms H which is logically equivalent to both (H1) and (H2).

4.3. Underdetermination and the Thesis of Strong Relativism

In *Science and Relativism,* Laudan critically discusses a strong form of the thesis of underdetermination or epistemic relativism which holds that (1)"*all* the rivals to any given theory are as well supported by the evidence as the theory in question" and (2) "we have no epistemic grounds for accepting or rejecting any theory rather than its contraries"(1990,p.54). Strong relativism holds that "evidence is always powerless to choose between any pair of [theories which are contraries or] rivals"(1990,p.56).

As Laudan presents the position of strong relativism, it leans heavily upon the claim that Nelson Goodman, in *Fact, Fiction, and Forecast,* showed that for "any hypothesis inductively supported by a certain body of evidence, there were indefinitely many contrary hypotheses which were as well supported by those instances"(1990,p.67).

Laudan criticizes strong relativism by claiming that Goodman's grue paradox, which assigns "equivalent degrees of support to contrary hypotheses"(1990,p.68), is merely a consequence of Goodman's failure to distinguish between confirming and conforming instances. Laudan's distinction between these types of instances is similar to that used in the models of confirmation described earlier in this paper. That is, a "positive instance of a theory or hypothesis is simply one of its empirical consequences which is true"(1990,p.61), but "a positive instance of a theory is a confirming instance for that theory only if it results from a test of that theory"(1990,p.62). Holding that "where there's no risk of failure, there is no test involved"(1990,p.20), Laudan adds "no test, no confirming instance—although one may have positive instances in the absence of tests"(1990,p.62).

The difficulty posed by Goodman's grue paradox is not resolved simply by noting that there is an important distinction between confirming and conforming instances. The strong relativist may accept a distinction between confirming and conforming instances and contend that the confirming evidence reports for 'all emeralds are green' equally well confirm 'all emeralds are grue'. Laudan has not given any reason to reject this contention.

Strong relativism cannot sustain its central thesis that *all* the rivals or contraries to *any* given hypothesis are equally well supported by the evidence. It can be shown that evidence is not always powerless to choose between any pair of rival theories. That is, it is demonstrable that, under some plausible conditions, some pairs of contrary hypotheses are not equally supported by the evidence.

Consider the *discontinuity hypothesis* HDS which says 'if anything is an M, then it is either inspected before *t* and is a G or it is not inspected before *t* and is not a G' and the *continuity hypothesis* HCN which says 'if anything is an M, then, whether it is inspected before *t* or not, it is a G'. HDS may be symbolized as '$(y)\{My \supset [(Iy \,\&\, Gy) \vee (\sim Iy \,\&\, \sim Gy)]\}$' and HCN as '$(y)(My \supset Gy)$'.

One of the difficulties in the evaluation of the central thesis of strong relativism is that the strong relativist provides neither a definition of nor a method for determining equal confirmation of contrary pairs of hypotheses such as HDS and HCN on the same evidence E. Moreover, as Laudan suggests, the strong relativist fails to describe E adequately.

Let us say that E confirms H' and H equally, given K, if and only if $P(H'/K\&E) - P(H'/K) = P(H/K\&E) - P(H/K)$. Let us adopt the model of confirmation not requiring certainty to supply the Bayesian method of confirmation and the description of the evidence E.

A further problem is that neither Goodman nor the strong relativist say what is allowable as background knowledge in the consideration of the central thesis of strong relativism. There are two characterizations of K which it would be wise to avoid. First, there is the danger of begging the question in favor of the continuity hypothesis HCN by the inclusion of a principle of the uniformity of nature within background knowledge K. Second, there is the danger of so impoverishing K that it fails to serve any useful purpose. In particular, if K fails to include or imply the claim that there exists at least one M which is not examined prior to time *t*, then K cannot serve as a basis for saying that HDS and HCN are contraries.

I propose to characterize K as including or implying '$(\exists y)(My \,\&\, \sim Iy)$'. That is, I take it as part of the background knowledge K of some community of inquirers that some individuals of type M are not inspected before *t*. For example, I take it as common background knowledge that there is coal deep in the ground which is not inspected for its sulphur content before an arbitrarily chosen time *t*.

Because K includes or implies '(∃y)(My & ~Iy)', the conjunction of HCN and K logically implies ~HDS. Hence, $1 \geq$ P(HDS/K) + P(HCN/K) and $1 \geq$ P(HDS/K&E) + P(HCN/K&E). If P(HCN/K) = P(HDS/K) $\leq 1/2$ and P(HCN/K&E) $> 1/2$, then P(HDS/K&E) $< 1/2$ and, contrary to the strong relativist, E confirms HCN more than it confirms HDS. There is an infinite number of cases in which the evidence provides adequate epistemic grounds for accepting a theory rather than its contrary.

In addition, using the previously described model of confirmation not requiring certainty, it is possible to describe evidence E so that (1) a clear distinction is made between confirming and conforming reports, (2) E confirms HCN, and (3) E does not confirm HDS. Since P(HCN/K&E) - P(HCN/K) > 0 and P(HDS/K&E) - P(HDS/K) $\leq$ 0, E confirms HCN more that it confirms HDS. This will provide further illustration of the falsehood of the central claim of strong relativism.

Adopting the distinction between confirming and conforming reports used in the model of confirmation not requiring certainty, we distinguish between (1) '#Mu&Gu' and '#~Gu&~Mu' which directly confirm HCN and (2) 'Mu&Gu' and '~Gu&~Mu' which merely conform to HCN. Let us suppose that evidence E is gathered before time *t* by looking exclusively among individuals known to have the property M on the basis of background knowledge K in order to discover only whether or not these individuals have the property G. That is, each of the evidence reports has either the form '#Mu&Gu' or the form '#Mu&~Gu'.

On the basis of Goodman's account (1965,p.73), E contains no reports which disconfirm HCN, but contains only reports that confirm HCN. Thus, E may be understood to contain only reports of the form '#Mu&Gu'. Given P(Mu&Gu/HDS&K)+P(Mu&~Gu/HDS&K) $\geq$ P(Mu&Gu/K)+P(Mu&~Gu/K) > 0, it is possible to assign values so that P(Mu&Gu/HDS&K) = P(Mu&Gu/K). That is, evidence reports of the form '#Mu&Gu' do not confirm HDS. Since E does not confirm HDS but does confirm HCN, P(HCN/K&E) - P(HCN/K) > 0 and P(HDS/K&E) - P(HDS/K) $\leq$ 0. E confirms HCN more than E confirms HDS. The evidence is not always powerless to decide between contrary hypotheses.

4.4. Confirmation in a Less Austere Language

The models of confirmation described in this paper may be extended to apply to lawlike subjunctive conditionals using operators such as the Fetzer-Nute fork operator '->-'. The lawlike subjunctive conditional '(y)(Zy ->- Qy)', on Fetzer's account (1981,pp.152-7), implies the extensional accidental universal conditional '(y)(Zy ⊃ Qy)'. Thus, if H is interpreted as '(y)(Zy ->- Qy)', the derivations given in sections 3.1 and 3.2 of this paper may be used to specify which evidence reports confirm, which disconfirm, and which merely conform to a lawlike subjunctive conditional. Although the models of confirmation described in this paper were developed within a fairly austere language, they may be used to articulate the notion of the confirmation and disconfirmation of hypotheses stated in some less austere languages.

References

Chihara, C.S. (1987), "Some Problems for Bayesian Confirmation Theory", *The British Journal for the Philosophy of Science* 38: 551-60.

Fetzer, J.H. (1981), *Scientific Knowledge*. Dordrecht: Reidel.

Glymour, C.N. (1980), *Theory and Evidence*. Princeton, New Jersey: Princeton University Press.

Goodman, N. (1965), *Fact, Fiction, and Forecast*. Indianapolis, Indiana: The Bobbs-Merrill Company, Inc..

Hempel, C.G. (1965), *Aspects of Scientific Explanation*. New York: The Free Press.

Horwich, P. (1982), *Probability and evidence*. Cambridge: Cambridge University Press.

Jeffrey, R.C. (1965), *The Logic of Decision*. New York: McGraw-Hill Book Co.

Lakatos, I. (1970), "Falsification and the Methodology of Scientific Research Programmes", in *Criticism and the Growth of Knowledge*, I. Lakatos and A. Musgrave (eds.). Cambridge: Cambridge University Press, pp.91-196.

Laudan, L. (1990), *Science and Relativism*. Chicago: The University of Chicago Press

Neurath, O. (1959), "Protocol Sentences", in *Logical Positivism*, A.J.Ayer (ed.). New York: The Free Press, pp.199-208.

Quine, W.V. (1960), *Word and Object*. Cambridge: MIT Press.

______. (1961), "Two Dogmas of Empiricism", in *From a Logical Point of View*. Cambridge, Massachusetts: Harvard University Press, pp.20-46.

______. (1969), "Epistemology Naturalized", in *Ontological Relativity and Other Essays*. New York: Columbia University Press, pp. 69-90.

______. (1981), "Five Milestones of Empiricism", in *Theories and Things*. Cambridge: Harvard University Press, pp.67-72.

______. (1990), *Pursuit of Truth*. Cambridge: Harvard University Press.

The Importance of Models in Theorizing: A Deflationary Semantic View[1]

Stephen M. Downes

University of Utah

1. Introduction

It is commonly acknowledged in science that model construction is one of the most important components of theorizing. Philosophers of science are gradually coming to acknowledge this situation, spurred on by holders of the semantic view of theories. In this paper I wish to defend a very deflationary version of the semantic view of theories, which is more or less a re-statement of the above commonplace. I reject the view encapsulated in the identity statement "scientific theories are families of models," although acknowledging the useful insights into science that holders of this strong position have given us. My position derives from a critique of various of the semantic views of theories, and further from a guiding presupposition that rather than providing necessary and sufficient conditions for what a theory is, philosophers should focus on the nature of scientific theorizing. Theorizing is carried out by practicing scientists, and we cannot say what scientific theories are unless we appreciate the myriad ways they are used and developed in all of the sciences.

The paper proceeds by investigating some key aspects of the semantic view of theories. I concentrate particularly on the notion of a model, and less on the various notions of "theory" that appear in the literature. Having cleared up some issues to do with the nature of models, I defend the claim that mathematical and meta-mathematical models are clearly different from scientific models. Next, I criticize the strong view that theories are families of models by looking at various examples that show the semantic view "merely" provides descriptions of particular cases of theorizing, rather than providing a general account of the nature of theories. Finally, I propose a more liberal or deflationary view, which is consistent with a naturalistic approach to the philosophy of science.

2. Working Definition of the Semantic View of Theories

In this section I introduce a working definition of the semantic view of theories. There are several formulations of the semantic view and many differences between each of these formulations. In the definition given here I am erring toward a strong version of the semantic view, but it is not a version held by any particular semantic theorist. The definition simply helps distinguish the view from other prominent views

---

PSA 1992, Volume 1, pp. 142-153

such as the "received view" (Suppe 1977), which is, roughly speaking, the view derived from logical empiricists that theories can be adequately reconstructed as sets of axioms and correspondence rules.

Here is the definition: Scientific theories consist of families of (mathematical) models including empirical models and sets of hypotheses stating the connections between the empirical models and empirical systems. Empirical models are models that specifically purport to have relations to an empirical system. There are many models in science that clearly do not purport to represent empirical systems and yet are still important in scientific theorizing. From here on I will follow Van Fraassen's (1980a) usage and refer to "empirical systems". Giere (1988) uses "real systems" in its place and I think this leads to misleading presuppositions about realism. The use of the phrase "empirical systems" allows for the discussion in this paper to remain mute about the realism/anti-realism debates (cf. Lloyd 1988).

The above definition is closest to Giere's (1988) and perhaps neither Van Fraassen nor Suppes would include the second clause. The important thing to note is that the models are the central feature. The view emphasizes semantic objects over syntactic objects, taking the lead from semantics in meta-mathematics. On the semantic view, whatever linguistic components a theory has, they are incidental to any serious understanding of that theory's nature. In the next section I give some examples of models to illustrate the diversity of the concept.

3. Theories and Models

In this section I will pay most attention to models, but a brief word is in order about the other side of the definition "theories". Predominantly philosophers have worked with a somewhat unexamined notion of theories. Although much work has gone into attempting to answer the question "What are scientific theories?" there is no real consensus over what the scope of the investigation is. Much work in philosophy of science has proceeded under the assumption that Classical Mechanics is a good example of a scientific theory, and so if we can explain what that is, we have a start on explaining what theories are in general. There are many axiomatizations of classical mechanics (e.g., Simon 1954) that under the received view count as attempts to answer the question.

What has been revealed by closer examination of the practice of science is that the term "theory" does not simply denote the finished product of years of research formulated in its most elegant fashion as served up in advanced text books. Many fields have general overarching theories, middle level, and low level theories. Biology provides an example of a field in which there are clearly several different levels of theorizing, each of which are crucially inter-related. Further, there is no clear candidate for "*the* theory of evolution" (cf. Hull 1988).

Another direction that discussion of theories has taken has been to promote anyone's account of anything as presupposing a theory. Examples abound in the philosophical literature on cognitive science of our folk theories of psychology, physics, and even middle sized every day objects (e.g., Churchland 1986). For example, on this account, I not only possess a theory of the cell, which I invoke in recognizing cells, but also a theory of chairs.

A middle course between the extremes of accepting only fully developed and formalized scientific theories, and promoting any set of concepts that guide perception seems the most sensible. How one identifies candidate theories should be derived from an investigation of scientific practice. There is no really clear guide in scientific practice to what counts as theorizing, but several rules of thumb can be adopted to recognize such practices. Obviously many theoretical presuppositions are invoked in even the most mundane types of scientific practice, but it seems clear that tissue slicing and staining are not theorizing. Predicting various matches between a mathematical model and a yet to be established experimental set-up clearly is theorizing. Most proponents of the semantic view take off from a close look at a particular piece of scientific theorizing when developing their account of the nature of theories in general, and this is a reasonable approach.

There are many referents for the term "model", and it is my contention that there are far greater differences between models in mathematics and logic and models in science than holders of the semantic view have been prepared to admit. Both Suppes and Van Fraassen have played down the distinction between models in logic and in science. Suppes says "I would assert that the meaning of the concept of model is the same in mathematics and the empirical sciences" (1960, p. 289). Van Fraassen makes a similar point: "... the usages of model in meta-mathematics and in the sciences are not as far apart as has sometimes been said" (1980, p. 44). Suppes and Van Fraassen's views derive from the way they approach the philosophy of science, and this is a topic I will return to later. First, I will elaborate what do seem to be some clear differences between models in science and models in meta-mathematics. I use some elementary examples to set out some of the differences, in the sections below I will introduce some more examples as I develop my criticisms of the strong version of the semantic view.

Let us look at two straightforward examples of models in mathematics and meta-mathematics. Consider the following set of postulates:

1. Any two members of K are contained in just one member of L.
2. No member of K is contained in more than two members of L.
3. The members of K are not all contained in a single member of L.
4. Any two members of L contain just one member of K.
5. No member of L contains more than two members of K. (Nagel & Newman 1958, p. 16)

A model for these postulates is a triangle with vertices K and sides L. The model satisfies any theorems derived from these postulates and shows us that the postulates are consistent. The model provides us with a semantics for the set of postulates.

Now consider a slightly more complex example. We can derive all of arithmetic from the Peano Axioms (let us just for arguments sake ignore Godel's result here). The set that provides a model for the Peano Axioms is the set of integers plus zero. Things are slightly more complex in this case, as we need various orderings on the set to satisfy theorems derivable from the Peano Axioms. One way to do this is by Tarski's method of sequences, so the set {2,2,4.......} satisfies "2+2=4". So a model for arithmetic is the set of sequences produced from various orderings on the integers.

These very elementary examples give us much of what we need to understand how models work in meta-mathematics. The crucial relation is a relation of satisfaction between a set of postulates, theorems, or axioms and a set of some type of objects or other. The latter set can be expressed as a more concrete geometric model as in the first case, or specified in a more abstract manner as in the second case. The one further notion that is important for our purposes is "embedding". Think of postulates 1 through 5 above as a describing a theory T; this theory is satisfied by the triangle, but the triangle can be embedded in the larger structure of a Euclidian plane because it is isomorphic to part of that plane. On this account our theory T can be embedded in the larger theory T', which is Euclidian Geometry, because the model for T is isomorphic with part of the model for T' (cf. Van Fraassen 1980a, p. 43). Embedding is a relation between models, and relies on the isomorphism of one model with part of another. In our other example the sequence {2,2,4....} can be embedded in the set of all sequences of integers. The notion of embedding works well in the meta-mathematical cases because isomorphism is so well defined. In the scientific case isomorphism is a harder relation to define.

Let us now consider a few models from science. Again these are very simple examples. The first is taken from Maynard-Smith by Lloyd (1988). The description of population growth in ecological theory can be expressed mathematically by "the logistic equation", $dx/dt = rx(1 - x/k)$. Here x is the population density at time t, r is the intrinsic rate of increase, and k is the carrying capacity of the environment. As Lloyd points out, Maynard-Smith observes the following: "[the equation] was not derived from any knowledge of, or assumptions about, the precise way in which the reproduction of individuals is influenced by density; it is merely the simplest mathematical expression for a particular pattern of growth" (1988, p. 15). The equation presents a set of relations between particular mathematical objects, and it is this structure that the equation defines that is a model.

We can go on to make comparisons between the model and actual experimental systems. If we take an experimental population and plot its growth, and the curve matches closely the curve for the logistic equation, then we can claim that the model is isomorphic to the real system. The term isomorphism is being used in a different sense here than in the mathematical examples, and I will comment on this below. What we need now to note is that the model is a mathematical structure defined by the logistic equation, and it may or may not have relations to particular empirical systems.

Giere introduces an example from classical mechanics. Consider the following equation: $m\, d^2x/dt^2 = -(mg/l)x$. This is the equation of motion for the horizontal component x of the motion of a pendulum with length, l, mass, m, where gravity is g. This equation is for a small angle of swing, a, $\cos(a) = 1$. Such a pendulum is a model. The equation describes this particular pendulum, or in the language of the meta-mathematical examples the pendulum satisfies the equation.

In this example there are no empirical systems corresponding to the pendulum. The equations describing the pendulum in my high-school physics laboratory, or any form of existing linear oscillator, require many added parameters for the resistance of the air and the size of the angle, a, and so on. The pendulum that satisfies the equation of motion above is an abstract system or a model that satisfies just that equation.

Finally, consider a typical biology textbook drawing of a cell. In most texts a schematized cell is presented that contains a nucleus, a cell membrane, mitochondria, a Golgi body, endoplasmic reticulum and so on. In a botany text the schematized cell will contain chloroplasts and an outer cell wall, whilst in a zoology text it will not include these items. The cell is a model in a large group of inter-related models that enable us to understand the operations of all cells. The model is not a nerve cell, nor is it a muscle cell, nor a pancreatic cell, it stands for all of these.

Many other models are presented in cell biology when one graphically zooms in on the inside of the cell. For example, when energy transfer is considered we look at a model of a mitochondrion. In the case of the cell there is no mathematical object, and there are no equations describing it, and yet the schematic drawing is not of any one particular cell; it is an idealized cell or model. Just as the model of the mitochondria is not a drawing of any particular mitochondria.

We have enough examples now to bring out some of the important differences between mathematical and the scientific models. There are two important differences, one centering on satisfaction and the other centering on isomorphism and the related concept of embedding. Let us take the satisfaction issue first.

In the mathematical case the triangle satisfies the postulates of theory T, and this is all the work it has to do. In the case of the logistic equation what we have is a system that satisfies an equation also, but this is not all it does. The difference between the mathematical and the scientific case can be put schematically as follows: Let ==> be "satisfies".

Mathematics:
MODEL ==> {Linguistic description of theory, Equations ,...}
Science:
{Empirical System}=?=MODEL==>{linguistic description of theory, Equations,...}

First, consider the cell example as an instantiation of the second schema. There is no clear place for the satisfaction relation in the cell biology case. A set of sentences may describe the cell model, but the cell model does not satisfy the description in any specifiable sense. The notion of satisfaction is a technical term from meta-mathematics with no correlate in many cases of scientific model construction.

Further, in the scientific case we are interested in more than just the satisfaction relation, for an account of theories that does any justice to scientific practice we need to say something about the relation between models and empirical systems they purportedly model. Of course holders of the semantic view have such a concept, my symbol =?= above is captured by isomorphism. Before turning to this a brief note is in order about the pendulum case.

We observed that in the case of the pendulum there was no real system that corresponded to the equation. In fact as Giere has pointed out (1988) there are few models in classical mechanics that bear a close relation to empirical systems. In this case perhaps the similarity between the scientific and the logical case is more apparent. Giere's weaker version of Suppes and Van Fraassen's claims above is that the terminology in classical mechanics "overlaps nicely with the usage of logicians, for whom a model of a set of axioms is an object, or set of objects, that satisfies the axioms" (1988, p. 79). But even in the case of classical mechanics there is some sense in asking the question "What is the relation between the pendulum satisfying the equation, and the pendulum in my high-school physics laboratory?" No such corresponding question can be asked in the meta-mathematical case.

Notice, also, that the notion of embedding partially captures the relations between the pendulum and other linear oscillators. The equation for any linear oscillator is a more general equation than the equation for the horizontal component of the pendulum, and a model for a linear oscillator satisfies the equation for the pendulum. The pendulum is embedded in the linear oscillator because the pendulum is in some sense isomorphic to a sub-structure of the linear oscillator.

Giere, following many other semantic theorists, separates out appropriate questions philosophers should ask about scientific theories: "'What are scientific theories?' and 'How do theories function in various scientific activities'" (1988, p. 62).

Proposing that theories in meta-mathematics are very similar to theories in science is to pursue the former structural issue unconstrained by the latter procedural issue. Most holders of the semantic view, despite their position on the relative status of meta-mathematics and science, do have a story to tell about the relation between models and empirical systems. The semantic view is an attempt to give a general account of the nature of scientific theories, and to give an account that only worked for classical mechanics would not be sufficient.

Isomorphism is a relation between mathematical structures. If there is a function that maps each element of one structure onto each element of another the structures are isomorphic. What is more useful is the idea that some structures can be isomorphic to a sub-structure of a larger structure. A case of this relation was introduced in the geometry example, where we saw that the triangle is embedded in the euclidian plane as it is isomorphic with a substructure of the theory of Euclidian geometry. When we turn from the mathematical to the scientific case there are some problems with the use of isomorphism.

Reconsider the following schema:

{empirical system} =?= MODEL ===> {theory, equation,...}

The operator =?= is generally taken by semantic theorists (Suppes 1967, Suppe 1977, Van Fraassen 1980a) to stand for isomorphism, but scientific cases strain the clear mathematical sense of isomorphism. Certainly when dealing at the level of the relations between parts and levels of a theory isomorphism is a useful relation. As long as the theory is expressible mathematically and has clearly delineable models, then its sub-theories and lower level theories can be shown to be isomorphic to sub-structures of the larger theory. This is clear in the classical mechanics case.

Where the semantic view needs isomorphism to do most of its work is between models and empirical systems. Lloyd, who is aware of the difficulties with the notion of isomorphism (1988, p. 14), gives an example of how the relation works in the logistic equation case. As we saw above she claimed that if the growth curve for an empirical system, say a population of yeast, was the same as the curve from the logistic equation, then the empirical system and the model are isomorphic. But here Lloyd herself notes "in practice, the relationship between theoretical and empirical model is typically weaker than isomorphism, usually a homomorphism, or sometimes an even weaker type of morphism" (1988, p. 14 fn. 2).

The kind of relation that exists between empirical systems and theoretical models is simply not the isomorphism of mathematics. There are a few reasons for the tendency to hang onto the term isomorphism. One stems from Van Fraassen's anti-realism. Van Fraassen (1980a) speaks of empirical systems rather than real systems, so that he can consistently argue for his view that theories need only be empirically adequate. Rather than theories mapping onto real systems, they map onto empirical structures derived from observations. So the isomorphism at work is always between well defined structures.

A further reason for the usage is that holders of the semantic view have concentrated on highly mathematical theories. For example, the empirical structures that Van Fraassen refers to are expressible entirely mathematically. If, by contrast, we consider what relation there is between a schematized cell model and what I observe of a stained piece of muscle tissue through a light microscope, we have a case where the application of the notion of isomorphism looks very strained. And yet this is closely analogous to the relation that holds between the curve for the logistic equation and the curve for the yeast population.

Giere (1988) believes he has an answer to these kinds of problems. He substitutes the term "similarity" for isomorphism. Models are similar in various respects and degrees to empirical systems (1988, p. 81). But Giere is also more sanguine about real systems. He invokes similarity between models and real systems, because he is a realist. Without passing judgment on whether models are similar to real systems or empirical systems, we can still assess what is gained by introducing similarity as the candidate relation for =?=.

Giere's introduction of the notion of similarity gets him out of the difficulties associated with the strong relation of isomorphism, at the expense of distancing his account of science from any account of logic or meta-mathematics. At best we can say that the use of models in mathematics inspired the semantic view of theories, but we are not entitled to any of the strong claims that models in science and mathematics are the same.

What appears to be the case is that on the semantic view one needs an account of the relation between theoretical and empirical models that can handle all the kinds of models that occur in science. In many cases neither the models nor the empirical systems are, or can be, expressed mathematically, so isomorphism will not do the trick. Giere's similarity relation appears to solve some of these problems, so for now let us assume that this is the kind of relation we are looking for.[2] I now turn to a specific set of criticisms that lead to a deflationary version of the semantic view of theories.

4. A Deflationary Semantic View

Several holders of the semantic view (Suppes 1960, Suppe 1977, Van Fraassen 1980a) propose a general account of the nature of scientific theories. My illustrations of the difficulties with the notion of models above leads to the following question: Is such a strong generalized version of the semantic view tenable? I will address this question by showing how different versions of the semantic view arise when different sciences are examined. I then ask whether holders of the semantic view should be satisfied with a more piecemeal and descriptivist version.

Suppes (1960), Van Fraassen (1980a) and Suppe (1977) share the view that an important part of the philosophical investigation of science is to give an account of the structure and content of scientific theories. Van Fraassen refers to this task as "foundational work" (1980a, p. 2). Much of what was at stake in the initial presentations of the semantic view of theories was to provide an analysis that would answer all the questions logical empiricists had about the nature of theories in a more coherent fashion. Marc Ereshefsky (1991) has clearly laid out the issues at stake here. There is a problem with emphasizing this motivation for the adoption of the semantic view, which is that the explanatory agenda is not being laid out by explananda in science, rather by explananda that were of interest to logical empiricist philosophers. When Ereshefsky assesses the semantic view he does so by contrasting it with the received view thus reinforcing these terms of contrast. For, although he is not a defender of the received view, one could conclude from his criticisms that a return to the received view is in order as the semantic view does not hold up.

An alternative motivation for the semantic view, emphasized by Giere (1988) and Lloyd (1988), and more recently by Griesemer (1990), is that the view provides a more adequate description of theories as they are represented in actual scientific practice. This motivation leads to a proliferation of views under the rubric of the semantic view. Philosophers of science closely following a particular piece of scientific theorizing present an account that most closely resembles that particular science. This is fine if no general account of the structure and content of scientific theories is required, but if it is, then there is a problem. Let us first consider some examples.

Lloyd (1988) presents a detailed analysis of evolutionary theory, specifically population genetics. Her aim is to demonstrate the logical structure of evolutionary theory by elucidating the structure of its models. In doing this Lloyd claims that she "demonstrate[s] the usefulness of a precise analysis of the structure of evolutionary theory" (1988, p. 23), and that she offers "further evidence for the appropriateness and utility of the semantic view of theories" (1988, 23).

The account Lloyd presents closely follows those in the biological literature. For example, her account of the general structure of population genetics closely follows Lewontin's (1974). Certainly such close attention to the detail of biological theorizing produces useful results, in that Lloyd provides new and useful insight into many problem areas in both philosophy and theoretical biology, especially the units of selection issue. The question is to what extent her approach is generalizable, and, more specifically, generalizable as a version of the semantic view of theories.

Lloyd emphasizes, as do Suppes and Van Fraassen, the importance of mathematical models, and the use of a precise formal approach. In the kinds of cases these authors consider this is a reasonable emphasis. The problem is that, although Lloyd (1988, p. 16) claims that the semantic approach does not attempt to delineate scientific from non-scientific theories, there is something of an implicit demarcation criterion at work here. As we saw in the discussion of models above, there are distinctly non-mathematical models in science. Elucidating theorizing involving these models cannot rely on a formal approach of the kind Lloyd presents, and even less on such a specific formal approach involving state spaces, laws, and parameters.

What is at stake here is not the value of Lloyd's contribution to understanding evolutionary biology, rather the integrity of a strong version of the semantic view of theories. Certainly she holds that model construction is crucial to our understanding of theorizing, but this is a much weaker position than one that says that all theories are families of mathematical models. It is my contention that gaining descriptive and explanatory insight into scientific theorizing is an adequate goal for philosophers of science, even in the absence of a general account of *the* nature of scientific theories. Let us now look at some more examples to try and strengthen the claim that there is no one strong semantic view of the nature of theories.

Giere's discussion of linear oscillators that we considered above is clearly a candidate for a formal analysis in terms of mathematical models. This example can be quite easily accommodated in a state space type account such as Lloyd's, this was shown previously by Van Fraassen (1980a) in his discussion of classical mechanics. If this was the only kind of case Giere (1988) examined it would perhaps lend support to the idea that there was one semantic approach, but Giere examines several different cases notably the revolution in geology.

Giere's (1988) version of the semantic view contains a twist that no previous semantic theorists endorse, which is a certain kind of naturalistic approach with an emphasis on cognitive science. This leads to a confusion over just exactly what models are for Giere. When he discusses classical mechanics he claims that scientific theories are families of models, and that models are constructed abstract entities. So far he directly follows Van

Fraassen. He goes on to say that models are "socially constructed entities" (1988, p. 78). Later, Giere introduces models as kinds of mental representations. So when he discusses the geological revolution he refers to the "mental models" various individual geologists possessed. Here he leans on accounts of representation derived from cognitive science. Certainly at this point we have a proliferation of views on the table.

Models can now be mathematical objects, which are socially, or otherwise, constructed entities, and mental representations of some sort. Now in Giere's geological case study there were two large scale crude models, the static model and the mobile model. Each of these consisted of families of more detailed models of various geological phenomena, for example on the mobilists account a clear pictorial model of sea floor spreading was developed (Giere 1988, pp. 249-270). These kinds of models, irrespective of whether they are mental models, are non-mathematical models, and Giere's investigation of them is not a formal analysis such as that of Lloyd, Van Fraassen or Suppes.

Giere is obviously working with a deflationary version of the semantic view. His notion of models extends over all manner of objects, including the mental models of the cognitive psychologists (e.g., Johnson Laird 1983). This is strategic on his part, as he not only wishes to dissociate himself from logical empiricist philosophy of science, but also from approaches that rely only on a formal analysis of the structure of theories. Giere's naturalism leads him to investigate scientific theorizing in all its forms. Although Giere claims that "theories are families of models", his view could be more adequately characterized as the deflationary: model construction is an important component of scientific theorizing. Far from following through the alleged similarities between scientific theories and those of meta-mathematics, Giere brings in as many interpretations of the term "model" as suit his ends. I will comment on his reference to mental models at the end of the paper, but first I provide a third example of a proponent of the semantic view who contributes to the proliferation of semantic views.

Griesemer (1990) used a case study of the naturalist Joseph Grinnell to "enrich the semantic conception of theories" (1990, 11). Griesemer adds the following two important model types to all the above models involved in scientific theorizing. The first are entities as models, of which one of the most famous is Watson and Crick's wire and bead model of DNA. These kinds of models are different from the schematized cell referred to above, because their representational capacities are enhanced by our ability to physically manipulate them. The second type of model, and the most important for Griesemer's paper, are "remnant models". An example of this kind of model is a museum specimen of an animal or plant. Griesemer's introduction of these further types of model are a result of his move away from specifying theories in terms of the kinds of logical structures they are, and towards accounts of scientific theorizing. This is the move I advocated at the outset of the paper, one that Giere makes to a certain extent, but Griesemer pushes further.

One puzzling aspect of Griesemer's account is his insistence that he is enriching the semantic view of theories. By the time we have admitted laboratory specimens and physical objects to the domain of models, the idea that theories are families of models becomes quite inclusive. The less liberally inclined might be moved to claim that the semantic view is a non-view, because just about anything counts as an extension to it, or an enrichment of it. My approach is to claim that there is no such thing as one semantic view of theories, but work on such an approach led us to a fruitful way to study scientific theorizing and that is to investigate model constructing practices.

This is almost the position Griesemer arrives at, but not quite. His emphasis on theorizing as a practice is evident in the following passage:

> We can ... distinguish two semantic routes to theory introduction, one in terms of abstract models and one in terms of material models. On the classical semantic view, a theory is specified by defining the mathematical structures that satisfy the propositions of the theory. On an extended semantic view, a theory can be specified by defining a class of models as the set of physical structures constructed according to a given procedure or tradition. (1990, p. 12)

A clear way of lessening the contrast between these two views, is simply to re-phrase the definition of the classical view in terms of socially constructed entities also. Then all investigations of scientific theorizing involve the investigation of the practice of constructing models. And also on this view all investigations would have to go beyond the individual theorizer as a locus, as Griesemer (1990, p. 5) has pointed out.

What we have seen from the three examples is not so much a series of extensions to one semantic view of theories, but a set of alternative proposals for the investigation of scientific theorizing. Lloyd's proposal is to pay close attention to the formal structures produced by theorists, Giere's is to develop an account of theorizing as the manipulation of mental models, and Griesemer's to account for theorizing that involves the practices of using physical objects as models. I claim that what is common to all these views is best understood as the somewhat weak claim that model construction is an important part of scientific theorizing. This claim is not weak when we consider the progress various versions of the semantic view have made over the received view. The key mark of this progress is the ability for philosophers to provide insight into detailed and specific cases of scientific theorizing. What could be considered the chief source of weakness is the lack of a general account of the nature of theories. I will close with some concluding remarks about the potential fruitfulness of the deflationary account for naturalized philosophy of science.

5. Conclusions

My criticisms are not an outright rejection of the semantic view of theories, in fact I embrace some of the insights of the semantic view, but only in so far as they are contained in the deflationary semantic view. That is that model construction is an important part of scientific theorizing. What lies behind my criticisms is a view of how to approach issues in the philosophy of science. Any reconstruction of a domain of scientific practice in terms of identity statements such as "theories are families of models" can only give us limited purchase on making sense of science. The deflationary version of the semantic view of theories is an attempt to leave room for accounts of types of scientific theorizing that may not fit the model building mould. The strong view has to provide an analysis of scientific theorizing that forces any and all types of theorizing into this mould. Although I happily acquiesce to the claim that much theorizing in science involves the development and testing of models, I do not agree that a sufficient account of scientific theorizing is allowed by the claim that all scientific theories are simply families of models.

What is implicitly rejected on my view is that there are a special set of issues, referred to by Van Fraassen (1980b) as internal issues, whose investigation supplies sufficient activity for philosophers of science. Once theorizing is focussed on, rather than theories as completed formal entities, distinctions between internal and external issues become hard, if not impossible to make. On the deflationary view, philosophers of science form a loose confederacy for studying scientific theorizing, gathered around the common insight that model building is one of the most important components of such theorizing.

This kind of approach, which I take to be naturalistic, opens up many avenues of inquiry. As Griesemer has pointed out, echoing many sociologists of science before him, the examination of scientific theorizing will require an examination of the social nature of scientific practice. Giere's focus on the notion of models as mental representations also deserves attention. It does not appear to be the case that an account of models as simply some kind of mental representation will do justice to scientific theorizing, as the Griesemer case should indicate, but Giere's emphasis on the psychological does point to an important set of problems about scientific theorizing. For example, it is not clear just exactly what the representational status of theories is, or if they are representational at all. Certainly models are representations in some sense, but even their representational status is unclear. This is emphasized by the difficulty of finding a suitable alternative to isomorphism, and whether, once arrived at, such an alternative will apply to mental models in the same way it applies to socially constructed abstract models. Further, it is an open question just exactly how much scientific theorizing can be usefully and fruitfully carried out by individual scientists without the assistance of their peers. This leads to questions about our individual cognitive capacities, and how they are enhanced by the addition of other theorists.

If one pays attention to only the fine detail of the formal construction of theories, I contend that the above issues remain unaddressed. It could be argued that the formal investigations are sufficient, but I propose that they will only be useful if they go hand in hand with the other kinds of investigation encompassed by a naturalistic approach to the philosophy of science.

Notes

[1]I am grateful to Arthur Fine and Todd Grantham for discussing the semantic view of theories with me, and to David Hull and Tom Ryckman for comments on drafts of the paper. Financial support was provided by a post-doctoral fellowship from Northwestern University.

[2]Space prevents me from going into the reasons for rejecting Giere's similarity relation, which is also inadequate for its proposed task. Suffice to say that similarity avoids some of the problems associated with isomorphism, but may bring with it many more of a different nature. Cummins' (1989) discussion of similarity as a representation relation provides a good introduction to some of these problems.

References

Churchland, P.S. (1986), *Neurophilosophy*. Cambridge, Mass: MIT Press.

Cummins, R. (1989), *Meaning and Mental Representation*. Cambridge, Mass: MIT Press.

Ereshefsky, M. (1991), "The Semantic Approach to Evolutionary Theory", *Biology and Philosophy* 6: 59-80.

Giere, R.N. (1988), *Explaining Science: A Cognitive Approach*. Chicago: University of Chicago Press.

Griesemer, J.R. (1990), "Modeling in the Museum: On the Role of Remnant Models in the Work of Joseph Grinnell", *Biology and Philosophy* 5: 3-36.

Hull, D. (1988), *Science as a Process*. Chicago: University of Chicago Press.

Johnson-Laird, P.N. (1983), *Mental Models*. Cambridge, Mass: Harvard University Press.

Lewontin, R.C. (1974), *The Genetic Basis of Evolutionary Change*. New York: Columbia University Press.

Lloyd, E.A. (1988), *The Structure and Confirmation of Evolutionary Theory*. New York: Greenwood Press.

Nagel, E. and Newman, J.R. (1958), *Godel's Proof*. New York: New York University Press.

Simon, H.A. (1954), "The Axiomatization of Classical Mechanics", *Philosophy of Science* 21: 340-3.

Suppe, F. (1977), *The Structure of Scientific Theories*. Urbana: University of Illinois Press.

Suppes, P. (1960), "A Comparison of the Meanings and Uses of Models in Mathematics and the Empirical Sciences", *Synthese* 12: 289-95.

_____. (1967), "What is a Scientific Theory?", in Philosophy of Science Today, S. Morgenbesser (ed.). New York: Meridian, pp. 55-67.

Van Fraassen, B. (1980a), *The Scientific Image*. Oxford: Clarendon Press.

Van Fraassen, B. (1980b), "Theory Construction and Experiment: An Empiricist View", in *PSA 1980*, volume 2, P.D. Asquith, and R.N. Giere (eds.). East Lansing: Philosophy of Science Association, pp. 663-77.

Towards an Expanded Epistemology for Approximations[1]

Jeffry L. Ramsey

Rice University

1. Introduction

Since "one can seldom directly deduce from a speculation consequences that are even in principle testable," a scientist must "articulate" a theory, mathematically altering a given speculation to bring it "into greater resonance with the world" (Hacking 1983, p. 214). When confronted with computational difficulties caused by analytically intractable equations, imprecise specifications of initial conditions, or the absence of required auxiliary theories, scientists commonly articulate a theory by applying approximations and idealizations to the theoretical equations they have at hand. Recognizing this, the following question immediately arises for both the scientist and the philosopher of science, "How does one judge the quality of a particular approximation? I.e., when is a given approximation valid?"

After viewing approximations "as mere distractions, complications to be explained away in some future, more developed theory of science" (Laymon 1989, p. 353), philosophers of science have recently begun to reflect more critically on the role of approximations in scientific activity. While we are wiser to the ways in which various philosophical arguments must be altered in response to the presence of approximate data and unsolvable equations, we have as yet few critical tools available to assess the validity of an approximation itself. To evaluate an approximation, most authors place a limit on the permissible discrepancy between theoretical and experimental values. In this paper, I argue that such a discrepancy is a necessary but by no means sufficient criterion for judging an approximation's validity. Other important criteria include: controllability;[2] the absence of errors which cancel each other; and the justification given for the approximation.

I also argue that authors who rely on the size of the discrepancy as the criterion of evaluation do so because they characterize approximation largely as a resemblance of something with something else. More concretely, an approximation is theorized as a relation of some structure A to another structure B. To develop a more complete understanding of how scientists employ and evaluate approximations—that is, to understand more fully the epistemology of this methodologically important feature of scientific activity—I advocate a focus on the practice of approximating. When performing mathe-

PSA 1992, Volume 1, pp. 154-164

matical computations, a scientist often needs to make an approximation. In other words, an approximation is often an act before it is a relation. This focus allows me to argue that judging the worth of an approximation requires one to take into account more than the size of the discrepancy. When the three criteria mentioned above are also included in the judgement, the result is an expanded epistemology for approximations.

In Section 2, I analyze current views on approximations. In Section 3, I diagnose more fully the second sense of approximation just mentioned and show how this relates to the additional criteria for calling an approximation "valid." In Section 4, I support my points with an elaboration of Ludwig Boltzmann's (1898) discussion of the kinds of approximations used in the kinetic theory of gases. Finally, I reflect briefly in Section 5 on what an expanded epistemology for approximations offers for the philosophy of science more generally.

2. Current Views of Approximations

Many philosophers interested in approximations rely on a single criterion for judging the worth of an approximation. Depending on who you read, the criterion is cashed out in metrical or non-metrical terms, but the central idea is that an approximation is considered a good one if the discrepancy between a theoretical prediction and an experimental result is at or below a certain limit. The application of this criterion depends heavily on characterizing an approximation as a comparison of previously generated theoretical and experimental results. That is, the appropriate sense of "approximation" is taken to be a resemblance between two independent results or structures. The discrepancy criterion is then applied to determine whether the two structures are similar enough for the purposes at hand. In this section, I survey views which rely on this criterion and characterization.

Structuralists (e.g., Balzer, Ulises-Moulines and Sneed 1987) and philosophers of science influenced by the structuralists (Laymon 1987, Lloyd 1988, Redhead 1975) have developed perhaps the most sophisticated view of approximations. They typify the focus on discrepancy as the criterion of validity for an approximation and on approximation as a comparison between two states of affairs. On this view of approximation, the comparison relation lies entirely outside the theoretical structure of the science as practiced.

Structuralists explicate the discrepancy in metrical or numerical terms. In their theory of approximations, they construct a numerical relation, $\in$, which assesses the amount of approximation between any two statements in a given language. The acceptability of a given statement as an approximate confirmation, explanation or representation of another statement depends on whether the $\in$ value for the two statements falls within an acceptable range. However, when it comes to specifying what counts as an acceptable range of discrepancy, structuralists resort to hand-waving and vague generalities. For example, Balzer, Ulises-Moulines and Sneed (1987, p. 347) claim that "the choice of the particular $\in$ will certainly depend on the particular application we [the scientists] have in mind and on some changing pragmatic factors," but they do not discuss any further what might constitute such dependencies or factors. They note only that any disparity beyond the upper bound of the relation will be "considered unbearable" (Ibid.).

As might be clear from this, the structuralist approach is committed to viewing an approximation as a static relation of comparison between two structures. The relation of approximation—not to mention the acceptability of a given approximation—is determined *post facto* after all calculations have been performed. Furthermore, the structuralist approach locates the approximation relation outside the structure of the theories involved. One speaks of the approximate reduction of one theory to another, the approximate explanation of laws by a theory, the approximate truth of a theory, etc. (Niinuluoto 1986).

In developing a version of the semantic view of theories, Giere (1988) also adopts a view of approximation which judges the quality of an approximation on the size of the discrepancy between theoretical and experimental results and takes approximation as a static relation largely external to theorizing and problem solving. In contrast to the structuralists, Giere explicates the discrepancy between theoretical and experimental results qualitatively. In his approach, the approximation is acceptable if the theoretical result—which for Giere is generated within a model—is similar enough in respects and degrees to the experimental result. While he differs from the strict structuralists in such respects as emphasizing physical rather than logical models as the locus of scientific activity, Giere thus also relies on a single criterion to discuss approximations. His criterion is similarity. "My suggestion, of course, is that the notion of *similarity* between models and real systems provides a much needed resource for understanding approximation in science. . . . (Similarity) reveals—what talk about approximate truth conceals—that approximation has at least two dimensions: approximation in *respects*, and approximation in *degrees*" (Giere 1988, p. 106, emphasis in original). What Giere's talk about dimensions conceals is the fact that, as Aronson (1990) rightly points out, the underlying characterization of an approximation remains a comparison between two states of affairs.[3] In this sense, he differs in inessential respects from more logically-minded semanticists.

Further, Giere and the structuralists are in agreement that an approximation is a comparison of previously generated theoretical and experimental results. While the structuralists offer only one dimension of justification for an approximation, Giere offers two dimensions along which the model and the data or world may be judged for their approximate correctness: similarity in degrees and in respects. Nonetheless—and this is the important point of similarity between the two approaches—the way one generates a judgement that something stands in a relation of approximation to something else is by comparing one existent structure with another existent structure. (One also uses this static comparison when deciding the acceptability of the approximate representation.) Thus, for Giere as for the structuralists, an approximation is a static relation which is discovered outside the theoretical structure of the science.

While the semantic theorists and Giere opt for a view of approximation which places the approximation external to the theoretical structure, Laymon (1983, 1984, 1987, 1989) and Cartwright (1983) adopt a view which gives a role for approximations in the theoretical structure or, more weakly, in the computational structure of the theory. However, both tend to rely too heavily on the comparison view to generate assessments of an approximation's validity.

Cartwright and Laymon are interested in providing an analysis of scientific practice which makes sense of the difficulties encountered by scientists. As a result, they see an approximation as a strategy which scientists employ to overcome computational problems of various sorts. "Because of analytical and computational shortcomings, and the absence of necessary auxiliary theories and data, idealizations and approximations nearly always must be used to obtain *real* computability" (Laymon 1987, p. 197, italics in original). He recognizes that approximations and idealizations are embedded in layers of argumentation. As a result, to judge whether the theory is confirmed or not, he focuses on the "piecemeal improvability" (Laymon 1989, p. 353) of idealizations and approximations rather than simply the size of the discrepancy between the theoretical prediction and experimental result. Similarly, Cartwright is less interested in the magnitude of the deviation and more interested in the ways in which approximations are used. She focuses on how approximations actually improve theoretical derivations from fundamental laws, arguing that such improvements can be made deliberately (as in the case of the design of amplifier circuits) or fortuitously (as in the case of the Weisskopf-Wigner and Markov approaches to exponential decay).

Although Cartwright and Laymon are far less likely to judge approximations solely in terms of the size of the discrepancy, both retain the comparison view in their assessment of an approximation's validity. Laymon's (1987) endorsement of the semantic theory (in roughly the same way Giere uses it) and his focus on "idealized and approximate data" (cf. Laymon 1984, 1987) suggest strongly that he relies on the comparison view of approximation. Cartwright does not talk about how to assess the validity of the approximation itself but rather about how to assess the worthiness of the explanatory conclusion drawn from the computation which employed the approximation. Her simulacrum account of explanation is nothing less than a full-blown comparison view.

In sum, "approximation" is often theorized as a comparison of something with something else to see if the two things resemble each other, are similar qualitatively, have similar quantitative values, etc. This characterization is encapsulated in the claim that "every theory is only 'approximate' in respect to nature itself" (Truesdell 1984). Whether the approximation is valid or not is determined largely by appeal to the magnitude of the discrepancy between the two things. In the next section, I will argue that talk about approximations is not exhausted by talk about how one structure compares to another and thus by talk about the magnitude of the discrepancy.

3. Another Sense of Approximation

I have delineated how a popular view of approximations relies on a single criterion for judging the worth of an approximation and how this criterion is usually used in conjunction with a particular characterization of an approximation. In this section, I follow Cartwright and Laymon in adopting a characterization which stresses the process of making the approximation. Once this alternate characterization is recognized, the need for additional criteria to judge the validity of an approximation becomes clear. The assessment of an approximation's validity thus becomes a matter internal to the theoretical structure rather than just a comparison of the theory with the data or a model.

An "approximation" can be an product or an act. The product sense of the word is linked to the adjectival use of "approximate"; the act sense to the use of "approximate" as a verb. Used as an adjective, one principal meaning of "approximate" is "closely resembling." The authors discussed in the previous section utilize this meaning. Used as a verb, "approximate" means "to carry or advance near; to cause to approach (to something)."[4] I will utilize this meaning since it allows me to emphasize the process by which an approximation is effected. An approximation is an act or process and not just a relation. An approximation thus becomes "any methodological strategy which is used to generate or interpolate a result due to underresolved data or deficits of analytic or calculational power" (Ramsey 1990b, p. 485).

Why adopt this alternate sense of approximation? My own reasons stem from my study of how scientists produce solutions to analytically intractable equations (see Ramsey 1990a, 1990b). More generally, this second sense is needed because the act of producing a structure which can be judged as relevantly similar to something else exists prior to a metric or a similarity judgement. One could dispense with an account of this act, but then we are left—literally—comparing apples and oranges. We could count properties, but we are stymied if we do not have an account of why some properties are relevant or important. More to the point, an account of the act is especially important in the mathematically-oriented sciences since there are few if any properties we can count to produce a judgement of similarity between an equation and a thing.[5]

Why does this sense require more criteria for judging the validity of an approximation? The reason is rather simple: by themselves, the deviation or consilience of two results tells us nothing about the worth or reliability of the procedure which produced the results. This can be illustrated quite easily by what I call the "bathroom scale" problem. Suppose you know that the scale in the upstairs bathroom is not reliable; one day it makes you three pounds lighter and the next day it makes you two pounds heavier. Now, suppose you go out and stuff yourself at a smorgasbord. You go home and weigh yourself, and the scale tells you that you weigh just what you have always weighed. What can you conclude? Nothing. You need to know something about the reliability of the instrument before you can make a conclusion. E.g., does the scale tend to underweigh you on humid days? Translated into the language of science, you need to know something about the reliability of the theory and its calculational structure before you make any conclusions about the worth of a theoretical result which matches the experimental result exactly or within a specified range of error.

Although the example is rather weak,[6] it illustrates the following thesis: although the nearness of a theoretical result to an experimental result may be a necessary condition for judging the worth of an approximation, it is not a sufficient condition in and of itself. Again, the magnitude of the error introduced by the approximation is by itself only a very weak criterion for judging the validity of the approximation. In addition to an assessment of the magnitude of the error, at least the following criteria are needed:

1) controllability

2) the effect of one approximation must not be canceled by another approximation or by other factors

3) the approximation must be appropriately justified

The purpose of adding the three additional criteria are, I think, somewhat obvious. (1) and (2) place restrictions on a calculation which is only fortuitously close to the experimental result.[7] (3) places restrictions on how the approximation can be employed; for example, if one approximation is applied in order to produce an *a posteriori* fit to a curve and another is made *a priori* on the basis of some presumed causal factor, we should not judge the two approximations equally.

Laymon (1983, 1984, 1987, 1989) has advocated criterion (1) as necessary for understanding the confirmation and testing of idealized theories and idealized data. He has argued persuasively that "a theory is confirmed if it can be shown that better approximations lead to better predictions, [and] a theory is disconfirmed if it can be shown that better approximations do not lead to better predictions" (Laymon 1987, p. 211). However, although criterion (1) is important, it is not sufficient by itself to secure the validity of the approximation strategy. Why? If the discrepancy between the theoretical and experimental values is small, and if better approximations do lead to better predictions, it is still possible that a set of approximations which can not be relaxed have counteracted each other to produce fortuitously the near agreement and the monotonic behavior.[8] That is, criterion (2) may be violated. One needs an independent argument that (2) is satisfied before assessing the worth of the approximation strategy.

Likewise, criteria (1) and (2) are not sufficient by themselves to secure the validity of an approximation strategy. If an approximation strategy is applied *a posteriori* to an equation in order to produce a fit to a curve, we are not likely to consider the strategy particularly satisfying even if it produces good agreement, does so in a controllable fashion, and we think there are no canceling errors. Although we may take the strategy's success as an indicator that we should look more closely at how it achieves that success—e.g., we may want to invest some of the mathematical terms with ontological import (Cartwright 1983)—the success itself says little about the validity of the strategy. We can say little about the worth of an approximation strategy when good theoretical motivations for its employment are absent.

These three criteria taken together with numerical accuracy may or may not form a sufficient set for judging the validity of any given approximation; I leave this as an open question. Nonetheless, they provide us with an understanding of the conditions for the possibility of using error magnitude as a criterion of the worth of approximations. As such, they provide a much richer language for studying how approximations are employed and judged. In short, they provide us with an expanded epistemology for approximations.

4. Boltzmann's Typology of Approximations

Ludwig Boltzmann's (1898) important but widely ignored typology of the approximation strategies used in the kinetic theory of gases at the turn of the century will allow me to illustrate the utility of the criteria and the characterization adumbrated in the previous section.

In developing the kinetic theory of gases, Ludwig Boltzmann (1898) divided approximations or omissions (Vernachlaessigungen) into two types or kinds (Arten): those which neglect terms of a different order of magnitude of the terms in the final result, and those which neglect terms of the same order of magnitude. That is, he differentiated approximations by the kind of variables omitted. For Boltzmann, an approximation was a procedure first and a relation second. The purpose of making an approximation was to solve equations exactly. Comparison of the theoretical and experimental results came later. Also, Boltzmann was clearly attuned to the fact that we judge the quality of an approximation by attending to more than just the magnitude of the error it introduces.

Boltzmann's first type (Type 1) involves the neglect of terms of a smaller order of magnitude of the terms in the final result. Examples in the kinetic theory of gases include assumptions that the duration of collisions between molecules will be small compared to the time between collisions and that the distribution law for molecular velocities is precisely correct. In this case, the results will be physically inexact since they rely on calculations whose boundary conditions are not strictly met. Since the mathematical terms are inexact, "(t)he exact validity of the results is not to be confirmed by means of experiment."[9] At best, the theoretical and experimental or observational answer are "close." However, the calculations will be mathematically correct since they provide a limiting case as the physical assumptions are realized more exactly, i.e., less approximately.

Boltzmann makes a clear appeal to the controllability of the approximation. One knows the distortions introduced into the equations, and the formulae become "more nearly exact" as the assumptions which cause the distortions are realized in a given physical system. Boltzmann actually makes only a secondary appeal to the magnitude of the error caused by the assumptions. While it is important that the theoretical and experimental answers are close, it is more important that we can perform the calculations and that the answers we get approach the ideal situation in a regular fashion.

The second type of approximation (Type 2) involves the omission of terms of the same order of magnitude as the final result. These omissions make the results mathematically incorrect in the sense they are not logical consequences of the assumptions made. In the kinetic theory of gases at the turn of this century, for example, calculation of a quantity Q (such as electrical conductivity, viscosity or heat) transported by a molecule through any given surface depended on the following approximating assumptions:

a) the velocity distribution is not altered by the nature of the quantity associated with the molecules,

b) all directions of the motion of a molecule are equally probable, and

c) the amount of the quantity transported by the molecule through the surface is independent of the molecule's direction and velocity.

In the mathematical equation used at the time, (a) and (b) were neglected and researchers assumed arbitrarily that the only relevant factor in (c) was the amount of the quantity the molecule possessed in the layer just prior to the collision.[10] If these effects were included in the mathematical expression, various differentials (such as the dependence of the quantity on the vertical displacement of the layer) could not be taken outside the integral sign. The resulting integral was intractable. When trying to use such complex equations, "terms which are of the same order of magnitude as those which determine the result" had to be neglected to produce a solution; as a result, formulas which took the extra factors into account were "not essentially better than the ones obtained here in a simpler way" (Boltzmann 1898/1964, p. 105).[11]

Here, tractability and simplicity are offered as justifications for preferring one calculation over others. In the face of impending computational complexity, more conceptual and numerical accuracy is not necessarily better. What one wants is a tractable model which, although it produces large errors, is faithful to at least some of the physical processes known to be at work in the system being studied. Such models are to be preferred over models with unsolvable equations, solvable models with terms that are not connected to any known processes, or models which contain correction terms or "fudge factors" which must be adjusted on a case-by-case basis. If the equations can not be solved, no empirical assertions can be generated and no judgement regarding the adequacy of the model is possible. If the parameters do not represent known processes, the model becomes purely instrumental and no explanation of the physical system can be offered. Finally, if the parameters of the model must be adjusted each time the model is employed, the approximation which produced the model is clearly not controllable.

The following example illustrates the need to ascertain that two approximations do not cancel each other out and produce a precise result fortuitously. Consider a viscous fluid in which molecules leave the collision layer with different directions and with different velocities; further, this fluid is subjected to a constant shearing stress. We will have to assume (falsely) that Boltzmann could produce exact solutions to the equations representing such a situation.[12] Because the viscous fluid is being deformed continuously, assumption (a) is not met; the velocity distribution deviates from the Maxwell-Boltzmann equilibrium. In fact, the mean velocity will be lowered somewhat slightly. Correspondingly, the amount of quantity Q transported through a specified layer will be lowered. However, because the molecules leave the collision layer with different directions and velocities, a greater amount of quantity Q than previously calculated will be transported from the collision layer.

The upshot is that we have a technical example of the "bathroom scale" problem described previously. Here, the calculations with and without the two original approximations could give the same value. Without an independent investigation into each approximation, we cannot claim to know how or why the calculation employing the approximations actually agreed with the experimental result. That is, we need independent arguments for the controllability and the justification of each approximation. Otherwise, the agreement produced in the original calculation is fortuitous.

5. Consequences of an Expanded Epistemology for Approximations

I take the aim of the philosopher of science to be both critical and interpretive. One hopes to intervene in scientific activity by diagnosing when and where judgements have been made correctly and incorrectly, and one hopes to understand how such scientific judgements are made.

One reason for moving to a dynamic view of approximations and an expanded set of criteria is that both the critical and interpretive tasks of the philosopher are advanced. This is particularly true when studying instances where scientists produce solutions to analytically intractable equations via the application of some kind of approximation strategy. As the discussion of Boltzmann's methodology shows, scientists judge the validity of a given strategy by appeal to the magnitude of the error in conjunction with a number of other issues. Magnitude is a necessary but not sufficient condition to consider when praising or blaming a particular strategy.

With that understanding of how the judgement has been made, we can then move to a discussion of whether the judgement has been made correctly. If we appeal only to the magnitude of the error and think of an approximation as a comparison of two structures, we cannot evaluate Boltzmann's arguments accurately. We are now in a position to withhold praise from a strategy because, for instance, it might produce a fortuitous result owing to the cancellation of errors in successive steps of a calculation.

The interpretive and critical aims are also advanced on a more general level. For instance, more depth and structure is added to our notion of scientific activity when approximations are characterized dynamically. Hacking (1983, pp. 212-215) has proposed a tri-partite division of science as speculation, calculation and experiment rather than the more traditional view of science as a collection of theoretical and experimental statements. The view advocated here supports Hacking's distinction by focusing on some of the salient moves which can be made in the realm of calculation. As a consequence, the computational structure of a theory a far more important element for analysis than usually admitted. In general, discussions of theory structure in the philosophy of science have simply passed over the computational structure of a theory in favor of analysis in terms of theory and evidence. Practicing scientists know better. In speaking of three different formulations of the law of gravitation, Feynman (1967) noted that Newton's law, the local field method and the minimum principle are equivalent scientifically but not psychologically. However, we can now go one better than just psychological difference. If we consider the computational structure of the theory as a place where approximations are made, and if approximations can be judged by a number of criteria, then the computational structure is more than just psychologically important. It is epistemically important in the most traditional sense of the term "epistemic." The computational structure can affect what we believe about our theories and about the world.

Second, by paying close attention to scientists' talk about and employment of approximations, we uncover a systematicity with regard to the talk used to judge the quality of approximations. Scientists do not talk about approximations solely in terms of whether they produce results which are close to the desired result. Unless we expand the criteria by which we examine the use of approximations, we will cover up the systematic nature of scientists' remarks. Reflection on such systematicity may be the only way we can create a workable "logic" of approximation. If so, it behooves us to pay attention to the practice of making approximations.

An understanding of this systematicity will be invaluable when discussing how approximations affect issues such as realism, confirmation, truth, etc. As it stands, discussion of approximations typically revolve around the effects of the approximation on traditional philosophical questions. A far less developed theme is the assessment of the quality or validity of the approximation itself. Surely the latter issue is a necessary propaedeutic to the former. For instance, if we do not know when an approximation has been applied properly, we will be hamstrung in our efforts to ascertain whether the ontological entities affected by the approximation have legitimately or illegitimately been introduced on to or wiped off the face of the earth.

In sum, when praising or blaming an approximate result, it is not sufficient to consider only the magnitude of the discrepancy between the theoretical and experimental (or observational) results. How we got to the result is just as important as the fact that we got pretty close to where we wanted to go. Only a discussion of the process of approximating will provide the means to determine if the approximation strategy has been employed legitimately or not.

Notes

[1]Much of this paper is based on material in my dissertation (Ramsey 1990a). Many thanks to Bill Wimsatt, R. Stephen Berry, and Dan Garber for helpful criticism and comments on the dissertation.

[2]Laymon has argued for such a criterion when discussing issues such as confirmation, truth, and approximate data. See Laymon (1984, 1987, 1989).

[3]Aronson (1990) argues that Giere's notion of similarity is unanalyzed and must be unpacked into various sub-types of similarity. This is a crucial point, but Aronson joins Giere and the structuralists in characterizing approximation as a comparison. He notes, "... I am convinced that Giere is right about approximation being a matter of similarity between a model and a real system..." (ibid., p. 13). Because he remains with a comparison view, Aronson's views on approximation remain deficient from the perspective identified in this paper.

[4]Webster's *New Universal Unabridged Dictionary*, 2nd ed. (1983). New York: Simon and Schuster.

[5]Symmetry is an obvious example of a property we can use. However, even if you notice a symmetry between a mathematical structure and a physical thing, you still have to calculate the mathematical result.

[6]One major disanalogy between the bathroom scale and an analytically insoluble theoretical equation is that, in the former, the scale generates the result immediately.

You do not have to make any approximations to generate an assertion; you simply have to discover enough about the behavior of the scale under given conditions. In contrast, scientists must often approximate analytically intractable equations to be able to generate an assertion. Until the equations can be solved, knowledge of the system under given conditions will be of little use computationally (as long as one remains with standard calculational procedures) or will be of use only in so far as it suggests ways to solve the equations.

[7]The term "fortuitous" was introduced, as far as I can determine, by Teller (1983).

[8]Laymon (1984, p. 122, fn. 11) notes this possibility but does not explore its consequences.

[9]"ihre exact Bestaetigung durch das Experiment ist nicht zu erwarten." Brush (Boltzmann 1898/1964, p. 106) mistranslates this as "their [the results] exact validity can only be determined by experiment."

[10]Thus, researchers neglected the fact that the amount of quantity Q transported "can differ for molecules that leave the layer in different directions and with different velocities." If these factors would have been taken into account, "(t)he amount of Q transported by a molecule . . . would then depend not only on the layer where it last collided, but also on the place where it collided the next to last time, and perhaps also on the place of the collision before that" (Boltzmann 1898/1964, pp. 104-105).

[11] See Boltzmann (1898/1964, p. 105) for references to competing accounts in the work of R. Clausius, O. E. Meyer, and P. G. Tait. Brush notes that the difficulties were resolved by Chapman and Enskog in the 1910s (Boltzmann 1898/1964, p. 109).

[12]Today, we can solve such equations exactly due to the work of Chapman and Enskog. See Brush's "Translator's Introduction" and his footnote on p. 109 in Boltzmann (1898/1964).

References

Aronson, J. (1990), "Verisimilitude and Type Hierarchies", *Philosophical Topics* 18: 5-28.

Balzer, W., Ulises-Moulines, C., and Sneed, J. (1987), *An Architectonic for Science: The Structuralist Program.* Boston: D. Reidel.

Boltzmann, L. (1898), *Vorlesungen ueber Gastheorie.* Leipzig: J.A. Barth.

_ _ _ _ _ _ _ _. (1964), *Lectures on Gas Theory,* (trans.) S.G. Brush. Berkeley: University of California Press.

Cartwright, N. (1983), *How the Laws of Physics Lie.* Oxford: Clarendon Press.

Feynman, R. (1967), *The Character of Physical Law.* Cambridge: MIT Press.

Giere, R. (1988), *Constructing Science.* Chicago: University of Chicago Press.

Hacking, I. (1983), *Representing and Intervening.* New York: Cambridge University Press.

Laymon, R. (1983), "Newton's Demonstration of Universal Gravitation and Philosophical Theories of Confirmation", in *Testing Scientific Theories*, J. Earman (ed.). *Minnesota Studies in the Philosophy of Science*, volume X. Minneapolis: Minnesota University Press, pp. 179-199.

________. (1984), "The Path from Data to Theory", in *Scientific Realism*, J. Leplin (ed). Berkeley: University of California Press, pp. 108-123.

________. (1987), "Using Scott Domains to Explicate the Notions of Approximate and Idealized Data", *Philosophy of Science* 54: 194-221.

________. (1989), "Cartwright and the Lying Laws of Physics", *Journal of Philosophy* 86: 353-372.

Lloyd, E. (1988), *The Structure and Confirmation of Evolutionary Theory.* New York: Greenwood Press.

Niinuluoto, I. (1986), "Theories, Approximations and Idealizations", in *Logic, Methodology and Philosophy of Science* VII, R. Barcan Marcus, et al. (ed.). Amsterdam: North Holland, pp. 255-289.

Ramsey, J. (1990a), "Meta-Stable States: The Justification of Approximative Procedures in Chemical Kinetics, 1923-1947", unpublished Ph.D. dissertation, Committee on the Conceptual Foundations of Science, University of Chicago, May 1990.

_______. (1990b), "Beyond Numerical and Causal Accuracy: Expanding the Set of Justificational Criteria", *PSA 1990*, volume 1, M. Forbes and A. Fine (eds.). East Lansing, MI: Philosophy of Science Association, pp. 485-499.

Redhead, M.L.G. (1975), "Symmetry in Intertheory Relations", *Synthese* 32: 77-112.

Teller, P. (1983), "The Projection Postulate as a Fortuitous Approximation", *Philosophy of Science* 50: 413-431.

Truesdell, C. (1984), *An Idiot's Fugitive Essays on Science: Methods, Criticism, Training, Circumstances.* New York: Springer-Verlag.

Part V

PROBLEMS IN SPECIAL SCIENCES

Sociology and Hacking's Trousers[1]

Warren Schmaus

Illinois Institute of Technology

For Hacking, the word "real" is one of J.L. Austin's trouser-words, taking its meaning from its negative uses in much the same way as the admittedly sexist expression "wear the trousers". In *Representing and Intervening*, Hacking proposes that at least some scientific realists can be interpreted as using the word "real" in this way. The word "real" is also substantive-hungry, he adds. Thus when a philosopher states that a type of entity is not real we need to know just what is being denied. The causal entity realist, for example, holds that the entities scientists postulate in their theories can be regarded as real if they have causal powers that can be manipulated to create observable and repeatable effects. Thus the causalist offers sociology a possible way to command respect: if social science entities could be manipulated, the social sciences would then be on a par with physics. By endorsing this causalist position, however, Hacking is not trying to be friendly to the social sciences. In fact, he believes that the social sciences have not yet yielded entities that we can use to create new phenomena (1983a, 248-9).

However, as I will argue in this essay, there is an alternative way to defend the equal status of the social sciences. In what follows, I will criticize Hacking's attempt to apply his causalist entity realism to the social sciences and argue that in the philosophy of the social sciences, the realism issue concerns systems for classifying people and not unobservable theoretical entities. I will then show that Hacking reaches a nominalist position regarding classes of people by considering them in abstraction from the social systems of which they form a part. Hacking, I will argue, gives us no reason to believe that the objects of study of the social sciences are any less real than the objects of study of the natural sciences.

1. Causalism and the Social Sciences

For Hacking, the issue of scientific realism in the natural sciences concerns unobservable entities not as particulars but as representatives of certain kinds (1989, 562). His argument for preferring his causalist brand of realism about entities over J. J. C. Smart's materialist entity realism is that causalism is more inclusive. For the materialist, an entity is not real unless it is one of the "building blocks" of the material world. In order for a type of entity to be a type of building block, it seems, one must

PSA 1992, Volume 1, pp. 167-173

be able to individuate or at least to count representatives of this type. Thus Smart accepts the reality of electrons but not of lines of magnetic force (1963, 34). For the causalist, on the other hand, an entity is real if it has real effects. From this point of view, lines of magnetic force are just as real as electrons (Hacking 1983a, 37).

Causalism not only explains what it means to say that lines of magnetic force as well as electrons are real, however. Causalism also encompasses the social sciences and explains what it means for the entities that they postulate to be considered real, Hacking argues. Smart's materialism, on the other hand, Hacking contends, can in principle make no sense of the reality of entities in the social sciences (1983a, 35, 38-40). Now I am not sure that I can accept what Hacking says against materialism in the social sciences. However, I will not try to defend materialism against causalism, as I think that the more fundamental question is whether the realism issue in the social sciences at all concerns unobservable entities.

According to Hacking, one distinguishes the real from the unreal among putative entities in the same way in both the social and the natural sciences: that is, the real ones have real effects due to their causal powers. Putative entities in the social sciences, he thinks, can be criticized on a case-by-case basis with regard to whether we understand the causal mechanism by which they produce their effects (1983a, 38). Thus, for example, Hacking claims that Max Weber criticizes Marxian forces and tendencies on the grounds that they lack causal powers, Stephen Gould rejects the reification of IQ, and we may accept Durkheim's collective consciousness while rejecting Jung's collective unconsciousness (1983a, 38-9). Presumably, for an entity like Durkheim's collective consciousness to be real for Hacking would be for there to exist a method for manipulating it in order to bring about changes in society.

Hacking, I am afraid, is less than fortunate in his choice of examples. According to Hacking, Weber's criticism of Marx reveals a negative use of a causalist attitude towards social scientific laws. That is, Hacking asserts, Weber rejects Marx's concepts of "forces" and "tendencies" on the grounds that these things have no causal powers. (1983a, 39.) This interpretation, however, overlooks the fact that for Weber, the goal of social science is not practical intervention through the manipulation of entities but making social and historical facts intelligible to a particular audience of a certain culture.[2] Weber criticizes Marx on the grounds that he fails to achieve these goals.

Unlike Weber's, Durkheim's conception of the goals of social science does include an explanatory role for theoretical entities with causal powers. However, the "social forces" that Durkheim invokes to explain such generalizations as the "law" that Protestants have higher rates of suicide than Catholics are not always taken seriously. In fact, these entities are sometimes considered an embarrassment: Robert Alun Jones, for example, dismisses Durkheim's use of terms like "social forces" and "suicidogenic currents" as so much "obfuscatory language" (1986, 114). Even Hacking, although he recognizes the explanatory role of social forces in Durkheim's sociology (1990, ch. 18, 20), fails to appreciate that social forces for Durkheim arise from another kind of theoretical entities, which Durkheim calls "collective representations". Durkheim, educated as a philosopher, was steeped in the tradition of the way of ideas. Collective representations for Durkheim are a special type of mental entity that, because of the social conditions under which they are formed, are more "lively" than other kinds of mental representations and thus override them and affect behavior. According to Durkheim, this superior psychological energy of collective representations is thus responsible for social forces.[3]

Even when social scientists continue to use Durkheim's theoretical terms, these terms no longer have the same referents. For example, although David Bloor continues to use Durkheim's term "collective representation," it refers no longer to a type of mental entity as it did for Durkheim but to a node in a Quinean web of belief. Although Bloor accepts the Durkheim-Mauss hypothesis that classifications of things in nature reproduce classifications of people in society, he rejects Durkheim's attempt to explain this hypothesis in terms of the causal powers of collective representations. For Bloor, the Durkheim-Mauss thesis now means that in modern societies, scientists will defend a system of natural classification that somehow legitimizes a social arrangement that they would like either to maintain or to bring about (Bloor 1982).

The point of these examples is that sociological generalizations may take on a life independent of the theories that account for them in terms of postulated causal entities. Unlike physics, the social sciences today simply do not seem to be concerned with unobservable entities. In his most recent work, Hacking concedes that the realism issue is not necessarily about unobservable entities but in fact differs for each style of scientific reasoning, according as each style introduces its own class of objects. Thus, he says, there are realist disputes about abstract mathematical objects, biological taxa, languages and rates of unemployment as well as about unobservable theoretical entities (1992b, 1992c).

When explaining this notion of a style of scientific reasoning, Hacking usually cites A. C. Crombie's list of six distinguishable styles. These include: (a) the postulational method of mathematics, (b) "the experimental exploration and measurement of more complex observable relations", (c) "the hypothetical construction of analogical models," (d) methods of comparison and taxonomic classification, (e) statistical methods, and (f) "the historical derivation of genetic development" (1982, 50; 1983b, 455; 1985, 147; 1990, 6; 1992a, 132; 1992b; 1992c). Hacking does not actually subscribe to this list other than as a starting point for further discussion. He has recently added two additional styles: (a*) an Indo-Arabic algorithmic style of applied mathematics, and (bc) the laboratory style, combining the experimental and hypothetical styles, "characterized by the building of apparatus in order to produce phenomena to which hypothetical modeling may be true or false" (1992b; cf. 1992c). Disputes about unobservable entities appear to be a product of the laboratory style of reasoning. The reality of classes of people turns on styles of reasoning other than the laboratory style. To take an example from *The Taming of Chance*, claims that certain types of people are "normal" or "deviant" are tied to a statistical style of reasoning in the tradition of Adolphe Quetelet.

2. Dynamic Nominalism

It is not clear, however, that Hacking would accept Quetelet's statistical classes as real. In general, Hacking is not a realist about social classes but instead defends a position he calls "dynamic nominalism". Acknowledging a debt to Michel Foucault, Hacking characterizes dynamic nominalism as the belief that systems for classifying people created by society may affect what people do, which in turn may affect our knowledge of them (1984; 1986a; 1986b; 1988; 1990; 1991). As he explains in *The Taming of Chance*, the way that we classify people "has consequences for the ways in which we conceive of others and think of our own possibilities and potentialities" (1990, 6). For example, we not only classify people as either normal or abnormal, but "we try to make ourselves normal, which in turn affects what is normal" (1990, 2). In an article on the concept of child abuse, he claims that this "looping" or "feedback effect" drives a wedge between the social and the natural sciences that is "substantial" and not "methodological" (1988, 62). Facts are constructed in the social sciences in a sense "well worth calling social" but different than the sense in which facts are constructed in the natural sciences, he argues, because this looping effect does not occur with inanimate objects (1988, 57).

Hacking's way of distinguishing the social and the natural sciences, however, is not convincing (cf. Bogen, 1988). In both the natural and the social sciences, I would argue, our classifications make new kinds of actions possible. Explaining the notion of "dynamic nominalism" in an essay on split personalities, he says: "if the classification is not made, then people will not adopt that as a possible way to be" (1986a, 79). Similarly, I would suggest, when we classify entities in the natural sciences in a certain way, we make possible the creation of new experimental effects. The analogy can be brought to light first by interpreting the slogan of *Representation and Intervening* as stating "If you can spray them, then they form a real kind" and then by arguing that "if the classification is not made, then people will not find a way to spray them". The truth of this latter statement may be made easier to see by considering not electrons, the philosopher's favorite kind of entity, but Hacking's other example concerning muons and mesons. According to Hacking, this classification grew up together with ways of manipulating these entities (1983a, 87-91). One could even argue that the way that we classify entities has consequences for the ways in which we think of our own possibilities and potentialities—that is, the possibilities opened us to us through new technologies made possible by techniques for manipulating entities.

So far, kinds of people seem to be created in the same way and thus just as real as kinds of entities. I do not consider it a serious objection to my analogy that, whereas in the natural sciences we are classifying unobservable entities, in the social sciences we are classifying people. Much as the social sciences may consider people as existing independently of our classifications of them, the natural sciences may regard entities as existing independently of our systems of classification. The issue then becomes one of realism about people or entities as representatives of categories or classes.

A more serious objection may be that the constraints on systems of classification are different in the natural than in the social sciences. Now Hacking is careful to distinguish his dynamic nominalism from what he calls "wishful-thinking nominalism" in the social sciences (1986a, 79). He studiously avoids providing a general account of categories of people in terms of nominalism and realism, suggesting that some categories may be more real than others. The class of homosexuals, he supposes, is more real than the class of split personalities, yet the class of waiters is even more real than the class of homosexuals (1986b, 233-4). However, he also says that classifications in the social sciences are "constituted by an historical process," whereas classifications in the natural sciences are constrained by the "world" (1984, 115). Explaining this distinction further, he says that although types of physical phenomena are "created" by us, they are nevertheless "timeless" in the sense that whenever the relevant conditions are brought into being, the effects will result. Social and political categories, on the other hand, he believes are "constituted in an essentially historical setting" (1984, 124).

I doubt, however, that Hacking's distinction between social and natural classifications matters or even holds up. His distinction looks plausible only when we consider a category of people such as split personalities or child abusers in isolation from the entire social system. A Durkheimian could say that such systems of social classification are both timeless in Hacking's sense and constituted historically. Durkheim's primitive classification hypothesis, for instance, could be interpreted either way.

The primitive classification hypothesis held at least two distinct meanings for Durkheim. According to one meaning, it is simply the historical claim that the practice of subsuming species under genera is rooted in primitive social classifications in which clans are subsumed under phratries. This interpretation arises from Durkheim's argument that we could not have arrived at the concept of subsuming species under genera simply by observing nature. Instead, he believes, we borrowed this idea of subsuming one category under another from our social organization (1912, 210-11). A survival of this primitive form of social classification can be found in universities, in which the faculty is classified by departments and colleges.

A second interpretation of the primitive classification thesis, however, is more specific. It says that primitive tribes use the same system of clans and phratries for classifying both people and things in nature. Under this second interpretation, one can distinguish a "timeless" and a "historical" sense. Under its historical interpretation, which is based upon Durkheim's discussion of the ethnographic record, the precise names and numbers of clans in each phratry will be emphasized. That is, it says that for any given tribe, the names of the different species and genera in a system of natural classification and the numbers of species in each genus are the same as the names and numbers of the various clans and phratries. Bloor's relativist interpretation of Durkheim is grounded in such an historical reading of the primitive classification hypothesis. Understanding this hypothesis in its timeless sense, however, we can say that wherever we meet with a society that satisfies Durkheim's definition of "primitive," whether it be the Northwest Amerindians or the natives of Australia, we will discover the use of a system of clans and phratries for classifying people in society and things in nature. For Durkheim, this appears to be an empirical fact about the way that people do things.

In sum, a system of social classification may be regarded either as a historical particular or as a representative of a type. Hacking's failure to consider social groups as parts of systems of classification, I think, has led him to drive a wedge between the social and the natural sciences. The kangaroo clan, say, and the class of people with split personalities may both be only historical entities. It is less clear that the kinds of social systems to which these categories belong are merely historical entities. In any event, I do not think that the goal of sociology is to find laws governing members either of the class of split personalities or the kangaroo clan, any more than the goal of biology is to find laws governing members of biological taxa. To push the analogy with biology a little further, the existence of timeless laws governing the evolution of social systems would not be affected by the fact that particular social groups may come and go.

3. Conclusion

I have tried to show that the social sciences are not inferior to the natural sciences due to the nature of their objects of study. Having removed Hacking's reasons for distinguishing the social from the natural sciences, we can conclude, if I may be permitted to use Austin's and Hacking's sexist expression, that social scientists have the same right to wear trousers as physical scientists.

Notes

[1]I would like to thank Ian Hacking for his comments on an earlier draft of this essay and for sending me offprints of some of his articles. Of course, I accept full responsibility for any remaining errors of interpretation.

[2]Stephen P. Turner, personal communication. Ironically, Turner explains, Weber denies causal reality to social entities as part of his rejection of the classical *mores* tradition in social theory. The causal powers of *mores* were defined in such a way that their existence made intervention in the course of social events impossible.

[3]I defend this interpretation of Durkheim at greater length in my forthcoming monograph, *Creating a Niche: Durkheim and the Sociology of Knowledge*.

References

Bloor, D. (1982), "Durkheim and Mauss Revisited: Classification and the Sociology of Knowledge", *Studies in History and Philosophy of Science* 13: 267-97.

Bogen, J. (1988), "Comments on 'The Sociology of Knowledge About Child Abuse,'" *Nous* 22: 65-6.

Durkheim, E. (1912), *Les Formes elementaires de la vie religieuse*. Paris, France: Presses Universitaire de France, 1960.

Hacking, I. (1982), "Language, Truth and Reason", in *Rationality and Relativism,* M. Hollis and S. Lukes (eds.). Cambridge, MA: MIT Press, pp. 48-66.

_______. (1983a), *Representing and Intervening: Introductory Topics in the Philosophy of Natural Science*. Cambridge, UK: Cambridge University Press.

_______. (1983b), "The Accumulation of Styles of Scientific Reasoning", in *Kant oder Hegel,* D. Henrich (ed.). Stuttgart, Germany: Klett-Cotta, pp. 453-65.

_______. (1984), "Five Parables", in *Philosophy in History: Essays on the Historiography of Philosophy,* R. Rorty et al. (eds.). Cambridge, UK: Cambridge University Press, pp. 103-124.

_______. (1985), "Styles of Scientific Reasoning", in *Post-Analytic Philosophy,* J. Rajchman and C. West (eds.). New York, NY: Columbia University Press, pp. 145-165.

_______. (1986a), "The Invention of Split Personalities", in *Human Nature and Natural Knowledge: Essays presented to Marjorie Grene on the Occasion of Her Seventy-Fifth Birthday,* A. Donagan et al. (eds.). *Boston Studies in the Philosophy of Science*, volume 89. Dordrecht: Reidel, pp. 63-85.

_______. (1986b), "Making Up People", in *Reconstructing Individualism,* T. Heller et al. (eds.). Stanford, CA: Stanford University Press, pp. 222-36.

_______. (1988), "The Sociology of Knowledge About Child Abuse", *Nous* 22: 53-63.

_______. (1989), "Extragalactic Reality: the Case of Gravitational Lensing", *Philosophy of Science* 56: 555-81.

______. (1990), *The Taming of Chance*. Cambridge, UK: Cambridge University Press.

______. (1991), "The Making and Molding of Child Abuse", *Critical Inquiry* 17: 253-288.

______. (1992a), "Statistical Language, Statistical Truth and Statistical Reason: The Self-Authentication of a Style of Scientific Reasoning", in *Social Dimensions of Science,* Ernan McMullen (ed.). Notre Dame: University of Notre Dame Press, forthcoming, pp. 130-57.

______. (1992b), "'Style' for Historians and Philosophers", *Studies in History and Philosophy of Science* 23 (forthcoming).

______. (1992c), "The Disunities of the Sciences", in *Disunity and Contextualism,* Peter Galison (ed.), (forthcoming).

Jones, R.A. (1986), *Emile Durkheim: An Introduction to Four Major Works*. Beverly HIlls, CA: Sage Publications.

Smart, J.J.C. (1963), *Philosophy and Scientific Realism.* London, UK: Routledge & Kegan Paul.

Neo-classical Economics and Evolutionary Theory: Strange Bedfellows?

Alex Rosenberg

University of California, Riverside

On the surface and to a first approximation, economic theory and evolutionary theory share salient features: a commitment to optimality and maximizing, a consequent similarity in mathematical formalism, an appeal to equilibrium as the preferred explanatory strategy. Moreover, both have at various times been stigmatized as empirically empty. However, the theory of natural selection has been widely accepted as the scientific cornerstone of modern biology, while microeconomics has remained open to repeated challenges as an explanation of human action and its aggregate consequences.

Because of their similarity, evolutionary theory looks like part of a powerful defense of the status quo in economic theory. I think that Darwinian theory is a remarkably inappropriate model, metaphor, inspiration, or theoretical framework for economic theory. The theory of natural selection shares few of its strengths and most of its weaknesses with neoclassical theory, and therefore provides no help in any attempt to defend, or even to understand the explanatory strategy of economic theory. In this paper I explain why this is so, by examining two classical appeals to Darwinian theory as a rationalization for orthodoxy in economic theorizing. The two appeals I explore retain the influence today that they had forty years ago, when they were new, and they have attained canonical status in the philosophy of economics.

1. Friedman's Ploy

Why should any one suppose Darwinian evolutionary theory will provide a useful model for how to proceed in economics. One apparently attractive feature of the theory for economists is the methodological defense it seems to provide neoclassical theory in the face of charges that the theory fails to account for the actual behavior of consumers and producers. Thus, Friedman (1953) offers the following argument for the hypothesis that economic agents maximize money returns:

> Let the apparent immediate determinant of business behavior be anything at all, habitual reaction, random chance, or whatnot. Whenever this determinant happens to lead to behavior consistent with rational and informed maximization of returns, the business will prosper, and acquire resources with which to expand; whenever it does not, the business will tend to loose resources and can be kept in existence

PSA 1992, Volume 1, pp. 174-183

only by the addition of resources from outside. The process of "natural selection" thus helps to validate the hypothesis or, rather, given natural selection, acceptance of the hypothesis can be based on the judgement that it summarizes appropriately the conditions for survival. (Friedman 1953, p. 35)

This argument does reflect a feature of evolutionary theorizing, though admittedly a controversial one. The natural environment sets adaptational problems that animals have to solve in order to survive. The fact that a particular species is not extinct is good evidence that it has solved some of the problems imposed upon it. This fact about adaptational problems and their solutions plays two roles in evolutionary thinking. First, examining the environment, biologists might try to identify the adaptational problems that organisms face. Second, focusing on the organism, they sometimes attempt to identify possible problems which known features of the organism might be solutions to. The problem with this approach is the temptation of Panglossianism: imagining a problem to be solved for every feature of an organism we detect. Thus, Dr Pangloss held that the bridge of the nose was a solution to the adaptational problem of holding up glasses. The problem with inferences from the environment to adaptational problems is that we need to determine all or most of the problems to be solved, for each of them is an important constraint on what will count as solutions to others. Thus having a dark color will not be a solution to the problem of hiding from nocturnal predators unless the organism can deal with the heat that such color will absorb during the day. On the other hand a color that will effect the optimal compromise between these two constraints may fail a third one, say being detectable by conspecifics during mating season.

Then there is the problem of there being more than one way to skin a cat. Even if we can identify an adaptational problem, and most of the constraints against which a solution can be found, it is unlikely that we will be able to narrow the range of equally adaptive solutions down to just the one that animals actually evince. Thus, we are left with the explanatory question of why this way of skinning the adaptational cat emerges and not another apparently equally as good a one. There are two answers to this question. One is to say that if we knew all the constraints we'd see that the only possible solution is the actual one. The other is to say that there are more than one equally adequate solutions, and that the one finally "chosen" appeared for non-evolutionary causes. The first of these two replies is simply a pious hope that more inquiry will vindicate the theory. The second is in effect to limit evolutionary theory's explanatory power, and deny it predictive power.

These problems have in general hobbled "optimality" analysis as an explanatory strategy in evolutionary biology. Many biologists find the temptations of Panglossianism combined with the daunting multiplicity of constraints on solutions to be so great that they despair of providing an evolutionary theory that contributes to our detailed understanding of organisms in their environments.

The same problems bedevil Friedman's conception, and limit the force of his conclusion. The idea that rational informed maximization of returns sets a necessary and or sufficient condition for long term survival in every possible economic environment, or even in any actual one is either false or vacuous. Is the hypothesis that returns are maximized over the short run, the long run, the fiscal year, the quarter? If we make the hypothesis specific enough to test, it is plainly false. Leave it vague and the hypothesis is hard to test. Suppose we equate the maximization of returns hypothesis with the survival of the fittest hypothesis. Then nothing in particular follows about what economic agents do, and how large their returns are, any more than it follows from the fitness-maximization hypothesis what particular organisms do and how many off-spring they have. However many off-spring, however much the returns, the results will be maximal, given the circumstances, over the long run. What we want to know is which features of organisms increase their fitness, what strategies of economic agents increase their returns. And we want this information both to explain particular events in the past, and to predict the course of future evolution. For the hypothesis of maximization of returns to play this substantive role, it cannot be supposed to be on a par with the maximization of fitness hypothesis. Rather, we need to treat maximizing returns as a specific optimal response to a particular environmental problem, rather like we might treat coat color as an optimal response to an environmental problem of finding a color that protects against predators, does not absorb too much heat, is visible to conspecifics, etc. But when we think of the maximization of returns hypothesis this way, it is clear that maximizing dollar returns is not a condition of survival in general, either in the long run or the short run.

Nature has a preference for quick and dirty solutions to environmental problems. It seems to satisfice, in Simon's phrase. But unlike satisficing, nature's strategy really is a maximizing one. It's just the constraints are so complicated and so unknown to us that the solutions selection favors look quick and dirty to us. If we knew the constraints we'd see that they are elegant and just on time. Learning what the constraints are and how the problems are solved is where the action is in vindicating the theory of natural selection. Because only that will enable us to tell whether the solution really maximizes fitness, as measured by off-spring. Similarly, in economics the action is in learning the constraints and seeing what solutions are chosen. Only that will tell us whether dollar returns are really maximized, and whether maximizing dollar returns ensures survival. To stop where Friedman does is to condemn the theory he sets out to vindicate to the vacuity with which Darwinian theory is often charged.

2. Alchain's argument

If the theory of natural selection is to vindicate economic theory or illuminate economic processes it will have to do more than just provide a Panglossian assurance that whatever survives in the short, medium or long run is fittest. What is needed in any attempt to accomplish this is a better understanding of the theory of natural selection. Such an improved understanding of the theory is evident in Alchain's (1950) approach to modeling economic processes as evolutionary ones.

Alchain's approach is not open to obvious Panglossian objections, nor does it make claims about empirical content which transcend the power of an evolutionary theory to deliver. Still, its problems reveal more deeply the difficulties of taking an evolutionary approach to economic behavior.

To begin with, Alchain's approach reflects the recognition that Darwinian theory's claims about individual responses to the environment are hard to establish, impossible to generalize, and therefore without predictive value for other organisms in other environments. Alchain recognized that the really useful versions of evolutionary theory are those which focus on populations large enough so that statistical regularities in responses to environmental changes can be discerned. And he recognized that Darwinian evolution operates through solutions to adaptational problems that are in appearances at any rate quick and dirty, approximate and heuristic, not rationally and informationally maximizing. Like the biological environment, the economic one need not elicit anything like the maximization of returns that conventional theory requires:

> In an economic system the realization of profits is the criterion according to which successful and surviving firms are selected. This decision criterion is applied by an impersonal market system... and may be completely independent of the decision

> processes of individual units, of the variety of inconsistent motives and abilities and even of the individual's awareness of the criterion.
> The pertinent requirement—positive profits through relative efficiency—is weaker than "maximized profits," with which unfortunately, it has been confused. Positive profits accrue to those who are better than their actual competitors, not some hypothetically perfect competitors. As in a race, the award goes to the relatively fastest, even if all competitors loaf.
> ...success (survival) accompanies relative superiority;... it may... be the result of fortuitous circumstances. Among all competitors those whose particular conditions happen to be the most appropriate of those offered to the economic system for testing and adoption will be "selected" as survivors. (alchain 1950, pp. 213-4)

Alchain also recognizes that adaptation is not immediate, is only discernable to the observer in the change in statistical distributions over periods of time, and what counts as adaptive will change as the economic environment does. Alchain uses a parable to illustrate the way that the economic environment shifts the distribution of actually employed choice strategies towards the more rational:

> Assume that thousands of travellers set out from Chicago, selecting their roads completely at random and without foresight. Only our "economist" knows that on but one road there are any gas stations. He can state categorically that travellers will continue to travel only on that road: those on other roads will soon run out of gas. Even though each one selected his route at random, we might have called those travellers who were so fortunate as to have picked the right road wise, efficient, farsighted, etc. Of course we would consider them the lucky ones. If gasoline supplies were now moved to a new road, some formerly luckless travellers again would be able to move; and a new pattern of travel would be observed, although none of the players changed his particular path. The really possible paths have changed with the changing environment. All that is needed is a set of varied, risk-taking (adoptable (sic)) travelers. The correct direction of travel will be established. As circumstances (economic environment) change, the analyst (economist) can select the type of participants (firms) that will now become successful; he may also be able to diagnose the conditions most conducive to greater probability of survival. (Alchain 1950, p. 214)

In order to insure survival and significant shifts in the direction of adaptation, several other conditions must be satisfied: To begin with the environment must remain constant long enough so that those strategies more well adapted to it than others will have time to outcompete the less well adapted and to increase their frequency significantly enough to be noticed. Moreover, the initial relative frequency of the most well adapted strategy must be high enough so it will not be stamped out by random forces before it has amassed a sufficient advantage to begin to displace competitors. And of course it must be the case that there are significant differences among competing strategies. Otherwise, their proportions at the outset of competition will remain constant over time. There will be no significant changes in proportions to report.

What kind of knowledge will such an economic theory provide? Even at his most optimistic Alchain was properly limited in his expectations. He made no claims that with an evolutionary approach the course of behavior of the individual economic agent could be predicted. Here the parallel with evolution is obvious. Darwin's theory not only has no implications for what will happen to any individual organism, its implications for large numbers of them are at best probabilistic:

> A chance dominated model does not mean that an economist cannot predict or explain or diagnose. With a knowledge of the economy's requisites for survival and by a comparison of alternative conditions, he can state what types of firms or behavior relative to other possible types will be more viable, even though the firms themselves many not know the conditions or even try to achieve them by readjusting to the changed conditions. It is sufficient if all firms are slightly different so that in the new environmental situation those who have their fixed internal conditions closer to the new, but unknown optimum position now have a greater probability of survival and growth. They will grow relative to other firms and become the prevailing type, since survival conditions may push the observed characteristics of the set of survivors towards the unknowable (to them) optimum by either (1) repeated trials or (2) survival of more of those who happen to be near the optimum—determined ex post. If these new conditions last "very long", the dominant firms will be different ones from those which prevailed or would have prevailed under the other conditions. Even if the environmental conditions cannot be forecast, the economist can compare for given alternative potential situations the types of behavior that would have higher probability of viability or adoption. If explanation of past results rather than prediction is the task, the economist can diagnose the particular attributes which were critical in facilitating survival, even though individual participants were not aware of them. (Alchain, 1950, p. 216)

As a set of conditional claims, most of what Alchain says about the explanatory and predictive powers of an evolutionary theory of economic processes is true enough. The trouble is that almost none of the conditions obtain, either in evolutionary biology or in economic behavior. But it is only these conditions that would make either theory as useful as Alchain or any economist needs it to be. Thus, the attractions of an evolutionary theory for economists must be very limited indeed. Alchain rightly treats the economy as the environment to which individual economic agents are differentially adapted. As with the biological case, we need to know what "the requisites of survival" in the environment are. In the biological case this is not a trivial matter, and beyond the most obvious adaptational problems, there are precious few generalizations about what any particular ecological environment requires for survival, still less what it rewards in increased reproductive opportunities. We know animals need to eat, breathe, avoid illnesses and environmental hazards, and the more of their needs they fulfill the better off they are. But we don't know what in any given environment the optimal available diet is, or what the environmental hazards are for each of the creatures that inhabit the environment. And outside of ecology and ethology, few biologists are interested in this information in any case. For its systematic value to biology is very limited.

Ignorance about these requisites for survival in biology make it difficult to predict even "the types of ... behavior relative to other possible types that will be more viable". It is easy to predict that all surviving types will have to subsist in an oxygen rich environment, where the gravitational constant is 32 feet/second^2, and the ambient temperature ranges from 45 degrees to minus 20 degrees Celsius. But such a "prediction" leaves us little closer to what we hope to learn from a prediction. The same must be true in evolutionary economics. We have no idea of what the requisites for survival are, and even if we learned them, they would probably not narrowly enough restrict the types that can survive to enable us to frame any very useful expectations for the future. Of course this is not an in-principle-objection to an evolutionary approach. But consider what sort of information would be required to establish a very full list of concrete necessary conditions for survival of say, a firm in any very specific market, and then consider the myriad ways in which economic agents could so act to satisfy those conditions. This information is either impossible to obtain or else if we had it, an evolutionary approach to economic processes would be superfluous. To see this, go back to Alchain's discussion.

Alchain notes that over time the proportion of firms of various types should change: the proportion of those which are fitter should increase while those less fit should decrease. If environmental conditions last a long time, "the dominant firms will be different from those which prevailed...under other conditions." True enough, but what counts as environmental conditions lasting a long time? In the evolutionary context "long enough" means at least one generation, and the duration of a generation will vary with the species. In addition the notion of "long enough" reflects a circularity which haunts evolutionary biology. Evolution occurs if the environment remains constant long enough for the proportion of types to change. "Long enough" is enough time for the proportions to change. Moreover, when the numbers of competing individuals are small, there may be change in proportions of types which is not adaptational but is identified as drift—a sort of sampling error. But what is a small number of individuals vs a large number? Here the same ambiguity emerges. "Large enough" means a number in which changes in proportion reflect evolutionary adaptation. The only way in which to break out of this circularity of "long enough, large enough," etc., is to focus on individual populations in particular environments over several generations. And the answer we get for any one set of individuals will be of little value when we turn to another set of the same types in different environments, or different types in the same environment.

Can the situation be any better in economics? In fact, won't the situation be far less promising? After all, the environment within which an economic agent must operate does not change with the stately pace of a geological epoch. Economic environments seem to change from day to day. If they do, then there is never enough time for the type most adapted to one environment to increase in its proportions relative to other types. Before it has had a chance to do so, the environment has changed, and another type becomes most adapted. But perhaps economic environments do not change quite so quickly. Perhaps to suppose that they do change so quickly is to mistake the weather for the climate. Day to day fluctuations may be a feature of a more long standing environment. The most well adapted individual to an environment is not one which responds best to each feature of it, including its variable features, but adapts best over all on an average weighted by the frequency with which certain conditions in the environment obtain. So, the period of time relevant to evolutionary adaptation might be long enough for changes in proportion to show up. For the parallel to evolution to hold up, this period of environmental constancy will have to be longer than some equivalent to the generation time in biological evolution.

But is there among economic agents any such an equivalent? Is there a natural division among economic agents into generations? With firms the generation time might be the period from incorporation to the emergence of other firms employing the same method in the same markets through conscious imitation; with individual agents the minimal period for evolutionary adaptation will be the time during which it takes an individual to train another one to behave in the same way under similar economic circumstances. But these two periods are clearly ones during which the economic environment almost always changes enough to shift the adaptational strategy.

The only way we can use an evolutionary theory to predict the direction of adaptation is by being able to identify the relevant environment which remains constant enough to force adaptational change in proportions of firms. As Alchain tacitly admits, this is something we cannot do: he notes, "Even if the environmental conditions cannot be forecast, the economist can compare for given alternative potential situations the types of behavior that would have a higher probability of viability or adoption". This is a retrospective second best. Suppose economically relevant environmental conditions could be forecast. Then, it is pretty clear we would not need an evolutionary theory of economic behavior. Friedman's rationale for neoclassical theory would then come into its own. If we knew environmental conditions, then we could state what optimal adaptation to them would be. And if we could do this so could at least some of the economic agents themselves. To the extent that they could pass on this information to their successors, Panglossianism would eventually be vindicated in economic evolutionary theory. Economic agents would conform their actions to the strategy calculated to be maximally adaptive, just as Friedman claims. An evolutionary theory of economic behavior is offered either as an alternative to rational maximizing or as an explanation of its adequacy. If rational maximizing is adequate as a theory, evolutionary rationales are superfluous, if it is not adequate, then an evolutionary approach is unlikely to be much better, and for much the same reason: neither economic theorists nor economic agents can know enough about the economic environment for the former's predictions or the latter's decisions to be regularly vindicated.

3. Equilibrium and information in economics and evolution

One of the features of evolutionary theory that make it attractive to the economist is the role of equilibrium in the claims the theory makes about nature. Equilibrium is important for economic theory not least because of the predictive power it accords the economist. An economic system in equilibrium or moving towards one, is a system some or all of whose future states are predictable by the economist. Equilibrium has other (welfare-theory relevant) aspects, but its attractions for economists must in part consist in the role it plays so successfully in physical theory and evolutionary theory. Evolutionary biology defines an equilibrium as one in which gene ratios do not change from generation to generation, and it stipulates several conditions that must obtain for equilibrium: a large population mating at random, without immigration, emigration, or mutation, and of course without environmental change. Departures from these conditions will cause changes in gene frequencies within a population. But over the long run the changes will move in the direction of closer adaptation to the environment—either closer adaptation to an unchanged one, or adaptation towards a new one. The parallel to economic equilibrium is so obvious that mathematical biologists have simply taken over the economist's conditions for the existence and stability of equilibria. If a unique stable market clearing equilibrium exists, then its individual members are optimally adapted to their environment, no trading will occur, and there will be no change—no evolution—in the economy. But if one or another of the conditions for equilibrium are violated an efficient economic system will either move back to the original equilibrium or to a new one by means of adjustments in which individuals move along paths of increased adaptation.

In evolutionary biology equilibrium has an important explanatory role. So far as we can see, populations remain fairly constant over time, and among populations the proportions of varying phenotypes remain constant as well. Moreover, when one or another of the conditions presupposed by equilibrium of gene frequencies is violated, the result is either compensating movement back toward the original distribution of gene frequencies, or movement towards a new level of gene frequencies. These facts about the stability of gene frequencies and their trajectories need to be explained, and the equilibrium assumptions of transmission genetics are the best explanations going. In addition they will help us make generic predictions that when one or another condition, like the absence of mutation, is violated, a new equilibrium will be sought. Sometimes we can even predict the direction of that new equilibrium. But in real ecological contexts (as opposed to simple text-book models) we can hardly ever predict the actual value of the new equilibrium level of gene frequencies. This is because we do not know all the environment factors that work with a change in one of these conditions, and among those factors which we do know about, we have only primitive means of measurement for their dimensions.

Now compare economics. To begin with we have nothing comparable to the observed stability of gene frequencies that needs to be explained. So the principle explanatory motivation for equilibrium explanations is absent. We cannot even appeal to the stability of prices as a fact for equilibrium to explain in economics, because we know only too well that neoclassical general equilibrium theory has no explanation for price stability. That is, given an equilibrium distribution and a change in price, there is no proof that the economy will move to a new general equilibrium. (For this reason general equilibrium theory has recourse to the Walrasian auctioneer and tatonnement.)

There is no doubt that economic equilibrium theory has many attractive theoretical features—mathematical tractability, the two welfare theorems—but it lacks the most important feature that justifies the same kind of thinking in evolutionary biology: independent evidence that there is a stable equilibrium to be explained.

One of the factors that gives us some confidence that equilibrium obtains with some frequency in nature is that changes in gene frequencies are not self-reinforcing. If something changes which has the effect of changing gene-frequencies, then hardly ever will the change in gene frequencies precipitate still another round of changes in gene frequencies and so on, thus cascading into a period of instability. Of course sometimes evolutionary change is "frequency dependent": if one species of butterfly increases in population size because it looks like another species that birds avoid, then once it has grown larger in number than the bad-tasting butterflies, its similar appearance and the genes that code for appearance will no longer be adaptive and may decline. But presumably, the proportions will return to some optimal level and be held there by the twin forces of adaptation and maladaptation.

In the game-theorist's lingo, evolutionary adaptational problems are parametric: the adaptiveness of an organism's behavior does not depend on what other organisms do. But we cannot expect this absence of feed-back in economic evolution. Among economic agents the problem is strategic. Economic agents are far more salient features of one another's environment than animals are features of one another's biological environment. Changes in agents' behaviors effect their environments regularly, because they call forth changes in the behavior of other agents, and these further changes cause a second round of changes in their behavior. Game theorists have come to identify this phenomenon under the rubric of the common knowledge problem. Economists traditionally circumvented this problem by two assumptions that have parallels in evolutionary biology as well. It is important to see that the parallels do not provide much ground for the rationalization of economic theorizing in the biologist's practice.

Both evolutionary equilibrium and economic general equilibrium require an infinite number of individuals. In the case of evolution this is to prevent drift or sampling error from moving gene frequencies independent of environmental changes. In the case of the theory of pure competition it is to prevent agent choice from becoming strategic. If the firm is always a price taker, and can have no effect on the market, then it can treat its choices as parametric. Where numbers of interactors are small the assumption of price-taking produces badly wrong predictions, and there is indeed no stability, and typically no equilibrium.

Is sauce for the biological goose sauce for the economic gander? Can both make the same false assumption with equal impunity? The fact is, though the assumption of infinite population size is false about interbreeding populations, it seems to do little harm in biology. That is, despite the strict falsity of the evolutionary assumption, populations seem to be large enough so that theory that makes these false assumptions can explain the evident facts of constancy and or stability of gene frequencies. In the case of economics, there are no such evident facts, and one apparent reason seems to be the falsity of the assumption of an infinite number of economic agents.

The other assumption evolutionary theory and economic theory traditionally make is that the genes and the agents are "omniscient". Genes carry information in two senses. First, they carry instructions for the building and maintenance of proteins, and assemblies of them that meet the environment as phenotypes. But they also indirectly carry information about which phenotypes are most adapted to the environment in which they find themselves. They do so through the intervention of selective forces which cull maladaptive phenotypes and thus the genes that code for their building blocks. And so long as the environment remains constant, the gene-frequencies will eventually track every environmentally significant biologically possible adaptation and maladaptation. In this sense the genome is in the long run omniscient about the environment. There are two crucial qualifications here. First the assumption of the constancy of the environment, something economic theory has little reason to help itself to. Second there is the "the long run"—another concept evolutionary theory shares with economic theory. Evolutionary biology has world enough and time empigj for theories that explain and predict only in the long run—geological epochs are close enough to infinite not to matter for many purposes. But Keynes pointed out the problem for economics of theories that explain only the long run. An evolutionary economic theory committed to equilibrium is condemned at best to explain only the long run.

We know only too well the disequilibriating effects of non-omniscience, that is, how information obstructs the economy's arrival at or maintenance of an equilibrium. Indeed, the effects of differences in information on economic outcomes are so pervasive that we should not expect economic phenomena ever to reflect the kind of equilibrium evolutionary biological phenomena do. Arrow has summarized the impact of information on equilibrium models succinctly:

> If nothing else there are at least two salient characteristics of information which prevent it from being fully identified as one of the commodities represented in our abstract models of general equilibrium: (1) it is, by definition, indivisible in its use; and (2) it is very difficult to appropriate. With regard to the first point, information about a method of production, for example, is the same regardless of the scale of the output. Since the cost of information depends only on the item, not its use, it pays a large scale producer to acquire better information than a small scale producer. Thus, information creates economies of scale throughout the economy, and therefore, according to well-known principles, causes a departure from the competitive economy.
>
> Information is inappropriable because an individual who has some can never lose it by transmitting it. It is frequently noted in connection with the economics of research and development that information acquired by research at great cost may be transmitted much more cheaply. If the information is, therefore, transmitted to one buyer, he can in turn sell it very cheaply, so that the market price is well below the cost of production. But if the transmission costs are high, then it is also true that there is inappropriability, since the seller cannot realize the social value of the information. Both cases occur in practice with different kinds of information.
>
> But then, according to well-known principles of welfare economics, the inappropriability of a commodity means that its production will be far from optimal. It may be below optimal: it may also induce costly protective measures outside the usual property system. Thus, it has been a classic position that a competitive world will underinvest in research and development, because the information acquired will become general knowledge and cannot be appropriated by the firm financing the research. ... if secrecy is possible, there may be overinvestment in information

gathering, each firm may secretly get the same information, either on nature or on each other, although it would of course consume less of society's resources if they were collected once and disseminated to all. (Arrow 1984, pp. 142-3)

If agents were omniscient these problems would not emerge. Genomes are omniscient so the parallel problems do not emerge in nature, and do not obstruct equilibria. There are no apparent economies of scale operating within species in the maximization of reproductive fitness. And besides, the information which the environment provides about relative adaptedness is costless and universally available. So there is no problem about appropriability. In the absence of secrecy and the need for strategic knowledge about what other agents know, there is no room in biological evolution for the sort of problems information raises in economics. Once biological systems become social, and their interactions become strategic the role for information becomes crucial. But at this point evolution turns Lamarckian. It is no surprise that when "acquired" characteristics are available for differential transmission, markets for the characteristics will emerge. But at this point Darwinian evolution is no longer operating. In fact, one good argument against the adoption of Darwinian evolutionary theory as a model for economic theory is just the difference made by information. Once it appears in nature evolution ceases to be exclusively or even mainly Darwinian. Why suppose that once information becomes as important as it is in economic exchange that phenomena should again become Darwinian?

References

Alchain, A. (1950), "Uncertainty, evolution and economic theory," *Journal of Political Economy* 58: 211-221.

Arrow, K.A. (1984), *The Economics of Information*. Cambridge: Belnap Press.

Friedman, M. (1953), *Essays in Positive Economics*. Chicago: University of Chicago Press.

Community Ecology, Scale, and the Instability of the Stability Concept[1]

E.D. McCoy and Kristin Shrader-Frechette

University of South Florida

1. Introduction

In the Preface to his recent philosophy of biology (1988) and in his earlier *Growth of Biological Thought* (1982), Ernst Mayr emphasized that recent progress in evolutionary biology is a result of conceptual clarification, not a consequence of improved measurements or better scientific laws. Although complete agreement on the meaning of key concepts is not essential for all communication in science (see Hull 1988, pp. 6-7, 513), we likewise believe that conceptual clarification is perhaps the most important key to progress in community ecology. In order to investigate some of the reasons that might explain why community ecology has been unable to arrive at a widely accepted general theory, in this essay we analyze "stability," one of its most important foundational concepts. After reviewing the stability concept and sketching its associated problems, we assess the epistemological status of four difficulties with the concept.

2. A Review of The Stability Concept

The idea that nature maintains itself in some sort of balance or stability is a very old one (see Egerton 1973, Goodman 1975, McIntosh 1985). Among traditional natural historians, nature was in balance when no changes could be detected in the identities or population sizes of the component species of a biotic community. The frame of reference for discerning lack of change was the period during which the community was observed, a period which usually encompassed a few years to decades. Interruptions of the balance resulted from "disturbances" (e.g., fire, timbering, cultivation) that promoted detectable changes in the environment and community. Often, such disturbances could be traced to human activities, giving rise to the common wisdom that nature would be in balance except for the meddlings of humans (see Allee et al. 1949).

Some influential ecologists in the early twentieth century held that communities possessed emergent and organismal properties (see, for example, Clements 1905). Their views reinforced the impression of nature as being in some sort of balance or homeostasis (i.e., no change in species composition or in population sizes of a community). Clements (1905; see also Egerton 1973, McIntosh 1985), for example, assumed that predictable groups of species replaced one another at a particular location

PSA 1992, Volume 1, pp. 184-199

until a "climax" was reached, in which one group was self-perpetuating. Under this view, any change in the community that did occur ultimately produced a static balance or stability, as long as the environment did not change. Some of Clements' contemporaries (e.g., Forbes 1880) also espoused a view similar to his. At least throughout the first half of the twentieth century (e.g., Dice 1952, Emerson 1954), the traditional, static, balance-of-nature concept refused to die.

The idea of a static stability or balance, however, was doomed by the weight of evidence that was beginning to accumulate during the same period. Elton (1930, p. 17), for example, stated that "the numbers of wild animals are constantly varying." Andrewartha and Birch (1954, p. 20), in response to the ideas of Nicholson (1933) and Nicholson and Bailey (1935)—who supported the classical (static) view of the balance of nature—stated that "it is not easy to understand what precisely is meant by the word 'balance'."

Also, as a result of expanding the time frame of reference, a more "dynamic" conception of stability emerged in the early twentieth century. The origins of this dynamic perspective were, in part, the claims of Clements (1905) and Forbes (1880) that species populations oscillated, but that the oscillations were kept within bounds, tending toward equilibrium (see McIntosh 1985). Within their view are the seeds of two modern ideas about stability, "dynamic balance" and "persistence." A dynamic balance exists when change in the environment is tracked by the community, but the community returns to some "normal" condition. A community exhibits persistence when its changes seem not always to be followed by return to some "normal" condition, but it never changes so drastically as to be unrecognizable by some set of established criteria. We shall return to these ideas later.

By about the middle of the century, the dominant conception of stability was that of a dynamic balance. For example, Allee et al. (1949, pp. 507-508) say that a community

> is a self-regulating assemblage in which populations of plants and animals hold each other in a state of biological equilibrium. This is an extension of the principle of biotic balance to embrace the whole community. This is not to say that communities are always in static equilibrium. Rather, they are in a condition of flux.

Kendeigh (1961, p. 196) echoes the same theme:

> even in entirely natural communities undisturbed by man, a strict balance of nature is probably never maintained for any appreciable period of time. It is characteristic for populations to vary in size.

Probably because of the dated connotations of the term 'balance of nature', it is rarely used and never defined explicitly in current ecological literature. The term appears typically only to hold it up to ridicule (e.g., Ehrlich and Birch 1967). The analogous, but acceptable, term in modern ecology is 'stability'. Yet, 'stability' seems to be no more precisely defined than 'balance'. In fact, by the late 1960's and early 1970's, use of 'stability' was so imprecise that several prominent ecologists (Lewontin 1969, Holling 1973, and Orians 1975, among others) found it necessary to try to define and categorize the various kinds of stability attributed to nature.

Lewontin (1969), in probably the best known of these reviews, discussed the meaning of stability in terms of particles represented in vector fields. He illustrated the tradition of borrowing ideas from other sciences, especially physics, and incorporating them into ecology, a tradition that began as early as the days of Lotka (1925), in part because laws and theory are not well established in ecology. Lewontin's vector fields include both a "position vector," which describes the succession of positions of an object (in the present case, a community), and "transformation vectors," which describe both the direction that the object would move at every point in hyperspace and the magnitude of the motion. For him, 'neighborhood stability' occurs when all of the vectors that are very near a "stationary point" (point at which the magnitude of the transformation vector is zero) aim toward that point (thus making it a "stable point"). He defines 'global stability' as the kind of stability in which "all of the vectors aim toward the stable point from all other points in the dynamical space. Neighborhood stability, therefore, describes arbitrarily small perturbations" (p. 16), while global stability describes large ones (Lewontin 1969, p. 16; see also, Pimm 1982, Wu 1977). Because neighborhood stability is mathematically more tractable than global stability, Lewontin (1969) says, and because random perturbations hamper detection of global stability,[2] it is understood better than global stability. Global stability, however, is likely to be the kind of stability of interest to ecologists, because they typically study the ramifications (for the community) of relatively large changes in the environment.

Holling (1973), in particular, emphasized the dichotomy between classical mathematical and ecologically relevant stability. He noted that traditional ecological models typically are "either globally unstable or globally stable," but that, under numerous realistic environmental conditions, systems may not show much evidence of typical neighborhood (point) stability. Rather, he claims that they may display a capacity simply to persist, something he (Holling 1973, p. 17) terms "resilience":

> resilience determines the persistence of relationships within a system and is a measure of the ability of these systems to absorb changes of state variables, driving variables, and parameters, and still persist.

'Stability', on the other hand, he defines (p. 17) as:

> the ability of a system to return to an equilibrium state after a temporary disturbance.

We made this same contrast earlier in this section, but used the terms 'dynamic balance' and 'persistence' rather than Holling's (1973) 'stability' and 'resilience'.

Orians (1975, pp. 141-143) recognized seven ways in which scientists described stability, as applied to ecological systems:

(1) the lack of change in some parameter (e.g., number of species, taxonomic composition) of a system ('constancy');

(2) the survival time of a system or some component of it ('persistence');

(3) the ability of a system to resist external perturbations ('inertia');

(4) the speed with which the system returns to its former state (e.g., number of species, taxonomic composition) following a perturbation ('elasticity');

(5) the area over which a system is stable ('amplitude'),

(6) the ability of a system to cycle or oscillate around some central point or zone ('cyclic stability'); and

(7) the ability of a system to move towards some final end point or zone despite differences in starting points ('trajectory stability').

The seven conceptions of stability identified by Orians (1975), however, are neither mutually exclusive nor even comparable, i.e., they do not all refer to properties of biological systems. More generally, the three reviews of the stability concept—by Lewontin (1969), Holling (1973), and Orians (1975)—suggest that in the late 1960's and early 1970's a wide range of concepts, with a variety of applications, could be found under the umbrella of ecological stability. At least two of the authors who reviewed stability realized that, because ecologists lack a solid understanding of what constitutes ecological stability, they could not advise resource managers on how best to ensure the long-term survival of ecological systems. Holling (1973), for example, pointed out the implications of adopting a "stability" view, as opposed to a "resilience" view. He said (p. 21) that proponents of stability would emphasize "harvesting of nature's excess production with as little fluctuation as possible," whereas proponents of resilience would emphasize persistence of species. Holling suggests that guaranteeing stability and thereby assuring a stable maximum yield might so change the deterministic conditions of an ecological system that resilience could be reduced or lost. Orians (1975), waxing equally pessimistic, warned (pp. 147-148) that "ecological advice on matters relating to community stabilities is highly intuitive," even though "scientific ecology should be able to provide better advice to decision makers who are responsible for...preservation."

Despite the imprecision of the stability concept and the difficulties associated with applying it to problems of preservation, many ecologists nevertheless believed (in the sixties and seventies) that, as a community became more diverse, in terms of number of component species (the so-called "richness" component of diversity), it also became more stable in relation to perturbation (e.g., Clements and Shelford 1939, Colinvaux 1973, Collier et al. 1973, Odum 1953). Paine (1966), MacArthur (1971, 1972), May (1972, 1973), van Emden and Williams (1974), Levins (1974, 1975), Goodman (1975), and others examined the supposed causal relationship between species diversity and community stability, but their pioneering work and subsequent mathematical analyses failed to demonstrate any causal link between species diversity and community stability (see Pimm 1984).

One outcome of these failures was a shift of emphasis away from diversity-stability relationships to complexity-stability relationships (e.g., Pimm 1984), a shift that seems actually to have been a return to earlier ideas about what attributes of a community might enhance its stability (see Goodman 1975). The complexity of a community encompasses more than just its diversity. While diversity is measured by species richness and species evenness (how near a relative abundance distribution of species comes to equal representation of all species), complexity includes both these factors as well as the components of "interaction strength" and "connectance." Pimm (1984) defines 'interaction strength' as "the size of the effect of one species' density on the growth rate of another species." He defines 'connectance' as "the number of actual interspecific interactions divided by the number of possible interspecific interactions" (see also Levins 1974).

The shift in emphasis from diversity-stability relationships to complexity-stability relationships was, in a sense, a liberal move. This is because complexity-stability relationships seem compatible with a great variety of conceptions of the community. Consider, for example, how they might mesh with the probabilistic conception of the community. Under this conception, interactions among the component species of a community may simply result from species' being in the same place at the same time. Even though, under the probabilistic conception, interactions may be fortuitous and temporary, they still can convey stability, by "spreading the risk" of extinction (see Hengeveld 1988, 1989, Reddingius and den Boer 1970). Hengeveld (1989, p. 129) explains that risk spreading occurs when

> the variance of fluctuation [of populations] is kept within bounds purely statistically: the greater the number of mutually dependent variables affecting a species, the smaller the variance.

That is, the larger the number of connections among the component species of a community, no matter how those connections come about, the greater the stability of the community. However, in situations of risk spreading, relationships among species are non-specific, and communities, as self-regulating feedback mechanisms (with resulting deterministic species composition), do not exist (Hengeveld 1989, p. 129).[3]

Despite numerous attempts to link some measure of complexity with stability, the matter remains poorly resolved at present. One problem is understanding how stability at the population level relates to that at the community level. Pimm (1984) suggests some possible relationships, and Strong (1986) reviews regulation at the population level. A second problem is measuring stability. Goodman (1975), Pimm (1984), and Williamson (1987) discuss methods of measuring stability, appropriate for various situations, and some problems in their use. We shall not deal further with these two problems, because they have been reviewed recently and ably. Instead, we shall concentrate on two other conceptual difficulties that have been addressed less frequently. These problems are (1) the large number of meanings that have been attributed to the stability concept and (2) the temporal and spatial scale over which stability is assessed. We shall discuss the second problem, next the first, and then evaluate their epistemological significance.

3. The Problem of Scale

Virtually every modern author, writing about stability at any level of organization, includes some mention of scale (e.g., Berryman 1987, Blondel 1987, Morris 1988, Pimm 1984, Ricklefs 1987, Shugart and West 1981, Williamson 1987). Particular examples of the importance of scale in understanding stability of communities may be found in Davis (1986) and Graham (1986). Connell and Sousa (1983) detail how scale may influence perceptions of stability in natural communities. They conclude, from their review of previous studies of stability, that virtually no evidence exists for conceiving of ecological stability as some variable remaining at equilibrium (what we have termed "static balance") or returning to equilibrium after perturbation (what we have termed "dynamic balance"), once difficulties like inadequacies of scale are resolved. They conclude (p. 808):

> There is no clear demarcation between assemblages that may exist in an equilibrium state and those that do not. Only a few examples of what might be stable limit cycles were found. There was no evidence of multiple stable states in unexploited natural populations or communities. Previously published claims for their existence either have used inappropriate scales in time or space, or have compared populations or communities living in very different physical environments, or have simply misconstrued the evidence...rather than the physicist's classical ideas of stability, the concept of persistence within stochastically defined bounds is...more applicable to real ecological systems. (Recall that this conclusion was also reached by Lewontin (1969) and Holling (1973).)

By "persistence within stochastically defined bounds," Connell and Sousa (1983) mean without extinction, or with extinction and rapid recolonization, within those bounds. They note that judgments of "persistence" are themselves dependent upon the scale of observation. For example, if a community is viewed on a relatively large scale, it may be easy to judge its persistence, because the criterion may be as obvious as determining, say, whether a grassland persists or gives way to forest. If, instead, a community is viewed on a relatively small scale, it may be more difficult to judge persistence, because the focus is on detailed changes, such as in the composition and relative abundances of species. Judgments of "persistence" also are dependent upon the time scale imposed by the organisms that make up the community. They are much more difficult for long-lived species because, to draw meaningful conclusions, observations must be made over at least one complete turnover of individuals. Problems of scale thus exist no matter what conception of stability one employs.

In sum, it appears that the modern conception of stability is imprecise. This imprecision probably derives in part from historical arguments about the relative importance in ecology of what we have termed "dynamic balance" (change in the environment is tracked by the community, but the community returns to some "normal" condition) and "persistence" (changes in the community are not always followed by return to some "normal" condition; communities possessing persistence never change so drastically as to be unrecognizable by some set of established criteria). Currently, most ecologists appear to favor the persistence viewpoint. Part of the imprecision also probably derives from the variety of meanings associated with the stability concept. These many meanings are specifically tailored to the wide variety of ecological situations that can arise. Finally, part of the imprecision likely comes from the fact that stability may be judged on a variety of scales, some of which lend themselves to precise measurement, and some of which do not.

4. The Problem of Terminology

The concept of stability currently appears to be as vaguely understood as it was when Orians reviewed it in 1975 (see Chesson and Case 1986, Connell and Sousa 1983, Rutledge et al. 1976, Santos and Bloom 1980). Pimm (1984, p. 322; see also Pimm and Redfern 1988), for example, lists five ways in which ecologists think about the concept, and his list reveals how modestly the vagueness has been reduced since Orians' (1975) analysis of the concept:

(1) the variables (his list of "variables" includes individual species abundances, species composition, and trophic level abundance) all return to the initial equilibrium (he seems to mean the conditions of variables before perturbation, although he does not say so explicitly) following perturbation ('stability');

(2) how fast the variables return towards their equilibrium following a perturbation ('resilience');

(3) the time a variable has a particular value before it is changed to a new value ('persistence');

(4) the degree to which a variable is changed, following a perturbation ('resistance'); and

(5) the variance ($(\text{animals})^2$/unit area) of population densities over time, or allied measures ('variability').

Part of the problem associated with precise understanding of stability is the existence of at least four closely related terms: 'stability', 'balance of nature', 'equilibrium', and 'homeostasis'. If one examines definitions of these four terms taken from classic secondary sources, each of which distinguishes two or more of the terms (Begon et al. 1986, Brewer 1979, 1988, Chesson and Case 1986, Colinvaux 1973, 1986, Collier et al. 1973, Connell and Sousa 1983, Grant 1986, Hall et al. 1970, Hanson 1962, Hubbell and Foster 1986, Jordan et al. 1972, Krebs 1972, 1985, Lotka 1925, MacArthur 1955, Odum 1963, Oosting 1958, Patten 1962, Pianka 1988, Pielou 1974, Ricklefs 1973, Smith 1980, 1986, Tansley 1935, Whittaker 1970, Williamson 1987), then one is able to trace some development of the stability concept.[4] It has changed recently, and in the manner that we suggested in the second section of this essay: namely, ecologists currently appear to favor the "persistence" viewpoint over the "dynamic balance" viewpoint.

Based on the wording they employed, the definitions found in the 28 sources (Begon et al. 1986, and so on; see endnote 4) seemed to fall naturally into three categories. The first category includes definitions whose wording suggests that some attribute of the community (e.g., species composition, relative abundances) exhibits *no change* in the face of disturbance or returns to initial conditions after disturbance (e.g., "constancy," "dampened oscillations," "recovery," "resistance"). The second category includes definitions whose wording suggests *persistence* of some attribute of the community (e.g., "bounded," "low chance of extinction," "persistence"). The third category includes definitions whose wording suggests *interaction* among components (usually species) of the community (e.g., "assembly," "checks and balances," "integration," "partitioning").

The proportions of definitions that fall into each of the three categories—(A) no change or balance, (B) persistence, and (C) interaction, respectively—when examined chronologically suggest a recent decline in emphasis on (A) lack of change in community attributes: Prior to 1900, during the 1960's, during the 1970's, and during the 1980's, respectively, the percents of (A) definitions (in the 28 sources) falling into each of the four time periods, respectively, are 0, 26%, 41%, and 33%. The proportions also suggest a recent substantial increase in emphasis on (B) persistence: The percents of (B) definitions (in the 28 sources) falling into each of the same four time periods, respectively, are 0, 0, 9%, and 91%. Admittedly, of course, these observations might change if one discovered other ecological texts containing comparative definitions of the four related terms (see endnote 4). Likewise, the proportions of definitions suggest that presuppositions about (C) interaction among community components was greater before and during the 1960's than after the 1960's. The percents of (C) definitions (in the 28 sources) falling into each of the same four time periods, respectively, are 58%, 25%, 4%, and 13%. Emphasis on interaction appears to have been least during the 1970's. We have no explanation for the apparent modest increase in use of such wording in the 1980's. Perhaps our sample size is not large enough and representative enough to have revealed a real pattern. If it was large enough and representative enough, then one possible reason for the decline in emphasis on interaction after the 1960's might be the general acceptance, at that time, of the importance of interactions in structuring communities. If ecologists supposed that interactions were integral to community structure, then they might also have supposed that interactions were integral to community stability. Thus, they might have seen no need to incorporate interactions explicitly into definitions of 'stability' and related terms. We also think that part of the reduced emphasis on interaction after the 1960's could be due to the increased emphasis on community persistence. Emphasis on persistence, rather than on lack of change in community attributes or on return to initial conditions of these attributes after disturbance, simultaneously reduces emphasis on interaction of components, at least in the sense of self-regulating feedback. This is because such feedback is not necessary to promote persistence.

5. Significance of the Uncertainties Surrounding the Stability Concept

Much of this discussion has focused on two main categories of evidence for the imprecision and uncertainty surrounding the stability concept: problems with determining the temporal and spatial scale over which stability is to be judged and problems with the variety of meanings ecologists attribute to stability terms. What is the epistemological significance of these and other problems with the ecological concept of stability?

Following Wimsatt's (1987; see also Wimsatt 1980) analysis of how and why some false models can often help us to find better ones, one might argue that some uncertain concepts in community ecology, like stability, can often help us to find more certain or precise ones. In other words, one might argue that, just as there are epistemological roles for false models, so also there are epistemological roles for imprecise concepts. After all, every science is uncertain, and no concepts or evidence in community ecology (or anywhere else) are unassailable. Given this unavoidable uncertainty, it may be that the problems we have uncovered in stability concepts are not so much evidence of obstacles to theorizing but, instead, are somehow heuristically useful for the progress of community ecology.

On Wimsatt's (1987) scheme, there are at least seven ways in which models can be false, four of which sometimes produce useful insights for error-correcting activity and three of which are rarely useful in such situations. Useful types of falsity, for Wimsatt, are cases in which the model may be: only locally applicable; an idealization never realized in nature; incomplete in leaving out causally relevant variables; and misdescriptive of the interactions of some variables. Rarely useful or not useful types of model falsity, for Wimsatt, are those in which the model is: a totally wrong-headed picture of nature; purely phenomenological; or erroneous in its predictions, that is, consistent with any of the preceding states of affairs.

If one assumes (as seems reasonable) both that Wimsatt's (1987; see also Wimsatt 1980) categorization of "useful" and "rarely useful" false models is approximately correct, and that this categorization yields analogous insights for distinguishing between "useful" and "rarely useful" conceptual imprecision in science, then his analysis provides one vehicle for interpreting the epistemological significance of the problems with the stability concept. Following these two assumptions, do the problems associated with the concept fit into analogous categories of "useful" or "rarely useful" instances of imprecision?

Problems (1) and (3)—respectively, that stability concepts at the population and community levels are not clearly related, and that stability may be judged present or absent, depending on the chosen spatial and temporal scales—are arguably instances of conceptual imprecision that could be useful in achieving greater precision. These two difficulties are, in large part, that the concept alleged applicable at one level or scale is not clearly applicable at another level. Wimsatt appears correct, that such problems of applicability are problems that can provide a basis for further scientific progress, because we can localize applicability problems and hence can use what he calls "piecemeal engineering" to improve the concept or model. For example, one could look for density independence or density dependence at different spatial scales, and then use the findings to determine the precise concepts that were applicable or inapplicable to different communities. Hence, problems (1) and (3) do not appear to be prima facie obstacles to theorizing in community ecology; indeed, they may well aid it, because they are able, at least, to pinpoint the precise areas in which applicability of a particular concept of stability is problematic.

The problem (2) that stability is difficult to measure, likewise, appears to be one that may be epistemologically useful. Although measurement difficulties may be indicative of conceptual problems of varying degrees of seriousness, the measurement problems associated with the stability concept appear to be difficulties associated with idealization in science. That is, the conditions of applicability of the concept are different in different situations in nature because different measures of stability are appropriate for different situations (Goodman 1975, Pimm 1984, Williamson 1987). Hence the concept itself is idealized. But, because different measures for the concept are apparently applicable to different situations, it appears possible to localize problems of idealization, given different ecological communities. For example, different temporal measures for stability-related concepts are applicable to marine, as opposed to terrestrial, communities, because of the longer cycles of many processes in marine communities. Because it appears possible to localize measurement problems and hence to modify, correct, or make more precise the idealized concept by working with the unrealistic parts—such as temporal measures that are too short—the typical sorts of difficulties associated with measuring stability seem to be of the "useful" type for theorizing. Hence, problem (2) does not appear to be a prima facie obstacle to theorizing in ecology.

The variety of meanings attributed to stability terms, problem (4) discussed earlier, however, appears to be a difficulty that is unlikely to be useful for further theorizing in community ecology. One reason is that the meaning variability is a manifestation of conceptual incoherence and perhaps even inconsistency. The five ways of thinking about the concept of stability, as outlined by Pimm (1984), for example, do not even all describe alleged characteristics of communities; some of them refer to the time during which community changes take place, rather than to the change itself, and some presuppose different spatial and temporal scales than other ones. Indeed, just as some scholars have argued about the species concept (see, for example, Ghiseli 1969, 1974, Gould 1982, Hull 1976, 1978, Rosenberg 1985; see also Kitcher 1985), there is no universally applicable specification of the characteristics that describe stability, no set of necessary and sufficient conditions for a community's being stable. What, in practice, is often described as stability is a disjunction of relationships, none of which is essential for all instances of stability. There is no homogeneous class of processes or relationships that exhibit stability, and no single, adequate account of what stability is. Hence, the conceptual incoherence surrounding stability terms and meanings appears more to block heuristic power and scientific progress, rather than to aid it, because the more plausible meanings associated with the concept do not appear to be isolable. Hence, it will likely be difficult to pinpoint problematic meanings and to correct them.

Second, the meaning variability associated with stability concepts seems unlikely to aid future theorizing in that scientists' meaning the same things when they talk about the same concepts appears to be a necessary condition for progress in clarifying those concepts. Scientific progress in conceptual clarification seems to presuppose the ability to isolate semantic from non-semantic problems, and such isolation does not appear currently possible, given the variability of meanings associated with the stability concept.

Third, because the apparently dominant meaning associated with stability, that of persistence, is compatible with all models of community structure, from mere species coexistence to feedback models, the concept of stability does not appear to provide useful information for theorizing. The concept seems to have little "cash value" for explaining community processes. Indeed, because the current concept of stability—emphasizing persistence—seems compatible with incompatible community models, it appears consistent with almost any state of affairs and hence seems incapable of explaining any of them. Thus, to the degree that the stability concept focuses only on persistence, it appears to have little epistemological or theoretical utility for progress in community ecology. Given the absence of potential explanatory power or theoretical utility and the presence of conceptual incoherence already mentioned, it is arguable that ecologists might do better not to employ stability terms (such as 'balance', 'equilibrium', 'homeostasis', or 'stability') at all. The one precise sense in which ecologists use the term 'stability' refers to "dynamic equilibrium" (Pimm 1984), but dynamic equilibrium is highly questionable and has largely been abandoned, as we argued earlier. As we explained in preceding paragraphs, it is a relationship supportable only if one misrepresents the ecological evidence (Connell and Sousa 1983). The alternative to employing an imprecise general concept of stability, or to using a precise but discredited term, 'stability', might be to abandon use of the concept and term altogether. Following our discussion of the various terms allegedly related to the stability concept (see Pimm 1984 and Orians 1975, for instance), community ecologists might do better to employ, analyze, and refine such specific terms as 'persistence', 'resistance', and 'variability'. For example, we might do better to clarify the temporal and spatial scales or applicability of such terms and to abandon, for the most part, use of stability terms such as 'equilibrium'.

Fourth, even if we abandon use of stability concepts and terms such as 'equilibrium' and 'homeostasis', it is not clear that pursuing the alternative—of employing precise terms such as 'persistence', 'resistance', and 'variability'—will necessarily get us "home free" in ecommunity ecology. The alternative merely appears less undesirable than to continue to use incoherent, vague, or discredited concepts. Indeed, the apparent evolution of the stability concept, away from emphasis on interactions among component species and toward emphasis on persistence of species, provides little basis for explaining patterns of stability. It appears merely to focus on presence/absence of species, independent of mechanisms that might account for such presence or absence. Hence, this evolution in meanings—toward the current emphasis on persistence—does not seem to provide a fertile ground for theorizing about what explains possible community characteristics. The most recent problems with the stability concept appear to indicate that it may be approaching heuristic bankruptcy, and that its problems may not be ones that are likely to lead us to greater conceptual clarification in community ecology.

Notes

[1]Work on this project was funded by NSF grant BBS-86-159533, "Normative Concepts in Ecology," although the opinions expressed are those of the researchers, not the NSF. The authors are grateful to Greg Cooper, Reed Noss, and Dan Simberloff for constructive criticisms of earlier drafts. See Shrader-Frechette and McCoy 1993.

[2]Random perturbations also may actually prevent stability. Lewontin notes that stability may be absent, yet species may persist, when a system is "dynamically bounded." In the terminology of physics, he says (p. 19):

> a system is 'dynamically bounded' in some interior set *S* if at all points in the neighborhood of the boundary set *B* the transformation vectors point into the interior set *S*.

A simple way to illustrate dynamic boundedness is to assume that the environment varies randomly in such a way that one species is favored on the average but a second species is very strongly favored occasionally. In this case, no stable equilibrium abundance of the two species (relative to each other) will be reached, but neither species will be eliminated.

[3]Currently, some researchers (e.g., Maurer 1987, Moore and Hunt 1988, O'Neill et al. 1986) propose that complexity is related to community stability, but in a different way than simply by risk spreading. They think that complexity derives from the arrangement of species into community subunits, and that stability derives from the inter-connections among these subunits. Their proposal requires further examination (see Winemiller 1989).

[4]We attempted to employ computer searches of databases to uncover books and articles in which definitions of the four "stability" terms appeared, but this effort proved not to be workable, for three reasons. *First*, the most important database, BIOSIS, originated in 1969 and was unable to provide us information on stability concepts prior to that date. *Second*, even during the years 1969-present, the number of books and articles listed in BIOSIS as addressing the four stability terms was too large (tens of thousands) to be examined. *Third*, there were no database entries for terms like "stability, concept of." Faced with too much material and no apparent way of circumscribing it in a wholly rational way, we compiled a list of explicit definitions of the four stability terms from a variety of sources including (1) classical ecological textbooks and dictionaries, (2) well known collections of readings in ecology, (3) "citation classics" in ecology, 80 of the most highly cited publications in ecology between 1947 and 1977, and (4) other books and articles known in advance to contain two or more definitions of the four stability terms. The result is our analysis of stability terms contained in 28 works. We realize that biases may have crept into our analysis, but we know of no alternatives to the approach we took, and we invite suggestions on ways to improve this approach.

References

Allee, W.C., Park, O., Emerson, A.E., Park, T. and Schmidt, K.P. (1949), *Principles of Animal Ecology*. Philadelphia, PA: W.B. Saunders.

Andrewartha, H.G. and Birch, L.C. (1954), *The Distribution and Abundance of Animals*. Chicago: University of Chicago Press.

Begon, M., Harper, J. and Townsend C. (1986), *Ecology: Individuals, Populations, and Communities*. Sunderland, MA: Sinauer.

Berryman, A. (1987), "Equilibrium or Nonequilibrium: Is That the Question?" *Ecological Society of America* 68: 500-502.

Blondel, J. (1987), "From Biogeography to Life History Theory: a Multithematic Approach Illustrated by the Biogeography of Vertebrates", *Journal of Biogeography* 14: 405-422.

Brewer, R. (1979), *Principles of Ecology*. Philadelphia: W.B. Saunders.

______. (1988), *The Science of Ecology*. Philadelphia: Saunders College.

Chesson, P. L. and Case, T. J. (1986), "Nonequilibrium Community Theories: Chance, Variability, History, and Coexistence", in *Community Ecology*, J. Diamond and T.J. Case (eds.). New York: Harper and Row, pp. 229-239.

Clements, F.E. (1905), *Research Methods in Ecology*. Lincoln, NE: University Publ. Co. Reprinted 1977, New York: Arno Press.

Clements, F.E. and Shelford, V.E. (1939), *Bio-ecology*. New York: Wiley & Sons.

Colinvaux, P. (1973), *Introduction to Ecology*. New York: Wiley & Sons.

_______. (1986), *Ecology*. New York: Wiley & Sons.

Collier, B.D., Cox, G, W., Johnson, A.W. and Miller, P.C. (1973), *Dynamic Ecology*. Englewood Cliffs: Prentice-Hall.

Connell, J.H. and Sousa, W.P. (1983), "On the Evidence Needed to Judge Ecological Stability or Persistence", *American Naturalist* 121: 789-824.

Davis, M.B. (1986), "Climatic Instability, Time Lags, and Community Disequilibrium", in *Community Ecology*, J. Diamond and T.J. Case (eds.). New York: Harper and Row, pp. 269-284.

Dice, L.R. (1952), *Natural Communities*. Ann Arbor, MI: University of Michigan Press.

Egerton, F.N. (1973), "Changing Concepts in the Balance of Nature", *The Quarterly Review of Biology* 48: 322-350.

Ehrlich, P.R. and Birch, L.C. (1967), "The Balance of Nature and Population Control", *American Naturalist* 101: 97-124.

Elton, C. (1930), *Animal Ecology and Evolution*. New York: Oxford University Press.

Emerson, A. E. (1954), "Dynamic Homeostasis: a Unifying Principle in Organic, Social, and Ethical Evolution", *Scientific Monthly* 78: 67-85.

Forbes, S. A. (1880), "On Some Interactions of Organisms", *Bulletin of the Illinois State Laboratory of Natural History* 1: 3-17.

Ghiselin, M. (1969), *The Triumph of the Darwinian Method*. Berkeley: University of California Press.

_______. (1974), "A Radical Solution to the Species Problem", *Systematic Zoology* 23: 536-544.

Goodman, D. (1975), "The Theory of Diversity-Stability Relationships in Ecology", *The Quarterly Review of Biology* 50: 237-266.

Grant, P.R. (1986), "Interspecific Competition in Fluctuating Environments", in *Community Ecology*, J. Diamond and T.J. Case (eds.). New York: Harper and Row, pp. 173-191.

Hall, D.J., Cooper, J.F. and Werner, E.E. (1970), "An Experimental Approach to the Production Dynamics and Structure of Freshwater Animal Communities", *Limnology and Oceanography* 15: 839-928.

Hanson, H.C. (1962), *Dictionary of Ecology*. New York: Philosophical Library.

Hengeveld, R. (1988), "Mechanisms of Biological Invasions", Journal of Biogeography 15: 819-828.

_________. (1989), *Dynamics of Biological Invasions*. London, UK: Chapman and Hall.

Holling, C.S. (1973), "Resilience and Stability of Ecological Systems," *Annual Review of Ecology and Systematics* 4: 1-23.

Hubbell, S.P. and Foster, R.B. (1986), "Biology, Chance, and History and the Structure of Tropical Rain Forest Tree Communities", in *Community Ecology*, J. Diamond and T.J. Case (eds.). New York: Harper and Row, pp. 314-329.

Hull, D. (1976), "Are Species Really Individuals", *Systematic Zoology* 25: 174-191.

____. (1978), "A Matter of Individuality", *Philosophy of Science* 45: 335-360.

____. (1988), *Science as a Process*. Chicago: University of Chicago Press.

Jordan, C.F., Kline, J.R. and Sasscer D.S. (1972), "The Relative Stability of Mineral Cycles in Forest Ecosystems", *American Naturalist* 106: 237-253.

Kendeigh, S.C. (1961), *Animal Ecology*. Englewood Cliffs: Prentice-Hall.

Kitcher, P. (1985), *Species*. Cambridge: MIT Press.

Krebs, C.J. (1972), *Ecology: The Experimental Analysis of Distribution and Abundance*. 3rd ed (1985). New York: Harper and Row.

Levins, R. (1974), "The Qualitative Analysis of Partially Specified Systems", *Annals of the New York Academy of Sciences* 231: 123-138.

_____. (1975), "Evolution in Communities Near Equilibrium", in *Ecology and Evolution of Communities*, M.L. Cody and J.M. Diamond (eds.). Cambridge: Harvard University Press, pp. 16-50.

Lewin, R. (1986), "In Ecology, Change Brings Stability", *Science* 234: 1071-1073.

Lewontin, R. (1969), "The Meaning of Stability", in *Diversity and Stability in Ecological Systems*, Brookhaven Laboratory Publication No. 22, G. Woodwell and H. Smith (eds.). Brookhaven: Brookhaven Laboratory.

Lotka, A.J. (1925), *Elements of Physical Biology*. New York: Williams and Wilkins. Reprinted 1956, New York: Dover.

McIntosh, R.P. (1985), *The Background of Ecology: Concept and Theory*. Cambridge: Cambridge University Press.

MacArthur, R. (1955), "Fluctuations of Animal Populations, and a Measure of Community Stability", *Ecology* 36: 533-36.

_________. (1971), "Patterns of Terrestrial Bird Communities", in *Avian Biology*, D.S. Farner and J.R. King (eds.). New York: Academic Press, pp. 189-221.

________. (1972), *Geographical Ecology*. New York: Harper and Row.

Maurer, B.A. (1987), "Scaling of Biological Community Structure: A Systems Approach to Community Complexity", *Journal of Theoretical Biology* 127: 97-110.

May, R.M. (1972), "Will a Large Complex System Be Stable?" *Nature* 238: 413-414.

______. (1973), *Stability and Complexity in Model Ecosystems*. Princeton, NJ: Princeton University Press.

Mayr, E. (1982), *The Growth of Biological Thought: Diversity, Evolution, and Inheritance*. Cambridge, MA: Harvard University Press.

____. (1988), *Toward a New Philosophy of Biology: Observations of an Evolutionist*. Cambridge, MA: The Belknap Press of Harvard University Press.

Moore, J.C. and Hunt, H. W. (1988), "Resource Compartmentation and the Stability of Real Ecosystems", *Nature* 333: 261-263.

Morris, D.W. (1988), "Habitat-Dependent Population Regulation and Community Structure", *Evolutionary Ecology* 2: 253-269.

Nicholson, A.J. (1933), "The Balance of Animal Populations", *Journal of Animal Ecology* 2: 132-178.

Nicholson, A.J. and Bailey, V.A. (1935), "The Balance of Animal Populations", *Proceedings of the Zoological Society of London*: 551-598.

Odum, E.P. (1953), *Fundamentals of Ecology*. Philadelphia: Saunders.

______. (1963), *Ecology*. New York: Holt, Rinehart, and Winston.

O'Neill, R.V., DeAngelis, D.L., Waide, J.B. and Allen, T.F.H. (1986), *A Hierarchical Concept of Ecosystems*. Princeton: Princeton University Press.

Orians, G.H. (1975), "Diversity, Stability and Maturity in Natural Ecosystems", in *Unifying Concepts in Ecology*, W.H. VanDobben and R.H. Lowe-McConnell (eds.). The Hague, Netherlands: W. Junk, pp. 139-150.

Oosting, H.J. (1958), *The Study of Plant Communities*. San Francisco: W.H. Freeman and Company.

Paine, R.T. (1966), "Food Web Complexity and Species Diversity",*American Naturalist* 100: 65-75.

Patten, B.C. (1962), "Species Diversity in Net Phytoplankton of Raritan Bay", *Journal of Marine Research* 20: 57-75.

Pianka, E.R. (1988), *Evolutionary Ecology*. 4th ed. New York: Harper and Row.

Pielou, E.C. (1974), *Population and Community Ecology: Principles and Methods*. New York: Gordon and Breach.

Pimm, S.L. (1982), *Food Webs*. London: Chapman and Hall.

______. (1984), "The Complexity and Stability of Ecosystems", *Nature* 307: 321-326.

Pimm, S.L. and Redfern, A. (1988), "The Variability of Population Densities", *Nature* 334: 613-614.

Pulliam, R. (1988), "Sources, Sinks, and Population Regulation", *American Naturalist* 132: 653-661.

Reddingius, J. and den Boer, P.J. (1970), "Simulation Experiments Illustrating Stabilization of Animal Numbers by Spreading of Risk", *Oecologia* 5: 240-284.

Ricklefs, R.E. (1973), *Ecology*. Newton, MA: Chiron Press.

________. (1987), "Community Diversity: Relative Roles of Local and Regional Processes", *Science* 235: 167-171.

Rosenberg, A. (1985), *The Structure of Biological Science*. Cambridge: Cambridge University Press.

Rutledge, R.W.; Basore, B.L. and Mulholland, R.J. (1976), "Ecological Stability: an Information Theory Viewpoint", *Journal of Theoretical Biology* 57: 355-371.

Santos, S.L. and Bloom, S.A. (1980), "Stability in an Annually Defaunated Estuarine Soft-Bottom Community", *Oecologia* 46: 290-294.

Shrader-Frechette, K. and McCoy, E. (1993), *Method in Applied Ecology*. Cambridge: Cambridge University Press.

Shugart, H.H., Jr. and West, D.C. (1981), "Long-Term Dynamics of Forest Ecosystems", *American Scientist* 69: 647-652.

Smith, R.L. (1980), *Ecology and Field Biology*. 3rd ed. New York: Harper and Row.

______. (1986), *Elements of Ecology*. 2nd ed. New York: Harper and Row.

Strong, D.R. (1986), "Population Theory and Understanding Pest Outbreaks", in *Ecological Theory and Integrated Pest Management Practices*, M. Kogan (ed.). New York: Wiley & Sons, pp. 37-58.

Tansley, A.G. (1935), "The Use and Abuse of Vegetational Concepts and Terms", *Ecology* 16: 284-307.

van Emden, H.F. and Williams, G.F. (1974), "Insect Stability and Diversity in Agro-Ecosystems", *Annual Review of Entomology* 19: 455-475.

Walker, D. (1989), "Diversity and Stability", in *Ecological Concepts*, J.M. Cherrett (ed.). London: Blackwell, pp. 115-145.

Whittaker, R.H. (1970), *Communities and Ecosystems*. New York: Macmillan.

Wiens, J.A. and Rotenberry, J.T. (1981), "Censusing and the Evaluation of Avian Habitat Occupancy", *Studies in Avian Biology* 6: 522-532.

Wimsatt, W. (1980), "Randomness and Perceived-Randomness in Evolutionary Biology", *Synthese* 43: 287-329.

_ _ _ _ _ . (1987), "False Models and Means to Truer Theories", in *Neutral Models in Biology*, M.H. Nitecki and A. Hoffman (eds.). New York: Oxford University Press, pp. 23-55.

Winemiller, K.O. (1989), "Must Connectance Decrease with Species Richness?" *American Naturalist* 134: 960-968.

Williamson, M. (1987), "Are Communities Ever Stable?" *Symposium of the British Ecological Society*: 353-370.

Wu, L.S.Y. (1977), "The Stability of Ecosystems—A Finite-Time Approach", *Journal of Theoretical Biology* 66: 345-359.

Part VI

DECISION THEORY

Kings and Prisoners (and Aces)

Jordan Howard Sobel

University of Toronto

What we make of information we come to have should take into account that we have come to have it, and how we think we have come to have it.

I relate this homily to several puzzles. In one, three cards, of which I know one is a king, lie face-down. After I select, without inspecting, a card and bet that it is a king, you reveal that a certain other card is not a king. I wonder what this does to my chances on that bet. In another puzzle I am one of three prisoners and learn that one of us will be released. Then I learn that a certain other prisoner will not be released. Again I wonder what this does to my chances. After demonstrating the utility of attending to that homily in these cases, I take up the question, What is it about them that makes this homily remarkable for them, and makes them puzzles? An appendix takes up second-ace problems.

I. Choosing Kings

1. There are three cards with numbers on their backs.

One is a king, and two have blank faces. You do, but I do not, know which is which. I pick a card, say card 1, without turning it up.

Then you turn up one of the other cards, say card 2, and reveal that it is not a king. You do not just happen to do this. Your business was to turn up a blank card.

PSA 1992, Volume 1, pp. 203-216

Should I now be more confident that I have picked the king? Before your turn I placed a bet on my card 1's being the king, a bet on which I stand to win \$2, or to lose \$1. [I care only about the money, so this bet was fair. I thought I had 1 chance in 3 of winning \$2, and 2 chances in 3 of losing \$1. I considered the expected monetary value of my bet to be 0—(1/3 X +2) + (2/3 X -1) = 0.] *If allowed should I now be eager to bet another \$1 at these odds?* And what about card 3? Should I now think better of its chances of being the king? Suppose that without changing other terms in my bet I can make it ride on card 3 rather than card 1. *Should I now be eager to do that?*

Answers to these questions depend on details that have not been fixed. After picking a card I come into new information, including the information that card 2 is not a king. But what I should make of this information depends on how I think it was generated; it depends in particular on what I take to be the rules and procedures that culminated in your turning up card 2. It is a general principle that what I should make of information that has come my way is identical with what I should make of its having come my way.

Two stories will be considered in which I have definite views concerning your rôle in Choosing Kings. Differences in these views lead to my making different things of the new information that card 2 is not a king, and revising differently my probabilities for cards 1 and 3 being the king. In the first plot offered below I am sure that my choice is not known to you before your turn; in the second I am sure of this, and that you take care not to turn card I pick.

What if I had no idea what you were up to? I think that then, if I really had no idea at all not even a probable conjecture, there would be no saying what I should do with my new information that card 2 is not a king. Even though I was sure that I had learned this by watching you turn card 2, if I had no views concerning why you turned it rather than the other blank card, I would not know what to make of the information gained by watching you that it is not a king.

The case in which I had no idea what you are up to would be, as far as my new certainty that card 2 is not a king goes, like the even more extraordinary case in which I had absolutely no idea whence this new certainty had come. In that case I would not view it as 'information', that is, as something I had 'learned': I would not view it as a conviction caused by "impingements...of realit...unmediated by cerebral processing" (Sobel 1990, p. 506), and so I would not view it as a conviction to be accommodated. Nor would I view it as something that had just popped into my head from nowhere, and so as an aberration to be questioned and perhaps expunged. In this extreme case I would be at a complete loss what to make the new certainty that card 2 is not a king. Similarly for the less extreme but still extraordinary case in which, though I knew that this new certainty had come from watching you turn up card 2, I had no idea, not even a probable conjecture, why you had done that.

2. First Scenario

Suppose I am sure that you will not know what card I choose and bet on, and sure that you will select at random a blank card to turn. I have, both in the beginning and after my choice but before your turn, the following equal probabilities for combinations of truth-values for propositions (and for six-conjunct corresponding conjunctions of these propositions and/or negations of them): 'K1' abbreviates 'card 1 is a king'; 'T1' abbreviates 'you turn up card 1'; and so on.

	K1	K2	K3	T1	T2	T3	P_1
Card 3 is the king.	f	f	t	t	f	f	1/6
	f	f	t	f	t	f	1/6
Card 2 is the king.	f	t	f	t	f	f	1/6
	f	t	f	f	f	t	1/6
Card 1 is the king.	t	f	f	f	t	f	1/6
	t	f	f	f	f	t	1/6

There are three equally probable possibilities for which card is the king, and in each there are two equally probable possibilities for which blank card you will choose randomly to turn. This is so before I choose card 1 to be 'my' card, and it is so after. Since I am sure that my choice is not known to you, I am sure that it makes no difference to your turn. In this scenario I am not sure that you will not turn the card I choose.

On seeing you turn card 2 my probabilities should become,

	K1	K2	K3	T1	T2	T3	P_1'
Card 3 is the king; you turn card 2.	f	f	t	t	f	f	0
	f	f	t	f	t	f	1/2
Card 2 is the king; you turn card 2.	f	t	f	t	f	f	0
	f	t	f	f	f	t	0
Card 1 is the king; you turn card 2.	t	f	f	f	t	f	1/2
	t	f	f	f	f	t	0

Upon seeing you turn card 2, I am sure you did not turn either card 1 or card 3 (lines 1, 3, 4, and 6). And seeing that card 2 is not the king, I am sure that exactly one of the other cards is the king, and I view these chances as equal (lines 2 and 5). Intuitively, I am sure that you turned at random a blank card, and think that it is as likely that you turned card 2 when you could have turned card 1 instead (line 2), as that you turned it when you could have turned card 3 instead (line 5).

For corroboration of and a more theoretical perspective on the entries on lines 2 and 5, since on seeing you turn card 2 I learn that, T2, and nothing else that is of further relevance to either K1 or K3 independent of that, my updated probabilities, P_1', for K3 and K1 should equal my *conditional probabilities* for K3 and K1 on T2,

$$P_1'(K3) = P_1(K3/T2) = P_1(T2 \& K3)/P_1(T2) = (1/6)/[(1/6) + (1/6)] = 1/2,$$

and,

$$P_1'(K1) = P_1(K1/T2) = P_1(T2 \& K1)/P_1(T2) = (1/6)/[(1/6) + (1/6)] = 1/2.$$

On seeing you turn card 2 I of course learn in addition to T2 that ¬K2, but I knew I would learn that in addition to T2, so it is not of further relevance to anything independent of T2,

$$P_1(T2 \rightarrow \neg K2) = 1,$$

so that,

$$P_1[(\neg K2 \ \& \ T2) \leftrightarrow T2] = 1,$$

and in particular, it is not of further independent relevance to either K1 or K3: interchanging probabilistic equivalents (¬K2 & T2) and T2 yields the identities,

$$P_1[K3/(\neg K2 \ \& \ T2)] = P_1(K3/T2),$$

and,

$$P_1[K1/(\neg K2 \ \& \ T2)] = P_1(K1/T2).$$

Reflection should convince that it is implicit in Choosing Kings, in every natural version of Choosing Kings, that nothing that I learn when I learn T2 is relevant to these things independent of its relevance, so that for updating purposes it can stand for everything that I learn that is relevant to them. [I let **s** be of possible evidential relevance to **q** at a time **t** for a person independent of the relevance of **p** if and only if, if P is this person's probability function for t, then $\neg(P[\mathbf{q}/((\mathbf{s} \ \& \ \mathbf{p})] = P(q/p))$.]

Let me now connect the analysis of this scenario with the homily that heads this paper. It is implicit in Choosing Kings that I am sure that I will 'learn' (i.e., be made certain by impingements of reality unmediated by cerebral processing) that a card is not a king if and only if you turn it, and that I will 'learn' that a card is not a king only if it is in fact not a king. Letting '**L**' abbreviate 'I will at the time of your turn learn that', it is in particular implicit that,

$$P_1[\mathbf{L}(\neg K2) \leftrightarrow T2] = 1,$$

so that **L**(¬K2) is not relevant to anything independent of the relevance of T2. And it is implicit that,

$$P_1[\mathbf{L}(\neg K2) \rightarrow \neg K2] = 1,$$

and thus that,

$$P_1([\mathbf{L}(\neg K2) \ \& \ \neg K2] \leftrightarrow \mathbf{L}(\neg K2)) = 1,$$

and so that,

$$P_1([\mathbf{L}(\neg K2) \ \& \ \neg K2] \leftrightarrow T2) = 1.$$

This means that pivotal probabilities that are conditional on T2 equal corresponding probabilities conditional on [**L**(¬K2) & ¬K2],

$$P_1(K3/T2) = P_1(K3/[\mathbf{L}(\neg K2) \ \& \ \neg K2]),$$

and,

$$P_1(K1/T2) = P_1(K1/[\mathbf{L}(\neg K2) \ \& \ \neg K2]).$$

Probabilities conditional on [**L**(¬K2) & ¬K2] reflect my views concerning whence information that ¬K2 would come, views which that homily cautions should be allowed to hold sway. These probabilities are authoritative indications of things I should make of the information that K2 has come my way because I do not expect, if and when I learn ¬K2, to learn at the same time anything of relevance to K3 and K1 independent of the relevance of [**L**(¬K2) & ¬K2], or, equivalently in my view, of T2, and because, in the event I am right, and do not then learn anything else of such independent relevance.

I have been relying, and will continue to rely below in this paper, on a general principle for updating endorsed in (Sobel 1990). Here for ease of reference is that principle slightly revised:

> For any person whose probability function for **t** is P and whose probability function for a subsequent time **t'** is P': if this person learns **p** for sure in the interval (**t**,**t'**), is sure 'just before' t' that **p**, and that he learned **p** for sure in that interval; then, if this person is ideally rational, for every **q** such that, for any **r** that this person has learned in the interval (**t**,**t'**) that is of possible evidential relevance to **q** independent of relevance of **p**, there is an **s** that entails **r** such that, (a), this person has learned **s** in this interval and is sure 'just before' **t'** that s and that he learned **s** in this interval, and, (b), **s** is not of possible evidential relevance to **q** independent of relevance of p, $[P'(q) = P_p(q)]$.

For this principle, let something be learned in an interval (t,t') if and only if it is learned at a time later than **t** and earlier than **t'**. Let a person be sure just before **t'** of p if and only if, for some time **t** that is earlier than **t'**, he is sure of **p** at every time that is later than **t** and earlier than **t'**.

In all versions of Choosing Kings, probabilities that are conditional on [**L**(¬K2) & ¬K2] are authoritative indications of what, as far as K3 and K1 are concerned, I should make of the information that ¬K2 should this information come my way. I note that in the present first scenario corresponding probabilities conditional simply on ¬K2 happen also to be authoritative indicators, since they equal corresponding probabilities conditional on [**L**(¬K2) & ¬K2]:

$$P_1(K3/\neg K2) = (2/6)/(4/6) = 1/2 = P_1(K3/[\mathbf{L}(\neg K2) \& \neg K2]),$$

and

$$P_1(K1/\neg K2) = (2/6)/(4/6) = 1/2 = P_1(K1/[\mathbf{L}(\neg K2) \& \neg K2]).$$

In this first scenario, that I have learned ¬K2, **L**(¬K2), which is in any case not of further relevance to anything independent of the relevance of T2, is also not of further relevance to these things, K1 and K3, independent of the relevance of ¬K2.

To sum up and answering the questions of Choosing Kings, I should, in the first scenario, on seeing you turn card 2 feel better about the chances of my card 1 being the king. I thought there was 1 chance in 3 for that,

$$P_1(K1) = 1/3,$$

but I should now think that there is 1 chance in 2,

$$P_1'(K1) = 1/2.$$

I should feel that the bet I made on K1 on which I stand to win \$2 or lose \$1 is now more than fair. If allowed, and if my expected utilities here are linear with my expected monetary returns, I should be pleased to double it. And I should not be eager to switch my bet to one that rides on card 3's being the king, for, though neither my probability for K1 nor my probability for K3 should be what it was in the beginning since they were both 1/3, they should still be equal—they should now both be 1/2.

3. Second Scenario

Suppose, in contrast with the first scenario, that I am sure that my choice of a card is *public* and known to you, that after my choice you consider as alternatives for turning only the other cards, that if both of them are blanks you select at random one to turn, and that if only one is a blank you turn it straightaway. In this scenario, though my probabilities, P_2, for each of the six combinations should be initially 1/6 as in the first scenario; after choosing card 1, and before you turn up a card, my probabilities for these combinations should be as follows:

	K1	K2	K3	T1	T2	T3	P_2'
Card 3 is the king;	f	f	t	t	f	f	0
I choose card 1.	f	f	t	f	t	f	2/6
Card 2 is the king;	f	t	f	t	f	f	0
I choose card 1.	f	t	f	f	f	t	2/6
Card 1 is the king;	t	f	f	f	t	f	1/6
I choose card 1.	t	f	f	f	f	t	1/6

I am in this second scenario sure you will not turn card 1, the card I have chosen (lines 1 and 3). For line 2, I still think, after I choose card 1, that there is 1 chance in 3 that card 3 is the king, so my probabilities for the combinations on lines 1 and 2 must still sum to 1/3. Similarly for line 4. And similarly for lines 5 and 6. Furthermore regarding these lines, probabilities for combinations on them should be equal, since I am sure that if, unknown to me but known to you, I have chosen the king, you will select at random one of the other cards to turn.

After you turn card 2 and reveal that it is not a king, my probabilities should be:

	K1	K2	K3	T1	T2	T3	P_2''
Card 3 is the king; I choose	f	f	t	t	f	f	0
card 1; you turn card 2	f	f	t	f	t	f	2/3
Card 2 is the king; I choose	f	t	f	t	f	f	0
card 1; you turn card 2	f	t	f	f	f	t	0
Card 1 is the king; I choose	t	f	f	f	t	f	1/3
card 1; you turn card 2	t	f	f	f	f	t	0

Seeing you turn card 2, I am sure you did not turn card 3 (lines 4 and 6), and I should think that it is half as likely that you could as well have turned card 2 (line 5) and that you thus turned card 2 at random, as it is likely that you had no choice but to turn card 2, since of the cards I did not choose only it was a blank (line 2). For corroboration of these last probabilities, we have that my new probabilities for K3 and K1 should be what were my corresponding probabilities conditional on T2:

$$P_2'''(K3) = P_2'(K3/T2) = P_2'(T2 \& K3)/P(T2) = (2/6)/(3/6) = 2/3,$$

and

$$P_2''(K1) = P_2'(K1/T2) = P_2'(T2 \& K1)/P2'(T2) = (1/6)/(3/6) = 1/3.$$

In all versions of Choosing Kings, these authoritative conditional probabilities equal probabilities conditional on my 'learning' that card 2 is not a king when it is not a king:

$$P_2'\,(K3/T2) = P_2'\,(K3/[L(\neg K2) \& \neg K2]),$$

and,

$$P_2'\,(K1/T2) = P_2'\,(K1/[L(\neg K2) \& \neg K2]).$$

But it is noteworthy that since, as in the first scenario, after my choice and before your turn,

$$P_2'\,(K3/\neg K2) = P_2'\,(K1/\neg K2] = 1/2$$

we have in contrast with the first scenario, the non-identities,

$$\neg[P_2'\,(K3/[L(\neg K2) \& \neg K2]) = P_2'\,(K3/\neg K2)]],$$

and,

$$\neg[P_2'\,(K1/[L(\neg K2) \& \neg K2]) = P_2'\,(K1/\neg K2)].$$

In this second scenario, in contrast with the first one, probabilities conditional merely on $\neg K2$ are misindications of what I should make of 'learning' that $\neg K2$. This is not surprising, since, given what I know you are up to in this second scenario, the further information that $L(\neg K2)$ is relevant both to K3 and to K1 independently of the relevance of $\neg K2$. In this second scenario, $P_2'\,(K1/\neg K2)$ would have me make too much of the information that $\neg K2$,

$$P_2'(K1/\neg K2) = 1/2 > P_2'\,(K1/[L(\neg K2) \& \neg K2]) = 1/3,$$

and $P_2'\,(K3/\neg K2)$ would have me make too little,

$$P_{'}(K3/\neg K2) = 1/2 > P_2'\,(K3/[L(\neg K2) \& \neg K2]) = 2/3.$$

Summing up for this second scenario, my final probability for my card 1' s being the king should be unchanged:

$$P_2''\ (K1) = 1/3.$$

But my probability for card 3's being the king should be enhanced:

$$P_2''\ (K3) = 2/3.$$

I should *not* after watching you turn card 2 feel better about the chances of card 1's being a king, or be eager to double my bet on it. But I *should* think better of card 3's chances and so, if allowed, be happy to switch my bet to it. Given my views concerning what you are up to, and of how my choice influences your turn, watching you turn a card provides information that is positively relevant to the unchosen and unturned card's being the king, which information is, however, quite irrelevant to my card's being the king:

$$P_2' (K3/T2) - P_2' (K3) = (2/6)/(2/6 + 1/6) - 2/6 = 1/3,$$

and,

$$P_2' K1/T2) - P_2' (K1) = (1/6)/(2/6 + 1/6) - 2/6 = 0.$$

II. The Prisoner's Predicament

Suppose that I am a prisoner, that there are two other prisoners, Jack and Jill, and that I know that one of us has been selected at random for release. I take my chances for release to be 1 in 3. It is then announced, and I believe it, that Jack will not be released. *Should I feel better about my chances after this announcement? If allowed, should I trade places with Jill, so that I will be released if and only if she was the one who had been randomly selected for release?*

As in Choosing Kings, the answers to these questions depend upon whence I think the information contained in the announcement is coming. Here are two scenarios that work like the ones for Choosing Kings. *First Scenario*: I am sure that the name of a prisoner who will not be released was selected at random from a hat that contained the names of both prisoners who will not be released. *Second Scenario*: I am sure that the hat did not contain my name; that it contained only the names of prisoners other than me who will not be released. I am sure that either I had been selected for release and the names 'Jack' and 'Jill' were both in the hat, or I had not been selected for release and only the name of the other prisoner who had not been selected was placed in the hat.

If my views are as in the first scenario, I should feel better after the announcement that Jack will be released, since, for all I still know, my name could have been announced. It wasn't, and so, though I do not know that this is because I am going to be released, I have more confidence in my release than I did before the announcement. If, on the other hand, my views are as in the second scenario, and I am sure that my name was not in the hat even if I am one of the two who will not be released, then I am sure that there had never been a chance that my name would be announced. In this case the announcement that Jack will not be released is only tantalizing. It should make no difference to my expectations for release, though I should now think better of Jill's chances, and be pleased to change places with her, just as she, if her views are, with suitable egocentric changes, similar to mine, should be pleased to change places with me.

III. What Makes These Problems Puzzles?

Confusion is possible in Choosing Kings and The Prisoner's Predicament as long as views concerning ways in which information will be, or was, generated are not made explicit. But it is not just their ambiguities that make these problems puzzles. Many 'story-problems', though seriously underspecified and ambiguous and thus strictly defective as problems, are so without being puzzles. Suppose that late one afternoon three persons walk into my clinic suffering from a form of influenza that is new to my part of the world. It is obvious that I should make one thing of this, if I think that no one knows of my special interest in forms of influenza, and another lesser thing if I think that this interest of mine is widely known, and that many physicians in my area are apt to refer novel influenza-cases to me. Settle this and the case is easy. Suppose, for another non-puzzle, that I watch as you lay out from left to right four spades from a five card poker hand that I have dealt to you. It goes without saying that I should make one thing of these spades, if I think you are disclosing your cards in the order that I dealt them, and that I should make less of it if I think that before laying out these cards you have looked at and arranged your hand.

What distinguishes our two problems and makes them puzzles is that in them confusions common as long as views regarding sources of information are left unspecified are apt to persist even when all relevant ambiguities are carefully resolved. Włodzimierz Rabinowicz has dramatic evidence of this in connection with a problem like Choosing Kings that is discussed in (Bar-Hillel 1989).

> The Cadillac Puzzle. Imagine that you are a participant in a TV game show. On stage are three large screens, one of which hides a brand new Cadillac. If you guess correctly which one it is, the car will be yours. You choose a screen. Tension in the studio mounts as the show's host approaches one of the two screens you have not chosen, and flings it open. There is no Cadillac behind this screen. He now turns to you, and says: 'Well, the choice of screens behind which the Cadillac might be has now been narrowed to just two. Do you want to stick to your previous choice, or do you want to change your mind? You re allowed to do that,' he tells you, 'for a $100 fee. What is your decision?' Everybody is waiting for your answer. What will it be? (p. 348.)

Rabinowicz reports that he,

> presented the second scenario [in which you know that the host opens a screen other than the one you pick, and opens a screen other than the one that fronts the Cadillac] to a group of my undergraduate students and asked them for an intuitive response (without calculation). Only one of the group was prepared to switch (out of about 30), and when I asked him about his reasons, it transpired that he didn't have any....I got the same response when I described the problem to ...participants in a workshop on belief-revision (!) and non-monotonic reasoning in Kostanz last year....Clearly, the right solution is difficult to arrive at by purely intuitive reasoning. (1990, personal correspondence.)

Indeed, this problem remains not merely difficult for intuition when made unambiguous in the manner of our second scenario; it still *boggles* intuition, and even *misdirects* intuition. Subjects are apt to fix on the fact (and it is a fact in the second scenario) that the host's act and its attendant non-Cadillac revelation do not change the chances of your having chosen the right screen, which chances remain 1 in 3. That can seem to imply that you have no reason to second-guess and switch screens. Attending to your screen's chances it is easy not to notice that the *other* unopened screen's chances for being the right one have improved to 2 in 3, so that while you should not feel worse about your screen's chances considering them in themselves, you should feel worse about them in relation to that other screen's chances. It is no longer as likely that your screen fronts the Cadillac as that that screen does. If it was your screen's chances would now be 1 in 2. But they are still only 1 in 3.

Maya Bar-Hillel, while stressing the importance "not only [of] *what* comes to be known but [of] *how* it comes to be known" (p. 352), adds that evidence she has gathered shows that:

> [I]n the Cadillac puzzle...the dilemma between an answer of 1/2 and an answer of 1/3... *derives not from ambiguity about the studio's screen removal procedure*, but from a conflict between two incompatible intuitive arguments. (p. 356)

I suggest, in partial disagreement with or supplementation of Bar-Hillel, that at least contributing to an explanation of her evidence are features of her puzzle that focus attention in ways that in the second scenario are apt to mislead. A subject can see after the card is turned or the screen opened in a second-scenario Choosing Kings or Cadillac puzzle that his chance is still 1 in 3, that is, that his chance is no worse than it was when he made his guess. And he may for this insufficient reason decide not to pay $100 to second-guess: thinking only of the constancy of the chance of his initial guess, he may fail to notice that, though still 1 in 3, it is now 1/2 of what would be the chance of a second-guess. Bar-Hillel, suggests that if a subject decides not to pay to second-guess in a second-scenario case this will always be because, and only because, he mistakenly thinks that his new chance is 1 in 2, and so, though different from what it was, still the same as the remaining alternative's.

The Cadillac Puzzle, and Choosing Kings and The Prisoner's Predicament, are compound challenges. They are not only seriously indeterminate in their initial statements, but possibly confusing even when disambiguated. They are apt, they are designed, even when their ambiguities are fully resolved to trick by distracting attention, as do conjurers, from parts of their settings. And it is not just their ambiguities, but especially these post-disambiguation capacities to dazzle that make them puzzles, and that give particular point to remarking in their connections the homily that heads this paper according to which what we make of information should take into account that we have come to have it, and can depend on how we think we have come to have it. Stressing that point is in these cases a prelude to working out, possibly with the aid of formal analyses, implications of alternative views concerning how featured pieces of information are generated.[1]

Appendix: Second Aces

1. There are affinities between my treatments of Choosing Kings and The Prisoner's Predicament, and Glenn Shafer's resolution of the Paradox of the Second Ace.

> I show you a deck containing only four cards: the ace and deuce of spades, and the ace and deuce of hearts. I shuffle them, deal myself two of the cards, and look at them, taking care that you do not see which they are....Now I smile and say, 'I have an ace.'....Now I smile again and announce, 'As a matter of fact, I have the ace of spades.'....Should my decision to identify a suit make any difference?
>
> You are puzzled about whether or not you should change your probability for [my having dealt myself two aces. Why? B]ecause we had not agreed to a protocol for what information I would communicate. Under some protocols, the change is reasonable; under others it is not." (Shafer 1983: "The puzzle of the two aces is...presented here in a form close to" the one in Freund 1965.)

Shafer states that his "resolution [of the puzzle]...is in agreement with Schrödinger (1947), Gridgeman (1963), and Faber (1976)."

2. One Ace or Two?

There are six two-card Shafer-hands.

	Ace of Spades	Ace of Hearts	Deuce of Spades	Deuce of Hearts
1	X	X		
2	X		X	
3	X			X
4		X	X	
5		X		X
6			X	X

The deck is shuffled and a hand is dealt. You learn that there is an ace in it. *What is your probability for the hand's containing two aces?* You learn that it contains the ace of spades. *What is your probability now for its containing two aces?*

If answers are to be numbers without qualifying discussion, this is a bad quiz, since what you should make of these pieces of information should depend on your views concerning how they were generated, and the quiz does not contain explicit indications of these views. Shafer contrasts two possible specifications that differ for the second question.

> If it had been agreed beforehand that I would tell whether or not I had at least one ace and then whether or not I had the ace of spades, then the second step would involve relevant information. (Had I said no... then your probability for my having two aces would have gone down to zero, and so it is reasonable that when I say yes this probability should go up.) On the other hand, if it had been agreed that I would first tell whether or not I had at least one ace and then, if I did have one, that I would name the suit of one I had [randomly, if I had two], then the second step would not involve relevant information.

[A protocol like these two of Shafer's in its first step, but different in its second step is:

> I say whether or not I have at least one ace, and then randomly name the face-value and suit of a card I have.

Additional protocols that differ in their first steps from Shafer's, and, in two cases, in their second steps are:

> I randomly name the face-value of a card I have, and then say whether or not I have the ace of spades.
> I randomly name the face-value of a card I have, and then, if I have an ace, randomly name the suit of an ace that I have.
> I randomly name the face-value of a card I have, and then randomly name the suit of a card I have.
> I randomly name the face-value of a card I have, and then say whether or not it is a spade.]

Dissolutions along Shafer's lines that feature alternative specifications concerning whence information is thought to come, contrast with solutions that would cast the problem as "a conflict between intuition...and a formalism" and take the form of simple formal lessons. For an example of the latter treatment we have: "Clearly, to resolve a paradox we must abandon something. And in this instance everyone, on reflection, accepts the results of the calculus and abandons intuition" (Burks 1977, p. 53).

It follows, Burks might have said, "by unimpeachable rules" (p. 53)—in particular, "the principle of counting equiprobable cases" (p. 52)—from admittedly "impeccable premises" (p. 53)—specifically, that hands are equally probable (that is, that the deck is not made), that 5 hands contain at least one ace, that 1 contains two aces, and that 3 (including this 1) contain the ace of spades—that the information that a hand contains the ace of spades is further relevant to its containing two aces, 1/3 being plainly greater than 1/5. But is that suit-information further relevant no matter where you think it is coming from—no matter what protocol you think I am following? Surely not. [Confession: I retailed a simple and unqualified lesson-view in my doctoral dissertation, and maintained it until I read Shafer's manuscript. (Sobel 1961, pp. 294-5.)]

Consider, for example, Shafer's second protocol, in which, it is agreed that if I have an ace, I name at random the suit of one I have. Let P, P', and P" be your probability functions respectively in the beginning, after learning that the hand contains an ace, and after learning that the hand contains the ace of spades. Numbering hands as above, we have:

hands	P	P'	P"
1	1/6	1/5	1/5
2	1/6	1/5	2/5
3	1/6	1/5	2/5
4	1/6	1/5	0
5	1/6	1/5	0
6	1/6	0	0

Under this protocol learning that the hand contains the ace of spades is not further relevant for whether or not it contains two aces: P"(hand-1) = P'(hand-1) = 1/5.

<<For P"-values, let 'S' abbreviate 'the hand contains the ace of spades', and, for example, 'hand-1' abbreviate 'the hand contains the ace of spades and the ace of hearts'. *Regarding P"(hand-1)*, it should equal P'[hand-1/L(S)], and thus P'[hand-1 & L(S)]/P[L(S)]. P'[hand-1 & L(S)] should equal (P'(hand-1) X P'[L(S)/hand-1]), which is [(1/5) X (1/2)]. P'[L(S)] should equal 1/2: P'[L(S)] should equal (P'(hand-1) X [P'[L(S)/(hand-1)]) + P'(hand-2) + P'(hand-3), which is 1/10 + 1/5 + 1/5. Therefore, P'[hand-1/[L(S)]) should be [(1/5) X (1/2)]/(1/2), which is 1/5. So P"(hand-1) should be 1/5. Regarding P"(hand-2) and P"(hand-3), they should be equal, and their sum should be [1 - P"(hand-1)]: recall that P"(S) is 1. So they should be 2/5.>>

Puzzles such as One Ace or Two are not occasions for simple lessons in which, without further ado, intuitions are corrected by equiprobable-case analyses. They are occasions first for disambiguations in which 'protocols' and sources of information are made explicit, and then for instruction by less simple analyses that while beginning with equiprobable cases may end with cases of several probabilities.

It is definitive of puzzles, and a feature of many problems that do not qualify as puzzles, that intuition is not a reliable guide to probabilities and evidential bearings. Nor does formal analysis—for example, analysis into cases, or Bayes' Rule analysis afford a 'quick-fix' panacea when intuition falters. Sometimes only analyses that are sensitive to nuances, and especially to what in given cases can be learned from the fact that certain things are learned, are not merely ways of coercion, but paths to enlightenment and understanding. Such analyses are not easy, and there are no prospects for algorithms for their assemblies.

Note

[1]William Seager teased with Choosing Kings. Gordon Nagel noted the similarity of The Prisoner's Predicament. Lloyd Smith, when I told him about my ideas, observed how my homily is obvious and unremarkable for cases such as the influenza and poker ones. Rabinowicz, when commenting on an early draft of this paper, cited (Bar-Hillel 1989), and called attention to a version (slightly defective, he noted) of The Prisoner's Predicament in (Kemeny, *et. al.* 1979, p. 115). Sandy Zabell relates to prisoners and aces a McCluanesque thesis somewhat like my homily.

> There are a number of probability paradoxes centered about the phenomenon that the informational content of a message may depend on the mechanism by which it is transmitted [and, Marshall McCluan might want to add, on the medium of its transmission]. Perhaps the most attractive of these is the *three prisoner paradox*....Another classic instance is the paradox of the second ace.... (1988, p. 334.)

I am grateful to Willa Freeman-Sobel for her comments, which included the reference to McCluan.

References

Bar-Hillel, M. (1989), "How to Solve Probability Teasers," *Philosophy of Science* 56: 348-58.

Burks, A.W. (1977), *Chance, Cause, Reason: An Inquiry into the Nature of Scientific Evidence*, Chicago: University of Chicago Press.

Faber, R. (1976), "Discussion: Re-encountering a Counter-intuitive Probability," *Philosophy of Science* 43: 283-85.

Freund, J.E. (1965), "Puzzle or Paradox," *The American Statistician* 19: 29-44. (See letters in 20 (1, 2, 5) and 21 (2,3,4).)

Gridgeman, N.T. (1963), "The Pit of Paradox," *The New Scientist* 20: 462-65.

Kemeny, J.G., J.L. Snell, G.L. Thompson (1979), *Introduction to Finite Mathematics*, Third Edition, New York: Prentice Hall.

Schrodinger, E. (1947), "The Foundation of the Theory of Probability—I," *Proceedings of the Royal Irish Academy* 51: 51-66.

Shafer, G. (1983), "Conditional Probability," *International Statistical Review* 53, 1985: 261-77.

Sobel, J.H. (1961), "What If Everyone Did That?", Ann Arbor, Michigan: University Microfilms, Inc.

______. (1990), "Conditional Probabilities, Conditionalization, and Dutch Books," PSA 1990: Proceedings of the 1990 Biennial Meeting of the Philosophy of Science Association, Volume One, Contributed Papers, edited by A. Fine, M. Forbes, & L. Wessels, East Lansing, Michigan: 503-516.

______. (1992), "On Conditional Probabilities," manuscript.

Zabell, S.L., "The Probabilistic Analysis of Testimony," *Journal of Statistical Planning and Inference*, 1988: 327-54.

Dutch Strategies for Diachronic Rules: When Believers See the Sure Loss Coming

Brad Armendt

Arizona State University

Diachronic Dutch book arguments, or Dutch strategy arguments, have been used to recommend the rule of conditionalization (*C*), Jeffrey's rule for fallible learning (*J*), and the principle of reflection (*R*). The arguments have been criticized in a variety of ways—more ways than can be discussed in this paper. Here we will look at two of the criticisms. As we proceed, it will become apparent how some criticisms not explicitly raised can be met; others will go unexamined. The criticisms are these: First, Dutch strategy arguments are said to fail as defenses of rules *J* and *C*, and principle *R*, because rational agents may violate each of those constraints, yet avoid vulnerability to a Dutch strategy; their knowledge of Dutch strategy arguments enables them to do this. Second, it is said to be no wonder that the arguments fail, because there are counter-examples to each of these purportedly general constraints on rational belief.

Though their history predates the best-known recent work on Reflection, the attention focused on Dutch strategy arguments has intensified since their use in defense of *R*.[1] It is interesting and significant that Dutch strategy arguments can be made for *R*, as well as for *C* and *J*. It has been argued that this is no accident: *R* and the rules *C* and *J* have been thought very intimately related, and their Dutch strategy arguments have been thought to stand or fall together (Christensen 1991; Maher forthcoming). But there are important differences between Dutch strategy arguments for, on the one hand, *R*, and on the other, *C* and *J*. Much of the recent debate about *R* as a principle of rationality relates to this difference. In this paper we pay enough attention to *R* to notice the difference, but we then sidestep the debate about *R* and concentrate on the use of Dutch strategy arguments to defend *C* and *J*.

1. Norms for belief and Dutch books

Given some norm *N* governing partial belief, whether *N* is synchronic or diachronic, a Dutch book argument for *N* is supposed to establish that a violation of *N* makes the agent vulnerable to a Dutch book. A converse Dutch book argument is supposed to show that compliance with *N* is sufficient for avoiding a Dutch book.[2] Both arguments assume an ideal, artificial *betting scenario*. In this scenario, the agent announces fair betting quotients for wagers on propositions, and commits himself to taking bets at those rates for stakes of any size. The scenario includes a cunning bettor,

PSA 1992, Volume 1, pp. 217-229

who knows the agent's advertised odds, but has, in some sense, no information about the world beyond the agent's. The agent is vulnerable to a Dutch book if the cunning bettor can offer him a set of bets, at the agent's own odds, such that however the bets end up paying off, the agent is bound to suffer a net loss. A sure loss is not a good situation for the agent; it is what we may call a *pragmatically defective outcome* (DO).[3] When the DO is produced by the agent's own fair betting quotients (and when other fair betting quotients would avoid it), we have a reason for judging the agent's betting behavior flawed. What has this got to do with defending a norm N for rational belief? The fair betting quotients are supposed to *be* the agent's degrees of belief; when their violation of N yields Dutch book vulnerability in the betting scenario, the stigma carried by DO is transferred to the betting quotients, *i.e.* the agent's beliefs. The stigma is only avoidable by complying with N.

There are plenty of points at which this story may be questioned: The betting scenario is a contrived one; why should the announced odds be the agent's beliefs? Bets may never pay off. There are not actually any cunning bettors. We could not come up with precise degrees of belief/betting odds even if we were required to. And so on. If there is anything to Dutch book arguments, there must be a better story than the one just given, or an improved way of presenting it.

In an often-quoted passage, Ramsey (1926) writes

> These are the laws of probability, which we have proved to be necessarily true of any consistent set of degrees of belief. Any definite set of degrees of belief which broke them would be inconsistent in the sense that it violated the laws of preference between options .... If anyone's mental condition violated these laws, his choice would depend on the precise form in which the options were offered him, which would be absurd. He could have a book made against him by a cunning better and would then stand to lose in any event.

Ramsey's suggestion that violations of the laws of probability (the norm N he has in mind) are tied to a kind of inconsistency[4] has been echoed and developed by Skyrms (1980, 1984, 1987a). The idea underlying a Dutch Book argument is that an agent whose beliefs violate the recommended constraint is making the mistake of evaluating the same option in two or more different ways. Since these evaluations involve (according to Bayesians) dispositions to choose and act, the distinct evaluations could be exploited by a bettor (who realizes what the agent is doing) in a way that is essentially quite simple: the bettor sells the option to the agent at the higher of the prices, and buys it back at the lower price.[5]

We should resist the temptation to think that a Dutch book argument demonstrates that the violations (violations of probability, for the synchronic argument) are bound to lead to dire outcomes for the unfortunate agent. The problem is not that violators are bound to suffer, it is that their action-guiding beliefs exhibit an inconsistency. That inconsistency can be vividly depicted by imagining the betting scenario, and what would befall the violators were they in it. The idea is that the irrationality lies in the inconsistency, when it is present; the inconsistency is portrayed in a dramatic fashion when it is linked to the willing acceptance of certain loss (the DO). The value of the drama lies not in the likelihood of its being enacted, but in the fact that it is made possible by the agent's own beliefs, rather than a harsh, brutal world. This improved account, recommended by Skyrms, is the one we endorse.

Abstractly put, then, the Dutch book argument connects violations of N with choices that yield a DO. It proceeds by drawing on the action-guiding character of the violators of N (degrees of belief, together sometimes with methods for changing them); by supposing a scenario in which those guides are operative; and by invoking (through the cunning bettor's offers) a particular pattern of guidance that yields the DO. When the DO is sure loss, derived from exchanges constructed only by reference to the agent's action-guides, we have this diagnosis of what goes on: The agent is susceptible to the argument because the agent displays pragmatic inconsistency—he gives conflicting evaluations to the same option(s). When the diagnosis is correct, the violation of N is then seen to be tied to the inconsistency. Hence when the inconsistency is objectionable, so is the violation of N.

What makes the diagnosis plausible? Partly the fact that the guidance pattern leading to the DO is ascertainable merely by examining the agent's guides—nothing further about the world is used. Mainly, though, the diagnosis is drawn from the particular guidance pattern that yields the DO. One looks at the recipe for the Dutch book and judges that the same goods are being evaluated differently, and are exchanged in different directions to yield the DO.[6] Examples: a bet on $p \vee q$, for incompatible p, q, can be made from a pair of bets, at appropriate stakes, one on p and one on q; a bet on p conditional on q can be made from appropriate bets on $(p \bullet q)$ and on $\sim q$.

Is the inconsistency, when it is present, objectionable? According to the present story, the arguments rely on a judgment that it is. Ramsey calls the inconsistency "absurd". I say it *is* a flaw of rationality to give, at the same time, two different choice-guiding evaluations to the same thing. This kind of synchronic (pragmatic) inconsistency we may call *divided-mind* inconsistency. It is difficult to find principles more fundamental, judgments more secure, from which to argue that it is a flaw. (We could make explicit that we are interested in rational belief of unified agents, rather than schizophrenic ones.) Perhaps, though, we should add this: In claiming that it is objectionable, we are not necessarily claiming that it is more objectionable than all other things. And in attributing divided-mind inconsistency to an agent, we need not be saying more than that he exhibits an imperfection of pragmatic rationality. It does not automatically follow that the agent is crazy, hopelessly confused, or blameworthy; it could be a minor inconsistency, or inconsistency in a complex situation, or inconsistency whose importance is overridden by other considerations. The point remains, however, that divided-mind inconsistency is a flaw in the agent's set of judgments and evaluations, if not necessarily in the agent.

Our interest lies with diachronic arguments. Since the passage of time is involved, another sort of inconsistency becomes relevant. The agent might give two different choice-guiding evaluations of the same thing, but give them at different times. Call this kind of diachronic inconsistency *change-of-mind* inconsistency. Is it objectionable? Our initial answer to this must be: *not necessarily*. We may well decide that some changes of mind are less reasonable than others, but it is not automatically a flaw of rationality to change one's mind. Should we find that some Dutch strategy argument shows only that violation of norm N produces change-of-mind inconsistency, it would be premature to regard it as a defense of N, without first examining the inconsistency. We would have to be convinced that the inconsistency is objectionable, in order to be convinced that the rationality of N has been defended. Christensen (1991) argues that all Dutch strategy arguments show only change-of-mind inconsistency, and also that we endorse no principle of diachronic consistency. So, according to him, Dutch strategy arguments have no force. But the first claim is incorrect: some Dutch strategy arguments, those for rules C and J, clearly point to divided-mind inconsistency.[7] There is a difference between those arguments and the argument for R. The second claim is controversial; we can agree that it is not a principle of rationality that we never change our minds, but there may be more limited principles of diachronic consistency that we do endorse. This is not an issue we will explore here, since we are mainly concerned with arguments for C and J, and since we are addressing criticisms alleging that the arguments point to no inconsistency at all.

2. Dutch strategy arguments

If my present degrees of belief (at time t_0) are given by the probability function P_0, and if P_1 is taken to be my beliefs at some later time t_1, then (a simple version of) the principle of reflection requires for each proposition A that

$$P_0(A \,/\, P_1(A) = r) = r \ . \qquad (R)$$

In an important sense, R is essentially a synchronic constraint, regulating at a fixed time t_0 the connections between partial beliefs about events, and partial beliefs about future beliefs concerning those events. It ties together present opinion and present anticipations of future opinion, and it ignores the process that is expected to produce, or does produce, the shift in opinion as time passes. In another sense, obviously, R is diachronic; it is a constraint that applies to opinion about change, *i.e.* change in opinion. The Dutch strategy argument for R says that a synchronic belief system that includes opinions about future beliefs is exploitable if R is violated. The system is not immediately exploitable, but its present configuration is such that whatever descendent system evolves from it, in whatever way, the pair is an exploitable one. The target of the Dutch strategy argument is the initial system, but the exploitation of the Dutch Strategy vulnerability takes time.

Rules C and J are essentially diachronic constraints. They regulate the evolution of the belief system at t_0 into a later belief system at t_1; they embody methods for such change.[8] The targets of the Dutch strategy arguments for C and for J are methods for belief change over time; more precisely (and this is important) the targets are *commitments to methods* that an agent might make. I assume the reader is familiar with the rules.

In each case, a Dutch strategy argument involves a specification of the possible ways of violating norm N, and a demonstration that for each of them a legitimate recipe exists by which a cunning bettor can offer the agent a set of acceptable bets that yield a sure net loss (Teller 1973; Armendt 1980; Skyrms 1980, 1987a; van Fraassen 1984). We will not rehearse the recipes for the various arguments here. Dutch strategy arguments are diachronic—their recipes will specify what bets to make at t_0, and also how to plan at t_0 to bet at t_1. Planning at t_0 for bets at t_1 requires possession at t_0 of information about the agent's beliefs at t_1. This is managed differently by the argument for R and the arguments for C and J.

A violation of R carries with it the sort of planning information about the agent's later beliefs that a recipe requires. A violation of R involves the agent's having a present belief about A, conditional on having some future belief about A. The agent's present belief about his (possible) future belief provides adequate information. (We treat conditional beliefs $pr(p/q)$ as quotients $pr(p{\bullet}q) \,/\, pr(q)$.)

A belief change that violates rule C or rule J, on the other hand, need *not* carry adequate information for the planning, unless it is anticipated. In arguments for those rules, the bettor needs to know at t_0 the agent's plans, in order to construct the Dutch strategy. So, of course, the agent must at t_0 *have* plans of some sort for changing beliefs on occasions to which the rule applies. A Dutch strategy argument for either rule assumes, as part of the betting scenario, that the bettor knows the agent's deviant rule for belief change well enough to plan the betting. It is a direct consequence of the need for early planning that the agent is assumed to have, and the bettor is assumed to know, a commitment to a pattern of belief revision that conflicts with the rule. That commitment yields Dutch strategy vulnerability, and the commitment is the target of the argument.

The commitments that link the agent's (possible) future beliefs to him at the earlier time t_0 are what justify the claim that his inconsistency is synchronic, divided-mind inconsistency. In the case of *R*, no such commitments are imposed by the presuppositions of the Dutch strategy argument, though it might be argued on other grounds (van Fraassen 1984, manuscript) that rational agents have them. But if the agent is viewed as merely having present beliefs about the beliefs he may wind up with in the future, without thereby endorsing those future beliefs, then his inconsistency need only be change-of-mind inconsistency. If change-of-mind inconsistency is not irrational, then Dutch strategy arguments alone (unsupplemented by further principles about our diachronic commitments) do not establish the rationality of *R*, nor do they show that all violations of *C* or of *J* are irrational. They can still show that among the commitments to rules for belief change that we might make, all those that conflict with *C* or with *J* are rationally flawed. It is claimed, however, that they do not even show this, that some agents can violate the norms without displaying inconsistency of either kind.

3. Knowledgeable agents

Concentrate for a moment on the Dutch strategy argument for *R*. A violation of *R* occurs when $P_0(A \,/\, P_1(A) = r) = s$, where $s \neq r$. Suppose for discussion that $s > r$. The bettor's Dutch strategy will consist in now selling the agent a called-off bet on *A*, called off if $P_1(A) \neq r$, and now planning to later buy back a bet (of similar magnitude) on *A*, should the condition ensue, *i.e.* if $P_1(A) = r$. Should both bets be made, the bettor gains, and the agent loses, the difference between *s* and *r*. The strategy is completed by a sale (now) to the agent of a side bet on the condition, with appropriately adjusted stakes. The agent will pay off on this bet if the condition fails, and he wins less if the condition holds than he loses in the exchange of the other bets. So the bettor gains and the agent loses in every eventuality.

Suppose, however, the agent who violates *R* knows all of the above. And suppose he considers this when the bettor's initial offers are made. He then reasons, "This bettor is offering me a conditional wager on A at a price that seems fair. But if I take that wager (and the fair side bet), he is going to execute the rest of his strategy if the condition holds, and I end up with a net loss. In pursuing his strategy, he is going to offer me a wager that I will then judge fair, but now judge unfavorable. The full consequences of accepting his present offer include not just the payoffs it explicitly carries, but also the value of the later wager I will make, which is negative (I now think). So really the wager he now offers is not fair—it does not pay off at rates that meet my advertised fair betting odds when its full consequences are considered. So I will not take the bet he now offers." This violator of *R* avoids the Dutch strategy. So we do not have an argument showing that all violations of *R* produce Dutch book vulnerability, leading to the DO in the betting scenario.

This kind of criticism is presented by Maher (forthcoming) and by Levi (1987).[9] Skyrms (manuscript) has recently given a reply that undermines their criticism. Skyrms points out that in making his full evaluation of the choices (take the initial bet on *A*, refuse it) the agent must apply his reasoning to both options. Let us call the thoughtful evaluation of the initial bet a *sophisticated* evaluation. Sophistication cuts both ways, and a sophisticated evaluation of refusing the bet also ought to include, as a possible consequence, the acceptance of the later bet. After all, the cunning bettor may offer the later bet whether or not the first bets are made. The apparent value of refusing the first bet is just as susceptible to discounting on this score as is the value of accepting it. Skyrms makes the further point that since the discounting applies equally, the cunning bettor can use part of the sure gains his Dutch strategy will yield, to induce the agent along the path of vulnerability. Maher and Levi fail to consider this, and their criticism must be fixed or abandoned.

There is a way of fixing the criticism. Assuming a sophisticated agent, the discounting of the two options will be *un*equal if the agent believes that the probabilities that the later bet will be offered differ. If he believes that the later offer is likely to follow his acceptance of the first bet, but unlikely to follow his refusal, then he will discount the *prima facie* value of accepting the bet more than he will discount the *prima facie* value of refusing it. He might then choose refusal, and avoid Dutch strategy vulnerability. If the disparity in the discounting is great enough, the cunning bettor's incentives will be unable to overcome it. So a sophisticated agent with these beliefs about the likelihood of the later offer can violate *R*, and avoid the Dutch strategy. The beliefs even make some sense: the agent may regard the bettor as having no incentive to make the later bet if the first bet is refused, but as having great incentive if the first bet is taken. Of course, a bettor might try to alter those beliefs by announcing his intention to offer the second bet even if, or especially if, the first is refused, but what if an agent simply does not believe it? The objection of the sophisticated agent is made more tenuous by Skyrms' response, but it does not collapse entirely.[10]

One might think that the sophisticated agent is contradicting himself by announcing his fair betting odds, and then judging bets at those odds unfair, but he is not doing that. He consistently applies his announced odds as standards of evaluation; he just takes the payoffs of wagers to be complex—to include offers of further wagers with non-neutral value. Recalling our abstract description of Dutch strategy arguments, we can characterize what goes on here this way: We are supposed to have an argument that says violations of *R* are tied to choices that yield a DO. Given any violation of *R*, the betting recipe given by the argument shows how to invoke a pattern of action-guiding beliefs that lead to choices which yield a DO. But this will not work when the agent is sophisticated; an agent who knows the argument will anticipate the DO, and redescribe the outcomes of the choices so as to avoid beginning exchanges that produce the DO. Notice that this is a strategy for criticizing arguments that seems workable against any that seeks to tie violations of a norm for action-guiding belief to some sort of DO, by way of choices guided by those beliefs. However the DO arises, an agent aware of the process can estimate the tendency of his choices to generate the process, and on that basis reevaluate the choices. Some such agents will at least sometimes avoid the process-producing choices, and avoid the DO. If that is correct, we might react with pleasure in the discovery of a general way of unmasking fallacious arguments for norms of rationality. Or we might begin to suspect that such objections (purport to) show too much, and do so too easily, to be the bottom line.

Agents do not actually have to know the Dutch strategy argument to be sophisticated; they just have to anticipate, for whatever reason, the future wagers. Knowledge of the Dutch strategy argument can explain that anticipation, but whatever explains it, the basis of the objection is the sophisticated evaluation of initial bets. So the above diagnosis might be said to rely on an inessential part of the objection. In any case, if we take the view that (divided mind) inconsistency, and not vulnerability to the DO, is the real issue, it would be better to attend to the former.

What the objection points out, then, is this: Take some norm N for action-guiding belief. We are presented an argument saying, "Whenever N is violated, the action-guiding beliefs in violation of N can be invoked to produce exchanges of goods (e.g., wagers) that yield a DO (e.g., sure net loss). The exchanges leading to the DO do so because violation of N involves inconsistent evaluations of the goods (e.g., divided-mind inconsistency). So we see that violation of N is bad." Now any such argument will proceed with some tacit assumption(s) about conditions of equivalence for the exchanged goods, and about how the values of choices to exchange depend on the values of the goods exchanged. All we need to do to undermine the argument, is find an agent whose evaluations are sensitive to features beyond those addressed by the assumptions. If we are not sure how to produce such an agent, we just look at the argument, see how it works, and imagine an agent who adds to the other factors at work in his evaluations, one more: What are the tendencies of his options to generate a pattern of choices like that appearing in the argument? The agent who is sensitive to this additional factor will not regard choices meeting the tacit equivalence assumptions as really equivalent, and he will not exhibit inconsistency.[11] So violating N need not be bad.

What is worrisome about this is its apparent generality. How are we to defend a norm for action-guiding belief? How, if not on the basis of actions to which we are guided by beliefs that violate the norm, and by beliefs that satisfy it?[12] But as soon as we fix the criteria of equivalence for actions and outcomes, and use them to produce a defense, we thereby produce a further criterion that may be invoked to undermine the claim of inconsistency. It is difficult to see how any norm except one can avoid this. The one secure norm is "Make your action-guiding beliefs such that your evaluations are consistent." The defense of this norm is secure against the objection, but only because the "defense" is trivial. The connection between violation of the norm and the flaw of inconsistency is immediate, and no reevaluation of choices and outcomes could undermine it. But this norm achieves maximum security by settling for minimum informativeness.

These observations suggest that we ought not be very impressed by the sophisticated agent objection to Dutch strategy arguments. At least, not if we are interested in finding *interesting* norms for subjectively rational action-guiding beliefs, and if we are indisposed to skepticism concerning such norms. To a critic who insists that we give the sophisticated agent objection some credit, we may recommend conjunctions of equivalence conditions and interesting norms, rather than norms *simpliciter*. And we may point out that sophistication is costly and avoidable: Maher likens the sophisticated agent's treatment of the future bets to that of a future tax on his present winnings. Notice that this tax, and the added complexity of deliberation, is self-imposed and unnecessary, at least from the perspective of the agent who complies with Dutch-strategy-immune norms.

It would be premature, however, to rest content here. Sophisticated agent objections have been supplemented by another criticism of arguments for the specific norms C, J, and R: There are cases alleged to be *counterexamples* to those norms.

4. Counter-examples

R is the most controversial of the norms; attention to the alleged counterexamples to R would require more discussion than we can give here. It would also take us away from our main subject, the defense of the rules. There are cases that look like counterexamples to both R and the rules, but their force against the rules should be judged independently of their force against R. Some have suggested that this is not so, that R and the rules are intimately connected, and that criticisms of Dutch strategy defenses of the former will automatically apply to defenses of the latter (Christensen 1991; Maher forthcoming). Maher claims that satisfaction of *R* entails, in the situations to which the rules *C* and *J* apply, satisfaction of the rules. These claims are mistaken.[13] Even if they were true, of course, it would not thereby follow that counterexamples to *R*, in the appropriate situations, must be counterexamples to the rules. The entailment justifying that inference would have to go in the other direction.[14] It is easy enough to see that, for some anticipated belief change, commitment to rules *C* and *J* need not entail satisfaction of *R*: The agent may doubt that either rule will apply to this belief change (the rule's conditions will not, he thinks, be met), and he may simultaneously hold beliefs that violate *R*. So a general commitment to the rules will not entail universal satisfaction of *R*, *i.e.* satisfaction of *R* even in cases to which the rules are not certain to apply.

We are interested in both *C* and *J*, but for simplicity, and because the differences between the arguments for the rules do not affect the points made, we discuss *C*. Here are two representative cases presented by Maher (forthcoming):

> You are sure there is a superior being who will give you eternal bliss, if and only if you are certain that pigs can fly: and there is a drug available which will make you certain of this. Let *d* be the act of taking the drug, and *q* the probability function you would have after taking the drug. Then we can plausibly suppose that ... in taking the drug you learn *d*, and nothing else. Consequently, conditionalization requires that your probability function after taking the drug be $p(\bullet/d)$, which it will not be. (With *F* denoting that pigs can fly, $p(F/d) = p(F) \approx 0$, while $q(F) = 1$.) Hence taking the drug implements a violation of conditionalization. Nevertheless, it is rational to take the drug in this case, and hence to violate conditionalization.
>
> Similarly for (the second example). Here (your current probability for heads is 1/2, and) you think there is a 90% chance that Persi knows how the coin will land, but you know that after talking to him, you would become certain that what he told you was true. We can suppose that in talking to Persi, you think you will learn what he said, and nothing else. Then an analysis just like that given for the preceding example shows that talking to Persi implements a violation of conditionalization. Nevertheless, it is rational to talk to Persi, because ... this maximizes your expected utility.

In the first example, the agent is offered an unusual option: choose an act guaranteed to bring about, in a manner that violates *C*, beliefs that are rewarded. Other examples involving Thought Police who punish beliefs brought about by *C* can be readily imagined. The claim that the belief change brought about by the drug violates *C* relies on the assumptions i) that the agent does learn exactly *something* for certain (Maher suggests *d*), and ii) that his prior beliefs (e.g., in *F*) conditional on that something differ from the beliefs produced by the drug. Let us accept those assumptions, and also the claim that taking the drug is the act that maximizes expected utility, hence is rationally preferred to not taking the drug. What does this show about the rationality of a commitment to rule *C*?

Just as the sophisticated agent objection was generalizable, so, it seems, is the superior being/Thought Police story. Given any norm for belief, we can imagine that superior beings reward violators, and make available easy means for violating it.[15] They might, for that matter, simply reward those who fall into pragmatic inconsistency, or who succeed in believing contradictions. It may well be a perfectly rational choice to employ the means they make available, and obtain the reward. Recall that we do *not* assume that pragmatic inconsistency, or some other flaw of rationality, is the only thing that matters: its importance might be outweighed by other considerations. In granting that, however, we are not thereby granting that examples in which one rationally chooses other benefits over pragmatic consistency, are examples in which pragmatic consistency lacks significance.

In the second case, the agent is confronted with a choice between acting so as to acquire information and undergo belief change (talk to Persi), or refraining. He does not take to be available a third choice that permits belief change conforming to C, but he does think that the beliefs produced by talking to Persi will be better calibrated than his present beliefs. Given that the improved calibration has value (he expects to use it in wagers), the act of talking to Persi is rationally preferred to refraining. Here the value of the act leading to violation of C comes not from an extraneous reward, but because the imperfect learning is better than no learning, and ideal learning is ruled out. Relatively uninformed consistency is worse than inconsistency that leads to valued information.[16]

Are these really examples showing that commitment to rule C is sometimes rationally flawed? Consider the agent who takes the drug, or who talks to Persi. Will he be guilty of the pragmatic inconsistency that Dutch strategy vulnerability is supposed to signify? He will be if he commits himself now to a pattern of belief revision that violates C. We then have cases where a flaw in rationality (simultaneous conflicting evaluations of the same thing) is outweighed by a more important payoff (bliss or information later). But this can arise for any punishable norm, and in any case it does not follow that the flaw is absent. *Must* the choice of the actions (take the drug, talk to Persi) constitute a commitment to a deviant pattern of belief revision? It is natural to say so: when we deliberately choose act A for the sake of obtaining outcome O, that choice reflects a present endorsement of O. Notice that the agent *might* take the drug, or talk to Persi, for the sake of both a deviation from C and the attached payoff. But an agent otherwise committed to C may well choose the acts for the sake of the payoffs alone, while regretting the means (the violation of C).

Does that sort of agent exhibit divided-mind inconsistency? He now endorses rule C, thereby now endorsing a (future) evaluation of F, in the future situation of having learned d, which agrees with his present evaluation of the conditional bet on F given d. Yet he also now judges instrumentally preferable a *different*, rewarded, future evaluation of F, in the same situation, and he acts so as to bring that about. There is this difference in the apparently conflicting attitudes toward the future evaluations of the wager on F: The agent has a present commitment to give one, but a present (derivative) desire to give the other. This is peculiar, but it is not inconsistency.

5. Conclusion

We have explored the objection that Dutch strategy arguments fail to support the norms they endorse, because knowledgeable, sophisticated agents may violate the norms while avoiding the Dutch strategy. We suggest that corresponding objections may be made against any defense of an informative norm for action-guiding beliefs, and this reflects badly on the objections.

We have also discussed alleged counterexamples to C (and claim our discussion applies to J as well). Contrary to what some may think, Dutch strategy arguments can consistently defend the following: The rules to commit yourself to, in order to change belief rationally are, or include, C and J. Some rationally chosen actions can lead to violations of those rules, so sometimes a belief change that violates these rational commitments can be the intended consequence of rational choice. This is not a paradoxical result, and it does not invalidate the rules nor their Dutch strategy defenses.

Notes

[1]The best-known work on reflection is van Fraassen (1984). See also Goldstein (1983, 1985), Sobel (1987), Jeffrey (1988), Skyrms (1987b), and van Fraassen (1989, manuscript). For Dutch strategy arguments for C and J see Teller (1973), who presents Lewis' argument for C, Armendt (1980), Skyrms (1980, 1987a), and Sobel (1987).

[2]Converse Dutch strategy arguments for C and for J are given by Skyrms (1987a).

[3]Sure losses may not be the only situations that count as pragmatically defective outcomes, but they are what Dutch books yield. Skyrms' (1987a) work on J includes some arguments which give *conditional* Dutch strategies, in which the agent may suffer net loss, and cannot obtain net gain.

[4]The inconsistency I refer to is the twofold evaluation of the same option. The inconsistency Ramsey explicitly mentions is the violation of what he takes to be laws of rational preference: the omitted part of the quotation says: "*... such as that preferability is a transitive asymmetrical relation, and that if* α *is preferable to* β*,* β *for certain cannot be preferable to* α *if p,* β *if not-p.*"

[5]A complication arises when diachronic Dutch Books are constructed: they involve extending a conditional sure loss for the agent, one that can be imposed should a particular eventuality occur, to an unconditional sure loss for the agent, by offering him an additional side bet against the eventuality.

[6]The judgment that the goods are the same is the source of much of the controversy about Dutch book arguments. This issue is important, and it arises in the discussion below, but our treatment of it is not a full one.

[7]See Skyrms (1980, 1987a, 1990), Jeffrey (1988), and Levi (1987). The idea is that Dutch strategy vulnerability arising out of commitment to deviant rules does yield synchronic inconsistency. Skyrms is particularly explicit in making this point. In the case of C, the inconsistency appears this way: the agent commits himself—via his conditional belief $P_0(A / E)$—to evaluating the (conditional) bet on A, whose value derives only from exchanges taking place should the condition E hold, differently from the evaluation he simultaneously commits himself to giving—via his commitment to the violation of C—to those exchanges should E hold. Levi doubts that this must always be so when commitments are properly distinguished and understood; he distinguishes commitment to a rule from a further commitment to maintain that commitment.

[8]Better, C and J are methods for *propagating* change; they take as input an alteration in some part of an initial belief system and they specify how this shall affect the rest of the beliefs to yield the later belief system. Each rule applies to a specific kind of input.

[9]Levi is more interested in C. The preceding sketch of the Dutch strategy argument and the objection would be very similar for C. The arguments for J are more varied, but the objection and the issues it raises are similar; see Armendt (1980) and Skyrms (1987a).

[10]Skyrms' point, and a response to it, must actually involve three parameters: the discounting depends on the likelihood of the later offer of the bad bet, should the condition of the initial bet hold, *and* the agent's initial degree of belief in the condition,

and the expected size of the stake of the later bet. The disparity needed between the likelihoods of the later bet in order to rationally refuse the first bet depends on the strength of the agent's belief in the condition. When the latter, and the deviation from the norm (R in this case) are both small, the discrepancy between those likelihoods will have to be large, in order to rationally refuse the initial bet. But large discrepancies can yield refusal. The amount by which the unfavorable later bet discounts the values of the first options also depends on how much is at stake in the later bet (this influences the size of the presently expected loss of that bet). Perhaps this parameter will be thought not to be an influence: the expected stake, whatever it is, might be thought the same whether or not the initial bet is taken. But perhaps, for some reason, the agent thinks larger later bets will be offered if the initial bet is taken (refused) than if it is refused (taken). This could make a difference, but there is no reason to think it will make the same difference in all cases.

[11]The assumption that the conditions of identity/similarity are so alterable by the sophisticated agent is innocuous, since we take our subject to be *subjective* rationality of action-guiding belief.

[12]There are other possibilities besides Dutch book arguments. When other approaches are sufficiently filled in to make plain their contact with *action-guiding* beliefs, avoidance of pragmatic inconsistency, and assumptions about equivalence conditions for choices, play a role for them as well. We will not explore them here.

[13]Maher's arguments that R entails C and J tacitly import assumptions of the rules into their premises. So they really demonstrate that R & C entails C; similarly for J. Let R_q be "at a later time $t+x$ I have (beliefs given by) probability function q." Take the case of C. Maher assumes that "you think you would learn E and nothing else, in shifting from p to q, just in case $p(\cdot/R_q) = p(\cdot/E)$." On his view, $p(\cdot/R_q)$ is what you think the result of what you learn will be, in going to q; when R is satisfied, $p(\cdot/R_q) = q$. But to say, as Maher does, that $p(\cdot/E)$ is what (you take) the result of learning exactly E to be, is tantamount to the assumption of C. If we assumed a different account of learning exactly E, then R and that assumption would together yield, not $p(\cdot/E) = q$ (*i.e.* C), but the other account. Maher makes the equivalent question-begging move in arguing that R entails J.

[14]I do not claim Maher is mistaken about this point. Actually, in spite of his emphasis on the alleged implication criticized above, Maher does support his claim that there are counterexamples to C and J by independently arguing that his counterexamples to R also happen to be counterexamples to the rules. But, as we will see below, a situation in which it is rational to choose an action that leads to a violation of C (or J) need not be a counterexample to the rationality of a commitment to C (J).

[15]A norm might be immune if it were somehow guaranteed to take *everything* into account (so that rewards and penalties are just more input for its application). One possible candidate, which Maher favors, is decision theoretic, with a recommendation to believe so as to maximize expected utility. We cannot pursue this here, but I doubt that such a norm is immune from the story.

[16]The difference between the two examples comes to the difference between expecting that the world will reward or punish the agent just for having a new belief, and expecting that the world will offer opportunities in which guidance by the new belief will be advantageous or disadvantageous. In the second example, the value of the improved calibration presumably rests on future offers of wagers that, from the point of view of the newly improved calibration, are favorable rather than neutral.

(Where is the reward in the new beliefs if we are only offered wagers that we will judge fair, according to our new, improved beliefs?)

References

Armendt, B. (1980), "Is There a Dutch book Argument for Probability Kinematics?", *Philosophy of Science* 47: 583-588.

Christensen, D. (1991), "Clever Bookies and Coherent Beliefs", *Philosophical Review* 100: 229-247.

Goldstein, M. (1983), "The Prevision of a Prevision", *Journal of the American Statistical Association* 78: 817-819.

________. (1985), "Temporal Coherence", in *Bayesian Statistics 2*, J.M. Bernardo et. al. (eds.). New York: Elsevier Science Publishers, pp. 231-248.

Jeffrey, R. (1988), "Conditioning, Kinematics, and Exchangeability", in *Causation, Chance, and Credence*, B. Skyrms and W. Harper (eds.). Dordrecht: Kluwer, pp. 221-255.

Kyburg, H.E. (1987), "The Hobgoblin", *The Monist* 70: 141-151.

Levi, I. (1987), "The Demons of Decision", *The Monist* 70: 193-211.

Maher, P. (forthcoming), "Diachronic Rationality", *Philosophy of Science*.

Ramsey, F.P. (1926), "Truth and Probability", in *Foundations*, D.H. Mellor (ed.). London: Routledge and Kegan Paul, 1978, pp. 58-100.

Skyrms, B. (1980), "Higher Order Degrees of Belief", in *Prospects for Pragmatism: Essays in Honor of F.P. Ramsey*, D.H. Mellor (ed.). Cambridge: Cambridge University Press, pp. 109-137.

______. (1984), *Pragmatics and Empiricism*. New Haven: Yale University Press.

______. (1987a), "Dynamic Coherence and Probability Kinematics", *Philosophy of Science* 54: 1-20.

______. (1987b), "The Value of Knowledge", in *Justification, Discovery, and the Evolution of Scientific Theories*, C.W. Savage (ed.). Minneapolis: University of Minnesota Press.

______. (1990), *The Dynamics of Rational Deliberation*. Cambridge: Harvard University Press.

______. (manuscript), "A Mistake in Dynamic Coherence Arguments?"

Sobel, J.H. (1987), "Self-doubts and Dutch Strategies", Australasian Journal of Philosophy 65: 56-81.

_______. (1990), "Conditional Probabilities, Conditionalization, and Dutch Books", in *PSA 1990*, Volume 1, A. Fine, M. Forbes, L. Wessels (eds.). East Lansing: Philosophy of Science Association, pp. 503-515.

Teller, P. (1973), "Conditionalization and Observation", *Synthese* 26: 218-258.

van Fraassen, B. (1984), "Belief and the Will", *Journal of Philosophy* 81: 235-256.

__________. (1989a), *Laws and Symmetry*. Oxford: Oxford University Press.

__________. (1989b), "Rationality does not Require Conditionalization", in *The Israel Colloquium Studies in the History*, Philosophy, and Sociology of Science, E. Ullmann-Margalit (ed.). Dordrecht: Kluwer.

__________. (manuscript), "Belief and the Problem of Ulysses and the Sirens".

The Collapse of Collective Defeat: Lessons from the Lottery Paradox

Kevin B. Korb

Indiana University

1. Introduction

Logicism as a philosophical enterprise died a sudden and unnatural death in the early 1930s. The logicist program was an attempt to secure our mathematical knowledge in the indubitable bedrock of our *a priori* logical intuitions. It was a program very much impressed by the remarkable achievements in formal logic and axiomatics in the early century. While that program is well dead and gone, a research program within artificial intelligence (AI) has come to be known by the same name, sharing with its homonymous predecessor at least that program's esteem for logic.

Logicism in AI, briefly, is the view that prerequisite to developing an artificially intelligent system we must find means for representing and reasoning with a vast store of propositional knowledge and that the only viable candidates for representing propositional knowledge are formal logics (and their associated formal languages). So succinctly stated, this may sound unobjectionable, yet within are the seeds of ideas that are objectionable: most notably, the exclusionary notion that *only* formal logic has a role to play in knowledge representation. This has been coupled with a tendency to consider deduction to be the only rational variety of inference. It is only in the last decade that any form of defeasible inference (i.e., nondemonstrative inference) has been widely acknowledged within AI to be legitimate.

Of course, there is an older approach to nondemonstrative inference within philosophy, namely the use of probabilities to represent degrees of belief and probability theory to reason about those degrees of belief. But logicism is explicitly at odds with probabilism. And so logicist attempts to deal with defeasible inference have been limited to extensions of formal logic, by introducing qualitative default rules, and so yielding default logic, for example.

For thirty years Henry Kyburg's Lottery Paradox (Kyburg 1961) has been thought by many philosophers and AI researchers to be a major obstacle to providing a probabilistic model of knowledge representation. John McCarthy and Pat Hayes, the two people most responsible for the emergence of logicism within AI, encouraged the use of qualitative models by attacking the adequacy of probabilities for knowledge repre-

PSA 1992, Volume 1, pp. 230-236

sentation (McCarthy and Hayes 1969). What was not noticed until recently was that Kyburg's paradox is as much a poser for qualitative systems of representation, such as default logic, as it is for quantitative systems.[1]

In his paper "Conjunctivitis" (1970), Kyburg demonstrated that a number of quantitative inductive systems, such as that of Hintikka and Hilpinen (1966), did not model anything recognizable as scientific induction: the price they paid to avoid the lottery paradox was to restrict acceptance to what were essentially redescriptions of the available evidence. Today I shall argue that John Pollock's qualitative inference scheme described in his recent book *How to Build a Person: A Prolegomenon* (1989; and earlier in 1987) fails in much the same way.

2. The Lottery Paradox

The Lottery Paradox requires a number of assumptions to get started. Perhaps most prominently, it needs a probabilistic rule of acceptance. An acceptance rule is supposed to warrant the belief in the conclusion of an inductive argument when that conclusion achieves some sufficiently high level of probability. Given an acceptance level of 0.9, say, and a million participants in a fair lottery, then I am warranted in concluding that my own ticket i will not win; i.e., if we call this conclusion '$\neg\Phi_i$', from the deductive argument that shows $P(\neg\Phi_i) = 1\text{-}10^{-6}$ (i.e., 0.999999), I can inductively conclude that $\neg\Phi_i$. While this may seem all right, it is clear that we can draw the same conclusion for every player in the lottery. Hence, $\neg\Phi_1, ..., \neg\Phi_{1,000,000}$. Having accepted one million assertions of failure, we can then conclude both

$\Lambda_j \neg\Phi_j$ (by conjunction)

$\neg\Lambda_j \neg\Phi_j$ (by fairness of the lottery)

that is, both that no one will win the lottery and that someone will win the lottery.

This argument has been thought to provide a conclusive reason—in the form of a *reductio ad absurdum*—to reject probabilistic acceptance and to turn instead to qualitative knowledge representations.[2] But the paradox turns on more than acceptance. As Kyburg pointed out in his paper "Conjunctivitis" (1970), the paradox requires the following assumptions as well:

(1) The Weak Deduction Principle. If a statement is accepted, then its direct consequences are acceptable as well.

(2) The Weak Consistency Principle. No self-contradictory statement is acceptable.

(3) The Conjunction Principle. If two statements are accepted, then their conjunction is acceptable.

Kyburg took his paradox not to require a rejection of acceptance, but as an argument from the principles of weak deduction and weak consistency to the inadequacy of the Conjunction Principle. The Conjunction Principle entails the equivalence of weak deduction and deductive closure. And it also forces the collapse of weak consistency, or any slightly stronger consistency requirement, into the strong consistency principle that all finite conjunctions of our beliefs be self-consistent. To some, the strong consistency principle is more obviously objectionable than probabilistic acceptance: minimally, we have no working examples of strongly consistent intelligences. It is clear enough that the Conjunction Principle is as central to the Lottery Paradox as acceptance.

3. Collective Defeat

The lesson that John Pollock draws from the Lottery Paradox is not that probabilities provide no reason to infer, but rather that what we have here is a case of collective defeat—that is, *none* of the conclusions of the form $\neg\Phi_i$ may be drawn because they collectively defeat each other. It is Pollock's stated objective to provide an account of inductive, defeasible reasoning and thereby to facilitate the construction of a real artificial intelligence. He believes that this objective can be fulfilled by laying down qualitative principles of reasoning, of which a principle of collective defeat is fundamental. That principle states roughly that if we have a set of defeasible conclusions where for each conclusion there is an argument from the other conclusions (and background knowledge) to its negation, then, regardless of how long or complicated such defeating arguments may be, none of the conclusions is warranted.

Specifically about the detached $\neg\Phi_i$ in the Lottery Paradox Pollock says (1987, p. 494; my italics): "Intuitively, there is no reason to prefer some of the [$\neg\Phi_i$'s] over others, so *we cannot be warranted in believing any of them unless we are warranted in believing all of them.* But we cannot be warranted in believing all of them," because of the inconsistency. This is intended to be an application of the principle of collective defeat; I shall call it Pollock's Rule. Pollock's Rule can be seen as derivative from the Conjunction Principle—and viewing it in this light perhaps makes it more plausible than otherwise. The contrapositive of the Conjunction Principle is: if a conjunction is not acceptable, then not all of its conjuncts are acceptable. Of course, to strengthen this contrapositive to allow, without restriction, the rejection of *all* of those conjuncts, as in Pollock's Rule, would be absurd.[3] But Pollock's Rule is restricted to cases where those conjuncts themselves have equal support. By the Contrapositive Conjunction Principle we must reject some of those conjuncts; by equal support we cannot prefer one conjunct over another; therefore, we appear to be constrained to reject them all.

This suspension of belief in the face of the Lottery Paradox has been found to be plausible before.[4] Yet Pollock's Rule, however plausible it sounds, fails to support a workable concept of justified belief and therefore it fails to provide an adequate response to the Lottery Paradox. This can be seen by observing that a reasonably close analogy can be drawn between the Lottery Paradox and everyday reasoning situations. That is, one can "lotterize" just about any inductive inference problem, and so, if using Pollock's Rule, one will *almost always* be constrained to indecision, even concerning the most ordinary, dull, unobjectionable inferences.

Consider a sequence of n coin flips of a fair coin. Any such sequence has a probability of occurrence of $(1/2)^n$, which is, of course, a very low probability for large n. It seems, in advance of flipping the coin, we should be able to state that the sequence that in fact turns up is highly improbable. And so, we would be in a position to infer that the actual state of affairs would not occur. But, naturally, we could say the same about any other sequence. Since *some* sequence must occur, by Pollock's Rule we cannot infer any of these conclusions. All that we need to trigger Pollock's Rule here, or anywhere, is a partition of the outcome space such that each member of the partition has the same low probability of occurrence.

But who cares whether or not we can draw conclusions about such artificial environments as lotteries or n-length sequences of coin flips? I will show that *we* ought to, because we can *typically* partition inductive decision problems in the way needed by Pollock's Rule. Consider the practical problem of whether or not to take an umbrella (for fear of rain, not sun) out on a walk tomorrow in the Sahara desert. We can partition the possible states of the weather so that each member of the partition has roughly the same, low, probability of occurring. Any continuous measurement scale that applies to the weather will do, for example the temperature. Thus, it may be that the probability of its being sunny and the high temperature being within 1/100th of a degree of 45 degrees Celsius is the same as the probability of its being rainy and the high temperature being within 3 degrees of 30 degrees. It is always possible to generate some partition with the cells having roughly equal, low probabilities of occurrence (see Figure 1). One minus this low probability is greater than our acceptance level, by stipulation; that is, we will create a fine enough partition that the probability of any member of the partition is sufficiently low. We are inclined to say then of any such event that it will not occur, and so conclude, for example, $\neg R_{30}$. But again, there *will be* weather tomorrow in the Sahara, so by Pollock's Rule we are held to indecision.

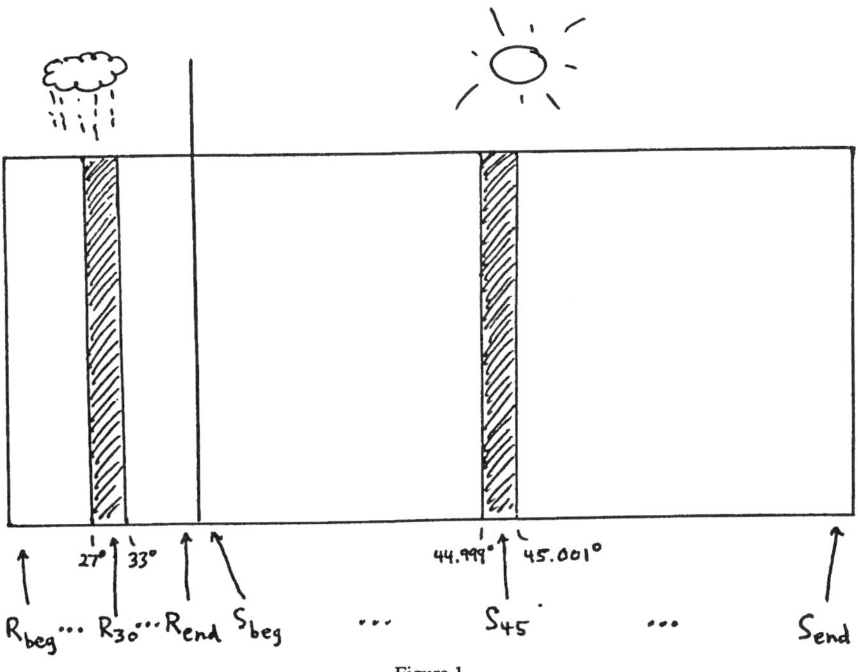

Figure 1

Clearly, we can play the partition game with just about any inductive decision problem. But perhaps that does not seem so bad; after all, the only conclusions we are obligated to avoid here are the denials of very particular states of affairs. Deciding whether we need umbrellas tomorrow does not hang on the temperature, but on the precipitation. That is, it seems open to us to forge ahead with the ordinary, boring conclusion that it will not rain tomorrow in the Sahara ($\neg R$) and so that we do not need our umbrellas. But there is a problem with this line of thought: if we can conclude that it will not rain tomorrow, then surely we can conclude—deductively this time—that it will not rain tomorrow with the temperature near 30 degrees (that is, $\neg R \vdash \neg R_{30}$; see Figure 2). Since any conjunction deductively implies its separate conjuncts (*this* conjunctive principle is not in doubt!), if we are obliged to *refrain* from inferring some conjunct ($\neg R_{30}$), we are equally obliged to refrain from inferring any conjunction containing it ($\neg R$). Therefore, we cannot conclude that it will not rain tomorrow—we cannot conclude $\neg R$; in general, we cannot reject any super-event for fear of rejecting its component subevents.[5] Likewise, since the affirmation of a set of events implies the rejection of the events excluded (since $S \vdash \neg R$), we also— absurdly—cannot *affirm* that it will be sunny tomorrow, no matter how likely that may be (short of certainty). In the end, we cannot draw any inductive conclusions at all.[6]

$R \equiv_{df} R_{beg} \vee \cdots \vee R_{30} \vee \cdots \vee R_{end}$

$S \equiv_{df} S_{beg} \vee \cdots \vee S_{45} \vee \cdots \vee S_{end}$

Pollock's Rule $\Rightarrow$ we must not conclude $\neg R_{30}$

But:

$$\neg R \equiv \neg(R_{beg} \vee \cdots \vee R_{30} \vee \cdots \vee R_{end})$$
$$\equiv \neg R_{beg} \wedge \cdots \wedge \neg R_{30} \wedge \cdots \wedge \neg R_{end}$$

So, $\neg R \vdash \neg R_{30}$

Thus, Pollock's Rule obliges us to refrain from concluding $\neg R$

Since $\neg R \equiv S$, we also cannot conclude S.

Figure 2

4. Diagnosis

As I argued above, Pollock's Rule is best seen as derivative from the Conjunction Principle, a specialized version of a Contrapositive Conjunction Principle. But why should we believe the Conjunction Principle? It's clear that if we believe that probabilities provide inductive warrant, then the Conjunction Principle cannot be right: the probability of a conjunction of independent non-trivial propositions is guaranteed to be less than the probability of any conjunct. The Conjunction Principle requires us to ignore this fact. Regardless of your attitude toward inductive probabilities, the Conjunction Principle is inimical to *any* variety of induction: any inductive conclusion will be less certain than its premises; those uncertainties will tend to *accumulate* through conjunction; the Conjunction Principle requires us to ignore this fact. (And, of course, these facts remain when the inductive support for each conjunct is equal, which is Pollock's special case.) It therefore should be unsurprising that inference systems founded on the Conjunction Principle, including Pollock's, end up rejecting induction.

In short, Pollock's attempt to dispatch the Lottery Paradox using a qualitative system endorsing the Conjunction Principle suffers from the very same defect that Kyburg identified twenty years ago in quantitative systems endorsing the Conjunction Principle: it eliminates induction. A philosophy which endorses the Conjunction Principle, therefore, can hardly serve as a framework for developing an artificial intelligence that learns inductively about its world.

5. An Exhortation

Perhaps my conclusion will surprise Mr. Pollock: in introductory moments, he quite sensibly urges that "a reasonable epistemology must accommodate both" deductive and inductive reasoning (1987, p. 481). One plausible reason that Pollock and others have been led to the extreme of rejecting inductive inference is an exaggerated concern with the *normativeness* of inference rules. Although I think naturalized epistemology goes too far in rejecting normative concerns, it is also possible to go too far by imposing utopian standards. Pollock is explicit in stating that his theory of warrant attempts to capture what an ideal reasoner would be justified in believing. His *ideal* reasoner is simply "unconstrained by time or resource limitations" of any kind (1989, p. 127). But it is not necessarily true that because such a reasoner is somehow an idealization, we or our AI systems can somehow approximate it: it is not clear how, or that, an epistemology for God has anything much to tell us about human or artificial epistemology. In particular, it may be plausible to those who have such intuitions that the (presumably infinite) conjunction of everything justifiably believed by God is itself justified for Him, but this is not merely implausible for humans, of humans it is known to be false.

The Preface Paradox (which is a relative of the Lottery Paradox) is a nice statement of our limitations: even though we may be justified in believing each individual sentence of a book that we write, we would also be well justified to insert into the preface the statement that at least one (other) sentence in the book is false. What we require is an epistemology of fallible beings, an epistemology that can deal with error and inconsistency. The impossibility of avoiding error was, after all, precisely the original stimulus within AI for investigating defeasible inference. To assume simultaneously that errors will never occur—as is required by the Conjunction Principle—is somehow inconsistent. It is high time that we recognized the global inconsistency that plagues the systems we are and the systems we build, and stop demanding that the next step in building a person be to scale up from human to godlike proportions. Artificial intelligence is hard enough already.

Notes

[1]See Perlis (1987) for a discussion of the lottery paradox for default logics. He demonstrates there that default logics suffer from the kinds of problems I find below in Pollock's system.

[2]Of course, rejecting probabilistic acceptance is *not* tantamount to endorsing qualitative methods; that is, probabilistic acceptance has frequently been rejected by *probabilists*, most notably by many Bayesians. But I will not be exploring the differences among probabilists here.

[3]For simplicity I am identifying the failure to accept with rejection here. Nothing essential in my argument turns on ignoring the difference between rejection and indecision.

[4]E.g., Cohen 1983, p. 249; Stalnaker 1984, pp. 91-2; and even Perlis 1987, p. 188, when he specifically addresses the Lottery Paradox, although he otherwise endorses the acceptance of tentative beliefs (in what he calls "jumps").

[5]Talk of rejecting and accepting events or subevents is meant here a convenient short-hand for talk of rejecting and accepting propositions asserting the occurrence of those events.

[6]Perhaps it is worth noting that normal statistical inference is also ruled out. In order to have reason to reject the null hypothesis, it is necessary that we be able to assert prior to a statistical test that the outcome will not lie in the critical region on the

assumption of the null hypothesis. But since we can partition the outcome space into regions each of which is as improbable as the critical region, Pollock's Rule obliges us to refrain from the normal statistical inference.

References

Cohen, L.J. (1983), "Belief, Acceptance, and Probability," *Behavioral and Brain Sciences* 6:248-49.

Hintikka, J. and Hilpinen, R. (1966), "Knowledge, Acceptance and Inductive Logic", in *Aspects of Inductive Logic*, J. Hintikka and P. Suppes (eds.). Amsterdam: North-Holland, pp. 1-20.

Kyburg, H. (1961), *Probability and the Logical of Rational Belief.* Middletown: Wesleyan University.

Kyburg, H. (1970), "Conjunctivitis", in *Induction, Acceptance, and Rational Belief,* M. Swain (ed.). Dordrecht: Reidel, pp. 55-82.

McCarthy, J. and Hayes, P. (1969), "Some Philosophical Problems from the Standpoint of Artificial Intelligence", *Machine Intelligence* 4:463-502.

Perlis, D. (1987), "On the Consistency of Commonsense Reasoning", *Computational Intelligence* 2:180-90.

Pollock, J. (1987), "Defeasible Reasoning", *Cognitive Science* 11:481-518.

Pollock, J. (1989), *How to Build a Person: A Prolegomenon.* Cambridge: MIT.

Stalnaker, R. (1984), *Inquiry*. Cambridge: MIT.

Part VII

REALISM: CAUSES, CAPACITIES AND MATHEMATICS

Cartwright, Capacities, and Probabilities[1]

Gürol Irzik

Boğaziçi University

1. Introduction

Nancy Cartwright's new book *Nature's Capacities and their Measurement* (1989)[2] attempts to achieve something remarkable: refute Hume's most crucial theses concerning causation and let in capacities largely through methodological arguments. The two central theses of Hume which Cartwright challenges are:

(1) Generic causal facts (e.g., causal laws) are reducible to regularities.

(2) Singular causal facts are true in virtue of generic causal facts.

Cartwright's arguments against (1) and (2) are methodological in the sense that she arrives at her conclusions by considering our best methods for studying the relationship between probability and causation. Cartwright rejects (1) on the grounds that if we look at the way in which probabilities are related to causal relations as expressed in her principle CC, or if we carefully study the methodology of causal modelling employed in the social, behavioral, and life sciences, we see that to get certain causal information from probabilities we must presuppose the truth of other generic causal facts. Therefore, a reduction of causal laws to laws of association (regularities) is impossible. Cartwright believes that (2) is wrong because "*the methods* that test causal laws by looking for regularities will not work unless some singular causal information is filled in first" (p.2, emphasis added). Hume's second thesis therefore must be turned on its head: it is the singular fact which is prior to the causal law rather than vice versa since "that is what causal laws are about" (ibid.). The methods Cartwright has in mind this time, however, are not those of causal modelling but based upon a revised version of the principle CC, called CC*. It is this revised principle which makes explicit the appeal to singular causal facts and thus refutes (2). Cartwright also claims that causal modelling strategies such as path analysis fail to establish the right connection between causes and probabilities to the extent to which they bypass singular causal information.

Cartwright's methodological arguments which challenge Hume's theses above are at the same time indirect arguments for taking causal claims as ascriptions of capacities. According to Cartwright, a typical causal claim such as "aspirins relieve headaches"

PSA 1992, Volume 1, pp. 239-250

says that aspirins, *in virtue of being aspirins*, have the capacity to relieve headaches, not that aspirin intake is regularly (or for the most part, or more often than not) followed by a relief of headache. This means that aspirins carry with them a "relatively enduring and stable capacity . . . from situation to situation; a capacity which may if circumstances are right reveal itself by producing a regularity, but which is just as surely seen in one good single case." (p.3). Admission of capacities as real things in nature is the strongest anti-Humean aspect of Cartwright's philosophy, despite her avowed empiricism. Capacities, Cartwright suggests, are metaphysical ingredients necessary to make sense of scientific practice, since our best methods in physics, econometrics, empirical sociology, etc. presuppose them. Therefore, their admission is a small price to pay for what is gained. Moreover, under right circumstances capacities manifest themselves empirically—either as regularities (deterministic or probabilistic) or as single cases. This means in Cartwright's language that they can be measured. That's why she finds them acceptable. Her empiricism is a liberal one which admits of any entity or property so long as it has observable manifestations that can be measured.

Being a realist, I believe that causal notions are metaphysically quite loaded notions, so any ontological framework which attempts to make sense of them must be rich enough to incorporate capacities, powers, propensities, or the like. Therefore I am not categorically against taking natural and social capacities as real, provided their acceptance is grounded not only on methodological arguments but also on ontological ones. It seems that Cartwright has placed too much "metaphysical weight" on methodology for it to carry and reached some hasty metaphysical conclusions from purely methodological arguments. She herself is aware of it and has signalled her plan to do something about it in the future, especially about the nature of capacities. Moreover, there are ontological arguments to support her theses without abandoning the framework of capacities, as I shall try to show. For the most part my problems with Cartwright's well-articulated philosophy pertain to some methodological issues concerning Principle CC*, her interpretation of causal modelling strategies such as path analysis, and how these bear on her views of capacities, especially mixed-dual ones.

To be more specific, I agree with Cartwright in rejecting both of Hume's theses. Concerning the first one, our reasons are similar; they come from examining causal modelling strategies. As for the second thesis, we both reject it but for quite different reasons. Her rejection is based on Principle CC*, mine on the propensity interpretation of causal probabilities, which she does not favor as an empiricist. I will argue that the propensity interpretation is tailor-made for endorsing capacities and can provide the sort of ontological support Cartwright needs for them.

Such a support becomes all the more important given the fact that CC* is the only major argument Cartwright has for refuting Hume's second thesis. So she painstakingly tries to show that no alternative account, including path analysis, which sidesteps the appeal to singular causal facts has been fully successful. I disagree with Cartwright's judgement concerning path analysis. Accordingly, I shall argue that the objections she raises against it can be met. That is why I urge for a propensity interpretation as an independent justification for endorsing capacities—independent, that is, from her methodological arguments.

2. Why causal laws are not reducible to regularities

Cartwright begins by noting that "in the context of causal modelling theory, probabilities can be an *entirely reliable instrument* for finding out about causal laws" (p.11). How? Relevant probabilities, when combined with certain background assumptions which include causal ones as well, yield causal knowledge. Since without presupposing the truth of some causal laws it is impossible to establish the truth of other causal laws from probabilities, it follows that causal laws are irreducible to probabilistic regularities and by implication to deterministic ones as well. In effect this amounts to claiming that causal modelling methods are essentially bootstrapping strategies and not hypothetico-deductive, because "causal relations can be deduced from probabilistic ones—given the right background assumptions" (p.22).

Cartwright sets to prove this point in the context of causal modelling as follows. Take a typical recursive set of equations from causal modelling:

$$x_1 = u_1$$

$$x_2 = a_{21}x_1 + u_2$$

•

•

•

$$x_e = a_{e1}x_1 + a_{e2}x_2 + \ldots + a_{ee-1}x_{e-1}$$

where x_e is the effect of interest. In each equation x's on the right represent causes, those on the left their effects. The set$\{x_1, x_2, ..., x_{e-1}\}$ is called a "full set of inus conditions" for x_e (inus conditions being understood in Mackie's sense). Each member of the set is a possible cause of x_e; that is, our background knowledge does not rule out any of them as a true cause. u's are called "error terms". They represent all other (unknown or unobservable) causal factors which are not explicitly incorporated into the model equations. There is an error term associated with each equation, and it is assumed that an error term is uncorrelated not only with other error terms but also with each of the causes in other equations.

Now just three assumptions are needed to relate causes to probabilities. These are:

(i) A generalized version of Reichenbach's principle, which says that every correlation has a causal source or explanation (p.29).
(ii) Transitivity of causality: if x causes y and y causes z, then x causes z.
(iii) Open back path assumption (OBP): "x(t) has an open back path with respect to $x_e(0)$ just in case at any earlier time t', there is some cause, u(t'), of x(t), and it is both true and known to be true, that u(t') can cause $x_e(0)$ only by causing x(t)" (p.33).

Given these assumptions, Cartwright shows that "if each of the members of a full set of inus conditions for x_e may (epistemically) be a genuine cause of x_e, and if each has an open back path with respect to x_e, all the members of the set are genuine causes of x_e" (p.33). Call this theorem T.

Consider now the following causal models to see how theorem T works:

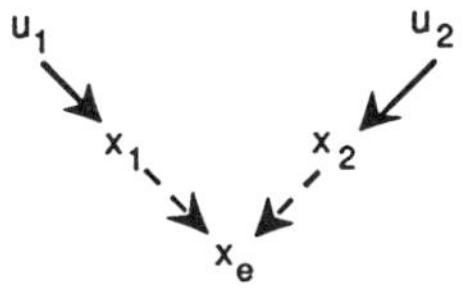

Model 1

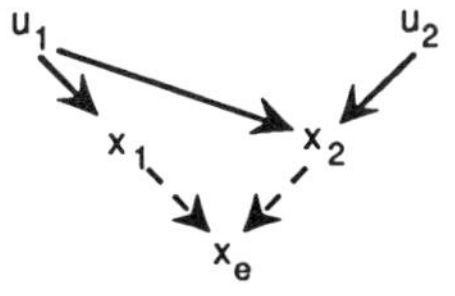

Model 2

In both models it is assumed that x_1 and x_2 are inus conditions for x_e and that each may (epistemically) be a genuine cause of x_e. This is indicated by the dotted-arrows. Solid arrows mean genuine causes. In model1 both x's are genuine causes of x_e. But in model 2 although x_2 is, x_1 may not be. The reason why we cannot say for sure that x_1 causes x_e in model 2 is that the OBP assumption is violated; the path u_1. x_1 is not an open path wrt x_e since u_1 causes x_2 which in turn may (for all we know) cause x_e. By transitivity of causation, u_1 may cause x_e via x_2 rather than via x_1, hence violating the OBP assumption. When the OBP assumption is violated, u_1 becomes a common cause of x_1 and x_e, and thus can be responsible for the correlation between them. Consequently, we cannot legitimately conclude that x_1 is a genuine cause of x_e.

By contrast, in model 1 we see that all assumptions (including OBP) are satisfied. So there can be no common causes to account for the correlation between x_1 and x_e. The only explanation left is that $x_1$is a genuine cause of x_e. (A similar argument applies to x_2.)

The general form of the argument discussed above is a familiar one: suppose x and y are correlated. If there is no common cause which can account for the correlation, the two variables are causally related given the generalized Reichenbach principle. Since the models are recursive to begin with, temporal precedence has already determined which one is the cause and which one the effect. In short, if the three strong causal assumptions stated above are met, one can deduce causal facts from probabilistic ones.[3]

So far I have outlined Cartwright's reasoning to the effect that in the context of causal modelling it is in principle possible to deduce causal laws from others with the help of probabilities. Because the deduction relies on background causal knowledge, causal laws cannot be reduced to purely probabilistic associations. Put differently, given the right background assumptions which are rich enough to include causal ones, certain causal relations can be deduced from the relevant probabilities or correlations.

This is the conclusion suggested by causal modelling methodology. Cartwright argues that the same conclusion can be reached through her principle CC:

$$\text{CC: C causes E iff } P(E/C \pm F_1 \pm \ldots \pm F_n) > P(E/\neg C \pm F_1 \pm \ldots \pm F_n),$$

where$\{F_1, \ldots, F_n, C\}$ is a complete causal set for E. To be a complete causal set for E means, roughly, to include all of E's causes. (p.56)

CC says that C causes E if and only if C increases the probability of its effect conditional upon all other causes of E. The idea is to look in those "causally homogeneous" populations where we hold fixed not only C, but also all other factors which cause and prevent E, and then compare $P(E/C{+}F_1{\pm} \ldots \pm F_n)$ with $P(E/\neg C{\pm}F_1{\pm} \ldots \pm F_n)$ for each possible combination of F's. ($\pm F_i$ means either F_i or $-F_i$, where i=1...n, but not both.) If C increases the probability of its effect in every such causally homogeneous population, then we can assert that C causes E since this is the only possible explanation of the increase in probability. Conversely, from the fact that C causes E it follows that the probability increase must occur in these populations. Cartwright points out that something like CC is at the basis of all of our reasonings from probabilities to causes and from causes to probabilities, but that it cannot be used to reduce causes to pure probabilities. The reason is obvious: the conditionalizing factors (F's) must be a complete set of *causal* factors. Thus, to infer from probabilities that C causes E, we must already possess (actually, an enormous amount of) causal information.

3. Why singular causal facts are not reducible to causal laws

Cartwright's second thesis I wish to discuss is this: "To pick out the right regularities at the generic level requires not only other generic causal facts but singular facts as well. So singular causal facts are not reducible to generic ones" (p.91). To see why singular causal facts are irreducible to generic ones, consider a simple example: $C \rightarrow F \rightarrow E$. Here C is a genuine cause of E via F. Applying principle CC, we get $P(E/C{\pm}F)=P(E/\neg C{\pm}F)$. So we wrongly infer that C does not cause E. A concrete example is provided by Eells and Sober (1983). Your dialling me (C) causes my phone to ring (F), and my phone's ringing causes me to lift the receiver (E). So, by transitivity, your phoning causes me to lift the receiver. But when F is held fixed per CC, this claim is not supported by the probabilities.

Cartwright also points out that not holding such intermediate causes as F fixed can be equally misleading. Imagine, she says, that you phone me in California every Monday from the east coast at time t. Imagine further that somebody else, who lives closer to me than you, also dials me every Monday at exactly the same time. Then he will be connected first, and you will fail to get through to me. So although it is not your phoning which causes me to lift the receiver, it may look that way from the probabilities when the other ringing is not held fixed.

The moral of these examples is that principle CC does not work in cases where a cause produces or prevents an effect through intermediaries. So Cartwright suggests that "the other factors relevant to E should be held fixed only in individuals for whom they are not caused by C itself" (p.95). Principle CC therefore must be revised to ensure that

> * Each test population of individuals for the law 'C causes E' must be homogeneous with respect to some complete set of E's causes (other than C). However, some individuals may have been causally influenced and altered by C itself; just these individuals should be reassigned to populations according to the value they would have had in the absence of C's influence (p.96).

Combining CC with * yields Principle CC*. Applying it to the examples above, we realize that we should divide F's into two groups (see figure below): those that are caused by C's and those that are not (the latter may occur spontaneously or due to some other cause, say, C').

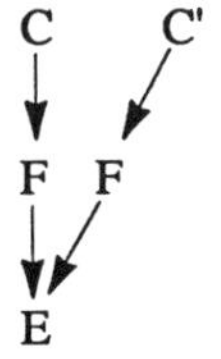

The former should be moved to the $\neg F$ population according to Principle * because the phone would not have rung were it not for the action of C (barring simultaneous causation). So the $\neg F$ population consists of all Monday afternoons on which the phone does not ring at all as well as those Monday afternoons my phone rings due to C's action. In this (hypothetical) population C causes E and the probabilities agree: $P(E/C)>P(E/\neg C)$. As for the F population, it includes those Monday afternoons in which my phone rings because of C', not because of C. Here C naturally does not cause E, and the probabilities agree again: $P(E/C)=P(E/\neg C)$.

So CC* works well, but it works by utilizing knowledge about individual causal relations. The right test populations are obtained by taking into account certain singular causal facts such as which F's are the results of C and which are not. Hence Cartwright's claim that it is singular causal facts which are primary.

It is true that *if* we know the true causal relations among your dialling me (C), my phone ringing (F) and my lifting the receiver (E), Principle * enables us to pick out the right test populations, and together with CC it gives the right probability relations. But coupling CC with Principle * is not without cost: we cannot make the converse inference from probabilities to causes anymore, for to know what to hold fixed we must know which F's are caused by C's and which ones are not in which individuals. Indeed, the whole point of appealing to probabilities in the first place is that we do not know whether C's in general cause F's or not, let alone which F's are caused by which C's, and that we are hoping that the probabilities will tell us whether C causes E via F or directly or not at all.

Notice that my complaint is not that CC* fails to reduce causation to probabilities. Rather, my point is that Principle * presupposes the very same causal knowledge which we are trying to establish from the probabilities (together with whatever other background causal information we might have). In other words, the CC part of the principle CC* should not be read as a biconditional statement anymore. It becomes a simple conditional statement which enables us to infer probabilities from causes, but not the other way around. This means that causal knowledge cannot be bootstrapped from probabilities and background causal information, and surely this is a costly price to pay for what is gained.

4. Mixed-dual capacities

Cartwright believes that an individual can carry truly mixed-dual capacities; that is, the same cause can both produce and prevent its effect. Hesslow's much discussed thrombosis example, she claims, is a case in point. According to this example, birth control pills cause thrombosis in women. But thrombosis is also caused by pregnancy. Since pills prevent pregnancy, by transitivity they also prevent thrombosis. In Cartwright's language pills carry both a positive capacity to produce thrombosis and a negative one to prevent it. The superiority of CC* (over other approaches such as Eells-Sober's and path analysis), Cartwright claims, is best seen by looking at such cases where individuals have such dual capacities.

As before, CC* works fine in such cases in getting the right probabilities given the correct causal structure, but is useless for the converse and obviously more important job of inferring the correct causes from probabilities. The problem is far more serious for Cartwright in Hesslow-type examples because it becomes virtually impossible to discover the existence of *mixed-dual* capacities from probabilities. If, as Cartwright seems to suggest, CC* is the only reliable way to establish a connection between causes and probabilities, and by implication, between mixed-dual capacities and probabilities as well, we can never know whether an individual has such capacities. All CC* does is to show how the probabilities would look *if* we already knew the mixed-dual nature of a cause, but it cannot tell us from the probabilities whether such a cause is at work or not.

Quite apart from these considerations, there is something peculiar about the notion of a mixed-dual capacity. In a joint article Dupre and Cartwright (1988) write:

> A basic assumption of this paper is that things and events have causal capacities: in virtue of the properties they possess, they have the power to bring about other

> events or states. (...) Capacities are carried by properties. That is, you cannot have the capacity without having one of the right properties. But the *same* property can carry mixed capacities. (p.521, emphasis added).

What troubles me here is not the nature of capacities in general, but rather the very coherence of the notion of *mixed-dual* capacities. What sort of things could they be? In particular, are they properties or not? If they are, clearly they will be contradictory ones. If they are not, we have a strange mixture consisting of properties carrying "mixed" things which are not properties. The idea of mixed-dual capacities becomes also problematic when we think of the transitive property of causal relations. Although it is plausible to think (for a regularity theorist, for instance) that if x causes y and y causes z then x (indirectly) causes z, it sounds strange to argue in the same way when we adopt a capacity interpretation for a case in which one of the causes is a preventive. Consider the thrombosis example again. Is it valid to argue that "pills have a negative capacity to prevent pregnancy, pregnancy has a positive capacity to produce thrombosis; therefore, pills have a negative capacity to prevent thrombosis"? I think not, for it seems possible to accept the premises as true but reject the conclusion as false. What is one committed to when one accepts a capacity interpretation for the conclusion? To the claim that pills have *some property* (or properties) *in virtue of which* they prevent thrombosis. Do Cartwright and Dupre really believe that chemical analysis will reveal the existence of such a property? Of course, the easiest way to test that claim is to administer pills to men and see what happens! This may sound ludicrous, but remember that properties carry with them "relatively stable and enduring capacities from situation to situation". And if pills do have such stable capacities to prevent thrombosis, I do not see why they should not reveal them in men.

The point is that it is the *context* in which capacities are exercised that is responsible for their apparent mixed-dual nature. Mixed-dual capacities are epiphenomena which arise only through the intermediate effects their causes give rise to. Take away those intermediaries, they disappear.

5. Defending path analysis

In section 2 we have seen that Cartwright's main argument against Hume's thesis that singular causal facts are true in virtue of generic causal facts is CC*. But Cartwright is well aware that this principle alone cannot compel her opponents to accept the primacy of singular causal facts for the simple reason that there might be an alternative strategy which works equally well without appealing to them. That is why Cartwright reviews the most promising alternatives to her account and tries to show that they are all defective. One of these is the method of path analysis. In this section I want to show that it is possible to defend path analysis against her charges.

First Criticism. Cartwright grants that path analysis works just fine in the thrombosis example provided all operations of causes are independent of all others. Her complaint is that "it is clearly unreasonable to insist that a single cause operating entirely locally should produce each of its effects independently of each of the others" (p.122). In the thrombosis example this amounts to the assumption that there is a correlation between the pill's action for preventing pregnancy (P) and for producing the harmful chemical (C'). Cartwright then shows that path analysis yields misleading results when this assumption is violated. The reason why she thinks the assumption of independent operation is a "clearly unreasonable" one is based on decay phenomena in the quantum domain where such correlations are commonly observed.

Reply: I agree that path analysis fails when the assumption of independent operation does not hold, but it is crucial to be clear on what the assumption says. As I understand it, the assumption is a denial that C screens off the two effects C' and P from one another *in spite of the fact that C is a genuine cause of them.* That there is a correlation between the pill's action for preventing pregnancy and for producing the harmful chemical means that C, C' and P form an interactive fork in Salmon's sense (see his 1984). So Cartwright's criticism is that path analysis fails in cases of interactive fork.

The obvious question is whether there are any genuine interactive forks at the level of macroscopic phenomena. As Glymour and et al (1991, pp.157-159) show convincingly, all alleged examples of macroscopic interactive forks discussed in philosophical literature turn out to be cases of conjunctive forks upon a careful and complete description of the common cause and its context, where the common cause does screen off its correlated effects. No wonder all of Cartwright's examples come from the quantum domain. The assumption of independent operation may be unreasonable there, but I think it is well-justified where path analysis is applied.

Second Criticism. "The path strategy works by pin-pointing some feature that occurs between the cause and the effect on each occasion when the cause operates" (p.125). But, Cartwright says, it is not always possible to find such intermediaries (a lesson which we learn from quantum mechanics again). Path analysis was successful in handling the thrombosis case because the cause operated through two intermediate effects, namely the harmful chemical and pregnancy. If, the criticism runs, there were no such effects, then path analysis would have failed to reveal the mixed- dual capacities of the cause.

Reply: I find this criticism puzzling because Principle * was endorsed to handle precisely those cases which involve such intermediaries. If there are no intermediaries, Cartwright's principle is totally useless to pick out the right test populations. Indeed, assume that C both causes and prevents E in such a way that there are no intermediaries between them. Let B represent other causes of E . Then it is in principle impossible to discover the mixed-dual capacity of C, a point already suggested in the previous section, for there is only one test population, and hence *only one* of the following probability relations can hold:

(1) (given B) $P(E/C) > P(E/\neg C)$, in which case C produces E but we cannot say that C prevents it; or

(2) (given B) $P(E/C) < P(E/\neg C)$, in which case C prevents E but we cannot say that C produces it; or

(3) (given B) $P(E/C) = P(E/\neg C)$, in which case C is irrelevant to E.

It is obvious therefore that the applicability of CC* depends on the existence of causal intermediaries just as much as path analysis does.

In the (1988) article Dupre and Cartwright recognize the difficulty. They admit that even in causally homogenous test situations "the converse conclusion, that there are no causal capacities when the probabilities do not change, does not follow" (p.534). What they seem to be saying is that when individuals which have mixed-dual capacities exercise their capacities in such a way that the two cancel each other out, there will be no change in the probability. Just because there is no change, they claim, it does not mean that there is no capacity. This may well be true, but then what independent evidence is there to believe that there *is* any causal relation at all, let alone that the cause has mixed-dual capacity? I find this hard to reconcile with Cartwright's empiricist maxim that capacities can and must be measured if they are to be acceptable.

Third Criticism. Cartwright complains that it is not clear what the strategy of path analysis is, and thus that there is some difficulty in formulating it (p.124).

Reply: Despite the popularity of path analysis among social scientists, philosophers are just beginning to take interest in it and other related techniques. Although these techniques are routinely used in the sciences, their underlying methodological and philosophical presuppositions and their implications are not made fully explicit and critically discussed. So Cartwright's criticism is right in this sense, and her own book is a valuable contribution to remedy this defect.

But in my opinion her complaint just points to the complexity of the method of path analysis rather than to an intrinsic difficulty in formulating it. The core of path analysis is to decompose a correlation between two factors into components corresponding to the direct and indirect (intermediate) causal paths through which they were related. Each coefficient, which measures the direct causal impact of one factor on another within the model, is expressed in terms of correlations by conditionalizing on factors from which there are direct paths to the factor of interest. One then solves for the coefficients. Once the coefficients are obtained, the correlations become decomposed, thus giving us the causal weights (which can be negative or positive) responsible for them. Contrary to what Cartwright says, one need not know at the beginning which paths carry positive and which ones negative capacities.

This is a rough summary, but I hope it nevertheless convinces people that the difficulty of formulation Cartwright complains of is not an insurmountable one. We must keep in mind that path analysis is just a version of the well-entrenched multiple regression analysis and "it differs from Ordinary Least Squares only because it uses correlations rather than covariances, so that all the estimated coefficients are standardized" (Hanushek and Jackson 1977, p.221). Part of the difficulty also lies in that these methods use correlations and covariations which are based on continuous variables, whereas those of philosophers use probabilities which are based on dichotomous variables. So more work needs to be done to understand and improve "the logic of causal modelling" and how it relates to philosophers' methods before we despair from them.[4]

Fourth Criticism. Cartwright believes that the difficulties of path analysis result from its confinement to the Humean program. The confinement is three-fold. Path analysis assumes that (a) causal paths can be delineated solely in terms of causal laws without appealing to singular causal facts; (b) causal laws can be bootstrapped from pure statistics; and (c) causal paths "represent the canonical routes by which causes operate" (p.125). Cartwright thinks that no strategy can succeed under such constraints.

Reply: I am not sure what Cartwright means by (b). If it means causes are reducible to pure statistics according to path analysis, then it is simply false as she herself has shown in section 1. On the other hand, path analysis does endorse (c) (and (a) of course). I believe that something like (c) must be presupposed by any strategy which aims to uncover the underlying social and behavioral mechanisms behind observed statistical associations. Postulation of intermediaries can be seen as a way of accomplishing just this and at the same time dealing with the problem of contrary effects without introducing mixed-dual capacities. Social phenomena are notoriously complex, and the social scientist is well advised to postulate and then test for intervening factors. Mechanisms by which a cause gives rise to its effect is usually more informative than the mere existence of a causal relation. That is why the assumption of transitivity is so crucial in causal modelling.

One must also recall the contexts in which causal modelling strategies such as path analysis are applied. Such contexts are stable equilibrium situations. In other words, it is assumed that the effects of various causes are more or less settled so that cross-sectional data, which consist of joint observations on several variables at a single time point, provide an adequate representation of social relations. Once the effects are settled as assumed, it is natural to expect that causes can be discovered through these "established routes" and that the routes will not change unless the equilibrium is disturbed.

6. Propensities as causal probabilities

I engaged in a lengthy defense of path analysis because I wanted to show the fragility of any argument for the primacy of singular causation, which is purely methodological. If my defense is adequate, this means there is an alternative to CC* after all, an alternative which does not appeal to singular causes. In that case, there is a pressing need for ontological arguments to support the primacy thesis. I propose the propensity interpretation of causal probabilities, which I find extremely conducive to that purpose.

Following Popper (1959), Mellor (1971), and especially Giere (1976, 1979), by propensities I mean "the inherent tendencies of a [physical] system for reaching the various possible final states" (Giere 1979, p.444). These tendencies are realized on each trial of the probabilistic physical system and give rise to a probability distribution over its possible final states. It can be shown that propensities obey the standard probability calculus and thus "have the logical structure of probabilities" (see Giere 1976). The greatest advantage of such an interpretation for our purposes is that propensities operate case by case. Thus the probability attaches to each individual outcome of the trial, rather than to a whole set of outcomes of similar trials, thereby eliminating the problem of single case and, its correlate, reference class. Of course, any empirical test of a propensity claim makes use of relative frequencies, but nevertheless the claim cannot be reduced to them. (Just like capacity claims can be tested by regularities but are not reducible to them.)

Now recall Cartwright's views about capacities. According to Cartwright, many laws of science are "laws about enduring tendencies and capacities" (p.1) and "are not reports of regularities but rather ascriptions of capacities, capacities to make things happen, *case by case*" (p.3, emphasis added). That is why singular causal facts are primary and generic facts secondary, being parasitic upon singular ones. Cartwright also points out that capacities are often probabilistic, but insists that probabilities should not be understood as tendencies. For she believes as an empiricist that "probabilities are just frequencies that pass various tests for stability" (p.59).

But frequencies are properties of classes of events, not of individual events, so it is not clear how probabilistic capacities in the sense of frequencies can operate case by case. If capacities in general operate case by case on the individual level, then it is natural to think that probabilistic ones also operate in the same way. This means they are nothing but single-case propensities in the full sense. In other words, to say that causes are real capacities which make things happen case by case is to say that *causation must be a natural relation which is wholly present in the single instance*. This in turn implies that probabilistic causation is best understood as propensities, that is, as real tendencies of the individual to behave in specifiable ways in each instance with a certain probability.

An ontological framework which admits of causes as real capacities can incorporate probabilistic capacities as real propensities without difficulty. Cartwright's empiricism is liberal enough to make this natural move. Such a move is aesthetically attractive as well since it will restore the symmetry between the irreducibility of deterministic capacities to strict regularities and the irreducibility of probabilistic capacities to relative frequencies. The standard (and arguably the strongest empiricist) objection to propensities that they are "too metaphysical" adds no extra burden on Cartwright's philosophy since she has already admitted capacities. These propensities can be measured (i.e., tested) by relative frequencies in the same way capacities can be measured by regularities.

Let me also point out that Cartwright has the option of not going all the way; that is, she does not have to maintain that all probabilities must be interpreted as propensities. Like Salmon (1984, p.203), she can distinguish between causal and non-causal probabilities and restrict the propensity interpretation to causal probabilities only. This way she can escape not only "undue" criticism (ibid. p.205), but can also exploit much of Salmon's account of singular causal processes to which she is apparently sympathetic (see p.9).

7. Concluding remarks

Taken as a whole Cartwright's book is an attempt to initiate a new research program for causation, which places scientific methodology at the center. However, Cartwright does not take scientific methodology uncritically. She meticulously examines it, reveals its hidden assumptions and presuppositions, criticizes them severely if she finds them wanting, and thus tries to improve upon it. Her research program is anti-Humean, but nevertheless only loosely empiricist. It proposes to treat causal laws as ascriptions of capacities, thus rendering capacities more fundamental and primary than regularities. But there is a far bolder aspect of her work which I did not discuss in this paper; ultimately, she wants to dispense with laws altogether, thus leaving us with capacities and singular causings only. Cartwright urges us to take up this project and develop it further since she believes that it is "the most powerful way to construct a reasonable image of science" and "a far more realistic picture" (p.181) than the alternatives provide.

As I pointed out in the introduction, Cartwright is well aware that this is a research program in need of more articulation and more support than she has provided. She obviously knows for instance that she owes her readers a detailed account of what capacities are, an account which should rely more on metaphysics than methodology because of the nature of the problem. Thus she writes:

> My aims in this book are necessarily restricted, then: I want to show what capacities do and why we need them. It is to be hoped that the subsequent project of saying what capacities are will be easier once we have a good idea of how they function and how we find out about them. (p.9)

Within this perspective then, Cartwright's work should be seen as an exciting exercise in drawing the boundaries of what we can accomplish by methodological arguments and insights and how far we can go with them. Cartwright has told us only half of her story; the rest should be even more exciting.

Notes

[1]The author gratefully acknowledges the support of the Boğaziçi University Research Fund.

[2]All references in the text are to this book unless otherwise indicated.

[3]It is to Cartwright's credit that she turns this informal argument into a formal one concerning the recursive linear causal modelling equations of the type introduced earlier above. The formal proof is included in the appendix to the first chapter of her book.

[4]See Irzik and Meyer (1987) for a work in this direction.

References

Cartwright, N. (1989), *Nature's Capacities and their Measurement*. Oxford: Clarendon Press.

Dupré, J. and Cartwright, N. (1988), "Probability and Causality: Why Hume and Indeterminism Don't Mix", *Nous* 22: 521-36.

Eells, E. and Sober, E. (1983). "Probabilistic Causality and the Question of Transitivity", *Philosophy of Science* 50: 35-57.

Giere, R. (1976), "A Laplacean Formal Semantics for Single-case Propensities", *Journal of Philosophical Logic* 5: 321-53.

_____. (1979), "Propensity and Necessity", *Synthese* 40: 439-51.

Glymour, C., Spirtes, P. and Scheines, R. (1991), "Causal Inference", *Erkenntnis* 35: 151-89.

Hanushek, E. and Jackson, J. (1977), *Statistical Methods for Social Scientists*. New York: Academic Press.

Irzik, G. and Meyer, E. (1987), "Causal Modeling: New Directions for Statistical Explanation", *Philosophy of Science* 54: 495-514.

Mellor, D. H. (1971), *The Matter of Chance*. Cambridge: Cambridge University Press

Popper, K. (1959), "The Propensity Interpretation of Probability", *British Journal for the Philosophy of Science* 10: 25-42.

Salmon, W. (1984), *Scientific Explanation and the Causal Structure of the World*. Princeton: Princeton University Press.

Objects and Structures in the Formal Sciences

Emily Grosholz

Pennsylvania State University

The priority of mathematics in the construction and deployment of the physical sciences has recently come under attack. Empiricism in the philosophy of mathematics, an offshoot of the trend towards naturalized epistemologies, subordinates formal disciplines to empirical facts (conceived variously as experience or experiment). (Kitcher 1983 and Maddy 1991) When allied with structuralism (Chihara 1973 and Resnik 1981, 1982, 1990), which denies to mathematics any specific subject matter of its own, it displaces intelligibility, heuristic force, and explanatory power away from mathematics into the realm of "fact".

In this essay, I will argue that mathematics, and mechanics conceived as a formal science, do have their own proper subject matters, their own proper unities. These objects, which are not merely structures, ground the characteristic way of constituting problems and solutions in each domain, the discoveries that expand and integrate domains with each other, and so in particular allow them, in the end, to be connected in a partial way with empirical fact. To introduce my arguments, I bring in as a foil Poincaré, whose interest in the formal dimension of science is refreshing in the current philosophical climate. As will become apparent, though his conventionalism inspires my position, I take issue with his early version of mathematical structuralism.

1. Poincaré and the Conventions of Mechanics

In his chapter on "The Classical Mechanics" in *Science and Hypothesis*, Poincaré points out that the principle of inertia is a convention. "If it be said that the velocity of a body cannot change, if there is no reason for it to change, may we not just as legitimately maintain that the position of a body cannot change, or that the curvature of its path cannot change, without the agency of an external cause?" (Poincaré 1905, p. 91) And in the same way, he argues that a generalized version of the principle of inertia is a convention: "The acceleration of a body depends only on its position and that of neighboring bodies, and on their velocities. Mathematicians would say that the movements of all the material molecules of the universe depend on differential equations of the second order" (Ibid., p. 92). For there are other thinkable possibilities: "In the first case, we may suppose that the velocity of a body depends only on its position and that of neighboring bodies; in the second case, that the variation of the acceleration of a body depends only on the position

PSA 1992, Volume 1, pp. 251-260

of the body and of neighboring bodies, on their velocities and accelerations; or, in mathematical terms, the differential equations of the motion would be of the first order in the first case and of the third order in the second" (Ibid., p. 93).

Roughly speaking, these three conventions correspond to Aristotelian physics, classical mechanics, and General Relativity Theory. This observation is particularly illuminating when it is conjoined with certain features of Poincaré's conventionalism. Experience, and experiment, says Poincaré, do not determine mathematical-physical principles, but rather guide our choice of them. "Experiment guides us in this choice, which it does not impose on us. It tells us not what is the truest, but what is the most convenient geometry" (Ibid., pp. 70-1). Experiment inclines us towards one convention rather than another when the convention results in the simplest formulation of laws. "All these conventions have been adopted for the very purpose of abbreviating and simplifying the enunciation" (Ibid., p. 91).

What counts as simplicity has changed over the course of the history of mathematics. And clearly, the simplicity Poincaré invokes here cannot be measured by algebraic degree, the order of a differential equation or indeed the degree of complexity of a logical formula. For if simplicity in that sense were paramount, we would have stayed with the Aristotelian choice of convention. Rather, Poincaré's simplicity has more to do with convenience or, more precisely, tractability. What are the mathematical items about which the age poses and solves problems?

For Aristotle, the simplest items were lines, circles and the rectilinear plane figures of Euclidean geometry. For Newton and Leibniz, they were the higher algebraic and transcendental curves amenable to Cartesian algebra and the differential equations of the infinitesimal calculus. For early twentieth century physicists, they were the manifolds and operations on manifolds of differential geometry. In each age, the progress of mathematics made these increasingly "complex" entities appear simple: useful, manipulable, tractable.

From this observation, we can draw a further conclusion that goes beyond Poincaré's own ruminations. That is, the development of new mathematics that involves new simplicities allows, and indeed in a sense drives, the choice of new conventions in physics. The availability of novel mathematical forms makes possible for the first time the Newtonian convention that "the acceleration of a body depends only on its position and that of neighboring bodies, and on their velocities," for it grants the new convention the required simplicity or convenience. Poincaré may be correct in saying that experiment guides, but does not determine, our choice of the simplest convention; it is equally true to say that the state of mathematics guides our sense of simplicity, and thus what the simplest convention may be.

To propose that mathematics drives physics, however, is an overstatement. For the role that physics plays in the formulations of mathematics increases as well over the centuries. The infinitesimal calculus, for example, arises out of a complex interplay among arithmetic, algebra, geometry and mechanics; moreover, the full story of that interaction cannot omit mention of the "guiding" role played by an accumulation of empirical astronomical data in the seventeenth century. All the same, the importance of mathematical considerations in the selection of physical principles must not be neglected. Current trends in American philosophy towards empiricism and a "naturalized" epistemology neglect the priority of mathematics in significant historical cases.

2. A Middle Ground between Empiricism and Structuralism

Poincaré regards his conventionalism about mathematics as the concomitant of his structuralism: "The aim of science is not things themselves, as the dogmatists in their simplicity imagine, but the relations among things." (Ibid., p. xxiv) "Mathematicians do not study objects, but the relations between objects." (Ibid., p. 20) These claims lie behind Poincaré's famous pronouncement, "There is no science but the science of the general." (Ibid., p. 4) To attempt to tie mathematics down to concrete particulars is the undertaking of dogmatists, and a futile project at that.

But a rejection of empiricism need not drive us into the arms of structuralism. I would like to retain some of the important insights of Poincaré's conventionalism and still maintain that mathematics and science have objects of study. I hold that we must recognize both structures and objects that instantiate those structures in the formal sciences in order to explain their intelligibility, development and indeed their relation to reality. Stated in slightly different terms, even formal sciences must have a semantics as well as a syntax.

Right away I should make two things clear. First, I do not claim that there is one true set of objects about which the formal and/or empirical sciences state truths. Rather, I prefer a tolerance for heterogeneity in ontological matters: different realms of study are occupied with different kinds of things. Second, I do not claim that the formal sciences have objects which are particulars. Rather, I want to show that they have objects which are formal unities in a strong sense, unities which cannot be reduced to mere place-holders in relational structures. Nor can they be reduced to mere sets of physical objects; they must be understood intensionally, not extensionally.

Thus, I will spend the remainder of this paper developing the view that only an account of the formal sciences which attributes to them objects as well as structure, proper semantics as well as syntax, can do justice to their intelligibility, heuristic force and explanatory power. My strategy depends in essential ways on examples from the history of mathematics and mechanics.

3. Intelligibility

The structuralist position holds that all alleged objects in mathematics can be unmasked as relational structures. I want to argue that relational structure by itself, utterly independent of objects to instantiate it, cannot be known. For what allows us to call relations "structures" or "systems" in the first place? These words point to an intelligible unity that the relations somehow constitute. But how do relations constitute a ratio, proportion, equation, group, sheaf, and so forth? The commonest way to write a proportion is a : b :: c : d. This is a simple and important kind of relational structure. While it is certainly true that a great variety of mathematical items can be substituted for a, b, c, and d, so that these terms are place-holders in the sense that any given substitution might be made otherwise, they are not "mere" place-holders in the sense that substitution can be omitted altogether. The proportion 2 : 4 :: 4 : 8 has the unity of a complete thought, indeed of a truth, and exhibits something important about the numbers 2, 4, and 8; whereas " : :: : " is either an accident of typography or nothing at all—it cannot be thought.

Let me make the same point in a different context. In the first pages of the *Elements*, Euclid posits a few basic kinds of formal geometrical unities: the point, which is a trivial unity; the line, which is a useful unity because lines can be used to measure themselves; plane figures, of which the simplest is the triangle; and the circle. The two latter kinds, triangles and circles, immediately generate deep and interesting problems, problems which in one way or another have kept mathematicians busy to the present day. Euclid's mathematical practice carefully distinguishes and segregates these kinds: points bound but do not compose lines; lines bound but do not compose plane figures; figures of different dimensions must be treated separately, as must straight and curved lines. (Euclid 1956, i pp. 153-55)

A triangle in the *Elements* is a formal unity which cannot be reduced to the points and lines which bound it, and which generates problems which do not arise with respect to points or lines. It is in no sense a structure of points, whose minimality might tempt one to regard them as "mere" placeholders, though that too would be a mistake. It combines intelligible unity (what Poincaré might call simplicity) with satisfyingly tractable mystery, due to its irreducible complexity. There is no more compelling display of this than the proof of the Pythagorean theorem, which reorganizes and crowns Book I. If mathematics were utterly transparent or utterly opaque, it would not be a human science, for it would provide nothing to think about. Since it is a human science, the philosopher's task is to give an adequate account of the middle ground between transparency and opacity that mathematics occupies.

To exist as an object in mathematics is to figure in a problem. Conversely, mathematical problems, simple enough to formulate and hard enough to solve, arise from problematic objects. A logician can take the domain of problems and problematic objects that is geometry and recast it as a set of axioms and theorems in a logical language, and even redefine its special geometrical terms in the language of "pure" logic, as logicists from Russell to Chihara have wanted to do. Then, as the logicist would have it, geometry appears as a set of theorems derivable from logical axioms that may remain and be considered as uninterpreted, that is to say, without any special subject matter.

What results from this undertaking, however, is unintelligible. There are two ways to argue the point. First, suppose that logic truly has no subject matter. Does this mean that logic, and along with it logicized geometry, can be understood as a purely syntactical structure, apart from any and all semantics? The answer to this question must be no. For the "structure" of logic in any of its important varieties, including predicate logic, has a shape inherent in its well-formed formulae and axioms, and in its metatheorems, that exhibit its power and limitations. This shape is the result of a two thousand year old tradition that has (not univocally) understood logic as about propositions about the world. What logic is taken to be about puts severe constraints on how logic can be formulated. If we suppose that logic has no object at all, then either the shape of logic is inexplicable and accidental, or a true logic would have no shape, standing under no constraints whatever. But such a logic would be unintelligible. (On this point, see also Jacquette 1991.)

Second, suppose that logic does have its own subject matter, propositions about the world, which constrains its systematicity and structure. Then within logic so construed there is no way of collecting the truths of logicized geometry; and even if they could be collected from the "outside," there would be no way to see how to go about extending them, except to generate an infinite and shapeless mass of theorems. No scrutiny, however long and thoughtful, of the way propositions behave will illuminate the side-by-sideness of things in space. A collection of truths about geometry, skewed and reduced in the first place by being pulled into an alien idiom, and then taken as a collection of truths about the sorts of things logic is about, has no reason to be a collection. Seen from within logic, it looks accidental; it has been stripped of intelligibility.

4. Heuristic Force

The foregoing argument should not be taken to mean that geometry and logic stand in no relation at all, or that there is nothing to be learned from juxtaposing them. (I study one related juxtaposition in Grosholz 1985.) My point is rather that logic and geometry are about different kinds of objects and problems, so different that geometry cannot be absorbed into logic without losing its intelligibility. But the relation between logic and geometry need not be proposed as one of total absorption. Indeed, once the heterogeneity of formal domains is admitted, the real interest of their mutual relations comes into focus.

Careful examination of historical cases reveals that heterogeneity must always be negotiated as partial unifications are established: shared structure is not easily located and once discerned cannot be taken for granted; and objects and problems behave oddly at the overlap of domains. The partial unification of domains involves objects as well as structures, and it is nontrivially problematic. How to set domains in relation to each other is neither transparent nor utterly opaque. The establishment of structural correspondences between different kinds of formal unities is difficult and surprising, and indeed often generates new kinds of objects embedded in new kinds of problems.

Thus this process is an important engine of discovery in mathematics and mathematical mechanics. Deductivists like Popper, inductivists like Carnap, conventionalists like Poincaré, all agree that empirical evidence cannot determine ideal form; an inferential gap always persists between the determinate ideal and the indeterminate real. This inferential gap also limits the degree to which empirical evidence can drive discovery in the formal sciences: all the more reason to pay close attention to heuristic processes that are inherently formal.

At the start of the seventeenth century, algebra, geometry, mechanics and natural philosophy were four quite separate disciplines. Algebra was a modest generalization of elementary arithmetic, geometry was Euclid's, mechanics was restricted to simple machines (that is, idealizations of human artifacts), and natural philosophy was a fundamentally Aristotelian treatment of nature. By the end of the century, all four of these had been impressively and complexly integrated in the scientific work of Leibniz and Newton. This conceptual unification is a central feature of the progress that took place during the one hundred years that bridge the Middle Ages and the Enlightenment.

A brief review of this historical development is in order. Problems arose in classical geometry which could be formulated but not solved in that domain; a famous example is Pappus' problem. And objects were generated there that seemed anomalous and without systematic relation to other parts of geometry; the best examples are transcendental curves, like the logarithmic spiral and the quadratrix.

Descartes brought geometry into relation with the realm of number by positing that operations on line segments could instantiate the algebra of arithmetic. This hypothesis allowed him to bring the resources of both algebra and geometry to bear on problems, and in particular to solve Pappus' problem far more completely than any of his predecessors. It also led him to exclude transcendental curves altogether from geometry. His exclusions were balanced by additions. Descartes was able to construct or schematically indicate a whole new hierarchy of higher algebraic curves. They appear as hybrids at the intersection of geometry and algebra, and would not have arisen from Euclidean geometry as it was classically constituted. Descartes was particularly intent upon a cubic curve that has come to be known as the Cartesian parabola, which he investigated as a point-wise construction, an equation and a curve traced by an apparatus that generalized ruler and compass (Grosholz 1991, ch. 1 and 2).

Descartes also took an important first step in integrating mechanics and natural philosophy. Characterizing nature as a plenum of material particles whose only interaction is collision, he put forward a set of rules that treat the collision of bodies in rough analogy to the classical treatment of the lever or balance. And he refers to all the possible configurations of matter as the machines of nature. However, as is well known, this part of his project of unification remained schematic and incomplete. Ironically, one reason it did so was because his reorganization of mathematics flattened and reduced geometry in a way that made it less apt for the mathematization of the machines of nature, especially once those machines were set in motion (Grosholz 1991, ch. 5).

Leibniz's style of unification is quite different from Descartes's, for it is motivated on the one hand by an intense interest in the combinatorial properties of numbers and on the other by a zest for the infinite. (Descartes never shows much interest in patterns of numbers, and stays away from the infinite as much as possible.) Fascinated by Pascal's treatment of problems of quadrature, Leibniz characteristically moved from the study of discrete patterns of numbers associated, as approximations, with the problems, to infinitary extrapolations that reclaimed the original continuity of the geometrical curve. The hybrid that recurrently claimed his attention was the curve considered as an infinite-sided polygon. At once a unified shape, the solution to a differential equation, the analogue of an algebraic equation, and an infinite array of numbers, it led him to formulate the algorithms of the calculus (Grosholz 1992).

Such problems of quadrature naturally give rise to transcendental curves, and Leibniz's treatment of them was designed to find a perspicuous and tractable expression for such curves. They in turn stand as hybrids linking mathematics with mechanics. The linkage ran in both directions. On the one hand, Leibniz used idealized mechanical constructions to extend mathematics for his own purposes, those of the infinitesimal calculus, beyond the Cartesian canon. On the other hand, the curve viewed as a solution to a differential equation was the obvious vehicle for the representation of a trajectory, which little by little had come to be understood as a nexus of forces couched in terms of times and distances (Bos 1988 and Meli 1986).

The combination of geometry and mechanics in Newton's *Principia* also hinges on trajectories taken as hybrids, though the combination is not mediated by algebra as it is in the work of Leibniz (and his followers the Bernoullis). Indeed, Newton's undertaking is so thoroughly and self-consciously geometrical that the contrasts with Euclidean geometry are particularly arresting. As Descartes' partial unification of geometry and algebra produces a strangely flattened geometry of line lengths, so Newton's partial unification of geometry and mechanics produces constructions that would have been meaningless to Euclid. Despite Newton's disclaimers, the proofs in Book I of the *Principia* are full of infinitesimalistic reasoning; and the diagrams that accompany them contain constructions that have no purely geometrical or even merely spatial significance, but can be understood only as representative of finite and infinitesimal configurations of force and time. For of course those diagrams are intended to show the paths of point masses acting under the influence of various abstract forces, whose relevance to the orbits of the planets of the solar system Newton will go on to argue. (Grosholz 1987 and De Gandt 1987)

Thus, when the Bernoullis begin the essentially eighteenth century project of developing Newtonian mechanics along the analytical lines of the Leibnizian calculus, they profit from a rich variety of ways of moving between arithmetic, algebra, geometry and mechanics, all of them partial and not all of them entirely compatible. The original domains retain their own objects and still give rise to problems independently, but their overlap with the other domains is marked by hybrids; and new sets of problems, sometimes elaborate enough to constitute new domains, arise around these ambiguous objects.

Descartes, Leibniz and Newton confronted certain difficult but promising problems about numbers, conic sections, algebraic and transcendental curves, and trajectories, and solved them by partially unifying certain domains in order to bring more conceptual resources to bear on them. This process did not make the original objects and problems disappear, though it distorted, extended, and reorganized them, but it did generate new hybrids and domains. And while Descartes, Leibniz and Newton posited correspondences between domains in order to solve problems, the correspondences, detached from their origins and considered as abstract structures, were not what those investigations were about.

A structuralist, looking back on these episodes, notices the sharing of structure between domains, and then jumps to the conclusion that structure is really all that is important or indeed, really all that is there. But this story of discovery cannot be told in terms of uninterpreted structures, which are either unintelligible or transparent, and so pose no problems. And it cannot be told in terms of the structures given to us by twentieth century logic, for Descartes, Leibniz and Newton were not trying to solve problems about propositions.

5. Explanatory Power

Towards the end of the foregoing story, Newton generates a geometrico-mechanical object, an ellipse-trajectory, that bears on reality. It stands in some sort of relationship to the sun and planets of our solar system and their motion with respect to each other; and that relationship, whatever it is, is somehow mediated by the data compiled by astronomers over a couple of thousand years. To bring this paper to its conclusion, I want to argue that however we characterize this relationship, it is unintelligible if the formal sciences are understood as pure structures, and intelligible if they are understood as domains of problems involving objects and structures. That is, the formal sciences must have their own proper semantics as well as syntax, their own proper ideal or intensional objects that constrain and shape the associated relational structures, in order to bear on reality.

Structuralism, because it wants to unmask all ideal objects as structures and because an utterly uninterpreted structure is incoherent, tends to borrow its objects from empiricism's "real" world. Empiricism, because it denies the gap between the indefinite real and the definite ideal, is all too happy to grant that the objects of its "real" world can be injected directly into empty formal structures. This bargain struck between structuralism and empiricism is usually called "instantiation." It is invoked by logicians who, in their expositions of predicate logic, wave their hands over the universe of discourse and call it "all the physical objects in the universe."

Instantiation, so understood, is impossible. For the objects of ordinary experience are ephemeral particulars whose unity and relations are indeterminate. They cannot have the features and relations that hold true of eternal, determinate formal unities. No accumulation of empirical data will ever make the realm of perceptual experience fit into any of the formal domains we wish to bring to bear on it.

All the same, it is undeniable that mathematics and mechanics bear on experience. But this relationship is postulated; it is a rational convention, as Poincaré would say, and it must be forged anew in every age. In every age it looks different, as the relationship among the formal sciences looks different. Amid all this difference, I might add, there is enough stability to allow human wisdom on the one hand and science on the other, for the world of existents and the objects of the formal sciences place severe constraints on how they can be set in relation to each other.

The rich, indeterminate world is encountered by us in experience which is "always already" organized. That organization is taken up in the rich, ambiguous discourse of natural language. Nominalist empiricism pretends that the world occurs to us as a collection of determinate particulars, and so assumes that they can have a determinate discursive structure. But the unity as well as the features of the things encountered in this world are indeterminate and fleeting; and the indeterminacies of natural language reflect experience.

Thus, I argue that when human beings have tried to bring the formal sciences to bear on experience, their project, variable and more and more extensive over the centuries, looks like the partial unification of two or more discourses. As in all the other cases of partial unification mentioned earlier in this essay, the domains unified are heterogeneous. Thus, they maintain their autonomy even as they are partially combined in the service of problem solving; and their characteristic objects and problems are distorted, rearranged and extended by the interaction. The correlation addresses questions that could not be answered from within ordinary experience alone: Why do the stars move as they do? Why does a lever, or a balance, or pulley, work? And at the overlap, hybrid objects and problems arise, that would never have arisen in the original domains considered in isolation.

This historical process thus looks like a series of mediations between the indeterminate real and the determinate ideal, none of which are definitive, all of which are severely constrained by the kinds of objects among which the mediations must hold. Natural language already brings the ephemeral particulars of the experienced world to stand in words, where they look more universal, precisely delineated, and eternal, despite the mortality of natural languages themselves. But natural language retains the ability, utterly foreign to mathematics, to name particulars, though this ability in a sense aggravates its indeterminacy.

It is more difficult to establish correspondences between natural language describing experience and mathematics than between, say, two mathematical domains. But the history of science reveals how this has been done in stages. The favorite hybrid objects for this mediation were initially human artifacts and the motion of the bodies in the solar system. That is to say, there are certain peculiar configurations of objects in experience that lend themselves to being correlated with formal objects, through the mediation of hybrids: simple machines, and idealized heavenly bodies. By the end of the seventeenth century, as we have seen, these hybrids become particles in collision, and the trajectories of mass points acting under abstract forces.

Human experience, mathematics and the hybrid domain of mechanics have all been altered by the interplay, yet we still distinguish among those heterogeneous enterprises. Science and mathematics bear upon, but have not fundamentally altered, the sources of human wisdom, birth, love, conflict and death. Mathematics continues to habitually detach itself from practical matters, and embroider its infinite complexities on the face of nothingness. And mechanics goes its own way through the labyrinth of experiment.

My final point is that the story of this process of bringing the formal sciences into relation with experienced reality cannot be told without speaking of formal objects as well as real objects. The correlations between domains of discourse which have historically provided the mediations depend on the recognition that different sorts of objects accede to, resist or exclude the imposition of common structure, so that new problems arise around the distorted, expanded and novel objects which the exploration of common structure precipitates. Mathematics has objects because unity is a condition of intelligibility; thought must have something to think about. The world has objects because unity is a condition of existence; Being, whatever else it is, is beings. In any case, the process I have so briefly indicated in this essay (though having elaborated on it elsewhere) ought to be of interest because of the light it throws on the relations between existence and intelligibility.

References

Bos, H.J.M. (1988), "Tractional Motion and the Legitimation of Transcendental Curves", *Centaurus* 13: 9-62.

Chihara, C. (1973), *Ontology and the Vicious Circle Principle*. Ithaca: Cornell University Press.

De Gandt, F. (1987), "The Geometrical Treatment of Central Forces in Newton's Principia", *Graduate Faculty Philosophy Journal* 12, 1 and 2: 111-51.

Euclid (1956), *The Thirteen Books of Euclid's Elements*. 3 volumes, Sir Thomas Heath, (ed.). New York: Dover.

Grosholz, E. (1985), "Two Episodes in the Unification of Logic and Topology", *British Journal for the Philosophy of Science* 36: 147-57.

______. (1987), "Some Uses of Proportion in Newton's *Principia*, Book I: A Case Study in Applied Mathematics", *Studies in History and Philosophy of Science* 18, 2: 209- 20.

______. (1991), *Cartesian Method and the Problem of Reduction*. Oxford: Oxford University Press.

______. (1992), "Was Leibniz a Mathematical Revolutionary?", in *Revolutions in Mathematics*, D. Gillies, (ed.). Oxford: Oxford University Press, pp. 117-33 .

Jacquette, D. (1991), "The Myth of Pure Syntax", pp. 1-14 in *Topics in Philosophy and Artificial Intelligence*, L. Albertazzi and R. Poli (eds.). Bolzano: Mitteleuropaisches Kulturinstitut.

Kitcher, P. (1983), *The Nature of Mathematical Knowledge*. Oxford: Oxford University Press.

Maddy, P. (1991), *Realism in Mathematics*. Oxford: Oxford University Press.

Meli, D.B. (1986), "Some Aspects of the Interaction between Natural Philosophy and Mathematics in Leibniz", in *The Leibniz Renaissance*, P. Mugnai, (ed.). Florence: Olschki, pp. 9-22 .

Poincaré, H. (1905; reprinted 1952), *Science and Hypothesis*. New York: Dover.

Resnik, M. (1981), "Mathematics as a Science of Patterns: Ontology and Reference", *Nous* 15: 529-50.

_______. (1982), "Mathematics as a Science of Patterns: Epistemology", *Nous* 16: 95-105.

_______. (1990), "Between Mathematics and Physics", *PSA 1990* Volume 2, East Lansing: Philosophy of Science Association, pp. 369-378.

Adding Potential to a Physical Theory of Causation[1]

Mark Zangari

La Trobe University

1. Introduction

Recently, several authors have attempted to characterise causation by identifying causal connections with physical processes. While this approach may not live up to providing a complete analysis of causation in all its contexts, it has raised some significant points regarding the causal role played by various entities described in natural laws. To date, much of the discussion has involved the attempted identification of physical terms or concepts that necessarily or sufficiently characterise causal processes, and has centered around three main contentions:

C1. Causal processes involve the presence of forces, and that forces are themselves a species of primitive causes (e.g., Bigelow, et al. 1988).

C2. Causal processes necessarily involve transfer of a physically measurable quantity, such as energy-momentum (e.g., Aronson 1982 and Fair 1979).

C3. Causal processes necessarily involve microscopic interactions between "fundamental" entities, such as is described by quantum field theory (e.g., Heathcote 1989).

In this paper, I wish to suggest that C1 and C2 cannot universally characterise physical causal processes because of the existence of a class of causal interactions that involve neither forces, nor energy/momentum transfer. In addition, C3 is usually assumed to reduce to C1 in the non-relativistic and classical (NRC) limits. As I wish to show, the behaviour of interacting systems in multiconnected regions shows that this view cannot account for the above kinds of phenomena. Furthermore, such effects suggest a fundamental ontology of interacting systems that is not captured in usual expositions of C3.

Section 2 contains a brief analysis of a physicalist account of causation based on interactions in the NRC limit. In section 3, I wish to argue that interaction theories, when further analysed, seem to lend at least implicit support to C1 through the principle of *gauge invariance*. However, as discussed in section 4, the analysis of certain well known results from quantum theory and the requirement of locality suggests that

PSA 1992, Volume 1, pp. 261-273

the correct characterisation of causal interactions is via the recognition of the role played by *potentials*.

2. Interactions

It has been suggested that causation should be fundamentally characterised in terms of time evolving state changes in physical systems.[2] In the non-relativistic, canonical formalism of physics, such state changes are signified by the presence of an interaction term in the particular Hamiltonian (H) or Lagrangian (L) that describes the dynamics of the system (In relativistic field theories, H and L are replaced by their densities, $\mathcal{H}$ and $\mathcal{L}$). In this scheme, a physical entity is considered to be subject to a causal influence just in the case that it is undergoing some kind of interaction process that results in a physical state change. In the particular case of two interacting objects (say the Earth and the Moon), a causal ontology based on *mutual interaction* overcomes problems in the earlier theses, particularly in C2.

Dieks (1981, p.106) criticised C2 on the grounds that transference of energy is an asymmetric process, whereas not all situations where causal explanations are employed share this property. Indeed, causal accounts of physical phenomena often possess an "explanatory symmetry", so that the cause and the effect can be interchanged, depending on what is being explained. As an example, he considers the situation where the placing of ice cubes in a glass of water results in both (a) the ice causing the water to grow colder and (b) the water causing the ice to melt. A causal ontology based on C2, it is claimed, would force (a) and (b) to be considered disjunctively, even if the requirement that energy necessarily flows from the cause to the effect is relaxed. Since there is an arbitrary choice of whether (a) or (b) characterises the causal link, and therefore whether the ice or the water is actually the cause, Dieks suggests that such a link may be an explanatory concept, rather than an ontological category.

However, Dieks is mistaken in assuming that the physical description of causal processes necessarily implies the ascription of an asymmetric cause-effect relation. This kind of approach applies, if anywhere, only to the determination of causal connections between temporally separated *events*. By contrast, the *evolution of states* (e.g., the state changes of the ice and water in Dieks' example) are causally connected in physical descriptions by the presence of a *mutual interaction* between the two systems. To make this a little more precise, consider a problem analogous to Dieks' glass of iced water, but simpler to represent physically: "Does the Earth cause the Moon to execute an epicyclic orbit around the Sun, or does the Moon cause the Earth to 'wobble' about its mean elliptic orbit?" This situation can be represented by the following Hamiltonian (a function of momentum **p** and position **x** that is equal to the kinetic plus the potential energy of the system):

$$H = \left[\sum_{i=1}^{2} (p_i^2 / 2m_i) + H_{\text{Sun and Planets}} \right] + m_1 m_2 V(\mathrm{x}_1 - \mathrm{x}_2)$$

where m refers to mass and the subscripts 1 and 2, refer to the Earth and Moon, respectively. The term in square brackets represents the "natural" or unperturbed motion of the Earth-Moon system (including the effects of the Sun and the other planets), and V is the potential between the Earth and Moon that represents the causal interaction between them. Notice that the description is symmetric in all quantities with subscripts 1 and 2 (up to a sign change, since V is antisymmetric under interchange of subscripts). In other words, the representation of the interacting system does not attribute the role of "cause" or "effect" to either the Earth or Moon, but merely indicates, by the presence of the interaction term $m_1 m_2 V$, that a certain causal link exists between their motions. One obtains the asymmetric causal explanations (e.g., that the perturbation in the Earth's orbit is caused by the Moon, or vice-versa) by deriving the equations of motion of the particle that displays the effect for which the cause is being sought. This is achieved by the application of Hamilton's equations to the above Hamiltonian:

$$\dot{\mathbf{p}}_i = -\partial H / \partial \mathbf{x}_i$$
$$\dot{\mathbf{x}}_i = \partial H / \partial \mathbf{p}_i \qquad (1)$$

Quite obviously, the symmetry in the roles of the subscripted quantities is broken in each of the 4 expressions obtained in this way (i.e., the two equations of motion for each of the bodies). Each equation of motion depicts a particular causal relationship between the kinematical state of the body that it describes and the external influences acting upon it, as in the above question concerning the orbits of the Earth and Moon (or, in fact, the ice and water). However, the ontological characterisation of the causal connection is in the interaction term itself, which possesses the desired symmetric properties.

There are certain situations where causal symmetry is not desirable. Consider the relation between the position of a barometer needle, and the state of the atmosphere. While it seems acceptable to assert that atmospheric fluctuation causes a change in the barometer reading, one would not wish to conclude that the barometer's state likewise "causes" the accompanying atmospheric change. A mutual interaction picture, however, is safe from such an unpalatable conjecture because, while the interaction term that describes the barometer's state evolution has large contributions from the atmospheric state, the atmosphere's state is dependent on the barometer only to a vanishingly small degree. When the equations of motion are applied, and the causal relations pertaining to each system are separately identified, it would be evident that the barometer readings are causally linked to the atmosphere, but the atmosphere's condition is practically independent of the barometer. In general, interaction terms determine the degree of "causal dependence" between interacting systems. In a loose sense, causal dependence can be quantified as a function of both the relative number of degrees of freedom from each system that contribute to the interaction, and the strength of the coupling between them. Roughly speaking, while an equal proportion of the Earth and Moon's degrees of freedom contribute to their interaction, only a very small number of the atmosphere's are involved in its interaction with the barometer.

3. A Case for Forces

In general, causal interactions are represented in relativistic field theories as non-linear couplings in the interaction terms of the densities $\mathcal{H}$ and $\mathcal{L}$. When the non-relativistic limit is taken for systems of particles however, 'interaction' is usually considered to become synonymous to the presence of forces and hence, forces become the means by which causation is physically characterised.[3]

To see how forces are introduced into such a scheme, let us now consider how the electromagnetic interaction of a single particle of charge q and mass m, with an "external" field is described in the NRC "interaction picture". One first writes the Hamiltonian and incorporates the interaction with the field by the inclusion of the vector potential $\mathbf{A}$ (via the substitution $\mathbf{p} \to \mathbf{p} - (q/c)\mathbf{A}$ to correctly conserve momentum) and the scalar potential ϕ:

$$H = (1/2m)[\mathbf{p} - (q/c)\mathbf{A}(x,t)]^2 + q\phi(x,t) \qquad (2)$$

H may be split into its kinematic (i.e., natural motion) and interaction terms:

$$H = H_0 + H_I = p^2/(2m) + q[\phi - (\mathbf{v}/c)\cdot\mathbf{A}] + (q/c)^2A^2/(2m) \tag{3}$$

where H_I is often referred to as the *generalised potential.* Applying Hamilton's equations of motion in equation (1) to equation (3) (and the judicious use of a few vector identities) yields the equations of motion and, in this case, the well known Lorentz force law:

$$\dot{\mathbf{p}} = \mathbf{F} = q(\mathbf{v}/c \times \mathbf{B} + \mathbf{E}). \tag{4}$$

E and **B** are the electric and magnetic fields which are related to **A** and ϕ in the following way:

$$\begin{aligned} \mathbf{E} &= -\nabla\phi - \partial\mathbf{A}/c\partial t \\ \mathbf{B} &= \nabla \times \mathbf{A} \end{aligned} \tag{5}$$

This formal development highlights some significant factors in the relation between interaction potentials, forces and fields. It can be seen that the specification of the interaction in H_I (equation (3)) yields a relation between forces and fields in equation (4) by the application of spatial derivatives. The usual (classical) interpretation of the above formalism is that the force and the fields in equation (4) are considered to be "real", whereas the potentials in the interaction term (equation (3)) are regarded as mere mathematical constructions defined by equation (5). While providing a convenient means of calculating the empirically observable equations of motion, the potentials do not themselves have any ontological significance. Thus, those degrees of freedom of the electromagnetic interaction that correspond to real dynamical processes are wholly attributable to the fields and the resulting forces, whereas the potentials are merely an alternative notational representation of such fields obtained by integrating them over space. Integration introduces an "arbitrary constant", and potentials therefore contain additional degrees of freedom over and above those present in the fields. In the classical picture, this overdetermines the observable motion, since the potentials can consequently be arbitrarily transformed by *gauge transformations* without any empirical consequences; i.e:

$$\begin{aligned} \phi &\rightarrow \phi - \partial\zeta/c\partial t \\ \mathbf{A} &\rightarrow \mathbf{A} + \nabla\zeta \end{aligned} \tag{6}$$

where $\zeta(x,t)$ is an arbitrary scalar function. Since equation (6) leaves **E** and **B** from equations (4) and (5) invariant, all the observable dynamics is fully accounted for by the degrees of freedom associated with the fields. The extra degrees of freedom in **A** and ϕ are therefore usually treated as a mathematical artifact that is not considered to correspond to anything physical. Quantities such as **E** and **B** that are immune to gauge transformations are said to satisfy the principle of *gauge invariance*, which acts as a criterion for ontological significance in physics:[4]

(GI) Only those mathematical terms that are gauge invariant are considered to be physically interpretable as descriptions of real entities.

The possibility that the potentials are actually ontologically significant, while the fields (and consequently forces) are derived from them via spatial differentiation is usually discounted, at least in classical physics, because potentials are not gauge invariant quantities. It is the gauge invariant forces and the fields that are related to them that are treated as primitive parts of physical causal interactions.

4. Non-Relativistic Quantum Mechanics and Potentials

4.1. The Aharonov-Bohm Effect

In this section, I wish to show that treating forces as ontologically primitive at the expense of potentials cannot be unequivocally upheld when certain results from quantum mechanics are considered. In 1959, Yakir Aharonov and David Bohm predicted the result of a proposed experiment in which a charged particle's state *could* observably change, even if all magnetic and electric fields (and therefore the corresponding forces) were completely shielded from the particle.[5] Their basic result involves a modification to the standard two-slit electron interference experiment in which an interference pattern is produced on a screen fitted with a suitable detector. If an impenetrable solenoid containing a completely confined magnetic field is placed somewhere between the slits and the detector, then the existence of the field will still observably affect the electrons, even though they cannot enter the region where the field is present. According to the Aharonov-Bohm effect, the phase of the interference pattern produced by the electrons will vary and the pattern will be shifted by an amount dependent on the magnetic flux ϕ_B inside the solenoid. This was an extraordinary result that had far reaching repercussions, particularly in the light of the discussion in section 3. Even though ϕ_B and therefore **B**, are completely confined to within the solenoid (from which the electrons are completely excluded), the electrons still undergo a type of causal interaction, despite being free of the local influence of the field. This would be impossible if the ontology of the interaction in the non-relativistic limit were correctly characterised by the Lorentz force law (as was thought to be the case), since there is no possibility that a force is exerted on the electrons in this situation. Furthermore, it has been shown that no energy transfer takes place between the electrons and the field (Peshkin and Tonamura 1989, pp. 25-6). So how can this particular causal interaction be explained? Because of the geometry of the magnetic field inside the solenoid, there is a vector potential **A** with zero curl in the region occupied by the electrons outside the solenoid. Can the modified wave function of the electrons, and the consequential observable shift in the interference pattern be explained as a local effect of **A**? To understand how this might occur, some further comments on the Aharonov-Bohm effect and a review of some of the peculiarities of the wave function are in order.

The effect that is caused in the Aharonov-Bohm effect does not involve any accelerations; it is of a nature that classical terms are ill equipped to describe. For, in a classical description, the dynamical state of a particle can be completely determined by specifying its position and momentum at a given time, just as was done in the Hamiltonian of section 3. In non-relativistic quantum mechanics, however, the dynamics can be completely described by the wave function, ψ. This introduces an extra "degree of freedom" in the state of a particle that has no counterpart in classical physics: the *phase* of the wave function. The phase has the significant property that it is *not* directly measurable (it is only inferred via the peculiarly "quantum" effects, such as the diffraction and interference patterns produced by particles). The phase endows ψ with degrees of freedom that are not gauge invariant so ψ cannot alone be physically interpreted in the standard picture. Only its modulus squared (which removes the contribution from the phase, and thus restores gauge invariance) is treated as meaningful, being the probability distribution for the observation of a particle in its possible states. The introduction of a curl-free vector potential into Schrödinger's equation amounts to a gauge transformation and results only in a phase shift in ψ. However, because gauge transformations produce no observable effects, such a phase shift cannot correspond to a real, physical process according to (GI).

Although **A** and the associated phase shift in ψ are not individually gauge invariant, the observable result of the experiment must be, if it is to satisfy (GI). While **A** is the quantity that explicitly appears in Schrödinger's equation, it is clear from equation (5) that a curl-free vector potential cannot be locally associated with any gauge invariant fields (and therefore, cannot produce any forces either). However, the measurable effect in the Aharonov-Bohm experiment depends only on the gauge invariant $\phi_{\mathbf{B}}$, rather than on **A** *per se*, so their result does not ostensibly violate (GI). This suggests an alternative ontological interpretation of the Aharonov-Bohm effect based on trying to identify a possible effect due directly to the field, rather than to the potential (see Olariu and Popescu 1985, pp. 423-426; references to original papers are included there). Such approaches can be broadly divided into two distinct groups. In the first, the potentials are re-expressed in terms of the fields by solving equation (5) for ϕ and **A**. These expressions can then be used to describe the details of the interaction without reference to the potentials, as they have everywhere been substituted with expressions containing only functions of **E** and **B**. However, since the strengths of **E** and **B** are always zero in all regions accessible to the electron, this approach has the consequence that the action of the fields must be inherently non-local. As I shall argue below, this non-locality is considerably more problematic than those previously encountered in physics and conflicts with the principles on which physical explanations are usually based.

The second type of argument for a field-based interaction ontology in the Aharonov-Bohm effect appeals to the return magnetic field that leaks from any solenoid of finite length. The complete confinement of all the flux so that none at all escapes, is an idealisation based on the case where the solenoid's length becomes infinite. While the total absence of any leakage flux is unrealisable in a real solenoid, a very small return field strength can be achieved near the middle of a sufficiently long one. It has been suggested that in any actual Aharonov-Bohm experiment, the return magnetic field through which the electrons pass on their way from the source to the detector can account for the observed phase shift (see Olariu and Popescu 1985, pp. 424-426). Thus, there is no need to depart from the picture of electromagnetic interactions painted in section 3, and in particular, no need to introduce ontologically significant non-gauge invariant potentials. This view, however, seems untenable (as is argued in Olariu and Popescu 1985, p. 426) because a direct, local coupling between the electron and the magnetic field implies that the electron's angular momentum will be quantised into integer multiples of $\hbar$. More importantly, it is possible that, for a suitable superposition of states, the electron wave packet will be sufficiently localised and close to the centre of a long (but finite) solenoid that the return field strength it encounters is negligible. Yet, even in this case, the phase shift is still dependent on the confined flux inside the solenoid, but is now attributed to interaction with the leakage field outside (but not necessarily in the region occupied by the electrons). Therefore, if the external return field is to have any influence, then it too must have some effect in a region where its strength approaches zero, and the problem is not significantly different from that of the completely confined field.

The above arguments are motivated by the fact that all observable quantities in the Aharonov-Bohm effect formally depend on only the field, and the respective authors attempt to physically vindicate this by showing that there is actually no need to suggest a causal role for the potential. However, one is then faced with the problem of how the field can influence the electrons, given that the local density $\psi^*(x)\mathbf{B}(x)\psi(x)$ is always zero. This problem does not arise if the potential is assumed to be the causal agent in the interaction, although simply promoting potentials to the rank of genuine ontologically significant entities challenges the very heart of (GI). It therefore appears that in order to provide an account of this odd interaction, one is forced to choose between violating gauge invariance and violating locality. As I now wish to argue, the violation of locality needed to attribute the causal role to the field is particularly severe. In fact, the Aharonov-Bohm effect highlights a fundamental discord between locality and gauge invariance.

4.2. Locality

The principle of locality has a fundamental bearing on physicalist accounts of causation, being a necessary condition for the possibility of establishing a causal relation between separated systems. It is imposed from the special theory of relativity (STR) as a consequence of a finite maximum velocity for signal transmission and the exclusion of action-at-a-distance.[6] It states that:

(L1) degrees of freedom must be completely determined by specifying the relevant physical quantities at *only one point*, and

(L2) systems depending on degrees of freedom $\alpha(x)$ and $\beta(y)$ can only be causally connected if, either $(x - y)$ is not spacelike, or

(L2a) $\mathbf{x}$ and $\mathbf{y}$ are infinitesimally close; i.e., $|\mathbf{x} - \mathbf{y}| < \varepsilon$ for arbitrarily small ε (italics and bold represent 4-vectors and 3-vectors , respectively).

I shall refer to theories that satisfy (L1), (L2) and (L2a) as *strictly local.* It was this requirement that led to the necessity of attributing electromagnetic effects to local field strengths, rather than to retarded integrals based on non-local charge distributions. The non-local field interpretation of the Aharonov-Bohm effect does not satisfy locality, but then other theories, such as Newtonian mechanics, do not either. As I suggest below, while the non-locality in Newtonian physics is ontologically benign, a similar thing cannot be said of that arising in the Aharonov-Bohm effect.

In non-relativistic, Newtonian mechanics, "instantaneous" causal interactions are permitted because signals are not restricted to finite velocities. However, the Newtonian picture, while often useful, is only an approximation to relativistic mechanics for the case where the time taken for interaction is negligible, compared to the relative velocities of the particles taking part. The locality conditions of STR are not threatened in such a picture just because it is understood as an approximation that one can explicitly remove by allowing for finite signal propagation times (for example, by replacing the Coulomb potentials with Liénard-Wiechert potentials—see Jackson 1975, p. 656). If this is done, then it can be shown that the relativistic picture, in the limit of small velocities, is formally identical to the non-relativistic one, which is indeed a valid formal approximation in that case.[7] Because it is arrived at by the *ad hoc* application of an arbitrary, formal limit, such an approximation does not challenge the ontological implications of the relativistic picture (assuming that the approximation renders the theory *less* empirically accurate—which is usually true). Additionally, there is a systematic, formal construction whereby the Newtonian non-locality can be reconciled with local interactions in relativity as the approximating limit is removed. However, no such construction is available for the type of non-locality that would arise were the Aharonov-Bohm effect seen as a non-local interaction between electrons and the confined field.[8] For, unlike the above case, the field can *never* actually reach the region where its effects are manifested, no matter how much time is allowed for the interaction, nor what velocity is assumed for the propagation of the influence, because *the field cannot escape the solenoid* under any circumstances (or, at least, it cannot reach the region where the probability density for finding the electron is not negligibly small). One would have to postulate an altogether new class of interactions, and a particularly mysterious one, were one to physically account for such non-locality.

It might be argued that an analogous problem already exists in quantum theory, because of non-local correlations arising in an EPR-type experiment.[9] However, these are of a qualitatively different and more restricted nature than what is emerging here and do not provide an analogous situation. In the EPR case, strict locality is violated because the determination of an eigenvalue (i.e., the fixing of a kinematical quantity by measurement) associated with a particle at one point simultaneously determines the eigenvalue of another particle at a distant point. Since no time passes for a signal to propagate between the two particles, the correlation cannot be maintained by local degrees of freedom that satisfy strict locality. Yet the difficulty here is over how the state of the first particle can be "transmitted" (if indeed it is) to the second one "instantaneously", and is associated with the generally problematic area of measurement and observation in quantum mechanics. By contrast, a non-local field-effect interpretation of the Aharonov-Bohm experiment has nothing to do with the measurement problem and would require the abandonment of locality, even while the system is "smoothly evolving" (i.e., described by Schrödinger's equation), rather than undergoing measurement.

For, in the EPR-type case, as in any other case, the two particle system can be described by a non-zero, configuration-space wave function Φ. When the particles are free and their probability distributions do not overlap, Φ can be written as the product of the free wave functions of the individual (distinguishable) particles, ϕ_1 and ϕ_2; i.e., $\Phi(x_1,x_2,t)=\phi_1(x_1,t)\phi_2(x_2,t)$. Note, however, that this is possible only if the two particles are completely separable and *do not interact*, i.e., that $\Phi(x,x,t)=\phi_1(x,t)\phi_2(x,t)=0$. In the EPR experiment, however, the particles do interact (at least momentarily) at time t_I (say). Thus, they are no longer separable and their probability distributions overlap, i.e., $\Phi(x,x,t_I)\neq 0$. This interaction, however, satisfies locality, as the probability distributions of the interacting particles spatially coincide. Non-locality arises only later because the particles remain correlated after the interaction, even though they appear to be in separable "free" states. Consider the analogous case in the Aharonov-Bohm experiment. The magnetic field **B** vanishes on the support of ψ, so the product $\psi(x,t)\mathbf{B}(x,t)$ is always zero for all t, *even when the two systems interact*. This demonstrates the qualitative difference between the non-locality associated with the EPR and the field-interpreted Aharonov-Bohm experiments. The EPR experiment highlights a non-local aspect of the role that *measurement* plays in quantum mechanics, while not challenging the locality of interaction. However, the non-local field interpretation of the Aharonov-Bohm effect, in addition to retaining the EPR non-locality, would also introduce a new kind of non-locality associated with the interaction pro cess. While Bell (1965) showed that the results of the EPR experiment cannot be attributed to local degrees of freedom, the same is not necessarily true for the Aharonov-Bohm effect, as I shall now discuss.

4.3. Potentials and Phase; Locality and Gauge Invariance

It seems clear that if the Aharonov-Bohm, and similar effects[10] are to be understood as the non-local manifestations of force fields, then the ontology of such interactions would be greatly at odds with currently accepted physical principles. Specifying the field strength at all points accessible to the electrons does not sufficiently determine the interactions that the electrons undergo. So let us examine the other option, namely attributing a causal role in the interaction to some local effect of the vector potential, **A**. It was suggested earlier that this is problematic because **A** does not satisfy (GI), and is therefore not a valid candidate in any ontologically realist, causal explanation. Of course, it is still perfectly acceptable as an abstract representation, or a theoretical construction, this being the role suggested for it in section 3. On closer examination, however, it can be seen that the way that **A** comes into play in describing the Aharonov-Bohm effect satisfies (GI), and suggests a new ontological picture for the fundamental processes that occur in physical interactions that differs from the account in section 3.

While the local field strength underdetermines the effects observed in the Aharonov-Bohm experiment, the potentials overdetermine them in the following way: while the formal result is shown to be dependent on the loop integral $\Delta\chi = q/(\hbar c)\oint A^{\mu}dx_{\mu}$, the observable effect is only dependent on $\Delta\chi$ up to modulo 2π. This is because it is actually the *phase factor* $\Omega = \exp(i\Delta\chi)$ that determines the change in the wave function due to the solenoid, rather than the phase $\Delta\chi$ itself, as a phase shift of $\Delta\chi$ is indistinguishable from one of $\Delta\chi + 2n\pi$, $n=1,2,...$ (this phase factor is extensively discussed by Wu and Yang 1975). It therefore seems that the term which correctly characterises the causal interaction in the Aharonov-Bohm effect is the phase factor Ω, and it is with the peculiarities and interpretation of this entity that I wish to conclude this discussion, along with the role that it suggests for the potential.

What is made clear by what has been said so far is that interactions are linked to changes in the phase of the wave function in a very fundamental way through Ω. The transformation of the phase of the wave function by Ω that accompanies the introduction of a potential (or a gauge transformation) in order to satisfy (GI) has led to a means of expressing fundamental interactions in terms of gauge transformation symmetries, known *as gauge field theories*. These are really a generalisation of the way **A** was introduced into the Schrödinger equation,[11] so that interactions other than electromagnetism can be understood in a similar way. If Ω is of the form indicated above, then the phase factor describes electromagnetic interactions. If it is replaced by the expression: $\Omega_W = \exp[(1/2)i\alpha_k\tau_k]$, where α_k are the components of a 3-vector and τ_k are the three generators of $SU(2)$ (specifically, the Pauli matrices), then the phase shift describes the weak interaction.[12] A similar move leads to a gauge theory of strong interactions based on transformations under the SU(3) group. All of these interactions are characterised by the introduction of potentials, although those of the strong and weak interaction are more complicated than the electromagnetic potential *A*.

Despite their ubiquitous application, the potentials still overspecify the dynamics of the interactions. Given this, can a C1-type account of causal interactions based on fields and forces suffice, at least in non-Aharonov-Bohm type cases (i.e., in singly connected regions)? It has been shown that the general answer is 'NO', although electromagnetism is a special case where it is 'YES', due to its simplicity (i.e., its algebra is Abelian, being that of complex numbers—see Wu and Yang 1975, p. 3856). In general, the field strength (and therefore, the force) at a point underdetermines the strong and weak interactions, even in simply connected regions, suggesting quite conclusively that C1 theories cannot be generalised beyond classical physics, and do not form a suitable basis for an "ontological" physicalist theory of causation.

Let us now return to considering locality. While an Aharonov-Bohm interaction based on potentials is strictly local, and that on fields is non-local, the phase $\Delta\chi$ and therefore the phase factor Ω, have a locality status that is intermediate between these extremes that I shall refer to as *global*. While there is a local overlap between the electron wave function ψ and $\Delta\chi$, the loop integral in $\Delta\chi$ is not shrinkable to a localised point because it encloses the excluded region bounded by the solenoid. Thus, while local connection of ψ and Ω can be established in accordance with (L2a), the interaction is not correctly characterised by local degrees of freedom (the potential) as these overdetermine the observable effects. Therefore, one cannot simultaneously preserve strict locality and (GI) because the two possible gauge invariant entities, Ω and the fields, imply the violation of (L1) and (L2a), respectively. In order to find an ontology of interaction that accommodates Aharonov-Bohm type effects, preserving (GI) at the expense of (L1) would require that Ω be ontologically irreducible. The potential is then cast back into its old light as a theoretical construction, although it is now treated as the "fictitious" phase of a phase factor, rather than the spatial integral of a field.

It has been remarked, however, that this approach undermines the reasons for which local field strengths were originally introduced to describe electromagnetism (Olariu and Popescu 1985, p. 427). Suppose that one is prepared to violate locality in order to accept a global phase factor as ontologically primitive. One does not then have sufficient grounds to disallow the non-local field interpretation because this was rejected by an appeal to strict locality. Now, if non-local field effects cannot be rejected as the basis for the interaction, the phase factor becomes redundant, because Ω is treated as primitive only to circumvent the need for the non-locality associated with the field (which is now acceptable). Therefore, a consistent approach based on preserving (GI) at the expense of strict locality, and therefore regarding the phase factor Ω as ontologically primitive does not seem possible. This has led to some investigation concerning how the potential may actually be ontologically significant.

In order for this to even be possible, (GI) must be weakened from a criterion for ontological significance, to a criterion for observability, where it is understood that the former does not entail the latter:

(GI*) Only those mathematical terms that are gauge invariant are considered to be physically interpretable as *observable* phenomena.

This implies that entities that are unobservable can now be considered to have ontological significance. In the case of the electromagnetic potential, various interpretations have been proposed.

Many authors have suggested that the electromagnetic vector potential **A** can be understood as the momentum associated with the electromagnetic field when it is interacting with particles, or, alternatively, that it is the current of the scalar density ϕ; i.e., $\mathbf{A}=\mathbf{v}/c\ \phi$ (Olariu and Popescu 1985, pp. 426-429). However, such identifications can only be made if a particular gauge is assumed (i.e., the function ζ from equation (6) is given, or a condition such as $\partial_\mu A^\mu=0$ is specified). Generally, different interpretations of the potentials require different gauges and there is no universal criterion for selecting a given gauge as "real".

Another interesting approach begins by taking the phase itself as ontologically significant, which is now permitted due to the weakened form of (GI*). The potential can then be directly interpreted as the local phase shift that occurs as a result of interaction, although again, a particular gauge must be assumed and can only be justified pragmatically. Both classical and quantum dynamics can be uniformly derived from this picture in a way that furnishes the formalism with considerable physical intuition (Wignall 1990). In a contrasting approach, the gauge potentials are treated as abstract geometrical objects using fibre bundle theory (Wu and Yang 1975, and Cao in Brown and Harré 1988, pp. 117-133). All of these approaches have interesting philosophical implications, but lie outside the scope of the present discussion.

5. Conclusion

While perhaps raising more questions than it answers, the ontology suggested by the above considerations implies that forces are not the fundamental causal agents that they appear to be in classical physics, nor in the usual non-relativistic limits.

Energy/momentum transfers are equally deficient, if not more so as, apart from sharing the shortcomings of force theories, they do not preserve the semantic relations that are required in causal explanations. Interaction itself should be understood as the ontological basis for a physicalist account of causal connection. If the Aharonov-Bohm effect is taken as a template for general interactions, classical forces, along with energy/momentum transferences, emerge as supervenient on a subset of more fundamental causal interactions involving phase changes in the quantum wave function that are characterised by potentials. In particular, electric and magnetic forces arise only as special cases of "classical limit" interactions where the potentials have the properties $\nabla \times \mathbf{A} \neq 0$ and $\nabla\phi \neq 0$. The Aharonov-Bohm effect conclusively shows that interactions still occur even if these conditions are not met, and further leads one to speculate on what the underlying mechanisms might be, regardless of the above constraints.

Given that one does not wish to adopt a variegated ontology, where the different types of fundamental interactions are segregated and Aharonov-Bohm type effects are treated separately from "normal" phenomena, it seems that causal interactions should be fundamentally understood as phase modulations in wave functions characterised by potentials. Indeed, this is exactly the approach that has been taken in recent research in physics where a new ontological framework unifying classical and quantum theory is being developed.[13] In more conventional physics, the prominence and great success of gauge theories of fundamental particles and their interactions also suggests that phase factors have a major role to play in a fundamental ontology of interactions extending beyond electromagnetism . However, this raises a question concerning the mutual compatibility of strict locality and the use of gauge invariance as a criterion for ontological significance. While relatively superficial considerations suggest a weakening of gauge invariance, at least as an ontological principle, a deeper analysis of such problems, along with the fundamental ontology of interaction processes, lies within the scope of future work.

Notes

[1]I am grateful to John Collier and Bill Wignall for many fruitful discussions and their helpful comments and criticisms.

[2]This is stated explicitly by Heathcote (1989, p. 77). A similar idea is alluded to by Horwich (1987, pp. 134-35). I am assuming here that the specification of co-ordinates in phase space, or of the canonical ensemble in statistical theories, sufficiently represent the physical "state" of a system.

[3]This line of reasoning is followed by Bigelow *et.al.* (1988, pp. 624-25) although relativity is not explicitly mentioned. Heathcote (1989, pp. 77-79) seems to make a similar claim.

[4]'Gauge invariance' actually refers to what was originally construed by Weyl (1921, pp. 121-123) as the independence of physical phenomena on local length standards; hence the term 'gauge'. See Cao in Brown and Harré (1988, pp. 117-133) for further discussion.

[5]Their original paper containing the theoretical calculation is Aharonov and Bohm (1959). This was supported experimentally a year later by Chambers (1960). A very comprehensive theoretical review is provided in Olariu and Popescu (1985), while Peshkin and Tonamura (1989) contains an excellent and up to date review of

experimental work on the Aharonov-Bohm effect, including Tonamura's definitive experiment.

[6]For the proof that interactions with non-local degrees of freedom violate Lorentz invariance see Currie et.al. (1963), and Kerner (1972) for a collection of discussions of related problems.

[7]Alternatively, the Newtonian approximation implies that $c\infty$. If this is applied to locality condition (L2), then this becomes $\lim_{c\rightarrow\infty} (|x-y|<c|t_x-t_y|)$ which is satisfied in Newtonian theory.

[8]There has been, to my knowledge, no fully relativistic treatment of the Aharonov-Bohm effect (i.e., one that has been second quantised).

[9]See Krips (1987, pp. 116-125) for an interesting account and analysis of this *gedankenexperiment*.

[10]The Aharonov-Bohm effect is actually the simplest case of a class of phenomenon where the introduction of a phase factor in the wave function in a multiconnected region leaves the dynamics unaltered, but still has observable effects. These were first consistently described by Berry (1984) where he showed that the general result was due to the quantum adiabatic theorem and the geometrical properties of the parameter space.

[11]Electromagnetic interactions are incorporated into Schrödinger's equation by transforming the momentum operator $-i\hbar\nabla$ to $-i\hbar[\nabla - (iq)/(\hbar c)\mathbf{A}]$, analogously to equation (2).

[12]Wu and Yang (1975, p. 3855) propose an interesting "weak interaction" generalisation of the Aharonov-Bohm effect in a thought experiment where the magnetic flux is replaced by an enclosed source of isospin, and the charged particles by an isospin doublet (e.g., protons and neutrons). Instead of a spatial interference pattern shift, the effect is a change in the mixing ratio between the two isospin states.

[13]See Wignall (1990). This work is in its early stages and while still rather speculative, has very interesting philosophical implications.

References

Aharonov, Y. and Bohm, D. (1959), "Significance of electromagnetic potentials in quantum theory", *Physical Review* 115: 485.

Aronson, J. (1982), "Untangling Ontology form Epistemology in Causation", *Erkenntnis* 18: 293-305.

Bell, J.S. (1965), "On the Einstein-Podolski-Rosen Paradox", *Physics*, 1: 195.

Berry, M.V. (1984), "Quantal phase factors accompanying adiabatic changes", *Proceedings of the Royal Society of London*, Series A, 39 2: 45-57.

Bigelow, J., Ellis, B. and Pargetter, R. (1988), "Forces", *Philosophy of Science*, 55: 614-630.

Brown, H. and Harré, R. (eds.) (1988), *Philosophical Foundations of Quantum Field Theory*, Oxford: Clarendon Press.

Chambers, R.G. (1960), *Physical Review Letters*, 5: 3.

Currie, D., Jordan, T. and Sudarshan, E. (1963), "Relativistic Invariance and Hamiltonian Theories of Interacting Particles", *Reviews of Modern Physics* 35: 350-375.

Dieks, D. (1981), "A Note on Causation and the Flow of Energy", *Erkenntnis* 16: 103-108.

Fair, D. (1979), "Causation and the Flow of Energy", *Erkenntnis* 14: 219-250.

Heathcote, A. (1989), "A Theory of Causality: Causality = Interaction", *Erkenntnis* 31: 77-108.

Horwich, P. (1987), *Assymetries in Time*, Cambridge, Massachusetts: MIT Press.

Jackson, J.D. (1975), *Classical Electrodynamics*, 2nd. Edition, New York: Wiley.

Kerner, E.H. (ed) (1972), *The Theory of Action-at-a-Distance in Relativistic Particle Dynamics*, New York: Gordon and Breach.

Krips, H. (1987), *The Metaphysics of Quantum Theory*, Oxford: Clarendon Press.

Olariu, S. and Popescu, I. (1985), "The quantum effects of electromagnetic fluxes", *Reviews of Modern Physics*, 57: 339-436.

Peshkin, M. and Tonomura, A. (1989), *The Aharonov-Bohm Effect: Lecture Notes in Physics,* No. 340, New York:Springer-Verlag.

Sakurai, J.J. (1985), *Modern Quantum Mechanics*, Menlo Park, California: Benjamin Cummings.

Weyl, H. (1952), *Space-Time-Matter*, Translated from the 4th German Edition (1921) by H.L. Brose, New York: Dover.

Wignall, J.W.G. (1990), "Maxwell Electrodynamics from a Theory of Macroscopically Extended Particles", *Foundations of Physics*, 20, 1 23-147.

Wu, T. and Yang, C. (1975), "Concept of nonintegrable phase factors and global formulation of gauge fields", *Physical Review*, D 12: 3845-3857.

Part VIII

KUHNIAN THEMES: SSR AT THIRTY

Theory-ladenness of Observations as a Test Case of Kuhn's Approach to Scientific Inquiry

Jaakko Hintikka

Boston University

1. What is Kuhn up to?

The overall character of the ideas Thomas S. Kuhn has offered concerning the nature of scientific inquiry has been generally misunderstood, or, rather, misconstrued. (See Kuhn 1957, 1970.) Kuhn's ideas do not add up to a fully articulated analysis of the structure of the scientific process. Kuhn does not offer a theory of science which should be evaluated in the same way as, e.g., the hypothetico-deductive model of science or the inductivist one. What Kuhn does is best viewed as calling our attention to certain salient phenomena which a philosophical theorist of science must try to understand and to account for. We do injustice to Kuhn if we deal with his views as if they were finished products of philosophical theorizing. They are not. Rather, they are starting-points for such theorizing; they pose problems to be solved by a genuine theory of science.

If we do not realize and acknowledge this, we run the risk of attributing to Kuhn a singularly shallow philosophy of science. As he uses them, several of Kuhn's central concepts can scarcely accommodate the theoretical traffic they have been put to bear. For instance, in an earlier paper (Hintikka 1988a) I have shown that the notion of incommensurability of theories does not behave in the way Kuhn assumes. Of course the incommensurability of two theories goes together with their conceptual alienation from each other, but it also goes together with the discrepancy between their respective consequences and hence can be characterized by reference to the latter, contrary to what Kuhn clearly assumes.

The attitude of most philosophers to Kuhn's work is all the more surprising as Kuhn has himself sought a deeper analysis of, and a firmer theoretical foundation for, his ideas. For instance, at one point Kuhn (1977) took very seriously the possibility that the so-called structuralist approach of Stegmüller and Sneed might provide a satisfactory theoretical framework for his ideas.

This perspective is also relevant to the evaluation of Kuhn's argumentation. If you measure it against what one is entitled to expect of a philosopher with a command of the conceptual and structural issues involved in understanding scientific inquiry, you

PSA 1992, Volume 1, pp. 277-286

will find Kuhn's argumentation disappointing. For one thing, his argumentation is often, not to say typically, negative. Instead of developing his own views constructively, he presents his point by criticizing other views, irrespectively of whether any major philosopher has ever actually held them.

Furthermore, some of Kuhn's arguments come close to being self-refuting. For instance, Kuhn cites empirical studies from the psychology of perception to buttress his claim that our observations are influenced by background beliefs, including theoretical ones. But such evidence is a double-edged sword. For if a rule-governed influence of beliefs on observations has actually been established, our awareness of the very rules governing it can in principle be used to eliminate the effects of our theories from our observations, by compensating against the bias these theoretical beliefs induce.

Even Kuhn's historical argumentation is frequently unconvincing. For instance, Kuhn's own prize specimen of a scientific revolution, the Copernican revolution, is as good a counter-example to his thesis of the theory-ladenness of observations as one can hope to find. For the actual observations used by the rebels of the astronomical revolution were theory-neutral. They were sightings of heavenly bodies, data as to which heavenly body was where on the firmament when. Indeed, what made possible the discovery of Kepler's Laws were Tycho Brahe's observations, notwithstanding Brahe's rejection of the Copernican system. (Which theory were his observations laden with?) Copernicus' picture of the world changed profoundly the way we humans have to think of our place in the cosmos, but his argumentation in no way presupposes understanding or interpreting observational data in a new way. It is couched mostly in time-honored terms of simplicity and naturalness. If there is a new way of looking at the situation, it is to extend to the heavenly bodies the same dynamic questions as can be raised about terrestrial bodies. (See Grant 1962.) For instance, this parity of the terrestrial and of the heavenly, rather than the new Copernican picture as such, is what motivates the denial of the immutability of the heavens. (See Grant 1991.) But this new way of looking at things did not affect the role of observations in Copernicus' argumentation.

2. Theory-ladenness and the interrogative model of inquiry

Yet the questions Kuhn is raising are not only very real but also of considerable subtlety. The right way of approaching them is not to focus on what Kuhn says, but to try to put the phenomena he is calling our attention to into a deeper perspective. In this paper, I shall use Kuhn's idea of the theory-ladenness of observations as a test case. This idea is important for Kuhn, for without the theory-ladenness of observations several of his other central theses, for instance the frequent incommensurability of theories, would be considerably weakened.

At first sight it might seem that Kuhn's theory-ladenness thesis is correct but quite shallow, and cannot bear the demonstrative traffic Kuhn loads it with. It may be a welcome correction to the positivistic picture, which Kuhn is using as his strawman. But as soon as a more realistic picture of the scientific process is adopted, Kuhn's thesis can be accommodated without much ado. As an object lesson, I shall here consider the issue of theory-ladenness of observations in terms of the interrogative model of inquiry which I have developed and applied in the last several years. (Cf. Hintikka 1988b.) Here only the most general features of the model are needed. In the model, scientific inquiry is conceptualized as a questioning game between an inquirer and nature (more generally, any suitable source of answers generically referred to as an "oracle", or even several oracles). The inquiry starts from some given theoretical premises (or set of premises) T. In the simplest case, the aim of the game is to prove a given potential conclusion C. In doing so, the inquirer can perform two kinds of moves, interrogative and logical. A logical move is simply a logical inference from the results (premises) so far reached. In an interrogative move, the inquirer puts a question to nature. (The presupposition of the question must of course have been established prior to asking the question.) If an answer is forthcoming, it is added to the inquirer's list of available premises. It is assumed that the set of available answers is fixed, and that it remains constant throughout the inquiry.

The answers can be thought of (in the application of the model considered here) as including observations and results of experiments—any factual data a scientist can lay her or his hands on. In the simplest case, nature's answers are all assumed to be true, but this assumption can (and must) be given up in other, more complex types of inquiry. If the inquirer can derive C no matter what true answers nature gives, C is said to be interrogatively derivable from T in M, in symbols

$$(1)\quad M{:}\ T \vdash C$$

where M is the model ("world") to which nature's answers pertain.

In the interrogative model, a scientist's observations can be construed as nature's answers (or as a subclass of nature's answers) to the inquirer's questions. They are among the premises that the inquirer can use in her or his logical inference steps, and they can also be used as presuppositions of further questions.

The basic fact about observations is that the role they play depends crucially on the rest of the inquiry. What can be done by means of a set of observations (for instance, what follows from them interrogatively) does not depend on these observations alone, but also heavily on theory the inquirer is relying on, that is, on his or her initial premises. If you change the initial premises, an observational answer will have entirely different interrogative consequences. This dependence of the role of observations on initial theoretical premises already amounts to a massive and clear-cut "theory-ladenness of observations".

Furthermore, in more realistic variants of the interrogative inquiry nature's answers are not assumed to be known to be all true. They can be accepted, rejected, reaccepted, and so forth, by the inquirer, in accordance with strategic rules which depend (over and above *a priori* knowledge about the probability of different answers' being true) on what happens in the rest of the interrogative inquiry, which in favorable circumstances can even be a self-correcting process. Among other things, these strategies depend on the given theoretical premises, thus revealing another kind of theory-dependence of observations in inquiry.

All this destroys the positivistic conception of observations as theory-neutral building-blocks more radically than Kuhn's arguments. At the same time, the interrogative model shows that the *theory*-ladenness of observations is not the only way they depend on their context in inquiry. The impact of an observation on inquiry does not only depend on the inquirer's initial theoretical premises. It normally also depends on the other answers the inquirer receives to her or his observational questions. This point should not come as a surprise to Kuhn. For instance, in discussing how earlier experience can affect our observations, Kuhn himself is not really talking about the theory-ladenness of observations. He is in effect talking about the observation-ladenness of observations.

The range of answers which nature will give often is an even stronger determinant of the role of an observation in interrogative inquiry than the set of its initial theoretical premises. In this sense, too, observations are thoroughly laden with (the totality of available) observations. This fact is one of the many reasons why I said above that on its *prima facie* construal Kuhn's theory-ladenness thesis is relatively shallow.

3. Interim conclusions

Thus in a sense the interrogative model vindicates splendidly Kuhn's thesis of the theory-ladenness of observations. In doing so, however, it puts Kuhn's ideas in a new light. For one important thing, we can now see that theory-ladenness does not have to be explained in the way Kuhn and others have tried to explain it, to wit, as being due to the dependence of the meanings of observational terms on the theory in which they occur. From the fact that the *meaning* (in the sense of significance) of an observation for a scientific inquiry depends on the theory on which this inquiry is based, one can only at the risk of the fallacy of equivocation infer that the observation *means* something different in different theories in any sense of "meaning" related to logical or linguistic meaning, e.g., in the sense that the linguistic meaning of observational terms depends on the theory in which they occur. Theory-ladenness of observations simply does not presuppose that the meaning of observational terms changes from theory to theory or from world to world. The way in which the interrogative model vindicates the idea of theory-ladenness is in many ways faithful to the ideas of the likes of Kuhn and Hanson, but it does not involve any dependence in the usual sense of the meanings of observational terms on the theory in which they occur.

Admittedly, there exist philosophical views on linguistic meaning which tend to make it theory-dependent. Frequently, however, such views are defended on the basis of ideas attributed to Kuhn, and hence cannot serve to defend Kuhn. And even on the most favorable perspective on their independent justification, the jury is still out on their justification.

In a different direction, the theory-ladenness of observations in the sense uncovered by the interrogative model does not automatically entail any unavoidable incommensurability of theories, either. (See Hintikka 1988a.) This incommensurability was thought to be a corollary to the dependence of the meaning of observational terms on the background theory, but (as was seen) there is no need whatsoever to assume the confused and confusing idea of meaning relativity in the first place. Theories can, of course, be more or less incommensurable also on the interrogative model, but this merely means that they have different (interrogative) consequences that can be tested against nature's answers. It even turns out, at least in oversimplified but representative cases, that this kind of incommensurability-without-meaning-variation is reconcilable with the idea that the incommensurability of two theories is due to conceptual discrepancies between them. Indeed, the observational incommensurability can be shown to be the smaller, the more conceptual ties there are between the terms occurring in the two theories in question. Hence once again we do not need meaning variation from theory to theory in order to do justice to the interesting phenomena to which Kuhn has called our attention.

4. On the logic of experimental science

Thus we seem to have reached a comfortable and comforting conclusion about what is true and what is false about Kuhn's claims. However, what is even more interesting than the conclusions so far reached is the possibility of pushing our analysis deeper. I suspect that Kuhn also saw deeper into the situation than I have so far brought out, even though he did not have the conceptual tools to articulate his insights. It turns out that theories are indeed involved in observations in a more basic way than we have diagnosed so far. It is not only that the consequences (and the other repercussions) of an observation are arguably theory-dependent. The very possibility of an observation may be contingent on theoretical assumptions.

In order to see what's what here, let us first register a *prima facie* objection to what I have said so far. It might seem that Kuhn could very well claim that the interrogative model tells only a part of the full story. The kind of inquiry the model codifies seems at first sight to match very well what Kuhn calls normal science. Indeed, Kuhn characterizes normal science as an exercise in puzzle-solving. In contrast, the interrogative model might not seem capable of handling the situations in which a revolutionary change takes place in a science or even, on a more modest scale, the kind of situation in which an entire new theory replaces an old one.

Nevertheless, it is precisely in the problem area of theory change that the interrogative model begins to show its real strength. A number of further insights will enable us to see what's what.

First, why is the interrogative model supposed to be incapable of serving as a paradigm for an inquiry which establishes a radically new theory, perhaps even without presupposing any strong initial premises? Here we come to an assumption which is weighty, widespread and wrong. It amounts to assuming that the only answers nature can give an inquirer are particular propositions. I have called this assumption the Atomistic Postulate. I have argued that it lies behind much of the traditional philosophy of science. Here I am suggesting that it also underlies Kuhn's thinking. It is because of this assumption that philosophers have thought that scientists cannot literally derive their theories from nature's answers without strong initial theoretical assumptions. I have also shown that this postulate is wrong in the sense that it does not adequately capture the structure of actual scientific reasoning. For instance, the answers nature gives in a successful controlled experiment express dependencies between two variables and hence has at least the quantificational complexity of a $(\forall x)(\exists y)$ prenex. More complex experimental setups can in principle yield even more complicated answers. (See Hintikka 1988b.)

From such answers the inquirer can logically infer complex laws and theories even without the help of strong initial theoretical premises and without the help of any inductive or other ampliative rules of inference. Even though I will not pursue the point any further here, this rejection of the Atomistic Postulate puts large segments of the philosophy of science to a new light.

5. Scientific inquiry as a two-level process

But even this analysis of the logic of experimental science is not enough to do full justice to the nature of experiments. For even though the results of an experiment typically enter into the reasoning in the form of dependence laws, the way in which those laws are reached must be capable of being analyzed.

The natural solution here is to consider scientific inquiry in theoretically sophisticated sciences as a two-level process. This two-level perspective is made possible by the double role of questions in interrogative inquiry. In more complex types of inquiry, the inquirer's goal is not to prove a predetermined conclusion (for example, to verify a hypothesis) but to answer a question. How the techniques of interrogative inquiry can be used to this end is not an easy question to answer. An answer covering all cases can nevertheless be given. In general, questions play two roles in interrogative inquiry. What happens is that the inquirer tries to answer a "big" (principal) question by means of a number of "small" (operational) questions. In any one inquiry, the two questions have to be distinguished from each other sharply. However, what for a higher-level inquiry is an operational ("small") question can for the purposes of a lower-level inquiry be the principal question of a complex inquiry in which it is to be answered by means of a number of lower-level operational questions. This, I find, is how we must view typical controlled experiments. For the purposes of a higher-level inquiry, the entire functional dependence (of the observed variable on the controlled variable) that is the outcome of the experiment is an answer to an operational question on the higher (theoretical) level. For the experimental scientist, in contrast, it is an answer to a principal question, and the experimentalist's operational answers are particular data brought to light during the experiment, for instance, instrument registrations.

What is interesting here is not just that experimental inquiry can be considered as a two-level process, but that the two levels have distinctly and characteristically different structures. The operational answers on the higher level can have a considerable logical complexity, whereas on the lower-level operational answers are typically formulated as particular propositions. Even more interestingly, the higher-level inquiry does not necessarily need strong initial premises, whereas the particularity of the lower-level operational answers presupposes that suitable general premises are available to the (experimental) inquirer.

Where do these general premises come from? The approximate answer is: The initial premises of a lower-level inquiry are theoretical laws established earlier on the higher level. Typically, they are not the ones that are being investigated or tested on the higher level; they are older and safer (and frequently only partial) generalizations.

I cannot argue here fully for this reading of the typical situation, but it should not be news to anyone familiar with experimental techniques in sciences like physics. It is also supported, it seems to me, by the recent studies of the role experiments in science by the likes of Alan Franklin (1986, 1990) and Peter Galison (1987).

If the view I have presented of the two-level character of scientific inquiry is right, it shows in some real detail how it is that an experimental inquiry depends on general laws and even on theories binding such laws together. In brief, it provides an account of what might be called the *theory-ladenness of experiments*. This account does not serve merely cosmetic purposes, either. It shows among other things what role induction does and does not play in actual scientific inquiry. On the higher level, induction is not needed, because the operational answers can have such a logical complexity that laws and even theories can in principle be derived from them deductively. The place of inductive inference is taken over by the kind of reasoning which is involved in extending a partial generalization and combining different partial generalizations with each other into a more sweeping generalization. As it happens, such reasoning was earlier in the history known under the very label "induction". (See Hintikka 1992.)

On the lower level, induction is needed because operational answers are usually particular propositions. But on this level the requisite general premises needed to back up so-called inductive inference are normally available, supplied by the results of earlier higher-level inquiry. Thus the (modest) role of induction in actual science receives a diagnosis in this way.

The two-level model can also be compared with the testimony of the actual history of science. Even though all the returns are not yet in, there already are several encouraging early results. For one important example, Newton's methodological pronouncements and methodological practice are not only put to a new light which strikingly vindicates some of them, for instance, Newton's claim of having "deduced" general truths from "phenomena"; what is more, Newton's theory and practice of methodology are found to be in an excellent conformity with each other (and with the interrogative model), current orthodoxy notwithstanding. (See here Hintikka and Garrison, forthcoming.)

6. The parallelism of experiments and observations

But how is this analysis of experiments relevant to the theme of this paper? I have argued for the theory-ladenness of *experiments*, but what does that have to do with the vaunted theory-ladenness of observations? The answer lies in a parallelism which I see there between experiments and observations. Such a parallelism seems to be to be implicit in Kuhn's and Hanson's thinking, and it can in any case be argued for in the same way as they argue for their views. From classics like von Helmholtz to contemporaries like Rock, most of the sophisticated psychologists of perception have recognized that unconscious cognitive processes, variously described as "unconscious inferences" (von Helmholtz), "a kind of problem solving" (Gregory, 1970, p. 31) or "hypothesis testing" (Rock, 1983), play an important role in perception. Assimilating these processes to experiments is merely to put a generic name to what these sundry descriptions have been attempts to capture. In so far as the parallelism between an experiment and an observation is an apt one, the theory-ladenness of experiments which we have discovered serves as an explanation also for the theory-ladenness of observations.

7. On the two-levelled character of experimental inquiry

The two-levelled character of experimental inquiry has interesting consequences even apart from the problem of the theory-ladenness of observations. Here I can mention only a couple of perspectives it opens. Clark Glymour has called our attention to the feature of scientific inquiry he calls "bootstrapping", which he describes as follows:

> ... the basic idea is clear enough: Hypotheses are tested and confirmed by producing instances of them; to produce instances of theoretical hypotheses we must use other theoretical relations to determine values of theoretical quantities; these other theoretical relations are tested then in the same way. (Glymour 1980, p.52)

Such explanations leave a multitude of questions unanswered. The problem is not just that there does not seem to be any guarantee that the process is free of circularity and that it therefore is likelier to succeed than the attempt by Baron von Münchausen which presumably lent Glymour's idea (attributed by him to Reichenbach) its name. The disturbing question is how there could be two different sets of theoretical hypotheses governing precisely the same phenomena. How can a blind hypothesis lead another blind hypothesis?

The most straightforward answer is to view the process which Glymour assumes can produce instances of theoretical hypotheses (hypotheses$_1$) by means of other hypotheses (hypotheses$_2$) as a lower-level experimental inquiry carried out to test hypotheses$_1$. Then hypotheses$_2$ are the theoretical assumptions on which the lower-level experimental inquiry is based. This perspective is obviously much more faithful to actual scientific practice than, e.g., a conventionalistic attempt to avoid the specter of circularity.

One can even raise the question as to what happens if the two sets of hypotheses or theories are incompatible. It has been claimed that this is the situation we encounter in quantum-theoretical experiments where the experimental situation apparently must be dealt with classically even though the target phenomena are quantum-theoretical. The perspective we have reached here promises new possibilities of conceptualizing such questions.

8. Theory-ladenness and the logic of identification

Another direction in which we can analyze the situation further relates to the alleged dependence of the meanings of observational terms on the underlying theory of the inquirer. Here, once again, Kuhn's argumentative strategy is unclear. For what is the meaning of the "meaning" which is supposed to vary from theory to theory? What is uncontroversially true is that when different theories are true in two different "possible worlds" (under two different possible courses of events), a term will apply to different cases in the two worlds. But in the normal understanding of the semantics of our language, this can be the case even if the meaning of the term does not vary from one world to another in the slightest. And if in some sense its meaning is different in the two scenarios, why should Kuhn or anyone else speak of the variation of the meaning of one and the same term, instead of two different terms? Clearly the issues have to be sharpened here before they can be resolved.

The most straightforward way of interpreting Kuhn is to take him to claim that the dependence of the reference (extension) of a term on the theory in which it occurs as well as on the world which is being investigated must be explained by saying that the meaning of that term changes when the theory is changed. In order to see whether such claims are true, we have to examine precisely *how* the extension (range of correct applications) of a term, say P, depends on the underlying theory, say T[P].

Once again, the interrogative model proves its mettle in answering this question. A good testing-ground is provided by the question: When does a theory T[P] containing a term, say a one-place predicate P, determine completely the range of correct applications of P? Here the interrogative model opens an interesting perspective in that it shows that this answer admits of two different kinds of answer. The difference between them is related to philosophers' and linguists' much abused distinction between *de dicto* and *de re* statements.

Speaking *de dicto*, the natural explication of the determinacy question is to ask: Does T[P] have among its interrogative consequences a quasi-definition of the following form:

(2) $$(\forall x)(Px \leftrightarrow D[x,a_1,a_2, ..., a_k])$$

Here $D[x,a_1,a_2, ..., a_k]$ must satisfy the usual requirements of a definiens. Furthermore, $a_1, a_2, ..., a_k$ are members of the domain of individuals of the model to which the inquiry pertains. (Naturally, (2) must be interrogatively derivable without answers containing P.) If the answer is yes, i.e., if (2) is so derivable, P is said to be *identifiable de dicto* in M.

The question whether P is identifiable in M depends of course on T[P], and so does the available quasi-definition (2). In this sense, the reference (extension) of P is indeed theory-laden. This observation is trivial, however, and can be considered an explication of the comment above to the effect that the reference of a term depends on the theory in which it occurs. The real question is: What else can be said here?

One interesting result here is the following: If P is *identifiable de dicto* (in the sense explained above) in every model of the theory T[P], then it is piecewise explicitly *definable* on the basis of T[P], that is, there is a disjunction of explicit definitions

(3) $V_i (\forall x)(Px \leftrightarrow D_i[x])$

logically implied by T[P]. This result follows from a well-known theorem of Svenonius. (See Rantala 1977, p.79).

What this observation means is clear. By the meaning-determination of a term like P by the theory T[P] in which it occurs we must of course mean a determination applicable to all models of T[P]. The differences between different models are due to contingent facts independent of T[P]. But our result shows that the only way in which a theory can determine the reference of one of the terms it contains for all its models —which presumably is the sort of meaning determination at issue here—is by way of logical consequence.

In other words, not all kinds of theory-ladenness of meaning make any philosophical difference. Complete meaning determination *de dicto* is possible only in the old-fashioned way, to wit, by way of logical consequence.

However, these consequences follow only if identifiability is construed *de dicto*. If it is understood *de re*, then P is identifiable on the basis of T[P] in a model M iff for each member b of the domain of individuals do(M) of M, b ε do(M), we have either

$M:T[P] \vdash Pb$ or
$M:T[P] \vdash {\sim}Pb$.

It can be seen that P can be identifiable *de re* in M and yet not captured by any definitory formula like (2). Then Svenonius' theorem does not apply, and even a complete determination of the extension of a concept by a theory, viz. in the sense of *de re* identifiability, does not reduce to logical consequence.

This throws some light on the subtleties of meaning determination by theories. But independently of all these subtleties, and independently of how we prefer to express ourselves, one thing is clear. Neither in the case of the *de dicto* identifiability, nor in the case of *de re* identifiability, does the dependence of the range of correct application of a term on the theory in which it occurs have absolutely anything to do with the relativity of the meaning of the term to the theory. On the contrary, this kind of meaning dependence is obscured it we do not represent the concept expressed by the term in different ways in different theories or think that statements made in its terms on the basis of different theories are incommensurable for that would make it impossible to talk about logical consequence relations here. There is a solid truth in the claim of the theory-ladenness of meaning, but this point is to be demystified and brought to the purview of sober logical analysis. In particular, it would be an extremely serious mistake to think that because of the theory-ladenness of meaning somehow our usual logic has to be suspended or modified.

References

Earman, J. (1977), "Theory-Change as Structure Change", in *Historical and Philosophical Dimensions of Logic, Methodology and Philosophy of Science*, R.E. Butts and J. Hintikka (eds.). Dordrecht: D. Reidel, pp. 289-309.

Franklin, A. (1986), *The Neglect of Experiment*. Cambridge: Cambridge University Press.

_______. (1990), *Experiment, Right or Wrong*. Cambridge: Cambridge University Press.

Galison, P. (1987), *How Experiments End.* Chicago: University of Chicago Press.

Glymour, C. (1980), *Theory and Evidence*. Princeton: Princeton University Press.

Gooding, D. (1990), *Experiment and the Making of Meaning*. Dordrecht: Kluwer Academic.

Grant, E. (1962), "Late Medieval Thought, Copernicus and the Scientific Revolution", *Journal of the History of Ideas* 23: 197-220.

______. (1991), "Celestial Incorruptibility in Medieval Cosmology 1200-1687", in *Physics, Cosmology and Astronomy 1300-1700,* S. Unguru, (ed.). Dordrecht: Kluwer Academic, pp. 101-27.

Gregory, R.L. (1970), *The Intelligent Eye*. London: Weidenfeld & Nicolson.

Hacking, I. (1981), *Scientific Revolutions.* Oxford: Oxford University Press.

Hanson, N.R. (1958), *Patterns of Discovery*. Cambridge: Cambridge University Press.

Hintikka, J. (1988a), "On the Incommensurability of Theories", *Philosophy of Science* 55: 25-38.

_______. (1988b), "What is the Logic of Experimental Inquiry?", *Synthese* 74: 173-88.

_______. (1991), "Toward a General Theory of Identifiability", in *Definitions and Definability,* J.H. Fetzer, D. Shatz and G. Schlesinger, (eds.) . Dordrecht: Kluwer Academic, pp. 161-83.

_______. (1992), "The Concept of Induction in the Light of the Interrogative Approach to Inquiry", in *Inference, Explanation and Other Philosophical Frustrations,* John Earman (ed.). Berkeley and Los Angeles: University of California Press (forthcoming).

_______. and Garrison,J.W. (forthcoming), "Newton's Methodology and the Interrogative Logic of Experimental Inquiry", in *Science in Context.*

Kuhn, T.S. (1957), *The Copernican Revolution.* Cambridge: Harvard University Press.

_______. (1970), *The Structure of Scientific Revolutions*. 2nd ed., enlarged. Chicago: University of Chicago Press.

Rantala, V. (1977), *Aspects of Definability*. Helsinki: Societas Philosophica Fennica.

Rock, I. (1983), *The Logic of Perception*. Cambridge: MIT Press.

Theory-ladenness of Perception Arguments[1]

Michael A. Bishop

Iowa State University

The theory-ladenness of perception argument is not an argument at all. It is two clusters of arguments. The first cluster is empirical. These arguments typically begin with a discussion of one or more of the following psychological phenomena: (a) the conceptual penetrability of the visual system, (b) voluntary perceptual reversal of ambiguous figures, (c) adaptation to distorting lenses, or (d) expectation effects. From this evidence, proponents of theory-ladenness typically conclude that perception is in some sense "laden" with theory. The second cluster attempts to extract deep epistemological lessons from this putative fact. Some philosophers conclude that science is not (in any traditional sense) a rational activity (Feyerabend 1975); while others conclude that we must radically reconceptualize what scientific rationality involves (Kuhn 1970; Churchland 1979).

This paper has two aims: First, to propose a framework for understanding the empirical cluster of theory-ladenness arguments; and second, to begin to assess them. Once we clearly understand the structure of these arguments—where they begin and what they must show—much conventional wisdom about the significance of the psychological data turns out to be false. In particular, the arguments from voluntary perceptual reversal, distorting lenses, and expectation effects often carry unexpected lessons for the theory-ladenness issue. The fourth empirical theory-ladenness argument, the argument from conceptual penetration, raises so many difficult issues that it cannot be adequately assessed here; however, the issues that need to be addressed in order for the argument to be evaluated are set in sharp relief.

1. The structure of empirical theory-ladenness arguments

Empirical theory-ladenness arguments typically employ some psychological phenomena to show that observations are in some sense "theory-laden." But the claim that observation or observation reports are "theory-laden" is ambiguous. On one reading, it is trivial—it amounts to no more than that observations are inevitably made in terms of some conceptual framework or other. On another interpretation, theory-ladenness involves the less trivial claim that all observations are *in principle* defeasible since the conceptual apparatus they employ might turn out to be faulty. And proponents of the theory-ladenness of perception seem to want to mean something more radical and epistemologically charged.

PSA 1992, Volume 1, pp. 287-299

The most productive step to take at this point is to put aside metaphorical talk about "theory-ladenness" and instead ask: What lesson are we supposed to extract from the above psychological phenomena? If the psychological phenomena are to be relevant to the breathtaking epistemological claims made by the parties to this debate, then they will support one or more of the following *theory-ladenness theses.*

> (T) Given a case of rational theory resolution, the
> psychological evidence **A** the set of observations S employs
> in deciding between **B** significantly different and competing
> theories consists of **C** theory-neutral observations.
>
> A: shows that inevitably, suggests that possibly
> B: all possible, all actual, most actual, some actual
> C: no, only some, primarily, only

Anyone wishing to take a position on the theory-ladenness issue needs to answer four questions. (A), is the psychological evidence supposed to show that the observations made by proponents of competing theories are *inevitably* neutral or non-neutral, or that there are features of our visual systems that *if* implicated in an episode of theory resolution *would* result in neutral or non-neutral observations?[2] (B), for what range of theories is the theory-ladenness thesis supposed to hold? (C), given that any experiment involved in the resolution of a scientific controversy always includes more than a single observation, which (if any) of these observations are (or are not) supposed to be theory-neutral? And (D), what is it for an observation to be theory-neutral?

With these questions in mind, we can formulate the various conclusions one might want to draw about the role of theory-neutral observations in the resolution of scientific controversies. For each definition of 'theory-neutrality' employed by the parties to this debate, there are at least 32 different possible theses they might be defending. Proponents of theory-ladenness mean to defend something close to the most radical theory-ladenness thesis, let's call it T-1.

> (T-1) Given a case of rational theory resolution, the psychological evidence *shows that inevitably* the set of observations S employs in deciding between *all possible* significantly different and competing theories consists of *no* theory-neutral observations.

Opponents of theory-ladenness want to defend something close to the most conservative theory-ladenness thesis, T-32.

> (T-32) Given a case of rational theory resolution, the psychological evidence *shows that inevitably* the set of observations S employs in deciding between *all possible* significantly different and competing theories consists of *only* theory-neutral observations.

These 32 theses do not exhaust the potential conclusions of the theory-ladenness argument since we must include conjunctions. One might believe, for instance, that the psychological evidence suggests that some actual cases involve no theory-neutral observations, and that most of the others involve only some such observations.

Now that we have a reasonably clear account of the theory-ladenness argument's conclusion, let us turn to the issue of what some piece of evidence would have to exhibit in order to support something close to T-1. It is not sufficient for the psychological evidence simply to show that our visual systems are malleable. If I shoot myself in the back of the head and don't die, my visual system will be radically transformed. But this fact can generate no epistemological magic. The defender of theory-ladenness needs to show that our visual systems are *capable of changing in such a way that lends support to his conclusion.* Ideally, the evidence would illustrate the following features:

(a) Subject adopts a new theory, T.
(b) Subject's visual system adapts to T.
(c) As a result of the adaptation, Subject sees the world differently, in the sense that it is now more difficult for Subject to resolve disputes on the basis of theory-neutral observations with those who have not adopted T.

The defender of theory-ladenness needs to find some psychological data that exhibit these three desiderata. Now let us turn to the data.

2. The conceptual penetration of the visual system argument

Proponents of the theory-ladenness of perception argument often adduce cases of expert-novice differences in perception in defense of their conclusions. These cases have a characteristic structure. The expert, well ensconced in a scientific theory, and the scientific novice view a scene. Nonetheless, what they see is supposed to be different. Paul Churchland dramatically illustrates this point by imagining a group of people who have expanded their "perceptual consciousness" by mastering at a young age our most sophisticated scientific theories.

> It is important for us to try to appreciate, if only dimly, the extent of the perceptual transformation here envisaged. These people do not sit on the beach and listen to steady roar of the pounding surf. They sit on the beach and listen to the aperiodic atmospheric compression waves produced as the coherent energy of the ocean waves is audibly redistributed in the chaotic turbulence of the shallows.... They do not observe the western sky redden as the Sun sets. They observe the wavelength distribution of incoming solar radiation shift towards the longer wavelengths... as the shorter are increasingly scattered away from the lengthening atmospheric path they must take as terrestrial rotation turns us slowly away from their source (1979, p. 29).

The conceptual penetration argument suggests that after a pair of individuals have mastered competing theories, the conceptual resources of their visual systems change in such a way that the observations they make are not, or perhaps cannot be, neutral between those theories. So when an individual masters completely contemporary optical theory, it is possible that she will no longer observe a red sky, but a sky that is reflecting solar radiation whose wavelength is between 650 and 700 nanometers.

In order to clarify the notion of theory-neutrality assumed here, we need at least a thumbnail sketch of the nature of scientific concepts. Let's begin with a description theory of terms, which defines a term (or a predicate) by some set of those descriptions believed true of it. More formally, it defines an expression, F, in a theory or belief-system, T, as follows (Russell 1919; Ramsey 1931; Lewis 1970).

a. Conjoin the sentences in T that contain F.

> We live on the earth, and the earth has a diameter of 7,918 miles, and the earth consists of seven continents, and the earth orbits the sun, and the earth rotates on its axis once a day, and the earth is the third planet in our solar system...

b. Quantify existentially (first or second order) over F.

There is an x such that we live on x, and x has a diameter of 7,918 miles, and x consists of seven continents, and x revolves around the sun, and x rotates on its axis once a day, and x is the third planet in our solar system...

c. Replace the existential quantifier with a definite description operator and define as F what satisfies the entire definite description.

The earth is defined as the unique x such that we live on x, and x has a diameter of 7,918 miles, and x consists of seven continents, and x revolves around the sun, and x rotates on its axis once a day, and x is the third planet in our solar system...

The use of a term expresses its *complete concept* when it is defined in terms of all the descriptions believed true of it. However, it is possible to define the term using less than all the descriptions believed true of it. The use of a term expresses one of its *incomplete concepts* when it is defined in terms of some (but not all) of these descriptions. So we could define 'earth' in a way that did not run afoul of either the heliocentric or geocentric theories.

The earth is defined as the unique x such that we live on x, and x has a diameter of 7,918 miles, and x consists of seven continents...

Now we are in a position to define a theory-neutral concept.

(NC) Term F expresses concept C which is theory-neutral between T1 and T2 just in case term F is defined by descriptions that are not incompatible with either T1 or T2.

The notion of theory-neutral observation employed by proponents of the conceptual penetrability argument is this.

(NO) An observation is neutral between competing theories T1 and T2 just in case its representational content is given only by concepts that are theory-neutral between T1 and T2.

An observation is theory-laden just in case it is not theory-neutral. Notice that this definition makes an observation theory-neutral or theory-laden only relative to a set of theories. It is perfectly possible for an observation to be neutral between T1 and T2, but not neutral between T2 and T3. (This account of scientific concepts is radically incomplete; it needs, among other things, a theory of meaning. Nonetheless, it will serve our present purposes.)

A curious feature of the conceptual penetration debate is that each side has tended to stake out extreme positions. Epistemological conservatives, such as Fodor (1984) and various foundationalists, hold that there are always theory-neutral observations that play a role in cases of rational theory-choice and that are essential to the objectivity of science. Epistemological radicals, such as Kuhn, Churchland, and Feyerabend, hold that the observations made by proponents of very different competing theories are inevitably theory-laden and, as a result, observation is so infected by theory that it interferes with the objectivity of science. What should hopefully be clear by now is that there are many intermediate positions that may well deserve our allegiance.

It is reasonably clear what the proponent of the conceptual penetration argument needs in order to make his argument plausible. The conceptual penetration argument will succeed if and only if the following three views can be defended.

(1) *A theory of representational content that allows for the possibility that the proponents of substantially different scientific theories can employ very different observational concepts.* Without (1), proponents of competing theories will always share the same observational concepts. It does not follow that these observational concepts will be neutral between all possible competing theories—some theories may be incompatible with some of our 'built-in' observational concepts. However, scientists' visual systems will not have the conceptual flexibility necessary to adapt to new theories. Thus, desideratum (b) (that subject's visual system adapt to the new theory) will be unfulfilled and a radical version of the theory-ladenness thesis will remain unsupported.

(2) *A psychological account of visual perception that allows for the possibility that after the appropriate sort of training, the proponent of a novel theory can visually perceive the world in terms of the theory's observational concepts.* Without (2), desideratum (b) cannot be satisfied because scientists' visual systems will be incapable of adapting to new theories.

(3) *The thesis that proponents of substantially different competing theories will often (or perhaps inevitably) employ very different concepts in historical episodes of theory-choice.* Let's suppose that the defender of the conceptual penetration argument succeeds in defending (1) and (2). He still needs to show that it is more than just a mere empirical possibility that proponents of competing scientific theories employ observations that are not theory-neutral. This case is most likely to be made on semantic, psychological, or historical grounds.

Perhaps a few (largely unsubstantiated) words are in order concerning these issues. The very best that a proponent of the conceptual penetration argument can reasonably hope to do is to support some moderate version of the theory-ladenness thesis. To see why, let's look at each issue facing the proponent of the penetration argument.

(1) Let's grant a theory of representational content that allows for the possibility that proponents of competing theories will employ incompatible, non-theory-neutral observational concepts. We should not, however, grant a theory that implies that proponents of competing theories must inevitably employ non-theory-neutral concepts. Such a theory fails to account for our ability to express many different concepts. For example, we can express a concept of light, or heat, or gene that is agnostic about what constitutes those things. Such concepts certainly seem to be neutral between a number of different theories about the structure of light, or heat, or genes. (For a fuller discussion of this point, see Bishop 1991.)

(2) There are good reasons to believe that experience and training can alter our visual systems so that we are capable of recognizing new things. Jerry Fodor (1983), no defender of the conceptual penetrability of our visual systems, offers a rough and ready test for determining whether a particular judgement about the visual surround is output by the visual system: A judgement is output by the visual system if its production is fast, mandatory, and automatic (non-voluntary). If my recognition of anything is fast, mandatory, and automatic, it is the recognition of my mother's face. And yet this representation is not a member of most other peoples' visual systems because they have had different experiences from my own. Consider the many things we can immediately recognize that Newton couldn't—planes and trains and automobiles, for instance. Again, it seems very likely that these representations penetrated our visual systems, and not Newton's, due to our constant contact with them. So it seems very possible that after lots of contact with Galapagos finches, proton traces, or Leyden jars, a scientist's visual system could expand to include such representations.

(3) We have granted a theory of representational content and an account of the visual system that allow for the possibility that both neutral and non-neutral observational representations were often available to proponents of competing theories. Which were actually employed in a particular episode of theory-choice is an open historical question. It seems plausible to suppose that the historical evidence will not show that in *all* actual cases of theory-resolution, only neutral categories were employed or only non-neutral categories were employed. If this is correct, then the conclusion to draw from the penetration evidence will be some subset of the following theory-ladenness theses.

(T) Given a case of rational theory resolution, the psychological evidence suggests that possibly the set of observations S employs in deciding between **B** significantly different and competing theories consists of **C** theory-neutral observations.

B: most actual, some actual
C: no, only some, primarily, only

It seems likely that even if the defender of the conceptual penetration argument wins the semantic and psychological debates, he can, at best, establish some reasonably moderate version of the theory-ladenness thesis.

3. Perceptual reversal of ambiguous figures

Familiar examples of ambiguous figures are the necker cube (ambiguous three-dimensional figures), the old woman / young woman and the duck / rabbit figures (ambiguous forms), and the face / vase figure (figure-ground ambiguity). These figures can be readily interpreted by the visual system in two ways; for example, the face / vase figure can be interpreted as either a pair of faces or a vase. Once we have recognized the faces and the vase, we are capable of perceptually reversing these—switching from recognizing the faces to recognizing the vase (or *vice versa*)—on command.

Some evidence suggests that perceptual reversal involves shifts in attention (Gale and Findlay 1983). For example, if a subject who is capable of reversing the old woman / young woman figure is asked to attend to a certain part of the figure (around the old woman's nose or mouth), she is more likely to report perceiving the old woman, whereas if the subject is asked to attend a different part of the figure (around the young woman's nose or ear), she is more likely to report perceiving the young woman. When a subject is asked to concentrate on the old or the young woman, she will tend to focus on those same parts of the figure.

It is clear that by itself this phenomenon implies nothing about how scientific controversies are resolved and so does not directly support any version of the theory-ladenness thesis. Voluntary perceptual reversal does not exhibit desideratum (a)—the subject does not adopt a new theory. Nor does it show that a subject's visual system is capable of adapting to a new theory (b).

It appears that this evidence is supposed to be suggestive: It is supposed to exhibit features of our visual systems that if implicated in an episode of theory resolution would result in no (or only few) theory-neutral observations being employed. The argument from perceptual reversal is an argument from analogy. The different hypotheses about the ambiguous figures are supposed to be analogous to the competing theories in an episode of theory-choice. Given this analogy, there are three potential lessons one might try to draw about theory resolutions in science.

Conceptual penetration. Although there is good evidence for the conceptual penetration hypothesis (see the discussion in section 2), one might advance the phenomenon of voluntary perceptual reversal as further support. It shows that it is possible for different individuals, or a single individual at different times, to perceive different things when their perceptual mechanisms are pointed in the same direction. The problem is that this phenomenon does not show the visual system's representational resources changing appreciably. Before and after acquaintance with the ambiguous old woman / young woman figure, a subject is capable of visually perceiving old and young women. There is no increase in the visual system's conceptual resources.

Voluntary reversal in historical episodes. The phenomenon of voluntary perceptual reversal suggests that individuals are capable of deciding which observational representations to employ in any given situation. So if a scientist has a number of different representations that she could bring to bear on an experimental situation, she can voluntarily decide which to employ. Although Churchland appears to endorse this consequence (1988, 170-1), Kuhn explicitly rejects it: "the scientist does not preserve the *gestalt* subject's freedom to switch back and forth between ways of seeing" (1970, p. 85). Why not? Why is the scientist's observational armamentarium so restricted by the theories she has mastered? No psychological evidence Kuhn adduces supports such a radically context-insensitive account of observation. In fact, voluntary reversal suggests a kind of *observational libertarianism*: many observational concepts are available to the scientist in any given experimental situation, sometimes including concepts neutral between the competing theories, and she is free to choose among them as she sees fit. This, in turn, supports the moderate conclusion we drew at the end of section 2. The psychological evidence so far examined provides no guarantee that only neutral or only non-neutral observations play a role in cases of theory-resolution; and given the availability of both types of observations, the historical evidence is not likely to be unequivocal.

The insufficiency of neutral observations. Given a complete hypothesis-neutral description of (say) the old woman / young woman (couched in geometrical terms), it is not possible to determine whether it is a figure of an old or a young woman. Therefore (one might argue) there may be analogous cases in which scientists are incapable of rationally deciding between hypotheses on the basis of theory-neutral observations alone. So even if there is a fund of observational concepts that are neutral between a pair of competing theories, such observations cannot be decisive (Kuhn 1970, p. 203). On one reading, this claim is obviously true. By themselves, observations or observation statements, whether neutral or not, cannot decide between theories. Non-observational (or theoretical) vocabulary is needed just to state most competing theories, much less to decide between them. The real issue is whether neutral observations are insufficient for rational theory adjudication, and therefore whether non-neutral observations are necessary. The lesson that the proponent of theory-ladenness needs to draw from voluntary perceptual reversal is that non-neutral observations are necessary in order to make rational theory choices. But voluntary perceptual reversal does not show this. The reason it is not possible to rationally decide whether the old woman / young woman figure really represents an old woman or a young woman is that the figure doesn't provide enough information. And no one denies that there have been many historical cases in which there is not enough available information to rationally decide between a pair of theories. What voluntary perceptual reversal doesn't show, and needs to, is that in cases in which there is enough information available to decide rationally between a pair of theories, non-neutral observations must be part of that information.

4. Inverting lenses

Both Kuhn (1970, p. 112) and Churchland (1988, p. 174) offer the fascinating literature on inverting lenses as evidence for theory-ladenness. An inverting lens turns the visual field upside down while reversing left and right. Subjects typically undergo three stages: the discrepancy, adaptation, and negative aftereffect stages. In the *discrepancy stage*, a conflict is registered between visual information and proprioceptive information (information about one's felt body position). For example, suppose you were staring at a blank screen and moved your right hand from left to right across your face. Your visual system would inform you that an inverted right hand (with the forearm disappearing toward the top of your visual field) was moving from right to left in front of your face, while your proprioceptive system would inform you that your right hand was being moved from left to right across your face. Perhaps the most interesting early effect of the inverting lenses is that head movements lead to the illusion that the entire visual scene is moving in the same direction as the head but twice as fast. This produces severe disorientation and sometimes nausea. In the *adaptation stage*, the subject undergoes a change that reduces or eliminates the registered discrepancies between the relevant perceptual systems (in this case, the proprioceptive and visual systems). Notice that this does not require any change to the subject's visual system. A change to the proprioceptive system (or some other system) might account for the adaptation. Adaptation to the inverting lenses takes a few days but is impressive. The nauseating illusion of motion goes away, and subjects are capable of engaging in quite complicated tasks, such as fencing, mountain climbing, skiing, and biking in heavy traffic (Kohler 1964; Taylor 1962). In the final, *negative aftereffect stage*, the lenses are removed, and the subject tends to make the same sorts of errors she made when the lenses were first put on, but in the 'opposite' direction. The negative aftereffect almost always disappears quickly (much faster than the original adaptation).

Like the argument from perceptual reversal, the argument from inverting lenses does not satisfy the first two desiderata on a theory-ladenness argument—the subject does not adopt a new theory, nor does her visual system adapt to a new theory. The argument from inverting lenses is an argument from analogy. Donning the lenses is supposed to be akin to adopting a new theory. When the new theory is adopted but not mastered, the subject's observations are not yet supposed to be theory-laden. Churchland makes this point forcefully.

> *Who ever claimed that the character of a scientist's perception is changed simply and directly by his embracing a novel belief?*... [Defenders of theory-ladenness have] emphasized the importance of long familiarity with the novel idiom,

> of repeated practical applications of its principles, and of socialization within a like-minded group of researchers (1988, p. 175).

Adapting to the lenses is supposed to be analogous to mastering the theory. Let's look into the nature of the adaptation. Kuhn and Churchland argue that the lensed subject's behavioral adaptation is subserved by the adaptation of her visual system: the visual fields of subjects flip over some time during the adaptation stage. Contrary to conventional wisdom, this finding should not give hope to the defender of theory-ladenness. It shows that after doing extreme violence to the visual system's input, *it adapts by going back to representing the world exactly as it did before*. The adapted lensed individual and the non-lensed individual will agree just as much as they ever did about the external world. Taking the analogy seriously, this outcome suggests that after a pair of subjects have mastered their competing theories, their visual systems have the profoundly conservative tendency to represent the world as they did before the adoption of those theories. It is hard to imagine a more damaging result for the proponent of theory-ladenness.

This point is striking when we ask what is supposed to be the analog of a theory-laden observation in the lens experiments. If adapting to the lenses is analogous to mastering a new theory, and the subject's visual field flips, where are the theory-laden observations? There aren't any. The subject now sees the world exactly as she did before donning the lenses.

As a matter of fact, the psychological data offer very good reasons to suppose that the visual fields of lensed subjects never flip over. Subjects who have adapted to the lenses often report that the world looks "normal," but reports that the visual world looks "upright" almost never survive close inspection and comparison with (memories of) pre-goggle perception (Dolezal 1982, pp. 227-8). Furthermore, the hypotheses that the visual field rights itself makes false predictions about the inverting lenses' negative aftereffects (Dolezal 1982, pp. 234-5).[3]

Regardless of what cognitive or motor mechanisms account for the inverting lens adaptation, it cannot lend support to the proponents of theory-ladenness. Recall the final desideratum on data that was supposed to support a radical theory-ladenness thesis.

> (c) As a result of the (visual system) adaptation, Subject sees the world differently, in the sense that it is now more difficult for Subject to resolve disputes on the basis of theory-neutral observations with those who have not adopted T.

When we take the proposed analogy seriously, the inverting lens data does not even exhibit this final feature. Proponents of theory-ladenness need to adduce visual system *adaptations* that increase observational disagreement between subjects. And although *donning* the distorting lenses increases observational disagreement (at least in the short run), the *subject's adaptations* to those lenses do not increase observational disagreement.

5. Expectation effects

How do expectations influence perception? Consider the findings of the Bruner-Postman card experiment (1949). On very short exposures, subjects identified anomalous playing cards (e.g., a red spade) as normal ones (a black spade or red heart) without apparent hesitation or doubt. However, most subjects were eventually capable of correctly identifying the anomalous cards on longer exposures. The Bruner-Postman card experiment (along with other experiments that demonstrate expectation effects, e.g., Warren 1970) suggests that there are circumstances in which an individual expects to see (or perceive) X, and when X does not occur, the individual will report having seen X nonetheless. Sidestepping the touchy issue of whether the individual actually saw X, it is clear that this phenomenon possibly could play an important role in the resolution of theory disputes in science. Can expectation effects give comfort to defenders of theory-ladenness?

5.1. The strong expectation argument

From the fact that subjects saw normal cards because they expected to see normal cards, one might suppose that what one sees is completely determined (or very nearly so) by one's theoretical expectations. So when a person masters a theory, he expects to see, and therefore does see, the predictions made by that theory. Define a theory-neutral observation as follows: *An observation is neutral between a pair of theories if and only if it is consistent with the predictions of both theories; otherwise, it is theory-laden.* (This is not the notion of theory-neutrality assumed in other empirical theory-ladenness arguments.) Given this interpretation of the Bruner-Postman card experiment, proponents of competing scientific theories cannot possibly adjudicate their differences on the basis of theory-neutral observations because their observations never conflict with the dictates of their theories. A very radical version of the theory-ladenness thesis follows from this interpretation of the data: For all possible cases of rational theory resolution, the psychological evidence shows that inevitably the set of observations S employs in making her choice consists of no observations that are theory-neutral.

Although the strong version of this argument supports the most radical version of the theory-ladenness thesis, it is demonstrably unsound. As a historical matter of fact, proponents of many theories have been surprised to see their expectations foiled. There is no psychological experiment that supports the thesis that theoretical expectations completely determine what one sees. In the case of the Bruner-Postman card experiment, just the opposite since most subjects were capable of identifying the anomalous cards on longer exposures.

No defender of the theory-ladenness argument, when thinking and expressing himself clearly, has defended the strong expectation argument. In fact, the existence of anomalies (perceived failures of expectations) is a fundamental tenet of Kuhn's philosophy of science. It is nevertheless important to be clear about the strong expectation argument. Foes of theory-ladenness have taken it to *be* the theory-ladenness argument and peremptorily rejected it as absurd. Consider the following critique of theory-ladenness by Jerry Fodor.

> [W]orking scientists indulge in every conceivable form of fudging, smoothing over, brow beating, false advertising, self-deception, and outright rat painting—all the intellectual ills that flesh is heir to... Nevertheless, it is perfectly obviously true that scientific observations often turn up unexpected and unwelcome facts, that experiments often fail and are often seen to do so... (1984, p. 42).

Fodor's argument is beside the point. The fact that experiments often violate expectations does not undermine any sane version of the theory-ladenness of perception argument.

5.2. The weak expectation argument

The weak expectation argument deploys the Bruner-Postman card experiment, and other experiments of the same vintage, to show that our perceptual judgments are sometimes determined by our theoretical expectations. So in the hurly-burly of scientific practice, theory adjudication on the basis of theory-neutral observation may be less common than usually supposed. Here then is a psychological phenomenon that, if at work in actual cases of theory resolution, makes it difficult for proponents of competing theories to resolve their differences on the basis of theory-neutral observations.

But we must be careful. What would an episode that exhibited expectation effects look like? Suppose S1 is a proponent of T1, and S2 is a proponent of T2, and T1 and T2 make incompatible observational predictions. Assuming S1 were experiencing expectation effects, he would not believe that his theory was in any observational trouble because he would perceive, or at least report that he was perceiving, what his theory predicted. Given this situation, what motivation would S1 have to reject his theory in favor of S2's theory (or *vice versa*)? After all, S2's theory makes predictions that conflict with what S1 is seeing, or reporting that he sees, and *vice versa*. So any time expectation effects are at work, it is very unlikely that either party will be motivated to resolve conflicts of theory, at least on the basis of those observations. Given that the presence of expectation effects in an historical episode undermines the likelihood of any resolution of theory conflict, it also undermines the likelihood of finding a case of theory resolution that has the extra property of not having been based on theory-neutral observations. Kuhn recognizes this point. He is not a proponent of the weak expectation argument. Kuhn employs the Bruner-Postman card experiment not to argue that theoretical disagreements are adjudicated on some basis other than theory-neutral observations, but to argue that in science "novelty emerges only with difficulty, manifested by resistance, against a background provided by expectation" (1970, p. 64).

6. Final words

The framework offered here for understanding theory-ladenness arguments allows us to see that there are many different arguments that can justly be called 'the theory-ladenness of perception argument'. Further, because the parties to this debate have not made clear what the evidence needs to show in order to prove their respective points, they have not always understood the significance of the psychological phenomena they advance. As a result, opponents of theory-ladenness rightly reject an argument that no one believes (the strong expectation argument), and proponents of theory-ladenness champion an argument that undermines their conclusions (the inverting lens argument).

The framework for understanding theory-ladenness advanced here should also make us wary of the extreme positions so often defended. For example, there are numerous moderate positions to take on the conceptual penetration issue. It may well be that scientists often have both theory-neutral and theory-laden observational concepts available to them, and which they employed in a particular episode of theory-choice is an open empirical question. If this is right, historians, sociologists and philosophers of science should not begin a study of an episode of theory-choice with the conviction that proponents of the competing theories must (or must not) employ theory-neutral observations. Doing so will likely keep closed what is, in fact, a very open and very interesting question.

Notes

[1]I would like to thank Paul Boghossian, Paul Churchland, Eric Gampel, Peter Godfrey-Smith, Patricia Kitcher, David Magnus, Joe Mendola, Sam Mitchell, Graham Nerlich, Stephen Stich and my colleagues at Iowa State University for very helpful comments on earlier drafts of this paper. I owe special thanks to Philip Kitcher for guidance that invariably led in fruitful directions.

[2]Kuhn recognizes that the theory-ladenness of perception arguments must deploy both psychological and historical evidence. The psychological evidence, alone, will not support a theory-ladenness thesis.

> [T]hough psychological experiments are suggestive, they cannot, in the nature of the case, be more than that. They do display characteristics of perception that could be central to scientific development, but they do not demonstrate that the careful and controlled observation exercised by the research scientist at all partakes of those characteristics. Furthermore, the very nature of these experiments makes any direct demonstration of that point impossible. If historical example is to make these psychological experiments seem relevant, we must first notice the sorts of evidence that we may and may not expect history to provide (1970, 113-4)

[3]Suppose the visual field had righted itself. When the subject removed the lenses, we would expect the visual field to (again) appear upside down and for the subject to (again) fall victim to the motion illusion. So when the subject moves her goggle-free head to the right, the world should appear to move in the same direction as her head. But this does not happen. In fact, the world appears to move in the opposite direction of her head motion (Dolezal 1982, p. 234-5).

References

Bishop, M.A. (1991), "Why the Semantic Incommensurability Thesis is Self-Defeating", *Philosophical Studies* 63: 343-356.

Bruner, J.S., and Postman, L. (1949), "On the Perception of Incongruity: A Paradigm", *Journal of Personality* 18: 206-223.

Churchland, P.M. (1979), *Scientific Realism and the Plasticity of Mind.* Cambridge, England: Cambridge University Press.

__________. (1988), "Perceptual Plasticity and Theoretical Neutrality: A Reply to Jerry Fodor", *Philosophy of Science* 55: p. 167-187.

Dolezal, H. (1982), *Living in a World Transformed: Perceptual and Performatory Adaptation to Visual Distortion.* New York: Academic Press.

Feyerabend, P. (1975), *Against Method.* London: New Left Books.

Fodor, J. (1983), *The Modularity of Mind.* Cambridge: MIT Press.

_____. (1984), "Observation Reconsidered", *Philosophy of Science* 51: 23-43.

Gale, A.G. and Findlay, J.M. (1983), "Eye Movement Patterns in Viewing Ambiguous Figures", in Groner (et. al.) *Eye Movements and Psychological Functions: International Views.* Hillsdale, New Jersey: Lawrence Erlbaum Associates, pp. 145-168.

Kohler, I. (1964), "The Formation and Transformation of the Perceptual World", *Psychological Issues* 3: 1-173.

Kuhn, T.S. (1970), *The Structure of Scientific Revolutions*. Chicago: The University of Chicago Press.

Taylor, J.G. (1962), *The Behavioral Basis of Perception*. New Haven: Yale University Press.

Warren, R. (1970), "Perceptual Restoration of Missing Speech Sounds", *Science* 167: 392-3.

The *Structure* Thirty Years Later: Refashioning a Constructivist Metaphysical Program

Sergio Sismondo

Cornell University

1. Introduction

I argue here that for the past thirty years there has been a persistent misreading of Thomas Kuhn's *The Structure of Scientific Revolutions*.[1] Not everybody has participated in this misreading, but its persistence is remarkable. So one of my tasks is to diagnose the reasons for its continuance. It is also an important misreading, because it attributes to Kuhn a view that is from most people's perspective highly untenable, that scientists in some strong sense construct the world by choosing a paradigm.

To avoid repeating work done by Gutting and many others (see Gutting 1980) I will focus on Kuhn's constructivist metaphysical program, as seen by commentators. I am interested in the relationship between paradigms and their ontologies, not in the rationality of science, its progressiveness, etc. As such, I will leave aside questions about incommensurability, about the dynamics of theory change, about the constitution of scientific communities, and other contentious issues stemming from *Structure*. And I use 'paradigm' in a loose way, to mean something like the "disciplinary matrix" that Kuhn describes in "Second Thoughts on Paradigms" (Kuhn 1974).

2. Idealist readings of Kuhn

The view that I think Kuhn does not hold, but which he is sometimes thought to hold, is that scientists, in adopting a paradigm, construct a world (in addition to the social and conceptual world in which they live) in which the basic tenets of the paradigm hold true. Readings of *Structure* that include this create an "idealist" or "Neo-Kantian" Kuhn; an early such reading was given by Israel Scheffler, who called Kuhn "an extravagant idealist" (1967). These readings of Kuhn's ontological program stem mostly from fairly literal interpretations of Kuhn's talk of different worlds existing before and after a revolution. The scientists in the changed discipline are taken to be not merely seeing the world differently, but living in a different world: When the "suggestion [that Herschel's new comet was a planet] was accepted, there were several fewer stars and one more planet in the world of the professional astronomer" (1970a, p. 115). For Carl Kordig, for example, it is this sentence and ones like it which lead him to the conclusion that Kuhn is an idealist, and therefore a relativist.

PSA 1992, Volume 1, pp. 300-312

> It must follow that before Lexell there were more stars in the world of the professional astronomer. But if this is true one can no longer say that these astronomers before Lexell were mistaken about the number of stars. . . . This is because from Kuhn's viewpoint there really were more stars in their world. It is not just that they believed there were more stars in their world. According to Kuhn there *really* was this number of stars. (Kordig 1971, p. 18)

This is elaborated by Richard Boyd. He claims that

> the realist denies, while the constructivist affirms, that the adoption of theories, paradigms, research interests, conceptual frameworks, or perspectives in some way constitutes, or contributes to the constitution of, the causal powers of and the causal relations between the objects scientists study in the context of those theories, frameworks, and so forth. . . . Realists affirm, and constructivists deny the *no non-causal contribution doctrine*. (Boyd 1990, pp. 182-3)

The "no non-causal contribution doctrine" is meant to distinguish Boyd's metaphysical position from that of Kuhn, his main example of a constructivist (see also Boyd 1992).

3. 'Different worlds' talk

It is clear that Kuhn means this talk of different worlds in a metaphorical sense. He says, for example, that there is a "*sense* in which [paradigms] are constitutive of nature" (p. 110, emphasis added). And, "in the absence of some recourse to that hypothetical fixed nature that he 'saw differently,' *the principle of economy will urge us to say* that after discovering oxygen Lavoisier worked in a different world" (p. 118, emphasis added). The chapter in which are the bulk of the references to the changing of worlds is entitled "Revolutions as Changes of World *View*," suggesting that changes in the "world" should not be read as more than changes in views. It begins:

> Examining the record of past research from the vantage of contemporary historiography, the historian of science may be *tempted* to exclaim that when paradigms change, the world itself changes with them. Led by a new paradigm, scientists adopt new instruments and look in new places. Even more important, during revolutions scientists see new and different things when looking with familiar instruments in places they have looked before. It is rather as if the professional community had been suddenly transported to another planet where familiar objects are seen in a different light and are joined by unfamiliar ones as well. Of course, *nothing of quite that sort does occur*: there is no geographical transplantation; outside the laboratory everyday affairs usually continue as before. Nevertheless, paradigm changes do cause scientists to see the world of their research-engagement differently. *In so far as* their only recourse to that world is through what they see and do, we may want to say that after a revolution scientists are responding to a different world. (p. 111, emphasis added)

Boyd, Scheffler, and others recognize that the "different worlds" talk is metaphorical, but they want to read it as a strong metaphor. In this they are surely right. Kuhn repeatedly says that paradigm changes alter the world of the professional scientist. Not only interpretations but also data change (p. 135). The affected scientists are faced with new problems, and have to use new techniques in order to solve them. Thus the world's changes are substantial enough to force a reaction.

4. Hoyningen-Huene's reading

Although many people assert that there is some idealism lurking in *Structure*, few people have argued the case at any length. One person who *has* done so is Paul Hoyningen-Huene, who sees some positive features in the idealism he finds (Hoyningen-Huene 1989a,b). He argues that there is an ambiguity in the use of "world" in *Structure*, an ambiguity between something "already perceptually and conceptually subdivided" and the more fundamental, un-humanly structured world. Much of Kuhn's discussion, claims Hoyningen-Huene, is about the constitution of the former world, which he likens to Kant's "totality of appearances" (Erscheinungswelt) (1989b). The latter world bears "great similarity to Kant's thing in itself although it is not identical with it" (1989b, p. 394), and is relatively unaffected by conceptual changes.

It isn't clear how this adds up to much idealism. Hoyningen-Huene claims that Kuhn's "world of appearances" is material (unlike a dreamt winged horse); its features are not completely dependent on its observers, but are shaped by "resistances" that the "world in itself" creates; and the "world of appearances" is not subjective in the sense that the individual observer shapes it, because it is a *socially* constructed world. The social construction comes about because appearances are shaped by held similarity-relations, which the community as a whole creates and keeps stable: communities hold the rules of classification fixed, in the same way that they hold languages fixed; individuals can influence a community's similarity-relations, but cannot single-handedly change them. What we are left with is that the world of appearances is partially shaped by our expectations and the frameworks that we try to impose upon it, and that these frameworks and expectations are social in nature. This is idealist only in opposition to a realism that maintains that there is a unitary, best description of the world potentially within our reach.

However, there is something in this reading of *Structure* that we can identify as incompatible with a sensitive realism. Kuhn the Neo-Kantian asserts that the second world, the world in itself, is not knowable. That is something that realism cannot accommodate, and it is something that Kuhn seems to say, as well. For example, when describing the sense in which paradigms are constitutive of nature, Kuhn says: "In so far as their only recourse to that world [the world in itself] is through what they see and do, we may want to say that after a revolution scientists are responding to a different world" (1970, p. 111). In addition, Kuhn is definitely skeptical of the notion of 'truth,' as he explains in the last chapter of *Structure*. Given his wariness about truth, we could guess that he would not be comfortable with a notion of description of the world that implied that we could really know it above and beyond knowing its appearance.

5. Resistances and appearances

The details of Hoyningen-Huene's analysis offer valuable insights, yet it creates some impressions that look wrong. I'm going to examine two problems that will lead us away from talk of worlds of appearances and away from the Neo-Kantian Kuhn. The first problem has to do with resistances, and the second with similarity-relations. The first is something that I think Kuhn would also see as a problem with this interpretation; the second is more a problem with Kuhn's emphases themselves.

As Hoyningen-Huene points out, Kuhn's world of appearances is very "object-sided." Resistances are everywhere. Nature plays an enormous role in *Structure*; no more than a couple of pages pass without mention of nature's intrusion into scientists' affairs. Through normal and extraordinary research we chart out the positions and the natures of the resistances, so it is not that science doesn't find out about the world.

The most important of resistances are those anomalies which may potentially lead to the overturning of paradigms. But Kuhn also mentions resistances which routinely face the scientist working unproblematically within a normal science tradition; when Kuhn describes fact-gathering activities, he is describing the contact of normal science with the resistances of the world. For example, there is

> that class of facts that the paradigm has shown to be particularly revealing of the nature of things. By employing them in solving problems, the paradigm has made them worth determining both with more precision and in a larger variety of situations. At one time or another, these significant factual determinations have included: in astronomy—stellar position and magnitude, the periods of eclipsing binaries and of planets; in physics—the specific gravities and compressibilities of materials, wave lengths and spectral intensities, electrical conductivities and contact potentials; and in chemistry—composition and combining weights, boiling points and acidity of solutions, structural formulas and optical activities. (p. 25)

This hardly reads like the writings of someone who thinks that we can't find out about the world. Kuhn's stance that scientists don't have direct access to the world seems to be held not because of a deep unknowability of that world, but because Kuhn sees the everyday side of scientists. They must live with interpretations, and within frameworks, but the world is not necessarily obscured. Kuhn's worry about the knowability of the world has to do with the absence of this higher authority, not with nature's obscurity. (see for example p. 114)

A second problem with Hoyningen-Huene's interpretation of *Structure* has to do with Kuhn's own emphases on appearances and classification. These emphases create much of the impetus for a reading of 'worlds' as 'worlds of appearances.' This I find misleading to the extent that we are apt to forget the actions of scientists in favor of their observations.

In "Second Thoughts on Paradigms" (1974), Kuhn discusses the dependence of classification and identification on similarity-relations. These are the features of objects that we take to be important in determining which similarities and differences arise because the objects are members of particular classes, and which are the result of individual idiosyncrasy. One gathers from this essay that Kuhn intends the notion of 'paradigm' to be replaced in part by 'learned similarity-relations.' Hoyningen-Huene takes this a step further and makes similarity-relations a central part of Kuhn's philosophy, seeing paradigm changes as changes in held classifications.

The limitations of the idea of paradigms as held similarity-relations becomes clear when we look at some actual examples of paradigm changes, and changes in the world, as described by Kuhn. One that looks as though it might fit the similarity-relation account is the discovery of Uranus. This was an object which was seen as a star, as a comet, and then as a planet.

> A celestial body that had been observed off and on for almost a century was seen differently after 1781 because, like an anomalous playing card, it could no longer be fitted to the perceptual categories (star or comet) provided by the paradigm that had previously prevailed. (pp. 115-6)

> Probably . . . the minor paradigm change forced by Herschel helped to prepare astronomers for the rapid discovery, after 1801, of the numerous minor planets or asteroids. Because of their small size, these did not display the anomalous magnification that had alerted Herschel. (p. 116)

New objects were potentially more similar to planets than to comets or stars. But the distinguishing features of planets, comets, and stars had changed little. Rather, a class which had previously been closed was now open. There had been resistance to the idea of new planets, but when this resistance was overcome the categories, and thus the similarity-relations, had not changed. What forced the "minor paradigm change" was the discovery of a *new* planet, not a different planet.

Other examples are more clear-cut. Roentgen's discovery of X-rays, while changing physicists' conceptual maps, was primarily interesting, and thus creating a new paradigm, because X-rays were a substantially new phenomenon. They were a new thing to be investigated, a new tool, and a new source of explanations (p. 59).

Changed similarity-relations make up only a part of what distinguishes paradigms. On the model of Herschel's observations of Uranus, it would seem that one thing that can separate paradigms is new examples of old phenomena. And following Derek de Solla Price (1984), I would see one of the most common factors separating paradigms to be techniques, in a broad sense, including technological changes, and the interests and directions of exploration that follow from the use of these techniques.

Despite these negative remarks on Hoyningen-Huene's reading of Kuhn the idealist, the details of his analysis are surely right: there are some Neo-Kantian streams in *Structure*; it is just unclear how much this amounts to. Kuhn uses "world" in an ambiguous way, and given his emphasis on appearances (gestalts, etc.) attributing to him talk of a world of appearances hardly seems unfair. I think that to some extent this can be blamed on Kuhn's over-use of perceptual metaphors and analogies to gestalt experiments, but it is unclear the extent to which he means these metaphors to define the limits of his inquiry. As well as issues about appearances, there is Kuhn's skepticism about the extent to which the notion of 'truth' can be applied in an inter-theoretic way, which sounds as though it entails a commitment to a dichotomy between the world of appearances and an underlying reality. None of this should be surprising, because *Structure* was written at a time when another Neo-Kantianism, namely positivism, was the dominant philosophical approach to science.

At the same time, something should be added to this reading in order for it to handle the strength of Kuhn's metaphor of changing worlds. While Kuhn places considerable emphasis on the way that scientists' worlds of appearances can change through revolution, he doesn't forget that scientists act on the world as well. They don't merely observe the world, they also carry out experiments. Revolutions don't just produce new views of nature, they produce new research programs. Thus talk of worlds of appearances is too limiting; it deals with only one aspect of paradigms. The two problems above can be directed against this type of talk. The limitations of the similarity-relation account of paradigms point to a limitation of seeing Kuhn's talk of paradigms as being primarily about categories of appearances, steering us away from worlds of appearances and towards worlds in which action and knowledge take place. The ubiquity of resistances takes us even further, questioning the extent to which Kuhn thinks that there is an unknowable realm beyond the worlds in which we live.

6. Putting aside idealism

If Hoyningen-Huene's characterization of Kuhn as a mild idealist doesn't seem to leave him saying much that a realist couldn't accept, what about the stronger readings by Boyd, Scheffler, and others? I think that their Kuhn doesn't bear much resemblance to the Kuhn of *Structure*; to argue this, I have to walk through Kuhn's discussion of changes in the world (primarily Ch. X), to show that, for him, the adoption of paradigms makes no "non-causal contributions" to underlying reality.

To start with, we should remember that Kuhn's discipline when he wrote *Structure* was history of science, so we should look for the motivations for his philosophical views in his experience as a historian. This is not to say that he was uninterested in philosophy of science; on the contrary, he was, and is, interested in philosophy. But at the same time, he claims that it was his dissatisfaction with philosophy of science in the light of historical work that prompted his change of fields from physics to history (v). And he claims that this historical work that led him to write *Structure*, as early as the first sentence of the book: "History, if viewed as a repository for more than anecdote or chronology, could produce a decisive transformation in the image of science by which we are now possessed" (p. 1). So, while his pictures of science may be informed by philosophical considerations, we should see them as compelled by historical ones.

It is historical considerations that could "tempt" one to say that the world changes with changes of paradigms (p. 111), something that Kuhn does not want to assert, except in a sense circumscribed by the the fact that scientists' "only recourse to that world is through what they see and do." If we take Kuhn seriously in his qualification, then "world" for him, when he is using it in this sense, means something like the social and perceptual context of human actions. To avoid confusing this new type of world with a Kantian world of appearances, and to avoid claiming that it is identical to the phenomenologists' life-world, I will call Kuhn's new worlds 'social worlds,' which will help to remind us that Kuhn is talking about professional science, which is a very social activity.

After the positing of social worlds, Kuhn wants to turn to their components, and our perceptions of them. Here his views are informed by gestalt experiments. His fundamental claim is that scientists' data are not the logical atomists' (and sometimes positivists') "here red now," but instead observations of planets and sunspots, the motion of pendulums, and even electrostatic repulsion and oxygen.

> Oresme sketched . . . [an] analysis of the swinging stone in what now appears as the first discussion of a pendulum. . . . [I]t was a view made possible by the transition from the original Aristotelian to the scholastic impetus paradigm for motion. Until that scholastic paradigm was invented, there were no pendulums, but only swinging stones, for the scientist to see. Pendulums were brought into existence by something very like a paradigm-induced gestalt switch. (p. 120)

Kuhn's controversial view of data, the historian's view, is prompted by very empirical concerns. What scientists refer to in their papers and notes are not patches in the visual field, but theory-dependent measurements and observations.

> The operations and measurements that a scientist undertakes in the laboratory are not "the given" of experience but rather "the collected with difficulty." They are not what the scientist sees—at least not before his research is well advanced and his attention focused. Rather they are concrete indices to the content of more elementary perceptions, and as such they are selected for the close scrutiny of normal research only because they promise opportunity for the fruitful elaboration of an accepted paradigm. Far more clearly than the immediate experience from which they in part derive, operations and measurements are paradigm-determined. (p. 126)

If we accept this view of data, which doesn't even deny the existence of traditional sense-experiences, then we can immediately see how the adoption of a paradigm can change the social world. The adoption of paradigms changes exactly those things that count as existing, the things that can be data. Once one has gained a great deal of experience with the new ontology, one might even only perceive *it*, and find it extremely difficult to perceive alternate entities.

Thus there is no non-causal contribution of paradigms to the world, either the social world or what lies behind it. The adoption of new paradigms makes definite causal contributions to social worlds, in that they define for the scientist what entities do and do not exist. That is, they are social causes of new social phenomena.

7. The worry about 'truth'

Besides Kuhn's 'other worlds' talk, probably the main reason why he has been seen as having a funny metaphysical picture has had to do with his attitude toward truth. The concept is pretty much avoided throughout *Structure*, introduced only to be dismissed as unnecessary (p. 170). In particular, it is unnecessary to see paradigm changes as bringing us closer to the truth. Science is an evolutionary process, Kuhn claims, but not one towards a goal. There is even progress, but not truth. For many, this view serves to make Kuhn an anti-realist, if not necessarily an idealist. Since realism seems to entail some commitment to truth, a commitment to the idea that scientific statements are sometimes true, it might be imagined that his apparently iconoclastic ideas about truth are connected to a more radical constructivist, or possibly even idealist, metaphysics.

There are a couple of reasons why Kuhn's disdain for truth need not be linked to idealism or even constructivism. First of all, skepticism about truth in no way implies constructivism, and if anything the two do not sit well together. Arthur Fine (see 1986), for example, is skeptical about truth, but doesn't hold anything like idealist or constructivist views. He simply wants to abandon truth altogether as a well-defined concept, and accept that the word 'truth' has different uses in different situations, uses that are not connected in any important way. But the constructivism that Kuhn is so often taken to espouse depends heavily on the notion of truth, because under that view paradigms establish what is true about the world. Constructivism's theory of truth is presumably not a correspondence theory, but it is a theory of truth. In contrast, Kuhn would rather not use the term at all. So he cannot both be a constructivist and deny truth any value.

Secondly, the position that Kuhn rejects is a scientific realism that today looks like a caricature. He objects to the position that holds that "there is one full, objective, true account of nature and that the proper measure of scientific achievement is the extent to which it brings us closer to that ultimate goal" (p. 171). Kuhn may have been objecting to that position for the same reasons that most realists today would: the idea that there is only one true account of complex phenomena is dependent on a reductionism (probably to the physics of elementary particles) that is outdated and almost certainly wrong.

Though Kuhn doesn't ever describe his views on truth very fully, there are short discussions in the "Postscript" to the second edition of *Structure*, and in "Reflections on my Critics" (1970b) that help us to understand his position, or at least his position at the end of the 1960s. Kuhn presents several problems with 'truth,' which lead him to reject it as a measure of progress in the sciences, at least over the long term. There is a historical problem:

> I do not doubt . . . that Newton's mechanics improves on Aristotle's and that Einstein's improves on Newton's as instruments for puzzle-solving. But I can see

> in their succession no coherent direction of ontological development. On the contrary, in some important respects, though by no means in all, Einstein's general theory of relativity is closer to Aristotle's than either of them is to Newton's. (1970a, pp. 206-207)

Kuhn's professional opinion as a historian leads him to believe that truth is not a measure of progress. This is partly because he is still thinking of a limiting truth, a single, best, scientific theory that science is said to be steadily approaching (1970b, p. 265). That type of truth is not to be found in history.

There are also at least three philosophical problems, though Kuhn does not separate them out in this way. The first one is closely related to the historical problem: in his opinion, there are no characterizations of truth adequate for application to whole theories, and thus none that allow one to say that one paradigm is more truthful than another. "Granting that neither theory of a historical pair is true, [many philosophers of science] nonetheless seek a sense in which the latter is a better approximation to the truth. I believe nothing of that sort can be found" (1970b, p. 265). The nature of paradigms, and large theories, prohibits talking about their overall truth value. They are not the sort of things to which truth can apply, being more like ways of life, or ways of establishing truth. Among other things, they set some of the criteria for establishing truth, but that doesn't allow them to be themselves true.

The second philosophical problem with truth boils down to an issue of language. For Kuhn, following Tarski, 'truth' is 'truth in a language.' Since scientists separated by paradigms operate in different languages, applications of the predicate 'true' across paradigms are possible only in so far as we can translate. Therefore 'truth' is a concept which can be applied locally, but not globally; it is reasonably well-defined intra-theoretically, but not inter-theoretically. To see this all we have to do is remember that translation is a difficult enterprise for Kuhn, and that paradigms are incommensurable. If we completely reject incommensurability, then we will reject this difficulty with truth, but it stands that when Kuhn wrote *Structure*, and even much later, he thought that there was at least some incommensurability between paradigms.

The third philosophical difficulty is an epistemological one. There is no way of escaping frameworks or paradigms; we are always within one or another. Therefore we cannot escape the paradigm we are in to establish an extra-theoretic truth to what we are saying. "There is . . . no theory-independent way to reconstruct phrases like 'really there'" (1970a, p. 206). Once again, this problem divides truth along intra- versus inter-theoretic lines, because when we attempt to talk about truths from another paradigm, we do so only from within a paradigm, ours. So there is a sense in which truth is inaccessible if we take it to be more than intra-theoretic truth. The fact that this is an epistemological problem means that we don't have to put too much weight on it. There is no difficulty in claiming that there are truths without our being able to establish particular truths in a non-question begging way. But for the practically-minded historian (and the pragmatist), this epistemological problem might mean that for practical purposes we should dispense with the notion of 'truth,' at least in its extra-theoretic and correspondentist flavors.

Some people may still find anti-realism in things that Kuhn says about his rejection of truth. The quote immediately above, for example, continues: "the notion of a match between the ontology of a theory and its 'real' counterpart in nature now seems . . . illusive in principle." Intuition would have it that there can be no realism without such a match. This seems right, but limited; Kuhn would be taking an anti-realist stance here were he talking about everyday, local applications of truth. But if we remember that Kuhn's disquietude is about truth as a theory-independent concept then this rejection is a simple corollary of Tarski's views on truth, or of Kuhn's appreciation of the theory-dependence of method. The fact that there can be no theory-independent truth in no way implies that there can be no truth at all. We are simply limited to a "partial perspective" (Haraway 1988).

If this is a reasonable reconstruction of what little Kuhn says on the topic of truth, then the Kuhn of these texts doesn't come out a scientific anti-realist. Kuhn agrees that science tracks reality; he just doesn't believe that that tracking of reality can be a measure of progress. He believes that some scientific statements are true; he just doesn't believe that it makes sense to talk about the truth of larger packages, and he doesn't believe that there is a God's-eye truth.

One can find other reasons why Kuhn, as a historian, might find 'truth' an unuseful concept, though these are not explicitly given. One of the targets of *Structure* is "Whig History," history that attempts to construct the past as a series of steps toward present positions. The introductory chapter, "A Role for History," claims that the book comes as a result of a small revolution in historiography. This revolution challenged several assumptions, the most important of which is the assumption that science is a steady accumulation of facts, theories and methods. Instead, history of science, particularly history of science after Alexander Koyré, has shown that science is anything but steady, and is in many senses not an accumulation. The "new," now old, history of science had new goals, goals that put an increased emphasis on specifying the causes of particular views and episodes. "Rather than seeking the permanent contributions of an older science to our present vantage, they attempt to display the historical integrity of that science in its own time" (p. 3).

Given these goals, 'truth' becomes less important as one of the historian's tools. The causes of a belief, for example, while they might include inputs from the world, do not include the truth of the belief. And a sequence of events in the history of science cannot be explained simply by the fact that they represent progress toward the truth. These are points that have been made repeatedly in recent years in the context of the "strong programme" and other sociology of science (see for example Bloor 1991). While they in themselves do not provide reasons to reject the idea of truth, they do show how history of science usually does not require scientific truth in order to go about its business.[2]

So there are many reasons why Kuhn may have been objecting to wholesale application of the idea of 'truth' that don't require him to be an anti-realist. And though I think that these *are* Kuhn's reasons, even if they are not the resulting anti-realism is not the constructivism which is sometimes read into *Structure*, because that constructivism requires 'truth.'

8. Re-interpreting Kuhn's 'constructivism'

With the collapse of an idealist interpretation of *Structure*, it seems that Kuhn doesn't really offer any interesting metaphysics. What is constructed is the social world, something we already knew. But although there may be no interesting non-realist metaphysics in Kuhn's philosophy of science, he can be read as saying something about the way the world is. On this reading, the term 'constructivist' becomes less misleading, since there is a metaphorical sense in which scientists construct the world.[3]

Many of Kuhn's remarks about his philosophical positions are ambiguous, some of them purposefully so. Part of the reason for this ambiguity lies in the fact that Kuhn may have other priorities from the ones that we normally attribute to him, and had have decided not to commit himself to definite positions on some issues that may interest philosophers. After all, some of his intuitions stem from his position as a historian and reader of history. Considerations such as the theory-dependence of what counts as data and good methodology (Ch. IX, X), a recognition of the strength of consensus about a field (III), and skepticism about uniform progressiveness (XIII) are part of what what guides positions in *Structure*. The importance of these factors, and ones like them, is also one of the things that we can learn from Kuhn.

The ambiguity comes out when Kuhn says things like: "In a sense that I am unable to explicate further, the proponents of competing paradigms practice their trades in different worlds." And in responding to a paper by Richard Boyd, he says:

> Both of us are unregenerate realists. Our differences have to do with the commitments that adherence to a realist's position implies. But neither of us has yet developed an account of those commitments. Boyd's are embodied in metaphors which seem to me misleading. When it comes to replacing them, however, I simply waffle. (Kuhn 1979, p. 415)

and

> Boyd's world with its joints seems to me, like Kant's 'things in themselves,' in principle unknowable. The view toward which I *grope* would also be Kantian but without 'things in themselves' and with categories of the mind which could change with time as the accommodation of language and experience proceeded. A view of that sort need not, I think, make the world any less real. (1979, pp. 418-9, emphasis added)

These quotes reinforce the position I take in Section 6, but they also serve to emphasize that the comments that follow represent only one possible development of the Kuhn's constructivist position.

If the natural world were tremendously simple in the number and type of different entities and causal relations it contained, then for each subject matter it would only take a little time before we constructed a theory that accounted for most things that we wanted it to. If, on the other hand, the natural world were spectacularly complex, then we would expect that each theory we constructed would be at best barely adequate for a short while, until our interests changed. We would probably lean toward nominalism, as each of our categories would turn out to be artificial and inadequate.

One option for a constructivist metaphysics is to claim that the natural world is likely to be on the complex end of the spectrum, but not so far that our science is nearly useless. Thus when we adopt a good position, or a paradigm, we would often be adopting it for its value in helping us to steer around the entities and relations that exist. It would tell us which entities to pay attention to and which ones not; it would even help us by convincing us that large numbers of entities don't, can't, exist. It would prescribe a way of life, by telling us which bits of machinery to put into our laboratories, and it would proscribe ways of life that don't involve manipulating, and finding out about, the preferred entities. In a very concrete way adopting a paradigm would construct a social world.

This is a picture of paradigms that is repeatedly given in *Structure*. For example:

> [The paradigm] functions by telling the scientist about the entities that nature does and does not contain and about the ways in which those entities behave. That in-

> formation provides a map whose details are elucidated by mature scientific research. And since nature is too complex and varied to be explored at random, that map is as essential as observation and experiment to science's continuing development. Through the theories they embody, paradigms prove to be constitutive of the research activity. They are also, however, constitutive of science in other respects. . . . Paradigms provide scientists not only with a map but also with some of the directions essential for map-making. (p. 109)

Nothing in this picture is antagonistic to realism. So the constructivism that I am proposing be explored becomes a species of realism, a realism committed to a certain amount of complexity in the natural world. This seems entirely appropriate as a description of Kuhn's views, given his self-description as an "unregenerate realist."

I call this position constructivist because it mimics the idealist position which holds that the world is constructed by our ideas or by social convention. This 'Kuhnian' constructivism sees the possibility of a number of highly successful, yet competing, world views. That is, it takes as a premise the notion that there is no unique, correct description of the natural world. It sees our interactions with nature partly determined by our world views, and our categories partly influencing what we find, even what we observe. In this way it takes seriously Kuhn's emphasis on the theory-dependence of observation and method.

9. A diagnosis

There don't seem to be good grounds to see in Kuhn the kind of idealism of which he is sometimes accused. Why is it that people have been eager to read some idealist anti-realism (or constructivist anti-realism) into *Structure*? Why is it that Kuhn's imagery has been often conflated with more genuine expressions of idealism? There are at least two reasons.

The first concerns the fact that at the time Kuhn was writing, realism was not always disentangled from positivist reductionism. Thus it may have been more like the type of position that Kuhn definitely does criticize, the position that there is a uniquely correct ontology that science is approaching. Therefore Kuhn would have become a realist only as scientific realism changed and was articulated. The old Kuhn would have seemed about as radical a 'constructivist' as they come.

The second reason concerns the lack of attention to what I call social worlds, the genuinely constructed worlds that scientists inhabit on a day-to-day basis, which includes the ontology they accept, the experiments that are permissible, and the institutions with which they must deal. For those not accustomed to thinking about these worlds, Kuhn's metaphors will seem strange. Philosophy of science has inherited a very small vocabulary with which to describe science. It includes such words as "observation," "theory," and "experiment," but tends not to include vocabulary that reflects the range of things that may be observations, theories, or experiments. For example, the concept of a protocol, a routinized sequence of manipulations designed to achieve a very specific end, such as isolating a particular protein, is not a part of the normal vocabulary of philosophy of science, even though it is a broad category of procedures that are a standard part of the laboratory scientist's day-to-day life. Protocols produce data or clean samples, though by means not necessarily understood by the people who follow them. They are interesting because they are, to a greater or lesser extent, taken for granted, and are a necessary part of the scientist's social landscape.

Without keeping in mind the importance of social objects, it is difficult to appreciate the potential value in saying that scientists' worlds change when paradigms do. We are apt to forget that a change in theory can result in the complete reorganization of the laboratory, and the lives that intersect with it. It can change the meaning and relative importance of different types of observations. Thus, even when a particular type of resistance is undoubtedly shared by two consecutive paradigms, for example a type of resistance that precipitates a change of paradigms, the lives of the users of those two paradigms might have changed sufficiently that they would give completely different meaning to those resistances. So instead of being something that unites their worlds, it might be something that divides them. It is this type of observation that might make it fruitful to talk of scientists inhabiting different worlds.

Notes

[1]References are to the 2nd edition (1970a), which, aside from the postscript, differs from the first in no way that is pertinent to this paper. Page references refer to this work unless otherwise noted.

[2]There are exceptions. Truth may be a very convenient shorthand for 'congruent observations' when discussing consensus. Consensus about a novel but highly repeatable observation might be explained in terms of the truth of the statements about it. We should note that this doesn't commit one to the global rather than the local notion of truth.

[3]There is also a sense in which scientists physically construct the world, stemming from the fact that scientists' objects of study are usually many-times purified versions of what occurs in nature, forced into situations that are distinctly unnatural.

References

Bloor, D. (1991), *Knowledge and Social Imagery*. 2nd ed. Chicago: University of Chicago Press.

Boyd, R.N. (1990), "Realism, conventionality, and 'realism about'", in *Meaning and Method: Essays in Honor of Hilary Putnam,* G. Boolos (ed.). Cambridge: Cambridge University Press, p. 171-195.

______. (1992), "Constructivism, Realism, and Philosophical Method", in *Inference, Explanation, and Other Philosophical Frustrations,* J. Earman (ed.). Berkeley: University of California Press.

Fine, A. (1986), *The Shaky Game: Einstein, Realism and the Quantum Theory*. Chicago: University of Chicago Press.

Gutting, G. (ed.) (1980), *Paradigms and Revolutions: Applications and Appraisals of Thomas Kuhn's Philosophy of Science*. Notre Dame: University of Notre Dame Press.

Haraway, D. (1988), "Situated Knowledges: The Science Question in Feminism and the Privilege of Partial Perspective", *Feminist Studies* 14: 575-599.

Hoyningen-Huene, P. (1989a), *Die Wissenschaftsphilosophie Thomas S. Kuhns: Rekonstruktion und Grundlagenprobleme.* Braunschweig: F. Vieweg & Sohn.

____________. (1989b), "Idealist Elements in Thomas Kuhn's Philosophy of Science", *History of Philosophy Quarterly*, Volume 6, Number 4: 393-401.

Kordig, C.R. (1971), *The Justification of Scientific Change.* Dordrecht: Reidel.

Kuhn, T.S. (1970a), *The Structure of Scientific Revolutions*, 2nd ed. Chicago: Chicago University Press.

_____. (1970b), "Reflections on my Critics", in *Criticism and the Growth of Knowledge,* I. Lakatos and A. Musgrave (eds.). Cambridge: Cambridge University Press: 231-278.

_____. (1974), "Second Thoughts on Paradigms", in *The Structure of Scientific Theories,* F. Suppe (ed.). Urbana, IL: University of Illinois Press: 459-482.

_____. (1979), "Metaphor in Science", in *Metaphor and Thought,* A. Ortony (ed.). Cambridge: Cambridge University Press: 409-419.

Price, D.J. de Solla (1984), "Notes Towards a Philosophy of the Science/Technology Interaction", in *The Nature of Technological Knowledge,* R. Laudan (ed.). Dordrecht: Reidel.

Scheffler, I. (1967), *Science and Subjectivity.* Indianapolis: Hackett.

Part IX

TOPICS IN THE PHILOSOPHY OF BIOLOGY

Additivity and the Units of Selection[1]

Peter Godfrey-Smith

Stanford University

1. Selection

Variation is the fuel of evolution, but as is well known, not all variation will drive Darwin's engine along. Within population genetics, "additivity" is one concept used to characterize the salient type of variation. The characterization of variation (variance) in terms of additivity originates with R.A. Fisher (1918), and Fisher's "fundamental theorem of natural selection" (1930) states that the rate of evolutionary change is proportional to the additive variance in fitness in the population at that time.[2] Evolution proceeds until the additive variance is exhausted.

It would not be surprising if a concept so important in the mathematical structure of evolutionary theory had philosophical significance. In two important works, William Wimsatt (1981) and Elisabeth Lloyd (1988) have claimed that additivity can be used to give a criterion for deciding the "unit of selection" in an evolutionary process. The units of selection debate began as a biological dispute (Wynne-Edwards 1962, Williams 1966, Lewontin 1970) but it has generated a large philosophical literature, as some of the biological issues seem to involve questions about the metaphysical status of biological entities (Hull 1980), reductionism (Wimsatt 1980) and causal explanation (Sober 1984). If additivity, a simple mathematical concept, could be used to clarify such imposing philosophical questions, this would be progress indeed.

It is necessary first to outline what additive variance is, and where it is found in a population genetics model. In a standard model of selection on viability, we suppose that at some locus the alleles *A* and *a* have frequencies *p* and *q*, where $p + q = 1$, and these genes form genotypes *AA*, *Aa* and *aa*, with fitnesses $W_{AA} = 1$, $W_{Aa} = 1 - hs$, and $W_{aa} = 1 - s$ respectively. With random mating and no other forces affecting a population of effectively infinite size, the frequencies of the genotypes before selection are in Hardy-Weinberg equilibrium. The frequency of genotype *AA*, or $f\{AA\}$, is p^2; $f\{Aa\}$ is $2pq$, and $f\{aa\}$ is q^2. These frequencies are multiplied by the genotypic fitnesses to derive the frequencies after selection, though these new frequencies must be normalized by dividing by $\overline{W}$, which is the total proportion of individuals left after selection, and also the population mean fitness.

PSA 1992, Volume 1, pp. 315-328

(1) $$\overline{W} = p^2 + 2pq(1\text{-}hs) + q^2(1\text{-}s)$$

The new frequency of *Aa*, for instance, is $2pq(1\text{-}hs)/\overline{W}$. From the new genotype frequencies it is trivial to derive the new gene frequencies. The new frequency of *A*, p', is the new frequency of *AA* plus half the new frequency of *Aa*. The rate of change in the frequency of *A*, or Δp, is $p' - p$.

(2) $$\Delta p = pqs(q + h - 2qh)/[1 - qs(2ph + q)]$$

So far we have talked of the fitness of genotypes, but it is possible to derive the same results using the fitnesses of the alleles *A* and *a*, sometimes called "marginal" fitnesses.

(3) $$W_A = pW_{AA} + qW_{Aa}$$

(4) $$W_a = pW_{Aa} + qW_{aa}$$

For in general,

(5) $$\Delta p = pq(W_A - W_a)/(pW_A + qW_a)$$

where the expression in the denominator is another way of writing $\overline{W}$. This formula gives us our first look at additive variance. When Δp is zero, evolution stops. There are two ways for Δp to be zero in formula (5). If p or q are zero, there is no variation in the population. If $(W_A - W_a)$ is zero, then though there may be variation in fitness, none of it is additive.[3] The expression $pq(W_A - W_a)$ is not itself a measure of the additive variance in fitness, but it is close to one.

2. Variance

The concept of variance is also due to Fisher, though it is simply a measure of the "spread" of a distribution and is the square of the standard deviation. The variance of a distribution is the average squared distance of values from the mean:

(6) $$Var(X) = \Sigma Pr(X_i)(X_i - \overline{X})^2$$

where $\overline{X}$ is the mean and the X_i's are the values X can take.

Turning to variance in fitness in a population, $\overline{X}$ is replaced by $\overline{W}$, each X_i is a genotypic fitness, and the probabilities of the X_i's are genotype frequencies.

(7) $$Var(W) = p^2(W_{AA} - \overline{W})^2 + 2pq(W_{Aa} - \overline{W})^2 + q^2(W_{aa} - \overline{W})^2$$

This can only be zero if genotypes have been lost or if the fitnesses are the same. The additive variance is a portion of this total. One way to measure it is as the variance of the marginal fitnesses:

(8) $$Var_{Ad}(W) = p(W_A - \overline{W})^2 + q(W_a - \overline{W})^2$$

It can be seen from (5) that $\overline{W}$ is the average of the marginal fitnesses. (8) reduces to

(9) $$Var_{Ad}(W) = pq(W_A - W_a)^2$$

an expression which resembles the numerator of (5).

There is a graphical way of making the concept of additivity intuitive, which Wimsatt and Lloyd refer to. In Figure 1, the horizontal axis shows genotypes ordered by their "dosage" of allele *A*, from 0 to 2. The vertical axis shows genotypic fitness. The points plotted correspond to the fitnesses described earlier with s=0.25 and h=0 (*A* is completely dominant to *a*).

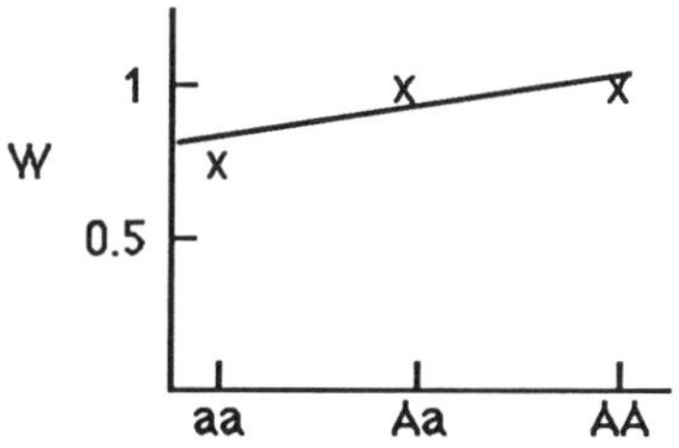

Figure 1

The line through the graph is a "least squares regression line," the straight line which minimizes the (weighted) average squared distance from the points to the line. The slope of such a line (squared and adjusted) measures the additive variance in fitness.[4] The term "additivity" should now become more intuitive; the additive variance measures the extent to which increasing the dosage of allele *A* increases the fitness of the genotype. Deviations from the regression line to the actual points measure the nonadditive variance. A third, equivalent way of thinking of additive variance is as the variance of the genotype fitnesses predicted by the regression line, which may not be the actual genotype fitnesses.[5]

The regression line is weighted by the genotype frequencies. In Figure 1 the line drawn is that for p= 0.5. As genotype *aa* becomes rarer, it exerts less pull on the line, so with time the slope approaches zero and additive variance declines. Given our assumption of infinite population size, even if q is very small there will be some even smaller proportion of *aa* genotypes, so there is always some additive variance while q is not zero. If on the other hand p is small so most of the *A* genes appear in heterozygotes, as long as there are some homozygotes *AA* there will be some nonadditive variance.

So with complete dominance, as long as p and q are nonzero, there is always some additive variance and always some nonadditive variance. Contrast this to the case where h=1/2, so the fitness of *Aa* is exactly between the homozygote fitnesses. Here the three points are on a straight line, and regardless of the (nonzero) frequencies of genotypes there is only one regression line, which hits all three points. Then all the variance in fitness is always additive, and this scheme of fitnesses is often referrred to as "additive." We will call it the case of no dominance.

Consider now a case in which the most fit genotype is *Aa* (heterosis):

$$W_{AA} = 0.75;\ W_{Aa} = 1;\ W_{aa} = 0.5.$$

When allele *A* is rare ($p = 0.1$), it generally occurs in the fit *Aa* combination. The *AA* point exerts little pull on the line, and there is additive variance where *A* is favored—a positive slope (solid line). If allele *a* is rare ($p = 0.9$) the regression line points the other way (broken line). There is additive variance, but favoring a different gene. In both cases the variance has a nonadditive component. At one frequency—$p = 0.67$—

neither gene is favored, and the slope is zero. When the population is at this equilibrium, there is variance in fitness but none of it is additive.

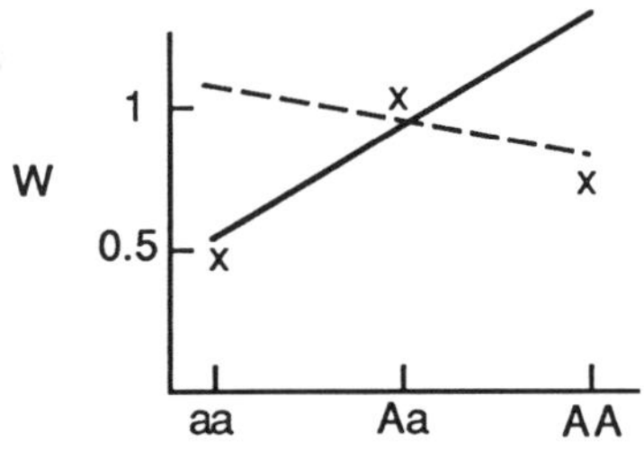

Figure 2

In this framework, additivity is not a property of genic fitnesses, but a property of the genic determination of genotypic fitnesses. Care must be taken here. It is not the marginal fitnesses which are added to get the genotypic fitnesses. Even with no dominance the marginals are frequency-dependent. The quantities added are the "average effects" of the genes.[6] The average effect of allele A is $W_A - \overline{W}/2$ and the average effect of a is $W_a - \overline{W}/2$. The average effects of the genes are added to get the predicted genotype fitness. With constant genotypic fitnesses if there is no dominance the average effects are constant. The average effect of A is 1/2 and that of a is $(1 - s)/2$. Otherwise the average effects are frequency-dependent.

In this discussion "additivity at the genic level" will mean "additivity of the genic determination of genotypic fitness." Asking about additivity at a level is asking about the relation between that level and the next level up. How might we understand additivity at the level of genotypes? Suppose a species consists of one large unstructured population whose individuals interact randomly, a scenario often associated with Fisher. Then we apparently lack a "next level up," at which there is variance in fitness, with respect to which additivity at the genotypic level can be determined. Perhaps the mating pair can be used. Then the problem reappears if we want to ask whether there is selection on mating pairs.

Wimsatt and Lloyd finesse this problem (Wimsatt explicitly, Lloyd implicitly) by having the property of *constancy* of genotypic fitness stand in for additivity some of the time. There is a motivation for this. Suppose a population is divided into groups but there is no group selection and constant genotype fitnesses. The size of the alteration in a group's productivity produced by an alteration in the group's "dosage" of a given genotype should be the same in all contexts. The genotype fitnesses function here as the average effects of genotypes upon group fitness. Different group productivities can be explained away in terms of the additive contributions of constant genotype fitnesses. In an unstructured population, groups could be stipulated arbitrarily, and this same simple relation between genotype fitness and group productivity would result. On the other hand, if there is group selection, then genotypic fitnesses will be frequency-dependent; the same alteration in genotype dosage will make a different alteration to group productivity according to genotypic frequencies. So there is a problem with the determination of the "additivity" of fitness at the genotypic level, though there is not a problem with constancy, and sometimes constancy suffices for additivity.

In summary: additivity is a property of functions or combinations of things. Secondly, measures of additive variance, like variances in general, are frequency-relative. Thirdly, it is hard for there to be no additive genetic variance at all in a scenario. If there is no variance in fitness at all, then there is no additive variance. But if there is variance in fitness, there is total loss of additivity only at equilibrium.

3. Wimsatt

I will now discuss Wimsatt's and Lloyd's appeals to additivity in the units of selection debate. First I will outline a shell common to both views, then discuss ambiguities in this shell, and lastly look to the texts for resolution of these ambiguities.

There are limitations to my discussion. This paper will focus on low-level issues, such as the conditions under which additivity criteria recognise genic selection as opposed to genotypic, in 1-locus systems. For Wimsatt, much of the motivation behind the invocation of additivity concerns the properties of multi-locus systems. In Lloyd's discussion, much of the motivation involves properties of structured populations, and the task of distinguishing different kinds of high level (kin, group, species) selection. Some of these questions are beyond my expertise, and my criticisms in this paper should not be taken to suggest that Wimsatt's and Lloyd's contributions to the analysis of selection in complex evolutionary systems, and the units of selection debate in general, are unimportant. On the contrary, I think both works are very significant advances. But though I may be displaying the additivity criteria in a domain where they are unlikely to shine, the criteria of Wimsatt and Lloyd can be applied to simple cases, and if they are intended as *general* solutions to the units of selection issue, they should give reasonable results in this context.[7]

Our question does not concern the "replicator" in evolution, but the "interactor" (Hull 1980), the type of entity whose interaction with the environment produces the differential replication of replicators. In some situations, fitness values can be assigned to a variety of entities. Entities to which fitnesses can be assigned are *prima facie* interactors, and when a variety of entities bear fitnesses, we face a choice between accepting a multitude of interactors, or recognising some as genuine and dismissing others. Wimsatt and Lloyd suggest that attending to the variance in fitness at different levels will decide the matter. If allelic fitnesses are additive, and suffice to predict the behavior of the system, then there is genic selection only. If allelic fitnesses are nonadditive but genotypic fitnesses are additive, there is selection on genotypes. If genotypic fitnesses are nonadditive but group fitnesses are, we have group selection.

	Frequencies: All	Frequencies: Some
Variance: All		
Variance: Some		

Table 1

This is a preliminary outline, but I will use it to present an ambiguity seen in more complex additivity criteria. Do these criteria demand that all variance at a level is additive, or just that some is? And must this additive component exist at all frequencies of the entities in question, or only some frequencies? So there are four distinct interpretations of the expression "there is additive variance at this level":

It should be apparent from the previous section that the claim that all variance at a level is additive is very strong, while the claim that some variance is additive is weak. Strictly, table 1 does not exhaust the possibilities — it omits criteria demanding "most" variance at "many" frequencies and other intermediate cases. Consideration of these would complicate matters enormously, and some of my arguments are intended to make such options generally unappealing.

In what follows, the claim that some or all variance is additive at "all" frequencies will mean that this variance is additive at all frequencies where there is variance in fitness at all.

We will begin with Wimsatt's criterion:

> A *unit of selection* is any entity for which there is heritable *context-independent* variance in fitness among entities at that level which does not appear as heritable context-independent variance in fitness (and thus, for which the variance in fitness is *context-dependent*) at any lower level of organization. (1981, p.144)

Wimsatt outlines the relation between context-independence and additivity in the following paragraph. Consider the case where the fitnesses of the three genotypes were 1, 1-(s/2), and 1-s. There is one regression line and all variance is always additive. Then the effect on fitness of substituting one allele for another in a genotype is context-independent; it is the same whether the the substitution takes us between *aa* and *Aa* or between *Aa* and *AA*, and it is the same regardless of genotype frequencies. So Wimsatt says: "Additivity is thus a special case of context-independence" (p.144).

The ambiguity outlined above is fostered by Wimsatt's switching between talk of context-independence and talk of additivity. One might think "context-independence" is an all-or-nothing affair. Additivity, however, is generally a property of part of the variance in fitness only. One would also not expect that context-independence would come and go with changes in frequency; context-independence should not be context-dependent, as it were. But additivity does come and go in this way. These considerations seem to place Wimsatt's criterion in the upper left of table 1. On this criterion, in a diploid system there is genic selection as opposed to genotypic only if the fitness of the heterozygote is exactly midway between those of the homozygotes. Any dominance elevates the unit of selection to the genotypic level.

The "all variance at all frequencies" option is a demanding criterion. Lloyd finds it too strong, and does not think Wimsatt holds it. Whether or not it is Wimsatt's view, it is worth discussing as it casts light on the relation between additivity and "context-dependence," and also has hidden resources.

Wimsatt claims that additivity is "a special case of context-independence." This suggests that additivity of fitnesses implies context-independence. But let us consider a well-known biological case often regarded as genic selection—segregation distortion by the *t* allele in mice (Lewontin 1962).[8] In this case male heterozygotes produce a larger proportion of sperm carrying one allele *t* than the other *T*. The fitnesses of the genotypes, and the genes, are frequency-dependent. If mating is random and *m* represents the proportion of the *t* allele found in the gametes of *Tt* males:

(10) $$W_{TT} = [f\{TT\} + f\{Tt\}(1 - m)]/p$$

(11) $$W_{Tt} = [(f\{Tt\}m + f\{tt\})(p - q) + q]/2pq$$

(12) $W_{tt} = [f\{tt\} + f\{Tt\}m]/q$

If "context-dependence" is frequency-dependence, this is not a context-independent scheme for the contributions of alleles to genotypic fitnesses.[9] But this context-dependence does not imply that variance is nonadditive. In fact, in this system all the variance is additive at all frequencies; the three points are always on a straight line. The slope of the line increases with $f\{Tt\}$, as the number of male heterozygotes determines the change in gene frequencies in a generation.

On the "all variance at all frequencies" criterion frequency-dependent genic selection of this sort can be recognised as genic selection. I, for one, was surprised that perfect additivity would emerge from formulae as disorderly as the fitnesses above. This result certainly supports the additivity approach.

There are other cases, however, in which the "all variance at all frequencies" criterion does not produce intuitive results. In the segregation distortion case, different regression slopes are produced at different frequencies but the slope has the same sign each time. If m is greater than 1/2 the slope is always negative; if m is less than 1/2 it is positive. The direction of selection is always the same, though its strength changes. Now consider a simpler scheme of frequency-dependent selection:

(13) $W_{AA} = f\{AA\}$

(14) $W_{Aa} = 1/2$

(15) $W_{aa} = 1 - f\{AA\}$

As in the previous case, all the variance is always additive; the three points are always on a straight line. Unlike the previous case, the *sign* of the regression slope changes with frequencies — the line "see-saws." Though the fitnesses depend on a genotype frequency, and though there is context-dependence with respect to which gene is favored, on the "all variance at all frequencies" view this should be understood as genic selection, because there is never nonadditive variance at the genic level. This is a counter-intuitive result, but it is uncertain how much force it has. First, the case is artificial. Secondly, the criterion could be supplemented with a clause requiring that the sign of the regression slope not change with frequencies.

There is a more general problem with the "all variance at all frequencies" approach however. Before we come to it, we must examine Wimsatt's stated reason for using the expression "context-independent" rather than "additive." The explanation is found in a footnote: "Speaking of additive variance implies the context of a larger unit in which more than one of the smaller units which contribute to fitness will co-occur, so that their contributions will add" (p.177). This is important. As noted earlier, the genotype is an obvious unit with respect to which the role of alleles can be assessed. If a population has no structure above the level of the individual then it seems we can note that genotypic fitnesses might be constant, or frequency-dependent, but there is no way to assess their "additivity." Constancy (context-independence) can stand in for additivity, at the genotypic level, but then there is no avoiding the result that in an unstructured population there cannot be frequency-dependent selection on genotypes, such as the advantage of a rare type, unless the nonadditive variance is always zero. Suppose the fitnesses are as follows:

(16) $W_{AA} = 1$

(17) $W_{Aa} = 1$

(18) $W_{aa} = 1 - f\{aa\}$

Here there is no frequency at which all variance is additive, though as long as there is some variance in fitness it will have an additive component. I am not sure how the "all variance at all frequencies" option classifies this case. Fitnesses are not constant, so it is not selection on genotypes. Fitnesses of mating pair types are not constant either. But neither is it group or genic selection, as there are no groups and there is not total additivity at the genic level. So this simple case is bad trouble for the criterion in the top left cell of table 1.

With constant genotypic fitnesses, if there is total additivity at some frequencies then there is total additivity at all frequencies. But if genotypic fitnesses are frequency-dependent, then at some frequencies all the variance can be additive while at others there is a nonadditive component.[10] This creates the distinction between the left and right cells in the top row. Schemes can even be devised in which all the variance is additive at some frequencies while there is *no* additive variance at others:

(19) $W_{AA} = f\{\mathrm{AA}\}$

(20) $W_{Aa} = 1/2$

(21) $W_{aa} = 1/4$

When $f\{AA\}$is 3/4, the fitnesses are perfectly additive. If $f\{AA\}$ is lower the additive component of variance is smaller. With f{AA} at 1/4 and Hardy-Weinberg equilibrium, we have a situation resembling heterosis and no additive variance at all. If genic selection exists when alleles determine genotypic fitnesses with perfect additivity, then there is genic selection in this case when f{AA} = 3/4, but selection on genotypes when the frequencies are different—a change in frequencies can produce a *switch* in the unit of selection. This is an unappealing result.

I will now turn to Lloyd's position. This is not because I take myself to have disposed of Wimsatt's, but because from this point we will be concerned with views in the second, "some variance" row of the table. Lloyd's discussion locates her more clearly in this row. It may be that Wimsatt's real view is found here too—perhaps he does not require that the size of fitness differences be context-independent, but just the direction of the inequality, the sign of the regression. However, at several points Wimsatt makes observations about additivity which some of my own arguments against the weaker, "some variance" criteria will resemble. Wimsatt claims, for instance, that additive variance at low levels of organization may often be not an "intrinsic" property of low-level processes of selection, but an "emergent" property of higher-level processes (1981, p.142). This is right, and as an instance of this, additive variance at the genic level need not be a sign of genic selection, as it can be a byproduct or "artifact" (Sober 1981, Sober and Lewontin 1982) of higher-level selection. It is a demanding condition that there *not* be additivity at low levels, if there is selection at higher levels. Thus Wimsatt's own arguments tell against weaker versions of the additivity criterion.

4. Lloyd

Lloyd's proposal is outlined explicitly in terms of additivity:

> *Additivity Definition:*
> A unit of selection is any entity-type for which there is an additive component of variance for some specific component of fitness, F^*, among all entities within a system at that level which does *not appear* as an additive component of variance in (some decomposition of) F^* among entities at any lower level. (1988, p.70)

Lloyd claims that, properly interpreted, Wimsatt's view is equivalent to her own (p.180). But Lloyd's definition, unlike Wimsatt's, seems clearly to be a "some variance" claim. This is seen in the phrase "an additive component." So she is apparently located somewhere in the second row of table 1. Further, though the definition does not spell this out, there is evidence that she requires additivity at all rather than some frequencies. She demands that a search for additivity include "consideration of a *range* of population compositions and environments" to reduce the chance of "misleading results" (1988, p.75). Perhaps the case I used against the "all variance at some frequencies" option, in which perfect additivity exists at only some frequencies, is a case of this kind.

This evidence is not enough to place Lloyd squarely in the lower left cell; perhaps "a range" of frequencies need not be "the *entire* range." Further, there is a footnote to her discussion which affects interpretation, which we will come to shortly. Of the options I am considering, however, the "some variance at all frequencies" option seems closest to the spirit of her work.

Earlier I agreed with Wimsatt that there is a problem with the concept of additivity at the genotypic level if there are no higher-level units in the system. Lloyd does not discuss this. In what follows I will continue to understand "additivity at the genotypic level" as constancy of genotype fitnesses. Lloyd does say "in the usual cases of frequency-dependence [of genotype fitnesses] there is nonadditive variance in fitness at the level of the individual organism" (p.76). If genotypic fitnesses are constant, this should be sufficient for "additivity," in whichever sense Lloyd intends.[11]

Before criticizing the "some variance at all frequencies" option we will examine the weakest "some variance at some frequencies" cell, to introduce the argument of this section. On this view, there is genic selection in a heterotic system, and with dominance, and in many cases of frequency-dependent selection. There is also selection on genotypes with heterosis or dominance. In fact, on this view most of the standard schemes of fitnesses involve *multi-level* selection. This is because there is an additive component at the genic level, and also *total* additivity (constancy) of the fitnesses of genotypes. Both levels have their own "additive" components, and both levels are then recognised as levels at which selection occurs. Selection at multiple levels does exist, but it is a special type of case, involving the interaction of several forces. On the "some variance at some frequencies" criterion, a multi-level verdict is brought in whenever a single force produces additive variance at more than one level.

The same argument applies to the cousin of this weakest view. Demanding that additive variance exist at all frequencies does not strengthen the criterion enough to deal with low-level additivity which is an artifact of higher-level selection. On the "some variance at all frequencies" view heterosis does not involve genic selection, as there is a frequency at which there is variance in fitness none of which is additive. Heterosis involves genotypic selection only. However, both partial and complete dominance must be regarded as multi-level, as there is additive variance at the genic and genotypic levels. So the "some variance at all frequencies" criterion also recognizes spurious cases of multi-level selection.

It is worth mentioning a real case of multi-level selection, to contrast it with the spurious ones. Earlier I discussed the mouse *t* allele, in which there is frequency-dependent genic selection. The *t* allele exists with its rival *T* in natural populations because the genic selection favoring *t* is balanced by viability selection against *tt* homozygotes. According to Lewontin (1962) there is a third force as well; group selection also works against *t*. This is a situation in which a multi-level selection verdict is appropriate. The problem with the "some variance" additivity criteria is that they see multi-level selection in too many other cases as well.

The "some variance at all frequencies" option is a relative of the view in Sober 1984 (Chapter 9). Sober finds genic selection if increasing the dosage of *A* alleles in a genotype raises fitness in at least one context and does not lower it in any. So heterosis involves genotypic selection but dominance may be understood as genic selection. The nonlinearity associated with dominance is considered insignificant—a change in the dosage of *A* always makes the same *kind* of difference. But the change in the regression's sign seen with heterosis is significant—increasing the dosage of *A* can raise or lower fitness, depending on context. Sober's view does not, however, regard dominance as multi-level. Sober sees the action of genes as causally "fundamental" in these cases (p.313).[12] Earlier I noted an interpretation of Wimsatt which holds that what must be "context-independent" is not the magnitude of fitness differences but the sign of these differences. On this view, Wimsatt is with Sober in seeing dominance as genic selection. On the stronger reading of Wimsatt, in which the magnitude of fitness differences must be constant, Wimsatt finds dominance to involve genotypic selection. It is only the "some variance at all frequencies" view which brings in a multi-level verdict here.

At this point we must discuss a complication in Lloyd's presentation, found in a footnote to her definition.

> "Additivity," in the additivity definition, should be understood as shorthand for "transformable into additivity," for the rest of this book.... Nonadditivity can arise from non-monotonicity or from simple non-rectilinearity; only the former relation is of interest here. (1988, p.180)

The solution, Lloyd says, is to allow as additive schemes of fitness which can be transformed into additive ones. This is apparently a major alteration to Lloyd's definition, but I will discuss it only briefly, as a "transformed additivity" criterion is probably less plausible than some of the untransformed criteria discussed already.

If the fitness of the heterozygote is between but not midway between the fitnesses of the homozygotes, it should be possible to find a re-scaling of the vertical axis which will produce a straight line between the three points. So if W_{Aa} is some small degree below W_{AA} and far above W_{aa}, then if the vertical axis is changed to represent something like the antilog of fitness, the differences between high values will be exaggerated relative to differences between low values. The small difference between W_{AA} and W_{Aa} can be swollen and the difference between W_{Aa} and W_{aa} shrunk until the three points lie on a straight line. Thus partial dominance can be transformed into additivity. This is not true of complete dominance however; if W_{AA} and W_{Aa} are exactly equal, there is no transform that will distinguish them, dragging W_{Aa} down to make a straight line. So suppose the transformed additivity criterion reads: there is selection at a level if all the variance can be transformed into additive variance at that level at all frequencies. Then we have the consequence that partial dominance results in genic selection but total dominance results in selection on genotypes. This is the probably the most artificial distinction made by any criterion we have considered.

The same result follows if the criterion demands total transformed additivity at some frequencies. Alternately, if the transform idea is applied to the "some variance" criteria, no difference is made at all. If some of the variance is transformable into additive variance, then some of the variance *is* additive, in the usual sense. All in all, appealing to transforms appears unlikely to take us further towards an acceptable criterion. An appeal to transforms also weakens the link between additivity criteria and the additivity of Fisher's theorem. Wimsatt and Lloyd appeal to this link for support, and Fisher's result is not a result about transformed additivity, but additivity simpliciter.

It might be helpful to summarise some consequences of the central criteria I have discussed. Asterisks signify an unwelcome result or problematic case.

Variance	Frequencies: All	Frequencies: Some
All	No dominance - genic Seg. distortion - genic Dominance - genotypic Heterosis - genotypic *Rare advantage with dominance?	No dominance - genic Seg. distortion - genic Dominance - genotypic Heterosis - genotypic *Unit can change with frequencies.
Some	No dominance - genic Seg. distortion - genic *Dominance - multi-level Heterosis - genotypic	No dominance - genic Seg. distortion - genic *Dominance - multi-level *Heterosis - multi-level *Unit can change with frequencies.

Table 2

Different versions of the additivity criterion lead to different verdicts in specific cases, and though some cases, such as segregation distortion, are dealt with impressively, all the versions I have considered have some unattractive consequences. On my own view, dominance should be regarded as genotypic, as top-row positions hold, but this must be reconciled with the possibility of frequency-dependent selection on genotypes, which bottom-row views deliver. These constraints are hard to satisfy simultaneously, as dominance affects the gene-to-genotype relationship in the same way frequency-dependence affects the genotype-to-group relationship: both make the average effects frequency-dependent.[13]

The additivity criteria I have discussed are designed to deal with spurious additive variance at a level higher than the real level of selection, but because of the bottom-up way in which these criteria are applied, additive variance at the genic level cannot easily be dismissed as spurious; there is no lower level which can explain it away.[14] My suspicion is that additivity is the wrong kind of concept to fashion into a criterion for a unit of selection, but I would not claim to have demonstrated this here. I have considered a restricted set of cases, and the significance of some of these could be questioned. In addition, there are no doubt other ways to refine the appeal to additivity in this debate, and the advance that would result from a successful analysis along these lines would be very substantial. I do think the following is clear: if an additivity criterion is to work, it must operate in such a way that sometimes additive variance can be recognized as a byproduct, when it appears above *or below* the level of the real unit of selection.

Notes

[1]I am very grateful to Sarah Otto for advice and comments. Thanks also to Richard Lewontin and Elisabeth Lloyd for helpful discussions.

[2]Fisher said simply "genetic variance" in his 1930 presentation (p.35).

[3]If $\overline{W}$ (the denominator) is zero the expression is undefined—everyone is expected to die before reproducing.

[4]The slope of the regression squared and divided by twice the variance in the number of *A* alleles possessed by individuals (4*pq*, conveniently) equals the additive variance in fitness as defined by (8). Equivalently, the square of the correlation $r^2 = 2Var_{Ad}(W)/Var(W)$. This formula relates the additive variance to the amount of variance "explained" by allele dosage (r^2).

[5]This is formula 9.30 on p.153 of Roughgarden 1979. Like Wimsatt, I recommend Roughgarden's presentation of these concepts.

[6]See Falconer 1981, pp.115-16. Roughgarden's "χ_1" and "χ_2" are average effects (1979).

[7]Brandon (1982, p.318) may go astray here. He says the empirical facts Wimsatt's rejection of genic selection requires are epistatic interactions among genes and linkage. If "bean-bag genetics" were true—independent assortment and one-gene-one-trait—Wimsatt would be a genic selectionist. But these are all multi-locus properties (unless Brandon includes dominance as "epistasis"), and it is not true that Wimsatt is forced to regard selection as genic whenever loci do not interact with each other.

[8]There are some, including Lewontin, who do not regard this as obviously genic selection. Godfrey-Smith and Lewontin (forthcoming) present a dynamically identical hypothetical case with a more clearly genic selectionist structure. Suppose the advantage enjoyed by *t* is the result of different sperm swimming speeds, rather than proportions.

If this causal difference makes a difference to the units of selection, then one dynamical model can represent two selective processes at different levels. This would be a fast argument against structuralist analyses of the units of selection, but I do not, myself, think the causal difference between these cases is significant.

[9]These are derived using the "method of weights" of Lewontin 1958.

[10]Sarah Otto suggested this line of argument to me.

[11]Lloyd needs constancy of genotypic fitnesses to not be necessary for additivity, or she would not be able to recognise frequency-dependent selection on genotypes, which she wants to recognise (p.77).

[12]Sober has other complaints about genic fitnesses: "[t]he allelic fitnesses vanish as soon as they appear, since they are only temporary reflections of the changing frequencies of the genotypes in which the alleles appear" (1984, p.311). However, this is true of marginal fitnesses (unlike average effects) even if there is perfect additivity at the genic level.

[13]Sober's 1984 view, which sees dominance as genic selection, would be rejected by these constraints.

[14]Lloyd herself notes a "bias" in her criterion towards seeing selection at the lowest levels (1989, p.412n).

References

Brandon, R. (1982), "The Levels of Selection", *PSA 1982*, volume 1. East Lansing: Philosophy of Science Association, pp. 315-333

Fisher, R.A. (1918), "The Correlation between Relatives on the Supposition of Mendelian Inheritance", *Transactions of the Royal Society of Edinburgh* 52: 399-433.

Fisher, R.A. (1930), *The Genetical Theory of Natural Selection*. Oxford: Clarendon Press.

Falconer, D.S. (1981), *Introduction to Quantitative Genetics*, 3rd Edition. New York: John Wiley and Sons.

Godfrey-Smith, P., and R.C. Lewontin (forthcoming), "The Dimensions of Selection."

Griesemer, J. and M. Wade (1988), "Laboratory Models, Causal Explanation and Group Selection", *Biology and Philosophy* 3: 67-96.

Hull, D. (1980), "Individuality and Selection", *Annual Review of Ecology and Systematics* 11: 311-332.

Lewontin, R.C. (1958), "A General Method for Investigating the Equilibrium of Gene Frequency in a Population", *Genetics* 43: 419-434.

Lewontin, R.C. (1962), "Interdeme Selection Controlling a Polymorphism in the House Mouse", *American Naturalist* 96: 65-78.

Lewontin, R.C. (1970), "The Units of Selection", *Annual Review of Ecology and Systematics* 1: 1-18.

Lewontin, R.C. (1991), Review of Lloyd 1988. *Biology and Philosophy* 6: 461-466.

Lloyd, E. (1989), "A Structural Approach to Defining Units of Selection", *Philosophy of Science* 56: 395-418.

Lloyd, E. (1988), *The Structure and Confirmation of Evolutionary Theory*. New York: Greenwood Press.

Roughgarden, J. (1979), *Theory of Population Genetics and Evolutionary Ecology: An Introduction*. New York: Macmillan.

Sober, E. (1981), "Holism, Individualism, and the Units of Selection", *PSA 1980*, P. Asquith and R. Giere (eds.), Volume 2. East Lansing: Philosophy of Science Association, pp. 93-121.

_____. (1984), *The Nature of Selection*. Cambridge: MIT Press.

_____. and R.C. Lewontin (1982), "Artifact, Cause and Genic Selection", *Philosophy of Science* 49: 157-80.

Williams, G.C. (1966), *Adaptation and Natural Selection*. Princeton: Princeton University Press.

Wimsatt, W.C. (1980), "Reductionist Research Strategies and their Biases in the Units of Selection Controversy", in *Scientific Discovery: Case Studies*, T. Nickles (ed.). Dordrecht: Reidel.

Wimsatt, W.C. (1981), "Units of Selection and the Structure of the Multi-Level Genome", *PSA 1980*, volume 2. East Lansing: Philosophy of Science Association.

Wynne-Edwards, V.C. (1962), *Animal Dispersion in Relation to Social Behavior*. Edinburgh: Oliver and Boyd.

Jacques Monod's Scientific Analysis and Its Reductionistic Interpretation

Spas Spassov

University of Montreal

Ever since the publication of his biophilosophical ideas in the early seventies, Jacques Monod has always been criticized quite severely for his old-fashioned philosophical reductionism. These numerous critiques have created a picture of an eminent scientist whose philosophical considerations are founded on the mechanistic world-view of Cartesian metaphysics, a view which is completely unacceptable to contemporary philosophers. Today it is the usual practice to cite Monod as one of the strongest and most radical reductionists in modern biology.

Monod's biophilosophical conception arose at a moment when vigorous discussions on reductionism followed the great successes of molecular biology. His ideas were criticized namely for their strongly reductionistic interpretation of biological knowledge, and for their mechanistic model of life which, as Monod believed, reflected the spirit of modern science. In the claim of the French scientist that contemporary biology has eliminated the possibility of any vitalistic or animistic interpretation of life, his critics revealed the echoes of a Cartesian approach which stipulated that vitalism and mechanism are the only possible alternatives.[1] This reductionistic methodology, grounded on Cartesian metaphysics, has led Monod to the conclusion that in vital phenomena there is nothing but molecules in motion, and to the rejection of the existence of different levels of organization, each with their specificity.

I do not intend here to discuss the extent to which these critiques are fair or exaggerated. I would rather distinguish between Monod's philosophical interpretation of biological phenomena, and the scientific analysis upon which it seems to rest. These two aspects of Monod's model have to be considered separately in order to reveal the non-necessary character of the relationship between them. My idea is to show that the philosophical interpretation Monod gives to the results of his scientific analysis is not the only possible or necessary one. In this way, one can better understand the meaning and the significance of his biophilosophical ideas, in spite of the controversies and the critiques it has raised. In fact, in his scientific analysis one can find an idea the real meaning of which both Monod and his critics have overlooked. This idea does not support his own mechanistic interpretation, but rather presupposes a different approach to the understanding of the phenomenon of life.

PSA 1992, Volume 1, pp. 329-334

The main aim of Monod's biophilosophical concept was to show that it is possible to give a full interpretation of the phenomenon of life, and first of all of its teleonomic properties, in terms of the principle of objectivity, which, as he believed, is consubstantial with modern science. One of the conclusions Monod derived from his analysis is that the understanding of the molecular mechanisms of living organization gives us the key to understanding the nature of teleonomy, in liberating it of all finalistic or subjectivistic implications. His thesis rests on the claim that the essence of this characteristic, that is the origin, the accomplishment, and the evolution of teleonomic structures and functions, lies in the specific properties and interactions of several biomolecules, and that it is possible to analyze and completely describe it in terms of these macromolecules. As a consequence, if we know the mechanism of the interactions between the macromolecules, as well as the chemical nature of their properties, we will be able to reveal the "secret" of teleonomy.

The molecular agents of teleonomic functions in the unicellular organisms are the proteins, which channel the chemical processes in the cell, assure their coherence, and build the biological structure. These three main teleonomic functions of proteins ultimately rest upon their stereospecific properties which consist in their capacity to recognize other biomolecules by their shape determined by their molecular structures. In other words, it is possible, in principle, to analyze all the teleonomic structures and functions of organisms in terms of this capacity of choice and elective discrimination of proteins (or, using Monod's expression, of their almost "cognitive" capacity), which is based on their stereospecific interactions. As a consequence, as the capacity of stereospecific recognition of the proteins depends on their three-dimensional forms and, therefore, on their structures, if one is able to describe the origin and the evolution of these structures, one can explain the origin and the development of the teleonomic activity linked to the protein (Monod 1970, pp.60-61).

The molecular mechanisms of teleonomic functions in the unicellular organisms are now accessible, thanks to several discoveries of molecular biology (and Monod himself played an important role in this progress). The principle of these mechanisms rests on the allosteric interactions between certain proteins, called regulatory, which are responsible for the coherence of the chemical activity in the cell. It is recognized that allosteric interactions play a fundamental role in the regulatory processes and are the most characteristic and essential components of cellular control systems. [2]Generalizing their role, Monod insists that it is impossible to understand the origin and development of regulatory systems in living beings without the idea of indirect interactions operating by the mediation of the allosteric enzyme (Monod 1970, 85).

I will not set out here the mechanism of allosteric interactions which have been very well described and analyzed, especially in several articles published in the early sixties by Monod in collaboration with F. Jacob and J.-P. Changeux. What is essential for their interpretation is the fact that these interactions rest entirely on the differential capacity of proteins in their different allotropic states to differentiating other biomolecules and to joint them by forming noncovalent stereospecific complexes. The interactions between ligands which prompt or inhibit the activity of the allosteric enzyme, as well as those between these ligands and the substrate of the enzyme, are totally indirect. Actually, each one of the ligands as well as the substrate interacts separately only with the allosteric protein. As a result, thanks to these properties, the proteins involved in these interactions are in indirect relation with each other, a relation the nature of which is not chemical, but which assures the transmission of different chemical signals:

> It would appear, in other words, that certain proteins, acting at critical metabolic steps, are electively endowed with specific functions of regulation and coordina-

> tion: through the agency of these proteins, a given biochemical reaction is eventually controlled by a metabolite acting apparently as a physiological "signal" rather than as a chemically necessary component of the reaction itself. (Monod, J., Changeux, J.-P. and Jacob, F. 1963, 306)

The chemical independence between the substrate of an allosteric enzyme and the ligands which influence its activity means that there is a chemical independence between the enzyme reaction and the character of the chemical signals which regulate it:

> The primary reason for considering allosteric proteins as essential and characteristic constituents of biochemical control systems is their capacity to respond immediately and reversibly to specific chemical signals, effectors, *which may be totally unrelated to their own substrates, coenzymes or products.*
> We have discussed several examples which illustrated this point (and need not to be recalled here), leading us to the paradoxical conclusion that the structure and *sui generis* reactivity of an allosteric effector is "irrelevant" to the interpretation of its effects. There remains no real chemical paradox, once it is recognized that an allosteric effect is indirect, being mediated entirely by the protein and due to a specific transitions of its structure. (Monod, J., Changeux, J.-P. and Jacob, F. 1963, 324)

In other words, the effect of the allosteric interactions is totally independent of the structure of the ligands because the structures of the allosteric enzyme in its different states determine it. The allosteric interactions are indirect, for they proceed by the virtue of the discriminatory property of the protein in the various states it is able to adopt, to recognize in a stereospecifical way other molecules. As a result, there is not any chemically necessary relation, structural or reactive, between the substrate of the allosteric enzyme and the ligands which influence its activity. This particularity of the allosteric interactions explains the directedness of the protein activity and the coherence of the chemical reactions in the cell, as well as the regulation of the process of macromolecular synthesis. What is more interesting is that Monod emphasizes the extreme biological significance of these mechanisms:

> It is hardly necessary to point out the critical role, indeed the physiological necessity, of such metabolic interconnections. ...
> The *absence* of any inherent obligatory chemical analogy or reactivity between substrate and allosteric effector appears to be a fact of extreme biological importance. ...
> Still, the arbitrariness, chemically speaking, of certain allosteric effects appears almost shocking at first sight, but it is this very arbitrariness which confers upon them a unique physiological significance, and the biological interpretation of the apparent paradox is obvious. (Monod, J., Changeux, J.-P. and Jacob, F. 1963, 306, 307, 324)

Another effect of the allosteric interactions with great biological importance, according to Monod, is the fact that they offer to molecular evolution almost limitless possibilities of exploration, and thus make possible the creation of complex systems of cybernetic interactions in living organisms. In fact, these control systems could not operate, and therefore could not have evolved, if their elementary mechanisms were restricted to direct chemical interactions. This restriction, however, is abolished in the allosteric interactions where there is no direct interaction between effector and substrates of the reaction, but only between each of them and the allosteric protein separately. As a consequence, an indirect interaction is accomplished between different metabolites without chemical affinity toward each other, an interaction which allows that the reaction becomes finally subordinated to the intervention of compounds chemically foreign and indifferent to this reaction. These compounds, therefore, can become subjects of a selection which operates through their effects upon the coherence and the functional efficacity of the organism (Monod, J., Changeux, J.-P. and Jacob, F. 1963, 324-325; Monod 1970, 91).

The biochemical mechanisms of these processes, revealed by molecular biology, serve as the basis of Monod's general conclusion concerning the teleonomic properties of living beings. The essential point in his conclusion is the idea that teleonomy rests, ultimately, on the discriminatory associative stereospecific properties of the proteins, and that we can analyze and describe it entirely in terms of these properties. As long as the steric structures of the molecules are responsible for their associative stereospecific properties, the "secret" of teleonomy lies in these spatial structures; therefore, to reveal this "secret", we must describe the mechanisms of formation and evolution of these structures (Monod 1970, 105). Now, the processes of formation and evolution of protein structures are well known. The mechanism of the molecular epigenesis is a distinct chemical process. According to Monod, there is nothing in it which can not be explained totally in physical and chemical terms. The determinism of the molecular epigenetic process, and therefore of the autonomous morphogenetic development of the organism, comes from the genetic information inscribed in the polypeptide sequences, screened by the initial conditions. Consequently, the teleonomic properties of living beings find their *ultima ratio* in the sequence of residues making up the polypeptide fibers:

> In a sense, a very real sense, it is at this level of chemical organization that the secret of life lies, if indeed there is any one such secret. And if one were able not only to describe these sequences but to pronounce the law by which they assemble, one could declare the secret penetrated, the *ultima ratio* discovered. (Monod 1971, 95-96)

This is the law of chance which molecular biology has revealed.

As we can see through this summary, Monod's fundamental thesis advances the idea that the very nature of teleonomic performance lies in the chemical specificity and in the molecular mechanisms of certain processes in the cell. Moreover, as the teleonomy represents a fundamental characteristic of a living organization, we can reveal the "secret" of life if we describe the molecular mechanisms of its fulfillment. This idea lies at the base of Monod's philosophical interpretation of teleonomy, and therefore, of his mechanistic view of the phenomenon of life. In the biochemical analysis of the molecular mechanisms through which teleonomic performance is accomplished, he believes that he has found the irrefutable scientific foundation of his biophilosophical thesis. Monod's mechanistic interpretation of life, however, is completely arbitrary in respect to the scientific analysis which underlies it. This analysis, instead of supporting the mechanistic view, suggests quite a different philosophical interpretation.

In general, teleonomic performance is accomplished through certain oriented processes which take place between the biomolecules, and in order to define their specificity with regard to other oriented chemical processes, we have to reveal the reason and the mechanisms of this orientation. Monod bases the explanation of the mechanisms of teleonomic performance on the molecular level, on the principle of allosteric interactions. In his scientific analysis he shows that this orientation is due entirely to the discriminatory properties of the proteins within the noncovalent stereospecific complexes, properties which lead to the establishment of a new relation, of a nonchemical nature, between these molecules. This relation between the proteins, which characterizes all allosteric interactions, represents a new type of relation between the biomolecules; its nature is nonchemical, in spite of the fact that it is accomplished through chemical interactions and can be explained entirely by these interactions. Its result leads to precise physiological effects concerning the conditions of survival of the cell. This new relation distinguishes the allosteric interactions from the other chemical processes between the biomolecules. We can say that the essence of the property called teleonomy lies in the specificity of the chemical interactions between the proteins, *and* in the new relation coming into being between the biomolecules in the process of their interactions, a relation whose nature is not chemically determined, but is chemically arbitrary or physiological.

For the philosophical interpretation of teleonomy this means that the teleonomic performance rests on this new relation between the proteins (which plays the role of a retroactive connection in the regulation of the chemical processes in the cell), and consequently, it is a totally "natural" or "objective" phenomenon. Teleonomy, interpreted in this way, receives its scientific explanation and its status as objective phenomenon without the necessity of invoking any vitalistic principle. At the same time, however, it is not a mechanistic interpretation in the philosophical meaning of the word, that is to say an interpretation which denies any specificity of the teleonomic performance as a biological phenomenon with respect to the ordinary chemical processes. In fact, in this interpretation the essence or the secret of teleonomy lies not simply in the specific chemical interactions between the proteins, but in the new relation of a nonchemical nature between the macromolecules (which are without mutual chemical affinity), a relation which comes into being in the course of their activity and which is something different, or to be more precise, something more, with regard to the ordinary chemical processes.

In this interpretation, the teleonomic properties are not reduced to stereospecific interactions between the proteins determined by their structures. In other words, teleonomic performance is not considered to be an ordinary chemical process, in spite of the fact that it rests on such processes, entirely determined by the structural stereospecific properties of the biomolecules, and that it can be explained by them. This interpretation, which is implicit in the context of Monod's scientific analysis, emphasizes the specificity of teleonomic performance, that is, the new quality which distinguishes it from other oriented chemical processes in nonliving matter. To put it another way, the specific nature of teleonomic processes of regulation in the cell arises from the new relation which comes into being between certain chemical compounds and is different from all chemical interactions between these compounds. This relation determines the biological functioning of the cell, with regard to its adaptation to the surrounding conditions. It rests entirely on and is realized through the stereospecific interactions between the biomolecules, interactions which are totally chemical; but its criterion is the survival of the cell, and its result is the adaptation of the cell to the environment, therefore it is a physiological relation which demonstrates the biological specificity of the regulatory biomolecular processes.

We can say that, in his scientific analysis of the molecular mechanisms of teleonomic properties, Monod reveals the specificity of these processes, which does not allow for their reduction to ordinary chemical interactions between the biomolecules. However, in his interpretative conclusions Monod identifies the biological specificity of the teleonomic performances with their chemical mechanisms, and as a result he gives them an inadequate mechanistic interpretation. His statements concerning the possibility of analyzing the teleonomic properties of the cell entirely in terms of stereospecific interactions between the proteins — interactions which are determined by the structures of these molecules — do not take into account their specific nature (which is nevertheless revealed in his scientific analysis). These assertions, which emphasize that the teleonomic processes rest, ultimately, on the stereospecific properties of the biomolecules, are quite true in the level of scientific analysis. Yet, for the philosophical interpretation it is not sufficient to say that the "ultimate source" of the teleonomy characterizing living beings in their performances lies in the molecular structures of the regulatory proteins. What is essential for this interpretation is the fact that out of the processes of regulation arises a new type of interaction between the biomolecules, an interaction of a nonchemical nature which governs and gives an orientation to the chemical reactions between these molecules, thus making possible the physiological functioning of the cell. In his mechanistic interpretation of these processes, Monod remains on the level of their chemical bases, neglecting the biological specificity which springs from these bases, and which is implicitly revealed in his scientific analysis. This incoherent approach is the source of Monod's mechanistic conception, and if we put his views in the right light, we can find in them at least the suggestion of a quite different philosophical interpretation of life.

Notes

[1]One of the main conclusions of the articles in the book *Beyond Chance and Necessity. A Critical Inquiry into Professor Jacques Monod's Chance and Necessity*, insists on the philosophical anachronism of Monod's conception.

[2]The idea that allosteric interactions play particular role in the regulatory systems of the cell, is the main subject of the article of Monod, J., Changeux, J.-P. and Jacob, F. (1963), "Allosteric Proteins and Cellular Control Systems", *Journal of Molecular Biology*, 6.

References

Lewis, J. (ed.) (1974), *Beyond Chance and Necessity. A Critical Inquiry into Professor Jacques Monod's Chance and Necessity*. New Jersey: Humanities Press.

Monod, J. (1970), Le Hasard et la nécessité. Paris: Seuil.

______. ([1970] 1971), *Chance and Necessity*. Translation. Originally published as Le Hasard et la nécessité (Paris: Seuil). New York: Alfred A. Knopf.

Monod, J., Changeux, J.-P. and Jacob, F. (1963), "Allosteric Proteins and Cellular Control Systems", *Journal of Molecular Biology* 6: 306-29.

A Kernel of Truth?
On the Reality of the Genetic Program[1]

Lenny Moss

Northwestern University

> The full understanding of the nature of the genetic program was achieved by molecular biology only in the 1950's after the elucidation of the structure of DNA. Yet, it was already felt by the ancients that there had to be something that ordered the raw material into the patterned system of living beings....What is particularly important is that the genetic program itself remains unchanged while it sends out its instructions to the body.
>
> One of the properties of the genetic program is that it can supervise its own precise replication and that of other living systems such as organelles, cells, and whole organisms.
>
> Ernst Mayr (1982)

1. Introduction

The notion of a genetic program, as described by Mayr, is ripe for an analysis inspired, at least in part, by recent debates in the philosophy of science taken under the heading of realism versus anti-realism. It is ripe for this because, as Mayr indicates, it is a notion with which contemporary biology orients itself to ancient and perennial problems. It is also pervasive and implicit in contemporary attempts at elaborating molecular level explanations of biological phenomena and so prerequisite to making sense of entities such as oncogenes. Yet while it is referred to as a kind of entity, i.e., a program, it is in some sense an object which is nowhere to be found.

Central to the current understanding of the genetic program is the governing force which is attributed to it. It is this characteristic, amply evident in the quotes from Mayr and widespread in current research literature, that determines the importance of this idea in conditioning all modes of biological explanation. One may thus be faced with the curious dilemma of accounting for biological events, much of which involves the discrimination of distinct components and attribution of their functions, on the basis of an entity invested with executive power, that doesn't exist.

PSA 1992, Volume 1, pp. 335-348

The reality of the genetic program will be pursued through exploring some of the history of developmental biology and genetics, attempting to locate the sources for such a notion. It will then be interrogated on a synchronic plane. What evidence is there for a genetic program in the findings of contemporary cell and molecular biology? What role does the genetic program play in the overall edifice of modern biology especially as regards the age-old question of teleology? In relation to questions of realism, the status of the genetic program will be considered with regard to the alternatives of an empirically derived hypothetical entity versus a heuristic principle, and its strength and warrant assessed in either case.

2. From Preformationism to the Challenges of Epigenesis

On a purely speculative plane, the traditional problem of how to account for the attainment of the mature form of an organism from its germ(s), i.e., the problem of ontogeny, can be approached in a variety of ways. The theory of preformationism, which held that the adult form was always already present in miniature in either the egg or the sperm, had the advantage of being able to defer the problem of form acquisition back to the first member of the species. However, emphasis on a detailed descriptive study of embryological development using the chicken egg as a model, carried out during the first half of the 19th century, served to rule out preformationism as a possibility. The doctrine of epigenesis advanced with these embryological studies which characterized the chick embryo in terms of three primary germ layers from which all subsequent tissues and organs derived (Coleman 1977). Eliminating preformationism, however, reopened the problem of form acquisition, and early epigenetic proponents were inclined toward some form of vitalism. Caspar Friedrich Wolff, for example, referred to an essential force, and comparative embryologist von Baer held a belief in a non-mechanistic essence which controls the development of the germ (Coleman 1977). Cell theory progressed concomitantly, contributing to an overall epigenetic picture, so that by 1855 Robert Remak could state that the development of the egg consists in the production of new cells derived from the division of "the egg-cell into morphologically similar elements" (Coleman 1977, p46).

In the latter half of the 19th century the understanding of epigenesis was structured by developments in the understanding of phylogeny. This received its most notable formulation in the pronouncement that "ontogeny is the brief and rapid recapitulation of phylogeny" (Coleman 1977, p47) by the Darwinian Ernst Haeckel in 1866. In fact, the ontogeny/phylogeny parallelism also had roots in the speculative naturalism derived from the Romantic era *Naturphilosophen* of F.W.J. Schelling. Driven by an idealist and teleological point of view, Johann Friedrich Meckel as early as 1821 argued that "The development of the individual organism obeys the same laws as the development of the whole animal series; that is to say, the higher animal, in its gradual evolution, essentially passes through the permanent stages which lie below it" (Coleman 1977, p50).

Whether driven by evolutionary materialist or idealist assumptions, the explanatory strategy of modeling epigenesis on phylogenetic history was linked methodologically to the descriptive practices of speculative morphology. By the late 19th century and within the intellectual horizons of the experimental physiology of Helmholz and others in Berlin, recapitulation could no longer suffice as an explanation of embryonic development, nor descriptive morphology as an adequate methodology. Wilhelm His, and more systematically Wilhelm Roux, initiated efforts at bringing experimentation oriented toward mechanistic explanation to bear upon the understanding of epigenesis. Roux's lifelong enterprise became that of the elaboration of a "developmental mechanics" (*Entwicklungsmechanik*).

The formative stage of modern experimental embryology became structured by the contrasting experimental results and interpretations of Wilhelm Roux and his competitor Hans Driesch. Roux proposed a *mosaic* theory of epigenesis based upon an unequal division and distribution of genetic material in the process of embryonic cell division. Working with frog eggs at the two cell (blastomere) stage, he could demonstrate that when one blastomere was punctured with a hot needle the embryo did not develop normally, and something like a half embryo resulted. Driesch, on the other hand, working with sea urchin eggs also at the two blastomere stage, could separate the two cells by an agitational method and demonstrate that each blastomere was capable of then going on to form a complete larva. Thus, rather than a process of progressive mosaicism, Driesch referred to his understanding of the embryo as that of an *harmonious equipotential system* "where all the parts are equivalent in their potential for producing a whole new organism" (Allen 1975, pp30-31). At stake in the difference between the thinking of Roux and Driesch were two very distinct basic conceptual strategies for approaching epigenesis. Roux understood an internally governed, programmatically driven, process of developmental unfolding consistent with the mosaic model. Driesch, by contrast, understood an equipotential system to be governed, not internally, but by external environmental influences, and progressive interactions of the developing parts. Driesch, however, eventually despaired of rendering this process in physio-chemical terms.

The perspective and problematics of experimental embryology arrived in the United States at the Woods Hole Marine Biology Laboratory in 1894 by way of Roux's manifesto of developmental mechanics in translation. Much of the work subsequently taken up there by a circle that included E.B.Wilson, T.H. Morgan, Ross Harrison, and Hans Spemann, was oriented toward resolving the basic differences at issue between Roux and Driesch. In the very particular matter of the developmental potential of the first two blastomeres of either frog or sea urchin, it was found to be the case that the equipotential view of Driesch was the correct one. Roux's results with the frog were shown to be based upon the effects of the residual tissue associated with the ablated blastomere. If the two frog blastomeres are separated, as performed with sea urchin by Driesch, then both frog blastomeres retain full developmental potential as well.

Further explorations of epigenetic development, based on the kind of interactive processes anticipated by Driesch, emerged from out of the Woods Hole circle. Most notable in this regard was the work of Hans Spemann, who went on to pursue most of his research career on the faculty at Freiburg. Spemann identified specific cases whereby the interaction of different embryonic tissues resulted in the transformation of one of the tissues into a new organ. Spemann introduced the term "embryonic induction" to describe this process. An important exemplar worked out during the 1920's, was that concerning the formation of the embryonic frog lens. The lens is derived by differentiation from ectodermal tissue, but in order for this differentiation to take place there has to be contact between the ectoderm and the underlying optic cup which is extended from the nascent brain. Quite consistent with the equipotential model suggested by Driesch, Spemann found that ectoderm from other regions of the frog embryo surface could be transplanted and induced by contact with the optic cup to differentiate into lens.

Following the lead of Spemann, and importing some language from physics, Paul Weiss introduced the idea of an embryonic *morphogenetic field* (Allen 1975, p122). This conceptualization can be contrasted nicely with the early notions of Roux in tracking the understanding of the processes associated with epigenesis. For Roux, each unit cell of the field would be internally constrained in its potential and already determined in its ultimate fate by its preordained share of the genetic material. In the field model by contrast, cells are intrinsically equal in their developmental potential but receive cues from their position in the field. The field as a whole is the developmental unit, and it is at this level at which the developmental fate resides. In this manner there can be plasticity and adaptability at the level of individual cells and groups of cells within the field, as was demonstrated to be the case by Spemann and others in transplantation experiments It should be pointed out, from the benefit of hindsight, that the progressive loss of developmental potential described by Roux, has itself been assimilated into the epigenetic model of development. Cells become progressively committed to a developmental pathway with a concomitant loss of their overall developmental potential. However, this process, unlike that envisioned by Roux, is one which necessarily relies upon a complex array of intercellular interactions, such as is thematized in the model of the morphogenetic field.

3. Sources of the genetic program-the rise of the 'Mastermolecule'

For all its success in elaborating an understanding of the processes involved in epigenesis, experimental embryology has been unable to realize the kind of physio-chemical level causal story about development expressed in the intentions of the *Entwicklungsmechanik* program. Had Roux's model panned out and each cell had turned out to be internally directed by its share of inherited material, then such a story which causally accounted for the behaviour of each cell according to the programmatic dictates of a chemical agent would presumably have been straightforward. Arriving at a suitably mechanistic account of a developmental process in which components mutually and interactively determine their fate proved to be far more difficult. Experimental embryology continued to suffer a singular rate of attrition from the ranks of its primary contributors. Even Spemann, who sought for the origins of the inductive cascade in the idea of a "primary organizer", came to disparage of securing a purely mechanistic account.

In searching for the origins of an understanding of a genetic program that governs, among other things the ontogenetic process, it is clear from its history that the findings of experimental embryology does not constitute such a source, as perhaps would have been the case if Roux's model had panned out. What will be attempted then, is to search elsewhere for those antecedents which could have contributed to the conceptualization of one component of the cell, namely the DNA, as the embodiment of a genetic program. The emphasis will be on the specifically governing sense, the imputation of an executive status to the role of DNA. Several areas of biology will be considered as possible contributors to this understanding.

(a) Cytogenetics and Transmission Genetics

Cytological studies in the late 1800's already called attention to the complicated and precise mechanisms involving the chromatin in cell division. By 1884-1885 "a number of leading cytogeneticists independently concluded that the nucleus, and doubtlessly the chromosomes were the all important physical bond between the generations" (Coleman 1977, p40). The combination of these studies with the "rediscovery" of Gregor Mendel's work led to the field of cytogenetics at the beginning of the 20th century. The "rediscovery" of Gregor Mendel's work, and the establishment of the classical understanding of the gene, grew out of, and was prompted by, controversies concerning whether natural variation occurred along discrete or continuous lines. Mendel's seminal work focused on the pattern of inheritance of certain traits of the pea plant, i.e., height, flower color, seed shape, and seed coat. Breeding experiments carried out by Mendel established the distinction between the "phenotype" of an organism, i.e., its embodiment of certain defined traits, and its "genotype" which accounts for the pattern of inheritance in terms of hypothesized units of heredity. The gene constituted as an entity which accounts for the pattern of Mendelian inheritance, was inferred to be solid and integral and not prone to mixing, and to come in pairs with each derived from one parent. Through hypothesizing the existence of genes with such characteristics, certain accurate predictions could be made about the appearance of certain traits in the progeny produced in breeding experiments.

What is of interest at present is a certain use of language that arose in the practice of classical (Mendelian) genetics. Cross-breeding experiments revealed that certain traits are preferentially expressed in progeny over others. We are generally familiar with this phenomenon in the case of brown eye color and blue eye color. Brown eye color is thus said to be *dominant* and blue eye color *recessive.* When any traits can be so distinguished according to their patterns of inheritance, then by extension one could say that the hypostatized gene for X (say brown eye color) is *dominant* and the hypostatized gene for Y (say blue eye color) is *recessive.* This sort of language is "natural" and parsimonious for the range of phenomena under consideration, but it must be kept in mind that all that classical genetics was addressing was the transmission of a very restricted set of traits. Classical genetics had nothing to say about how phenotypes were produced at all, that is, how genes, whatever they happened to be, interacted chemically, physiologically, developmentally, in the cell, the organism, etc. The only strong claim that classical genetics could make in this regard was that without some certain unit of transmission, i.e., some specific allele, some characteristic of certain members of a species, i.e., a trait, would not obtain. Nonetheless, I would suggest that the language generated by classical genetics, referring to genes as *dominant* and *recessive*, and even referring to the lack of a fully predictable appearance of a certain trait given the appropriate genotype in terms of "*penetrance*" and "*expressivity*", contributed to a perception of a passive cell/organism phenotype and active determining genes.

(b) Quantum Physics and the Informationist Turn

A significant influence on the genetic program idea, may have derived from an unlikely source. Niels Bohr had raised doubts about the possibility of explaining living creatures on a purely mechanistic basis. He suggested a principle of complementarity taken on analogy with that used to address the problem of indeterminacy in quantum mechanics. In studying a living system the problem would arise because one could not simultaneously observe the living system, as a living system, and observe the composition and state of its constituents. The limits of a purely mechanistic viewpoint was also a theme for Bohr's student Max Delbrück, who had attempted to turn away from the emphasis upon chemical mechanism and toward a conception of information as a new kind of biological commodity. Delbrück initiated the efforts of the famous phage group. They sought to exploit the simplicity of the bacteriophage for experimental study but oriented toward the information idea, tended to place more emphasis on the phage coat proteins than on the analysis of the phage nucleic acids. Delbrück's thoughts about the limits of a mechanistic account were acknowledged to be influential on the thoughts of friend and colleague Erwin Schrödinger. Schrödinger's own reflections on the matter, in his little book *What is Life?(1944)* was well read amongst the generation of the "romantic phase" including being one of the few points of commonality in the intellectual backgrounds of Francis Crick and James Watson. Following the lead of Bohr and Delbrück, Schrödinger went on to suggest that in order to understand life at the molecular level, new laws of nature may need to be discovered.

Schrödinger argued in his book that physics finds predictability at the molecular level based upon probability. What is rendered predictable in cases such as diffusion and radioactive decay given large numbers of molecules becomes decreasingly predictable with fewer and fewer molecules. Biology appears to veer off from the conditions of the inanimate world in its apparent ability to rely upon the stability and predictability of single molecules. "We are here obviously faced with events whose regular and lawful unfolding is guided by a 'mechanism' entirely different from the 'probability mechanisms' of physics. For it is simply a fact of observation that the guiding principle in every cell is embodied in a single atomic association existing only in one copy (or sometimes two) - and a fact of observation that it results in producing events which are a paragon of orderliness (Schrödinger 1944, pp84-85). Schrödinger lends a great deal of prestige to the genetic molecule in an interesting way. He believes that some new principles will have to be uncovered in order to account for the ability of something like an aperiodic crystal to stay aloof of the heat motion of the cell and thus retain its high level of order. Given the expectation that the understanding of the nature of the hereditary substance will not, and could not, be constrained by the limits of available models of physio-chemical mechanism, and perhaps showing the influence of Delbrück's informationist intuitions, Schrödinger lends an uninhibited voice to the glorification of the genetic molecule to be. "Since we know the power this tiny control office has in the isolated cell, do they not resemble stations of local government dispersed through the body, communicating with each other with great ease, thanks to the code that is common to all of them? (p84)."

With the victory of the new field of molecular biology and its conquest of the structure of the DNA molecule within the framework of a classical physio-chemical mechanistic account, went all the spoils of a privileged status, born of the speculative minds of the anti-reductionist quantum physicists, to the genetic molecule. There is perhaps a certain irony in the fact that the kind of stability which Schrödinger felt would require new basic laws of nature to account for, may have contributed to the apotheosis of a molecule, in part, for the sake of its *conventionally inert* chemical structure.

(c) Simplicity and Molecular Virology

Prior to the breakthrough of the Watson and Crick model, Hershey and Chase proved that in the case of the bacteriophage, it was the nucleic acids and not the proteins of the virus which entered the target cell and were thus responsible for transforming the bacterium. Their experiments entailed the use of radiolabelled phosphorus as a marker for nucleic acids, and radiolabelled sulphur as a marker for protein. The ease with which phage could be cultured and radiolabelled, the simplicity of its structure consisting of only protein and nucleic acids, and the native ability of the phage to penetrate the cell and notably effect its behaviour, were all factors in making the phage a compelling model system for study. Viruses, not just bacterial, but avian and mammalian as well, have continued to be prime tools of the molecular biologist. The highly touted identification of a growing number of cellular genes as "proto-oncogenes" for example, arose through the use of viruses as a kind of model molecular system.

The concluding point to be made with regard to the sources of the idea of the genetic program, is that the extensive use of viruses in molecular biological research has itself been a major factor. The realization of the potent effects that a virus can bring to bear upon a cell/organism, owing to the penetration of its DNA (or RNA) alone, carries with it a strong message. Adding to this, is the experience of numerous molecular virologists who have become accustomed to effecting their own consequences upon test cells, through altered constructions of the viral genome. What becomes easy to overlook in the midst of such apparent power and efficacy is that viruses are molecular parasites whose ability to act entirely presupposes a living system, in relation to which the virus is a kind of trigger or perturbant. Shooting DNA constructs into a cell and shouting "Now dance!" does not constitute an explanation of the mechanisms by which "the genetic program informs and instructs ontogeny" or "supervise(s) its own precise replication and that of other living systems such as organelles, cells, and whole organism" (even if the cell dances).

4. Searching for the Genetic Program at the Lab Bench

Biological research oriented toward achieving a mechanistic account of fundamental life processes at the cell and subcellular levels has become a massive enterprise. Specialized subdisciplines form in order to investigate even a single type of molecule, such as N-CAM the neural cell adhesion molecule, or a very specific process such as the targeting of lysosomal enzymes to the lysosome, or the "splicing out" of introns from newly synthesized messenger RNA, or the transcriptional regulation of a certain gene. Across the bounds of numerous subdisciplines, there has been a general tendency toward increased use of the techniques of recombinant DNA technology, both for the experimental power that they provide and for the sake of elucidating the role that some sequence of DNA may play in relation to the particular process or molecule of interest. To be sure, it could *not* be said at this time that the methods for analyzing DNA or the desire to explain biological processes in terms of the central importance of DNA are lacking. However the reality of the genetic program cannot be assessed merely on the basis of the influences of this idea upon the norms of experimental method and research strategy. If the genetic program is to warrant a claim to some form of real empirical existence then there must be evidence for it playing some role within the explanatory models which are the fruits of the cell and molecular biology research enterprise.

While a thorough review of the state of all biological research is not presently possible, an attempt will be made to offer the explanatory models of several biological subdisciplines in order to begin to frame the question as to whether the reality of the genetic program can be fleshed out, as well as to pose some specific challenges.

(a) Cytodifferentiation of Alveolar Mammary Cells

Contemporary studies in developmental biology have recourse to cell isolation and culture methods which were not available to the investigators discussed above. When possible, the isolation and culturing of animal cells allows for the analysis of events at the biochemical/molecular level not possible when working with the whole embryo. On the other hand, cell culture is an artificial environment, and as Driesch had long since postulated, cells do not undergo developmental maturation individually and independently but only by means of a mutually efficacious succession of interactions. On the whole, animal cells are not easy to sustain in culture[2], the more advanced they are along a certain developmental path, the less likely they are to be amenable to culture, and the general tendency is for cells that can adjust to tissue culture to lose whatever developmental specifications they had acquired *in vivo*. Various cell culture systems have been adapted for use as experimental models, and the sources of these have spanned the developmental distance from early embryo to adult. The developmental model that will be presently discussed is of the latter variety.

The biology of the mammary gland has certain unique features for developmental study. The epithelial cells of the alveolus which produces milk in the lactating gland first arises during pregnancy. As a result the morphogenesis of the milk producing apparatus and the biochemical differentiation of the alveolar cells into milk producers occur in fairly close proximity. A tissue culture model had been successfully established using mammary epithelial cells extracted from late pregnant mice in order to try to understand the mechanisms involved in the induction of milk synthesis and secretion (Emerman & Pitelka 1977). Although it was found that while these cells could be cultured on plastic petri dishes, the ability to induce a lactating phenotype was highly sensitive to the character of the substratum employed. Cells cultured on plastic secreted negligible if any quantities of the milk protein component ß-casein into the culture medium. A certain amount of ß-casein could be detected if the cells were cultured on a layer of the extracellular matrix molecule collagen, but the greatest increase in ß-casein secretion occurred when the collagen substratum, with cells on top, was released from its attachment to the plastic and allowed to float in the culture medium (Lee et al. 1984). In this situation the cultured cells caused the gel to contract in a concave fashion (from the perspective of looking top down at the dish). Scanning electron microscopy and transmission electron microscopy indicated both that the cells on released gel had assumed a rounded-up morphology and that the internal organization was typical of secretory epithelia, as where cells on plastic retained a flat morphology and a fairly undifferentiated internal organizational structure (Emerman et al. 1979).

Two points of interest were suggested by these results. The first was that the presence of collagen may have played an inductive function and the second that conditions affecting the shape of the cell may have been of significance in the induction of the lactating phenotype (or some approximation of a lactating phenotype). Collagen (type I) is deposited by fibroblasts of connective tissue which is ultimately derived from the mesodermal germ layer. The ability of collagen to effect epithelial differentiation can be taken in the light of studies on mesodermal/ectodermal inductive interactions that go back to Spemann. Subsequent studies tended to corroborate the understanding that substratum was a crucial factor in inducing cells to take on a secretory phenotype. Complex matrices consisting of numerous molecules characteristic of an *in vivo* stromal-epithelial tissue interface were even more effective (Parry et al.1987, Li et al. 1984), being capable of inducing a milk protein (whey acidic protein) that had never been found on collagen alone (Chen & Bissell 1989). When studies probed to see whether the affects of the substratum were based on changes in gene expression, it was found that the differences in the transcriptional activation of milk protein genes, that is the level of *de novo* milk protein messenger RNA production accounted for only part of the difference in milk protein secretion, with differences in the half-life, i.e., the rate of degradation, of milk protein also accounting for a share of the difference (Lee at al. 1985).

From the purview that this model system of cellular differentiation provides, one can see a number of external-to-the-cell factors influencing the internal state of a cell including the transcriptional state of some of its genes. Now one might want to argue that all of these factors can be tracked back to DNA somewhere along the line, and so therefore the genetic program, while not visible in this micro-view is still a tenable proposition on a global level. This question clearly cannot be adjudicated on the basis of the mammary cell model, but the mammary cell model can be used to point out the kind of questions which are endemic across the board. In this model the DNA is a touchpoint in the loop of causal events to the extent that transcriptional activation of certain genes is required. The signals that lead to activation are coming from outside the cell, and the means by which they are being transmitted to the DNA are unknown (although lots of models have been proposed, see for example Bissell et al [1982]). In addition, changes in the rate of degradation of RNA appear to also be significant. This difference may be due again to the external factors inducing the transcriptional activation of an enzyme that degrades RNA, or an enzyme that through some means inhibits the degradation of RNA, or through the "turning off" of the transcription of the enzyme that degrades the RNA, or it may be due to events in the cell that do not interface with DNA at all. In any model system that one examines, a chain of biomolecular events may or may not include DNA "in the loop." Alternatively, if one zero's in on the role of DNA in some process, such as the transcription of a certain gene into messenger RNA, the desire to track the mechanisms responsible for activating transcription will take one back into the state of the cell as a whole, and beyond. This latter point will be further illustrated by a glimpse of the research findings from work that's been done directly on the mechanisms of transcriptional regulation.

(b) Transcriptional Initiation and its Regulation

The Mendelian gene, while confirmed in a certain sense by the model of the DNA structure, does not in any straight forward fashion map one to one onto the DNA (see Hull 1974). The first distinction one would want to make in this regard, which does not readily correlate with the Mendelian model, is that between stretches of DNA which are destined for transcription into RNA (and of these sequences that are transcribed into RNA, only a fraction are then used for translation into protein) versus those that are involved in the regulation of transcription and are themselves never transcribed (nor need it be the case that most genomic DNA even fall into either of these two categories). Early studies on the regulation of transcription in bacteria found that "upstream" and close to the site of transcriptional initiation is necessarily an "A-T" rich region which became dubbed the promoter sequence. Transcription requires a complex of initiation factors (proteins) as well as the RNA polymerase enzyme which "reads" the relevant sequence and catalyses the polymerization of a complementary sequence from RNA precursors. Initiation of transcription requires initiation factor proteins to latch onto the DNA in the appropriate region, separate complementary strands, and in this manner set the stage for the catalysis. In as much as "A-T" base pairs are linked more weakly than "G-C" base pairs, owing to the presence of only two hydrogen bonds in the case of the former, as compared with three for the latter, the "A-T" rich promoter appears to provide a structural opening for initiation to take place.

In eukaryotic cells there are generally at least two promoter sequences whose relative spacing is moderately variable. These are recognized by initiation factors (proteins) which are themselves recognized by other initiation factors and the RNA polymerase enzyme complex. It appears that the composition of proteins in a eukaryotic initiation complex can be quite variable from one promoter to another, and even at the same promoter. In addition to the role of promoter sequences, and all the proteins that become associated with it in the regulation of transcription, another class of DNA sequences which effect the transcriptional state of a "gene" is that of so-called "enhancers". These are generally compound modular elements comprised of repeated sequence motifs and/or mixtures of different sequence motifs. A salient feature of enhancers as regulatory stretches of DNA is the context dependence of their role in transcription. "Enhancers have no intrinsic activity. Their effects on initiation frequency depend upon multiple contexts, including the promoter itself, the presence and activity of other nearby promoters and enhancers, cell type, physiological state (Yamamoto 1988).

An example of the explanatory regress that one faces in attempting to elucidate even a single transcriptional activation event at the level of the the DNA molecule itself can be illustrated by a recent paper concerned with hormonal induction of transcription. Glucocorticoids are steroid family hormones that circulate systemically and bind to specific glucocorticoid receptors in cells. These hormone-receptor complexes then exert their influence on the activity of a cell through binding directly to the DNA at enhancer regions known as glucocorticoid response elements (GRE's) which are close to hormone responsive promoters. "As with many transcriptional regulators, the receptor stimulates gene expression in some genetic and cellular contexts and represses expression in others" (Diamond et al. 1990). In the interest of exploring the mechanisms associated with the hormone associated regulation of gene transcription,

Diamond et al., constructed a highly simplified 25 base pair segment of a GRE that they found to be capable of binding glucocorticoid receptor and mediating both positive and negative regulation of a hormone sensitive gene. The initiation factor AP-1 components "c-jun and "c-fos" have also been implicated in the GRE modulated expression of the gene "proliferin." Using the simplified GRE construct in a model system, Diamond et al., explored the relationships of the glucocorticoid receptor complex and the c-jun and c-fos proteins in the regulation of proliferin transcriptional activation. From their findings in this regard, they offered the following model. In the absence of c-jun glucocorticoid has no effect on transcription as the receptor will not bind to the GRE. In the presence of c-fos, with no hormone, c-fos will bind to the GRE and mildly induce proliferin transcription. If hormone is then added, the receptor complex will then bind to the GRE (in the presence of bound c-fos) and strongly induce proliferin transcription. C-jun alone does not bind to the GRE and does not effect induction, but c-jun in the presence of c-fos and in the absence of hormone will bind to the GRE and strongly induce transcription. If hormone is then added to the c-jun-c-fos complex, the receptor will bind to the GRE and now inhibit the induction of proliferin transcription. It can thus be seen that even in this artificially simplified model of transcriptional regulation by a hormone, any mechanistic account must immediately refer to the presence or absence of at least three other factors (c-jun, c-fos, and the hormone receptor).

Attempts at securing a discrete causal account of even the most proximate events in the transcriptional activation of a certain gene can thus be seen to quickly devolve into an array of antecedent conditions which are exponentially more complex than the event one was trying to account for. Explorations of the mechanisms involved at the level of the DNA molecule itself, have not led to any privileged point of causal origins, but rather immediately refer back to the complex state of the cell/organism as a whole as the causal basis of the activity of the genes.

(c) Cellular Self-Assembly and the Genetic Program

Eukaryotic cells (i.e., not bacteria) are highly complex structures, consisting of a vast array of internal compartments (organelles) partitioned by phospholipid bilayer based membranes. The phospholipids that constitute the membrane are amphipathic molecules consisting of a hydrophilic polar head group and hydrophobic tails. Constructed in a bilayer fashion with the hydrophobic moieties buried within, they provide a significant thermodynamic barrier to the movement of anything other than small hydrophobic molecules. Taken as a whole, the membranous system partitions the cell into two topological domains, luminal i.e., within the organelles, and cytoplasmic (which includes the nuclear domain). In addition to the phospholipids (and other lipids), the membranes contain proteins that are either peripherally or integrally associated with the membranes, and which are typically conjugated to carbohydrate groups. With the exception of the mitochondria and plastids (in plants), all of the other organelles of the cell, although not physically continuous, constitute a unified system in as much as inclusion into either the internal (luminal) domain of the organelles or into the membrane bilayer itself, requires original entry at a common port. This common port of entry is the endoplasmic reticuluum, and it is mediated by a system consisting of indigenous docking proteins, cytoplasmic signal recognition particles, ribosomes and other factors.

The organization of the cell and the processes involved in its self-replication, its targeting of newly synthesized enzymes, for example, to their proper location within a particular organelle membrane, presents a challenge to the genetic program idea, above and beyond that which has already been discussed. The three dimensional structure of the cell, especially the specificity of each of the boundary forming organelle membranes, are essential features of any cell, and are the means to their own replication. Whereas the DNA constitutes a kind of template of one-dimensional information for the amino acid sequence of proteins, the organelles are the only template for their own replication, the only source of the "information" for their own three dimensional structure. Nascent proteins may be directed by template derived structures toward the membrane "port of entry", but an intact system of membranes with its port of entry are always the prior condition of cellular self-replication. The cell as an organized three dimensional entity is the prior condition by which the template function of DNA "can make sense." Information is not indigenous to DNA, it is constituted in the specific conditions of a cellular environment which is always already there. And if it wasn't there, there is not an iota of evidence to think that the sequence of base pairs of nucleic acids in the DNA could provide the direction for reconstructing it.

Gunther Stent has been one of very few prominent investigators in the field to reflect critically upon the idea of a genetic program (Stent 1986). It is probably no accident that his own career has spanned the ideational space from the time of his involvement as a junior member of Max Delbrück's phage group, in a pre-Watson and Crick environment, to that of trying to bring the experimental fruits of molecular biology to bear on understanding the epigenesis of the nervous system. Stent recalls the point soon after the Watson and Crick model where it was thought that perhaps the genes contained "a circuit diagram" of the nervous system. When the untenability of this notion became apparent, Sydney Brenner in the early 1970's suggested that what developmental neurobiology was about was revealing the genetic program that governs neural development. Stent argues against this position on grounds that are relevant to the kinds of problems that arise in considering the findings of the subdisciplines reviewed above. Stent argues that the idea of a genetic program, of any program, suggests something isomorphic, i.e., the program, with that which it is a *program of.* Clearly, the inspiration for this type of metaphor, comes out of the informationist turn and the euphoria of the disclosure of the DNA model. However, while the kind of template relationship that DNA has with RNA and protein may be assimilatable to a program metaphor, there is no indication of anything having an isomorphic relationship with development. (This argument in effect was already made above with regard to the organization of the cell). Quoting Stent "As George Szekely has pointed out, we know enough about neuronal development already to make it most unlikely that neuronal circuitry is, in fact, prespecified; rather, all indications point to stochastic processes as underlying the apparent regularity of neural development. That is to say, development of the nervous system, from fertilized egg to mature brain, is not a programmatic but an historical phenomenon under which one thing simply leads to another" (Stent 1986, p212).

5. Conclusions

This attempt at assimilating to a "genetic program" the processes of development, which from all that we have learned going back to Driesch entails a complex interactive epigenetic unfolding in which the cells and tissues are the source and soul of their own ontogenesis, is likened by Susan Oyama to an attempt at returning to a kind of preformationism. Form, she argues, emerges historically, and contingently, in the sequence of interactive processes. "Our mistake is to project it, not into those processes, but into an originating entity that stands in a creator-creation relation to them" (Oyama 1985, p138). In this light, the case is made that the self-organizing and self-developing capacity of living things have been projected into an abstraction called the genetic program which, as an informational analogue of the preformed homunculus, is housed in the nuclear germ of every living cell.

The objective of this undertaking has been to evaluate the claim to reality of the concept of a genetic program that was so readily put forward by Ernst Mayr. Having attempted to review at least some of the relevant history and some characteristic and salient features of contemporary biological understanding, we have to conclude that, as an empirical entity, as a model which is subject to the force of evidence and argument, the existence of a genetic program is indeed highly problematic. Yet, the ease with which one may come to this conclusion, juxtaposed to the pervasity and tenacity in its belief, suggests that there is more at stake then just another hypothesis.

Biologists for over 200 years have been grappling with the problem of explaining organismic development in the light of an understanding of matter that does not readily lend itself to the task, resulting in an oscillation between some form of preformationism that defers the problem and an epigeneticism which finds a need to supplement the properties of living matter with special vital forces. Immanual Kant addressed this problem for his day through arguing for the need to distinguish between constitutive and regulatory (heuristic) principles. Kant motivated his reflections on nature with the example of a man coming ashore and finding a hexagon drawn in the sand (Kant 1987). To perceive the hexagon is to know it by its principle and to recognize that it existence cannot have been determined by the mechanistic forces of nature. The judgement of the hexagon as being governed by an idea, a principle of reason, is such as to cause one to perceive its presence as a reflection of some purpose (i.e., that of whomever drew it). Like the hexagon, for Kant, we can comprehend a living entity, a tree for example, only on the basis of an idea. This idea is that of an entity which is both the cause and effect of its own existence. Kant cited several ways in which the tree is a cause and effect of itself. It is the self-producing cause and effect of itself in that it brings elements in from its environment, water, nutrients, etc, and these become in-formed by the tree. It is cause and effect of itself as a produced and reproducing member of a species. And it is possessed of its own cause and effect in terms of the mutual relationship of its parts to each other and to the tree as a whole. The judgement of an entity as being its own cause and effect, Kant referred to as a "natural purpose". For Kant, this judgement is not one based upon an understanding which can be evidenced scientifically, i.e., not a constitutive principle. It is, rather a principle of our judgement of nature i.e., a regulative principle without which we could not make sense of living things. An anatomist, for example, who cuts open an animal can only begin to make sense of the chaos inside with the *prior* understanding that things are there for a purpose. Kant did not set *a priori* limits on what could be eventually explained mechanistically, although he doubted that a causal account of a self-producing entity would ever be accomplished on the basis of constitutive principles. In so doing Kant set out the guide-lines through which biologists working within the confines of the understanding of matter of their times could advance a mechanistic account of developmental processes as far as possible under the guidance of a regulative teleological heuristic (see Lenoir 1982).

In his discussion of teleology, Ernst Mayr offers the notion of teleonomy as the modern scientific alternative. Teleonomy is the appearance of natural purpose, but it is natural purpose based upon the presence of a genetic program. Life is distinguished from non-life for Mayr, and thus biology from the other sciences, because living things are invested with a genetic program. From Mayr's perspective biology has thus accomplished what Kant thought could not be done. It has explained natural purpose by way of the mechanism of the genetic program.

However, the genetic program, as has been argued, is not such an entity. Rather than overcome the limitations that Kant presaged for biology, Mayr (1988) and others have fallen prey to the error that Kant cautioned against. If the genetic program has any status as a tool for understanding living things, it is not as a constitutive principle but rather as a regulative or heuristic principle. But, granting this, the floodgates of doubt are opened anew. If the genetic program is a modern heuristic of our understanding of nature, it is then on an equal epistemological plane with that of natural purpose. The two must then be weighed one against the other, on the basis of their respective merits. In other work I have argued that the conceptualizations of autopoietic systems developed by Maturana and Varela consist essentially of a rediscovery of the heuristics of natural purpose. Further, in considering the last 20 years of research in the area of the molecular biology of cancer, I have attempted to show that the idea of a genetic program has become a significant impediment to biological understanding and that the guidance of an autopoietic/natural purpose heuristic offers better prospects for further advance (Moss 1992).

Notes

[1]This paper is an outgrowth of Arthur Fine's Northwestern University graduate seminar on "Realism and Anti-realism." I thank Professor Fine for his encouragement, and Professors Fine, David Hull, Thomas Ryckman, Susan Oyama, and Elizabeth Lloyd for their comments.

[2]Most experimental work on cells in culture use "cell lines". These cells are the descendants of cells extracted from an animal, which underwent a kind of "crisis" in culture. The survivors of this crisis have undergone a kind of adjustment to life in a petri dish, and typically retain certain of the original properties of the ancestral cells *in vivo* while losing others. Cell lines offer great advantages to the experimentalist owing to their stability, longevity, and ease of handling. The perceived downside is that they no longer perfectly reflect the nature of cells *in vivo* and so pose doubts with regard to their applicability to the *in vivo* situation.

References

Allen, G. (1977), *Life Science in the Twentieth Century*. New York: John Wiley & Sons.

Bissell, M.J., Hall, G. and Parry, G.(1982), "How does the extracellular matrix direct gene expression?" *Journal of Theoretical Biology* 99:31-68.

Chen, L.H. and Bissell, M.J. (1989), "A novel regulatory mechanism for whey acidic protein gene expression". *Cell Regulation* 1:45-54.

Coleman, W. (1977), *Biology in the Nineteenth Century*. Cambridge: Cambridge University Press.

Diamond, M., Miner, J., Yoshinaga, S. and Yamamoto, K. (1990) "Transcriptional factor interactions: selectors of positive or negative regulation from a single DNA element" *Science* 249:1266-1272.

Emerman, J.T. and Pitelka, D.A. (1977), "Maintenance and induction of morphological differentiation in dissociated mammary epithelium on floating collagen membranes". *In Vitro* 13:316-328.

Emerman, J.T., Burwen, S.J. and Pitelka, D.A. (1979), "Substrate Properties influencing ultrastructural differentiation of mammary epithelial cells in culture", *Tissue Cell* 11:109-119.

Hull, D. (1974), *Philosophy of Biological Sciences*. Englewood Cliffs: Prentice-Hall.

Kant, I. (1987), *Critique of Judgement*. Indianapolis: Hackett Publishing Company.

Lee, E.Y., Lee, W.H., Kaetzel, C.S., Parry, G. and Bissell, M. (1985), "Interaction of mouse mammary epithelial cells with collagen substrata: regulation of casein gene expression and secretion", *Proceedings of the National Academy of Science* 82:1419-1423.

Lee, E.Y., Parry, G. and Bissell, M.J. (1984), "Modulation of secreted proteins of mouse mammary epithelial cells by the collagenous substrate", *Journal of Cell Biology* 98:146-155.

Lenoir,T. (1982), *The Strategy of Life*. Chicago: University of Chicago Press.

Li, M.L., Aggelar, J., Farson, D.A., Hatier, C. and Bissell, M. (1987), "Influence of a reconstituted basement membrane and its components on casein gene expression and secretion in mouse mammary epithelial cells", *Proceedings of the National Academy of Science* 84:136-140.

Mayr, E. (1982), *The Growth of Biological Thought*. Cambridge: Harvard University Press.

_ _ _ _ . (1988), *Toward a New Philosophy of Biology*, Cambridge: Harvard University Press.

Moss, L. (1992), "Oncogenes, tumor suppressor genes, and the nature of cancer", preprint.

Oyama, S. (1985), *The Ontogeny of Information*. Cambridge: Cambridge University Press.

Parry, G., Cullen, B., Kaetzel, C., Kramer, R. and Moss, L. (1987), "Regulation of differentiation and polarized secretion in mammary epithelial cells in culture. extracellular matrix and membrane polarity influences", *Journal of Cell Biology* 105:2043-2051.

Schrödinger, E. (1944), *What is Life?* Cambridge: Cambridge University Press.

Stent, G. (1986), "Hermeneutics and the analysis of complex biological systems" in *Evolution at a Crossroads*, Depew & Weber (eds.). Cambridge: The MIT Press.

Yamamoto, K. (1988), Unpublished lectures on gene regulation. University of California, San Francisco.

A Defense of Propensity Interpretations of Fitness[1]

Robert C. Richardson
University of Cincinnati

Richard M. Burian
Virginia Polytechnic Institute and State University

1. Introduction: Probabilities and Propensities in Fitness

That there was *some* chance element to evolutionary change was clear even to Darwin (cf. Hodge 1987; also Hodge & Kohn 1986), but chance has assumed a more central role in evolutionary theory since the early decades of this century. It was certainly clear to theoreticians such as Dobzhansky, Wright and Fisher that an adequate evolutionary theory would have to take a stochastic form, though they disagreed as to the significance of chance in evolution (Gigerenzer et al., 1989, ch. 4; Turner 1987). As Beatty explains, the central difference is that in more modern treatments, not only is the *origin* of variations a matter of chance, their evolutionary *fates* are also matters of chance (Beatty 1984; cf. Beatty 1987). Probabilistic theories of evolutionary change, as we shall be concerned with them, emphasize this dependency of evolutionary fates on chance, and treat evolutionary change as irreducibly probabilistic.

In recent philosophical literature on the topic, the dependence of evolutionary fates on chance has often led to a treatment of fitness as probabilistic and, in particular, as a propensity. On propensity interpretations, "... the fitness of an organism is its *propensity* to survive and reproduce in a particularly specified environment and population" (Mills and Beatty 1979, p. 42). Alternatively, "... the (relative) expected fitness of a type of organism as compared with specific competitors in a specified environment is its propensity to manifest a certain (relative) rate of reproductive success as compared with those competitors" (Burian 1983, p. 301). Propensities are natural dispositions which can be manifested in long term frequencies; they help explain the actual observed frequencies within a domain. As a propensity, fitness would be a natural disposition manifested in long term reproductive differences, useful for explaining actual differences in survival and reproduction. With fitness, understood as *expected reproductive success*, natural selection becomes differential reproduction due to fitness differences. So understood, propensity interpretations of fitness are critically tied to an acknowledgment of the probabilistic character of evolutionary change, and also to a particular interpretation of probability.

Representing fitnesses, or fitness differences, as probabilistic quantities has several *prima facie* advantages. Any treatment of fitness will need to maintain these advan-

PSA 1992, Volume 1, pp. 349-362

tages, whether or not the account is probabilistic. First, such representations allow us to distinguish fitness from actual reproductive success, or what we often will call *realized fitness*.[2] Actual reproductive success can deviate significantly from what would be expected on the basis of fitness assignments. Equally fit organisms can differ in their actual reproductive success, and less fit organisms can out-reproduce more fit organisms; indeed, if we consider an array of cases with intermediate fitness values, we should *expect* less fit organisms to prevail at least sometimes. Any given level of reproductive success is, except in the limiting cases, consistent with any fitness assignment. One principled way of motivating the distinction between expected and realized fitness is embedding fitness within a probabilistic framework.

Second, probabilities seem to be pitched at an appropriate level of generality for evolutionary explanations. Phenotypic or genotypic properties which are otherwise quite different can result in the same level of fitness; as Rosenberg puts it, "... there are many different ways in which the same level of fitness may be possessed, and there is consequently no one-to-one relation between a given level of fitness and a manageable set of its causal conditions" (1983, p. 459). This dependence of fitness on phenotypic or genotypic properties is what he means by "supervenience." There is no manageable characterization of the phenotypic or genotypic properties on which a given level of fitness supervenes; that is, no disjunction of phenotypic or genotypic properties, however heterogeneous, will yield a general characterization of fitness differences. In this respect, fitness assignments abstract from phenotypic and genotypic details, while depending on them. Even though phenotypic differences can explain differences in fitness within populations and even though fitness supervenes on phenotypic differences, there can be no general characterization of fitness in terms of phenotypic differences (cf. Brandon 1990, ch. 1).

Third, probabilistic analyses of fitness promise to clarify the crucial distinction between the effects of selection and those of drift in evolution. Genetic drift–essentially, sampling error due to small population size–can result in differential reproduction and therefore in evolutionary change. We will see that this issue is vexed in a number of ways. Brandon (1990, p. 9) says that though natural selection and drift both occur in actual populations, "drift and natural selection are alternatives", each capable of explaining differential reproduction. Similarly, Falconer (1989) treats selection and drift as different "agencies" through which the genetic properties of a population may be changed. An adequate account of fitness must provide a conceptually satisfactory distinction between drift and selection.[3] This is a corollary of the first point above; for, as Beatty explains, if fitness is identified with *actual* reproductive success, "... the distinction between random drift and natural selection of the fittest dissolves, and along with it dissolves the issue of the relative evolutionary importance of the two supposedly different sorts of processes" (1984, p. 191).

We shall emphasize the interpretation of fitness within a general account of evolution. We will deal only with a limited fragment of evolutionary theory, neglecting a variety of factors affecting evolutionary change, including developmental constraints and factors affecting the origin of variation. There are a number of important issues even within the narrower fragment which we will not address directly, but which are clearly relevant to evaluating the adequacy of an interpretation of fitness. First, we will not discuss the level at which selection acts (cf. Brandon 1982; Kane et al., 1990; Sober 1992). For convenience, we shall treat selection as if it occurred at the level of organisms, acting on the phenotype. While this is obviously a common case, we do *not* presuppose that this is the only level at which selection acts, or that there is only one such level. Indeed, a general account of fitness should apply at many levels, but we will set this issue aside. Accordingly, we will concentrate on selection over a generational time scale, changing the means, variances, and covariances of phenotypic distributions. Second, it is important to discriminate different propensities, distinguishing a propensity for reproduction from, for example, a propensity for survival, or within the former, distinguishing mating success from viability and fecundity. It is theoretically important to distinguish the various components of fitness (cf. Arnold and Wade 1984a, 1984b; also Prout 1971a, 1971b); an account that did not allow one to discriminate between them would be inadequate. Finally, we will assume the relativity of fitness to both the biotic and the abiotic environment. It is allowed on all hands that fitness is relative to the environment, even though there is in point of fact no entirely general way of accommodating this relativity in most models. We believe that a propensity interpretation can handle these issues.

Many discussions of the propensity interpretation fail to be responsive to the sources of probabilistic components in evolution and the various problems concerning the interpretation of probability. We intend to take such issues seriously in discussing propensity interpretations. In Part 2I, we will turn to the *probabilistic* character of evolutionary change under selection and drift, and will isolate the peculiar role of fitness values in probabilistic models of evolutionary change. In Parts 3 and 4, we will turn to the commitment to *propensity interpretations* as opposed to other interpretations available for understanding fitness as probabilistic, including in particular appeals to limiting frequencies.

2. Fitness as a Probabilistic Quantity

Realized fitness cannot suffice to *define* fitness: "... in any *single* run organisms of the relatively fittest type may not out-reproduce their competitors; indeed, there are occasionally cases in which none of the fittest organisms survive" (Burian 1983, p. 301). Sober uses the following analogy to explain the role of statistical information in estimating fitness levels. Tossing coins provides relevant *evidence* concerning whether the coins are biased. If three coins are thrown, one a million times, one three times, and one never, they will obviously differ in the *actual* frequencies of heads thrown even though they may have the same *probability* of coming up heads. As Hacking says,

> If we take seriously the notion of objective possibility, feasibility, proclivity, propensity, or whatever we call it, such degrees of feasibility may themselves be objects of knowledge, to be known with varying degrees of precision. (1975, p. 128)

Fisher, in his work on statistical theory, thought of frequencies as defined in "hypothetical infinite populations", and the actual data as a random sample from those infinite populations (1938; cf. Hacking 1965). Objective probabilities are objects of knowledge, known more or less completely, supported by evidence with varying degrees of sensitivity; the actual data provide some evidence to assess probabilities. The fitness of phenotypes does not depend, as such, on how often they occur, any more than the bias of a coin depends on how often it is tossed; neither can the fitness of phenotypes be defined in terms of the actual changes of frequencies within populations. "Populations of different sizes and compositions may be characterized by the same set of fitness values, just as coins that differ in the number of times they are tossed may have the same bias" (Sober 1984a, p. 43). From such considerations, Sober concludes that fitness is a "probabilistic quantity", and Brandon that "natural selection is a statistical phenomenon" (1978, pp. 69-70).

The precise sense in which fitness is a probabilistic quantity, or a statistical phenomenon, often has been misunderstood or misdescribed. It is not always transparent what advocates take the proposed propensities to consist in. Beatty suggests that the environment is specified by a "range of circumstances, each weighted according to the likelihood of its occurrence" (1984, p. 193). In each environment, a phenotype will have a definite fitness, and the phenotype can be regarded as having a range of offspring contributions depending on the specific circumstances it encounters. Different probabilities for reproduction, or survival, are a consequence of the different probabilities of encountering specific circumstances, and this alone is the source of the probabilistic character of fitness functions. In May's (1973, ch. 5) deterministic models, it is possible to find constant equilibrium values for, say, two-species systems, and these equilibrium values will be points; but when the environment varies temporally, equilibrium values will vary as well and these values must be represented as probability distributions. The probabilistic character of fitness in such cases is a consequence of the environmental relativity of fitness, and of the uncertainties of environmental variation. In other interpretations of fitness as a propensity, what seems to be contemplated is more radical, taking fitness itself as a probabilistic function, so that, even in a uniform environment and in the absence of drift, the consequences of selection would be "statistical" or "probabilistic" (e.g., Brandon 1990, Burian 1983, Sober 1984a). That is, a fitness function would itself be a function from a phenotypic frequency to an array of frequencies which describe the possible outcomes after selection.[4]

The role of fitness within a probabilistic evolutionary framework can be clarified by examining the relationship of random drift and selection. Let us grant for the sake of argument that fitness is a supervenient property; that is, that fitness depends only upon "the manifest properties of organisms—their anatomical, physiological, and behavioral properties—right down to their molecular constituents and their interaction with physical properties of the environment" (Rosenberg 1985, p. 165).[5] Fitness will be "objective" in the sense that it is a real and measurable feature of organisms. Yet there will be no "manageably small" number of manifest properties in terms of which a particular fitness level can be defined. Fitness of *types* may still be a statistical property; for example, we may identify the fitness of a phenotype or genotype with the mean value of the realized values for individuals. This is suggested by Sober's comment that

> Evolutionary theory understands fitness as a probabilistic quantity. An organism's fitness ... is its *chance* of surviving. A genotype's fitness is the average of the relevant probabilities attaching to the organisms who have that genotype. The genotypic fitnesses are survival probabilities, which is to say that they represent the average chance an organism of a given type has of surviving from egg to adult. (1984a, p. 43)

Sober's claim must be understood with care. His discussion is not altogether transparent. It could be understood as defining fitness by a simple frequentist averaging procedure (Horan 1991); however, we think it is better read in terms of models of fitness in infinite ensembles of populations, as described below.

Evolution is certainly a matter of chance. Genetic drift is simply the "error" in transmission of types from generation to generation, arising from finite population size. Drift is standardly treated using models incorporating infinite, or effectively infinite ensembles of finite populations. Given a single gene with two alleles, in the absence of selection ensembles of populations initially polymorphic at that locus will tend to disperse across a wide range, from populations fixed for one allele to populations fixed for the alternative allele (see Falconer 1989 or Roughgarden 1979). Once a population is fixed for one allele, it will not change unless there is significant mutation. The extremes are therefore absorbing states, from which populations will not deviate. The end result will be that, as the ensemble disperses, each population will tend to be monomorphic. In the limit, the ensemble of populations will bifurcate into a bimodal distribution at the two extremes. (For experimental confirmation, see Kerr and Wright 1954.)

The rate at which this dispersion occurs will depend critically on N_e, the effective population size. The overall effects will be diffusion of populations in the ensemble, differentiation between populations, and uniformity within populations. This is a *neutral* case, in which there are no relative selective advantages. Selection, by contrast, is treated in these models as a *deterministic* process; that is, if selection were the *only* force operating, then a given frequency within a population at one time will yield a unique distribution of frequencies in the next generation. This is expressed mathematically by supposing that we have infinite or effectively infinite populations, in which sampling error, and therefore drift, could not occur. In these cases, the change in $\bar{z}_t$, the mean value of some phenotypic trait z at time t, will be

$$\Delta\bar{z}_t = \bar{z}_{t+1} - \bar{z}_t = h^2\sigma^2(s\bar{z}_t - \bar{z}_t),$$

where $\bar{z}_t$ is the mean value of z in generation t (before selection), s the selection coefficient, and $h^2\sigma^2$ the heritable variance. In an infinite population, these changes would *always* be a deterministic function of fitness and will *always result in an increase in fitness as a function of the selection differential and heritable variance in fitness*. Changes in fitness will be a function of the local adaptive topography. When confronted with finite populations (cf. Lande 1976; Wright 1931, 1932), drift is incorporated as the random exploration of adaptive zones; that is, drift is expressed as changes in frequencies that are *un*correlated with differences in fitness, or random with respect to selection (Beatty 1984). Drift would change the frequencies within populations in an infinite ensemble, but should leave the mean value of the distribution in the ensemble unchanged from the initial value, in the absence of directional selection.[6] Thus, the *variance* of the ensemble would change, though the *mean* would not. Fitness values will determine the strength and location of the central tendency within the ensemble, and the amount of drift will be seen as the amount of dispersal around that mean value. It is *only* in this sense that the fitness of a genotype or phenotype is given by the average chance of survival: drift alone will not change the *mean* value in an ensemble of populations, though it *will* change the variance.[7]

The general point is then rather simple. Fitness, as modeled in infinite populations with uniform environments, is strictly a deterministic function. Understood as a consequence law (cf. Sober 1984a), in infinite populations it allows a simple deterministic prediction of frequencies from one generation to the next, and even in ensembles of finite populations allows a unique prediction of the expected mean value (though the variance is a function of stochastic variation). Selection is *not* strictly a probabilistic process, but a deterministic component *within* a stochastic process. Fitness functions yield unique mappings from prior to subsequent frequencies, with absolute fitness increasing in proportion to differences in fitness when the population is not yet at an equilibrium state. There may be stochastic effects due, for instance, to drift and the result is an evolutionary process that is stochastic, but the effects of selection are not probabilistic. Likewise, the fitness of a phenotype may be represented as an array of values, as it is by Beatty and Finsen, depending on the likelihood of encountering various environments, and thus be a probabilistic function in a heterogeneous environment. In this case, the stochastic component to the evolutionary process follows directly from the grainy character of the environment, and the environmental relativity of fitness.

The case is more complicated in finite populations, for a number of reasons. Since there cannot be continuous variation of frequencies in such populations, there is generally significant rounding error; this means that the mathematical models discussed above have no univocal application. This effect will be compounded by the inevitability of drift and associated phenomena such as variance in heritability. It might be tempting to think that once we descend from the level of mathematical abstractions, we will find more fundamental sources for the probabilistic character of fitness. Perhaps so, but neither of us knows of any cases in which this is clearly so. In either case, the view that simply shifting to a finite population requires a probabilistic treatment for fitness is a mistake.

The point can be seen by turning to some simple models of population growth to explore the consequences of finite population size. All assume a homogeneous environment. In deterministic exponential models,[8] population increases as a simple exponential function over time. With stochastic effects introduced—for example, from environmental variance—the population dynamics yields an array of values. An actual growth curve may lie above or below the value predicted on the simple exponential models. The exponential curve reflects what is commonly called the *intrinsic rate of increase* or the *Malthusian parameter*. This represents what would be predicted on the basis of reproductive rates alone. The stochastic array is a function of the variance induced by the environment. Analogously, consider a case of simple directional selection, with complete dominance and selection against one homozygote. In the deterministic case, we have a simple exponential decay representing the changing expected frequencies. With stochastic effects introduced—for example, drift with a small population size—we obtain an array of values. An actual, finite, population may experience changes under selection which lie above or below the value predicted on the simpler model. The effect of selection, however, is still to define the *expected* value, or the *intrinsic* effects due to selection. Again, this is what would be predicted in the absence of drift. The variance is not due to selection, but to drift. By itself, the shift to finite population size provides no motivation for treating fitness as a probabilistic function. When N_E is finite, the *realized* fitness will vary stochastically and the variance will be inversely related to N_E. As N_E becomes very large, the variance will approach zero, and evolutionary change will approximate the deterministic limit. As N_E becomes smaller, the effects of drift become more pronounced, and may dominate selection. Nonetheless, even in this finite case, fitness is properly represented as a deterministic function, defining the *expected reproductive success*. It is a deterministic component in a stochastic model. Assuming that, in finite cases, fitness must be probabilistic risks conflating the effects of selection with those of drift, and collapses the distinction between expected and realized fitness. Fitness values tell us what the *expected* values are, even though we know that, in finite populations, drift, rounding error, and the like will disperse *actual* values around the mean this provides. Alternatively, the fitness of a phenotype becomes the mean value in a sufficiently large population. In any finite population, or any finite set of populations, even the average of the realized fitnesses will deviate from what would be expected if fitness were the only determinant of evolutionary change. Similar consequences follow when environmental heterogeneity induces variance: environmental variance results in fitness across the ranges of environments being a probabilistic distribution.

3. Why Propensities Rather than Limiting Frequencies?

Defenders of propensity interpretations offer a variety of reasons for treating the probabilities involved as propensities. Having dismissed the interpretation of fitnesses as actual frequencies, there are two options: fitnesses can be identified with objective propensities or with limiting frequencies. One standard interpretation of probability depends on the frequencies in the "long run" (Hacking 1965, ch. 4). For our purposes, it does not matter whether the "long run" is infinite or not, so long as it is effectively infinite. Given an environment or a distribution of environments, the limiting frequency of a genotype or phenotype is the frequency that will occur if the sequence is continued over the "long run". The question is whether, assuming a probabilistic interpretation of fitness, these probabilities should be interpreted as propensities rather than as limiting frequencies or frequencies in the "long run". Alternately, the question is whether, within a probabilistic framework, fitness should be interpreted as providing the likely expected values rather than the values expected in the "long run".

The problem is that limiting or "long run" frequencies are not always well defined; whether they are depends on whether the frequency in question "settles down" to a stable limit. As is now commonplace from the work of May (1972, 1973) and others, for some biologically plausible parameters, we encounter periodic oscillations in frequency—that is, limit cycles—or chaotic variation in relative frequency, destabilizing the system until it "crashes". Additionally, there are biologically plausible cases in which there are stable limiting frequencies whose value depends on the actual sequence of environments. Such is the case, for example, when two competing types are at or near an unstable equilibrium such that whether one or the other predominates depends on the actual sequence of environments; for example, one or the other may predominate depending on whether a series of relatively dry or relatively wet seasons occurs first. This is one aspect of the notorious role of historical accident in evolution. Lewontin (1985) holds that such sequence dependence is common in evolution. Cohen (1976) has shown that a number of biologically realistic cases can be simulated by sampling with replacement; this means that, again with reasonable parameter values, each run has a stable limit even though the limits of a series of runs are randomly dispersed. There is no unique and well defined limit in such cases. The risks of sequence dependence and of the non-existence of the relevant limits faced by "long run" interpretations of the probabilities show that they are in principle unsuited for defining fitness because they yield mistaken values.

The point is that attempts to provide a probabilistic definition of fitness by use of "long run" frequencies will often yield mistaken values, or no univocal value, for the relevant probabilities. They will yield no value whatsoever where there are limit cycles or chaotic regimes. They will yield no univocal value in cases of sequence dependence. Yet, fitnesses are generally regarded as well defined in such cases. We think they *are* well defined. So much the worse for this class of attempts to define fitness.

It is, perhaps, worth remarking on one way in which one might attempt to employ limiting or "long run" frequencies counterfactually to define or evaluate fitnesses. Some of the problems we have discussed depend in important ways on environmental variation. Since fitness is relativized to environment, it might appear that these problems are simply irrelevant. We might fix the environment exactly as it currently is—for example, by holding the relative number of competing organisms constant, or by fixing the range of temperature and rainfall excursions—and then evaluate the "long run" frequencies within these fixed regimes.[9] In fixing the biotic environment we exclude all forms of frequency dependent selection and limit cycle phenomena; similarly, in fixing the abiotic environment, we exclude interactions between organism and environment that in fact play a role in evolutionary dynamics. The most significant problem here is the systematic underestimation of the importance of variation in biotic interactions from one generation to the next. This introduces feedback relations which are crucial in assessing fitness. As Richard Levins and Richard Lewontin comment,

> The simple view that the external environment changes by some dynamic of its own and is tracked by the organisms takes no account of the effect organisms have on the environment. The activity of all living forms transforms the external world in ways that both promote and inhibit the life of organisms. (1985, p. 69)

For organisms, most strikingly in small populations, variation in population size from generation to generation by itself represents *variation in biotic environments* that may significantly affect fitnesses. The same may be said for changes in the relative frequency of alternative phenotypes (Maynard Smith 1982). Fixing the environment, whether biotic or abiotic, in calculating fitnesses is biologically unrealistic, and badly distorts the notion of fitness that is to be explicated. One can obtain stable long run frequencies by requiring a stable environment, but only at the expense of a realistic account of fitness. This maneuver cannot save an interpretation of fitness that depends on frequencies over the "long run".

4. Which Propensities?

Beatty and Finsen (1989) raise a number of difficulties for propensity interpretations of fitness. Some of their concerns seem to be mistaken. The central difference between their position and ours turns on the recognition that fitness is a relation between an entity in an evolving lineage (for example, an organism) and *its* environment. Specifically, we think it is important to insist on a careful analysis of *which* environments are relevant to an organism in assessing its fitness. This underscores questions of scale. Our analysis eliminates many of the propensities that Beatty and Finsen consider to be relevant measures of fitness, though more than enough remain as contenders.

The importance of issues of scale arise, for example, in Beatty and Finsen's claim that among the many propensities that might measure the fitness of a (type of) organism are the "expected time to extinction" (ETE). They (1989, pp. 20-1 and 26-7) also consider the use of long term effects as the relevant measure of fitness, evaluating these, for example, by reference to the relative numbers of distant progeny. Our assessment of the mutual relevance of long and short term fitnesses turns crucially on the issue of scale. Differential multiplication of organisms alters the selective environment (cf. Brandon 1990; Antonovics, et al., 1988); accordingly, long term *extrapolations* on the basis of current fitnesses are not appropriate measures of *current* fitnesses. Moreover, increases in fitness can trap lineages in evolutionary dead ends. Such traps are sometimes predictable. Rather than demonstrating that the resultant low long-term fitness makes the choice of an appropriate measure of current fitness more difficult, however, this argument shows the irrelevance of ETE and of expectations about long term fitness as measures of the fitness of *organisms.* It may be that ETE is an appropriate measure of the fitness (or a component of the fitness) of higher level taxonomic units such as species or genera (cf. German 1991; Sepkoski 1991). It cannot provide an appropriate definition or estimate of fitness at the organismic level, and therefore does not provide a useful *general* definition of fitness. Fitness is a measure of the relations of a unit to (expectable) *current* environments. Indeed, this is precisely what fitness must be for it to be *contingently* related to long-term evolutionary outcomes. Long-term survivorship propensities, even if well defined and calculable, are not fitnesses.[10]

It is important to recognize the fundamental move in the preceding argument. For a vast number of biological situations, the salient aspects of the selective environment are biotic. In cases of frequency dependence, for example, a new generation typically faces a different selective regime from preceding generations. Thus, proper delimitation of the biotic aspects of the selective environment is crucial to any measure of fitness. Since the biotic and abiotic environments exhibit significant change (cf. van Valen's 1973 red queen hypothesis), in the vast preponderance of biologically realistic cases any sound measure of fitness is restricted to the current range of (expectable) environments. This refocuses the difficulties in achieving propensity accounts of fitness to the problem of delimiting the appropriate range of current or currently expectable environments with which the organism (or, generalizing, the unit of evolution) must interact.

These considerations suggest a principle in assessing what constitutes a relevant environment, and the relevance of "long-term" considerations to fitness. In cases in which an organism (say, an insect with a two-week generation time) employs environmental cues to make facultative adjustments to long-term (say, seasonal) cycles, the interrelation between the organism and the environment justifies the inclusion of the range of environmental excursions over *that* particular length of time in evaluating *current* fitness. Where there is no such 'informational feedback' between environmental changes and the properties of the organism, the long-term considerations are not appropriate in evaluating current fitness. We do not exclude dynamic patterns, but insist that one be sensitive to matters of scale.

The notion of 'informational feedback' provides the basis of an account of relevance. It also accounts for cases where fitnesses are assessed across two or more generations. One example is Fisher's (1930) well-known argument explaining the rough equality of the sex ratio in most sexual species: to maximize representation in the F_2 generation, it pays to produce minority sex offspring in the F_1 generation. What is crucial to the assessment of relevance here is the informational feedback rather than the physical details on which it supervenes; the organism systematically abstracts from these physical details in fixing on features that it uses as cues. To assess the putative relevance of environmental features to selection and fitness, one must ascertain whether there is a feedback at the appropriate level. Where no environmental information is available about the relevant features of the environments of, for example, the distant future one can exclude long term measures of fitness. This provides the proper content for the metaphor of 'blind evolution'.

These points do not remove all the difficulties that Beatty and Finsen raise for the refinement of propensity definitions or measures of fitness. In particular, we offer no solution here to the problems posed by the fact that in very similar circumstances distinct statistical measures (sometimes variances, sometimes means, etc.) correlate with the expected contribution to the next generation of organisms or other units. As Beatty and Finsen show, such problems arise naturally when organisms (units of evolution) can produce varying numbers of offspring (descendant units) with some sort of probability distribution.[11] But though we have not handled *these* problems, our argument helps to separate difficulties which are important from those that are spurious.

To summarize, the best interpretations of fitness as a propensity make fitness a *causal consequence* of the features of an organism in a relevant range or distribution of (expectable) environments. The most sensitive issue that this leaves open is the delimitation of the relevant environments and the interpretation of relevance. Fitness for an organism is then a matter of the likely number or distribution of offspring given the relevant environments. Because organisms are members of potentially continuing lineages, fitness can predict evolutionary outcomes, although it is a weaker and more fragile indicator than many optimistic selectionists have believed.

5. Conclusion and Prospects

We have focussed mainly on issues regarding the concept of fitness in evolutionary theory, but our position has larger implications. Since fitnesses are measures of evolutionarily relevant interactions between organism and environment and since fitnesses cannot be analyzed solely in terms of the mechanistic underpinnings on which they depend, various versions of reductionism are misguided. In particular, we reject the consequences that Rosenberg, for one, draws from supervenience:

> ... everything that the theory of natural selection can explain about what is happening in a well-controlled laboratory experiment can be explained more deeply, more directly, and in greater detail by physiological and biochemical principles that *do not mention the supervenient evolutionary concept of fitness*. When enough theoretical and experimental detail has been gathered to make a prediction that specifically confirms the claims of evolution about the maximization of fitness, the theory of natural selection and the notion of fitness become *superfluous*: ... The prediction that can be extracted from [evolutionary] theory in such cases is at best generic, and the explanation it provides will be qualitative at most. (Rosenberg 1985, pp. 173-4, his emphases)

We need not dispute the reality of supervenience; however, an account restricted to the "vast and heterogeneous class of determinants of fitness" fails to capture the patterns and processes which evolutionary theory aims to explain. It is simply false that what is explained by fitness can be explained by "physiological and biochemical principles". For example, a 50:50 sex ratio in peas might be explained by cytological mechanisms, but it may require shifting to an evolutionary model such as Fisher's to explain the general prevalence of 50:50 sex ratios. The patterns and processes which are the core of evolutionary theorizing are intelligible only at a higher level of abstraction. The physical principles themselves provide an inadequate account of causal relevance. What supervenes on physical differences will be the actual, realized, fitness, and there will be a systematic difference between fitness understood as a propensity—that is, fitness as it is employed in evolutionary theory—and the values resulting from physical differences. Indeed, in any given finite population, even if the expected mean value for an evolving trait is a deterministic function of prior values, given selection and heritable variance, the *actual* mean value is more likely to deviate from the expected value than not, and the actual value is what will be determined by the physical "determinants of fitness". Expected fitness is indispensable, and cannot be defined in terms of the actual values achieved. An account which emphasizes supervenience fails to respect the distinction between expected and realized fitness. Of course, we could always pick out some subset of the phenotypic and genotypic properties which would yield appropriate fitness values, but there is no principled means, absent an account of the selective regime, for reading out the correct subset of physical characters from the "physical and biochemical principles" on which fitness is supposed to supervene. Not all the physical features of the organism or its environment are relevant to evolutionary processes or patterns, and the "physical and biochemical principles" on which fitness supposedly supervenes cannot tell us *which* features matter for evolutionary processes and patterns. Selection operates at a level between the chaos of detailed interactions particular to each organism and one that ignores the complexities of environmental variation and biotic interaction. The relevant selective environment, or range of expectable environments, is that in which there is informational feedback between the organism and the environment, both biotic and abiotic. This cannot be given a purely physical analysis.

Notes

[1]The names of the authors are in reverse alphabetical order; the order has no other significance. RMB was supported by a residential fellowship at the National Humanities Center, funded by the National Endowment for the Humanities, and by a Study-Research leave from Virginia Polytechnic Institute and State University. RCR was supported by grants from the National Science Foundation and enjoyed the Taft Committee at the University of Cincinnati, and a residence at the Department of Philosophy and Religion at North Carolina State University. We gratefully acknowl-

edge this support. We have benefitted from discussions with Janis Antonovics, John Beatty, Robert Brandon, Greg Cooper, Rebecca German, Donald Gustafson, Brent Mischler, W. E. Morris, and Kelly Smith.

[2]Unless otherwise specified, we use *fitness* to mean *expected fitness* as opposed to *realized fitness*. Brandon (1990) calls expected fitness *adaptedness*, distinguishing it from fitness. The significance of the distinction is the same.

[3]It is another matter, and equally important, to be able to discriminate the effects of drift and selection empirically. This turns out to be a difficult matter for a number of reasons, though it is one that can be done in some cases. It is clear in any case, that if the effects of selection and drift are collapsed and cannot be distinguished conceptually, any empirical discrimination would be impossible.

[4]There are a number of possible sources of stochastic effects that are not central for our discussion. If, for example, heritability is not perfect then developmental noise may change the distribution from what would be predicted on the basis of fitness differences alone. This clearly does not imply that fitness is itself a probabilistic quantity. Similarly, in finite populations there will be effects of averaging from theoretical selection values comparable to rounding error. Such effects are not stochastic; they do not support the claim that fitness *per se* is a probabilistic function.

[5]We trust that Rosenberg intends to include encounters with other organisms under "interaction with physical properties of the environment." A great deal turns on the way in which biotic interactions enter into fitness functions; such formulations as Rosenberg's underemphasize their importance.

[6]As Brandon has reminded us, drift will change the variance and also the mean within finite sets of populations. This is easily seen, since a finite set of finite populations will be a finite population, and thus subject to some sampling error.

[7]Even in the purely stochastic (neutral) case where there is no selection, the mean value of the ensemble would not be expected to change, despite the fact that there will eventually be no populations that are polymorphic.

[8]There are more realistic models incorporating density dependence, but that does not affect our central point.

[9]Brandon suggests averaging across environments. This would have the same effect.

[10]There is a difficulty regarding the characterization of 'long-term' here; consider the case of annual seasonal cycles in the case of short-lived organisms that adopt different morphs or switch from asexual to sexual reproduction on a seasonal basis.

[11]Roughly the problem is that the relative contributions of different organisms in a given distribution of expectable environments depends, in various circumstances, on the *variance* of the expected distribution of offspring number and on other statistical differences between them even when the *mean* number of expected offspring is identical as between the types (cf. Lacey, et al., 1983). We view this as a problem concerning the appropriate *measure* of fitness, not its *definition*.

References

Antonovics, J., Ellstrand, N., and Brandon, R. (1988), "Genetic Variation and Environmental Variation: Expectations and Experiments", in *Plant Evolutionary Biology,* L. Gottlieb and S. Jain (eds.). London: Chapman and Hall, pp.275-303.

Arnold, S.J., and Wade, M.J. (1984a), "On the Measurement of Natural and Sexual Selection: Theory", *Evolution* 38: 709-719.

________________. (1984b), "On the Measurement of Natural and Sexual Selection: Applications", *Evolution* 38: 720-734.

Beatty, J. (1984), "Chance and Natural Selection", *Philosophy of Science* 51: 183-211.

_____. (1987), "Dobzhansky and Drift: Facts, Values, and Chance in Evolutionary Biology", in *The Probabilistic Revolution,* volume 2, L. Krüger, M. Gigerenzer and S. Morgan (eds.). Cambridge: MIT Press/Bradford Books, pp.271-312.

_____. and Finsen, S. (1989), "Rethinking the Propensity Interpretation: A Peek Inside Pandora's Box", in Ruse (1989), pp.17-30.

Brandon, R. (1978), "Adaptation and Evolutionary Theory", *Studies in History and Philosophy of Biology* 9: 181-206; as reprinted in *The Nature of Selection,* E. Sober (ed.). Cambridge: MIT Press, 1984b, pp.58-82.

______. (1982), "The Levels of Selection", in *PSA 1982,* volume 1, P. Asquith and T. Nickles (eds.). East Lansing: Philosophy of Science Association, pp. 315-323.

______. (1990), *Adaptation and Environment.* Princeton: Princeton University Press.

Burian, R. (1983), "Adaptation", in *Dimensions of Darwinism,* M. Grene (ed.). Cambridge: Cambridge University Press, pp. 287-314.

Cohen, J. (1976), "Irreproducible Results and the Breeding of Pigs (or Nondegenerate Limit Random Variables in Biology)", *BioScience* 26: 391-394.

Dupré, J. (ed.) (1987), *The Latest on the Best.* Cambridge: M I.T. Press.

Falconer, D.S. (1989), *Introduction to Quantitative Genetics,* third edition. Essex: Longman.

Fisher, R.A. (1930), *The Genetical Theory of Natural Selection.* Oxford: Clarendon Press.

Fisher, R.A. (1938), *Statistical Theory of Estimation.* Calcutta.

German, R.Z. (1991), "Taxonomic Survivorship Curves", in *Analytical Paleobiology,* N. Gilinsky and P. Signor (eds.). Knoxville: University of Tennessee Press, pp. 175-184.

Gigerenzer, G., Swijtink, Z., Porter, T., Daston, L., Beatty, J., and Krüger, L. (1989), *The Empire of Chance*. Cambridge: Cambridge University Press.

Hacking, I. (1965), *Logic of Statistical Inference*. Cambridge: Cambridge University Press.

Hacking, I. (1975), *The Emergence of Probability*. Cambridge: Cambridge University Press.

Hodge, M.J.S. (1987), "Natural Selection as a Causal, Empirical, and Probabilistic Theory", in *The Probabilistic Revolution*, volume 2, L. Krüger, G. Gigerenzer and M.S. Morgan (eds.). Cambridge: MIT Press/Bradford Books, pp.233-70.

Hodge, M.J.S., and Kohn, D. (1986), "The Immediate Origins of Natural Selection", in *The Darwinian Heritage*, D. Kohn (ed.). Princeton: Princeton University Press, pp.185-206

Horan, B. (forthcoming), "The Statistical Character of Evolutionary Theory".

Kane, T.C., Richardson, R.C., and Fong, D. (1990), "Cave Fauna as Natural Model Systems: Cave Organisms as Model Systems", in *PSA 1990* Volume 1, A. Fine, M. Forbes and L. Wessels (eds.). East Lansing: Philosophy of Science Association, 151-64.

Kerr, W. and Wright, S. (1954), "Experimental Studies of the Distribution of Gene Frequencies in Very Small Populations of Drosophila melanogaster: I. Forked", *Evolution* 8: 172-77.

Kohn, D., ed. (1986), *The Darwinian Heritage*. Princeton: Princeton University Press.

Krüger, L. Gigerenzer, G., and Morgan, M.S. (1987). *The Probabilistic Revolution*. Cambridge: M. I. T. Press/Bradford Books.

Lacey, E.P., Real, L., Antonovics, J., and Heckel, D.G. (1983), "Variance Models in the Study of LIfe Histories", *The American Naturalist* 122: 114-131.

Lande, R. (1976), "Natural Selection and Random Genetic Drift in Phenotypic Evolution", *Evolution* 30: 314-334.

Levins, R., and Lewontin, R. (1985), *The Dialectical Biologist*. Cambridge: Harvard University Press.

Lewontin, R.C. (1985), "Population Genetics", *Annual Review of Genetics* 19:81-102.

May, R. (1972), "Limit Cycles in Predator-Prey Communities", *Science* 238: 900-902.

May, R. (1973), *Stability and Complexity in Model Ecosystems*. Princeton: Princeton University Press.

Maynard Smith, J. (1982), *Evolution and the Theory of Games*. Cambridge: Cambridge University Press.

Mills, S. and Beatty, J. (1979), "The Propensity Interpretation of Fitness", *Philosophy of Science* 46: 263-288; as reprinted in *Conceptual Issues in Evolutionary Biology*, E. Sober (ed.). Cambridge: MIT Press, 1984, pp.36-57.

Prout, T. (1971a), "The Relation Between Fitness Components and Population Prediction in Drosophila I. The Estimation of Fitness Components", *Genetics* 68: 127-149.

_____. (1971b), "The Relation Between Fitness Components and Population Prediction in Drosophila II. Population Prediction", *Genetics* 68: 151-167.

Rosenberg, A. (1983), "Fitness", *The Journal of Philosophy* 80: 457-473.

Rosenberg, A. (1985), *The Structure of Biological Science*. Cambridge: Cambridge University Press.

Roughgarden, J. (1979), *Theory of Population Genetics and Evolutionary Ecology: An Introduction*. New York: MacMillan.

Ruse, M. ed. (1989), *What the Philosophy of Biology Is*. Dordrecht: Kluwer.

Sepkoski, J.J. (1991), "Population Biology Models in Macroevolution", in *Analytical Paleobiology*, N. Gilinsky and P. Signor (eds.). Knoxville: University of Tennessee Press, pp.136-156.

Sober, E. (1984a), *The Nature of Selection*. Cambridge: MIT Press.

______. (1984b), *Conceptual Issues in Evolutionary Biology: An Anthology*. Cambridge: MIT Press/Bradford Books.

______. (1992), "Screening-Off and the Units of Selection", *Philosophy of Science* 59: 142-152.

Turner, J.R.G. (1987), "Random Genetic Drift, R.A. Fisher, and the Oxford School of Ecological Genetics", in Krüger, Gigerenzer and Morgan (1987), volume 2, 313-354.

Van Valen, L. (1973), "A New Evolutionary Law", *Evolutionary Theory* 1: 1-30.

Wright, S. (1931), "Evolution in Mendelian Populations", *Genetics* 16: 97-159.

______. (1932), "The Roles of Mutation, Inbreeding, Crossbreeding, and Selection", *Proceedings of the Sixth International Congress of Genetics* 1: 356-366.

Part X

QUANTUM THEORY II

Locality, Complex Numbers, and Relativistic Quantum Theory[1]

Simon W. Saunders

Harvard University

1. Introduction

What is relativistic quantum theory? How does it differ from non-relativistic quantum mechanics? Is there something more involved than the transition from the Galilean group to the Lorentz group?

These questions are evidently of great importance to the understanding of relativistic quantum theory (RQT), in particular to the distinction between the basic principles of quantum theory, and the characterization of a particular class of dynamical evolutions. Within non-relativistic quantum mechanics (NRQM) a reasonably precise response is possible (differences concern the number of degrees of freedom, the existence of internal symmetries, and the form of the Hamiltonian as a function of the canonical variables); what appears characteristic of RQT is the appearance of a dynamics of a qualitatively different character (involving the creation and annihilation of particles), involving new kinds of constraints (renormalization theory; anomalies). Nevertheless, it seems we must know how to pass from the inhomogeneous Galilean group (IHGG) to the Lorentz group (IHLG), if only because we have successfully constructed a RQT; surely, then, we know what it is that we hold invariant in this transition.

But RQT is not quite an open book, even for the simplest of couplings. It is a remarkable fact that there is still no mathematically well-defined interacting theory in 3+1 dimensions. This fact is all the more inscrutable in view of the relative ease and simplicity with which interacting theories can be formally defined, and the phenomenological accuracy of perturbative calculations. But in the simplest of all such theories, $\lambda\Phi^4$, triviality appears inevitable (Fröhlich 1982). An understanding of this situation, as conjectured by Wilson and Kogut in 1974, must be sought in the structure of the renormalization group: it seems that there is no non-trivial fixed point for the coupling constant. Evidently the traditional phrasing of dynamics, in which perturbative techniques yield a simple and phenomenologically accurate description, badly misleads. Rather than persist with such a dynamical picture, in which perturbative techniques can be easily applied only to be called somehow illusionary, it may be that other features of the dynamical setting should be held suspect. In particular, one might do better to seek a mathematical framework in which it is at once apparent

PSA 1992, Volume 1, pp. 365-380

that a non-trivial dynamics is problematic, and the conventional perturbative methods inapplicable.

That such a theory should exist is somewhat remarkable. At its core stands a certain perspective on two of the most elementary features of RQT. The first, indeed, is that very feature which blocks any extension of mechanics to the relativistic regime (forcing a field-theoretic approach), and that is the negative energy difficulty. Recall that even when one does not seek a representation of the full IHLG (e.g., giving up time-or space-translation invariance), there is no positive energy solution to the Klein-Gordon or Dirac equation, not even when the external potential is smoothly varying and constant in time (the Klein paradox).

The other relevant feature of RQT is considered more trivial, and that is the problem of localization: there is no satisfactory configuration-space Born interpretation. To be precise, there is no covariant commuting triple of operators which can be interpreted as position operators (and correspondingly, no velocity operators either). Here, unlike the negative energy difficulty, one is inclined to a certain indifference; one has, after all, a reasonable physical understanding of the impasse: on the one hand, precise particle locality implies an uncertainty in the momentum (and hence energy) which is so large that one can no longer be sure that one deals with a single particle. On the other hand, demanding only an approximate localization (within a region of the order of the Compton wavelength), the particle is thus localized at a given time, and relative to a different frame of reference it will not be localized to the Compton wavelength at any time. From the latter perspective it is neither surprising that Newton and Wigner were able to introduce non-covariant position operators, nor particularly interesting: such quantities pick out the center of mass of approximately localized particles. But here two considerations which should give us pause. First, it is not clear that one has a notion of "approximate" particle locality available, unless one has the precise statement to hand. Second, the non-covariance is irrelevant to the non-relativistic limit of the theory. Since, in particular, something in RQT must go over to the position operators in NRQM, Newton-Wigner locality should figure in any fundamental account of the relationship of RQT to NRQM.

In practice the problem of localization has long been brushed under the carpet of what used to be called the correspondence principle. Particle trajectories, as revealed by ionization chambers, are a micron or more in extent, whilst their curvatures (in the case of charged particles) are orders of magnitude more pronounced. In the laboratory one simply applies classical estimates to deduce particle momenta, and these directly correlate to the momentum-space Born interpretation (which is, of course, well-defined in Minkowski space). (One tends to forget that it was exactly to eliminate appeal to the correspondence principle that Dirac introduced a quantum theory of fields in the first place.) The maneuver is particularly inapposite when one considers space-times of low or vanishing symmetry. Quantum mechanical variables such as momentum and energy are derived from global symmetries. In the absence of these, one has only the coordinate-space densities.

In what follows I wish to consider a somewhat unorthodox point of view on RQT. I am not so optimistic as to suppose that it is a point of view that will triumph, but it is one that involves rather deep aspects of the mathematical structure of NRQM, leading to a simple, elegant, and compelling formulation of virtually all the curious features of relativistic kinematics. The foregoing difficulties are not solved; rather, they are presented in a new light. Further, many of these insights can be translated back into more conventional formalisms, particularly those of Section 7. This approach is due to Irving Segal, and was developed in the mid '60s; it is, I assume, because it was not successfully extended to non-linear fields, that it has not been more widely appreciated. But, in line with my earlier remarks, that it becomes at once apparent that a dynamics presents formidable difficulties, may be considered a virtue of the approach.

For simplicity I shall treat only the scalar field; for the application of the theory to spinor fields see Bongaarts (1972). For the further background, and the relationship with the Dirac vacuum, see Saunders (1991).

2. The Plane-Wave Expansion

When we attempt a direct assault on a dynamical problem, using perturbative techniques, one must make use of the plane-wave expansion, of the form:

$$\Phi(x) = \int (e^{-ip\cdot x}a(p) + e^{ip\cdot x}b*(p))d^3p \,/\, p_0$$

In the case of the real scalar field, a(p) is the annihilation operator for a particle of momentum p, and b*(p) a creation operator for the same particle. When the field is complex, the latter is the creation operator for an antiparticle, and the adjoint field involves the creation operator for a particle and the annihilation operator for an antiparticle. Since the antiparticle has opposite charge to the particle, annihilating a particle and creating an antiparticle have exactly the same effect as goes the total charge present: it is reduced by one unit. Therefore $\Phi(x)$ is the annihilation operator for charge, and $\Phi^*(x)$ the creation operator for charge. And this can be taken as the rationale for this particular interpretation of the plane wave expansion.

There is a more constructive basis to this interpretation (Weinberg 1964, Novozhilov 1975). One begins from the basic idea of creation and annihilation operators, appropriate to a particular irreducible representation of the IHLG; from these one attempts to define spacetime fields, which transform covariantly under the group representation, and which satisfy the principle of microcausality (that the fields commute or anticommute at spacelike separations). This is not sufficient to force the interpretation above, but it does require that the plane-wave expansion contains both creation and annihilation operators. What is sufficient is the further assumption that the fields transform under a U(1) gauge symmetry, $\Phi \rightarrow e^{iw}\Phi$, and that the corresponding conserved quantity is the charge. It is no longer possible for a and b* to be operators on the same Fock space (if b* = a*, and a transforms as $a \rightarrow e^{iw}a$, then $b^* \rightarrow e^{-iw}b^*$); they must act on different Fock spaces. Further, from the relation of the U(1) gauge to charge, if the one creates particles of charge e, the other annihilates particles of charge -e. Since the group theoretic properties of these particles are the same, they have identical mass and spin (and their representations are unitarily equivalent). We say the one is a particle, the other an antiparticle.

Evidently these gauge properties of the field are foreign to NRQT. The only analog concerns the global choice of gauge in defining a state vector, but to construe the U(1) gauge in this way would require that the particle and antiparticle states are *antiunitarily* equivalent. Further, whilst the distinction between the two parts to the plane-wave expansion is related to the duplexity in the sign of the relation between frequency and wave-number (energy and momentum in NRQM), these have nothing to do with positive and negative particle energies, which are defined rather through the group representation. The energy here is positive for both particle and antiparticle alike; the relation of antiparticle states in the mechanics to those in field theory is left hanging (correspondingly the charge conjugation operator is antiunitary in particle theory, unitary in field theory).

Nevertheless, the two kinds of U(1) gauge can be intimately related—as we shall see in detail in what follows. To accomplish this it is necessary to *identify* the antiparticle states with the negative-frequency one-particle solutions. The negative-energy difficulty is dealt with by the remarkable strategy of introducing a new (essentially geometric) imaginary unit. At the same time, this imaginary unit acts non-locally on the solutions to the wave equation. The locality and negative-energy difficulty are now related at the level of the action of the complex numbers. As for the usual imaginary unit, this too has a fundamental role: the action of the creation and annihilation operators defined with respect to this, expressed in terms of the those defined by the former, is exactly the content of the plane-wave expansion.

It is important to realize that something more than mathematical niceties are involved. With so much in place, the basic phenomenology of relativistic quantum dynamics -pair creation and annihilation processes—follows immediately: gauge invariant couplings in the field (of the form $\Phi^*V\Phi$) involve terms of the form a^*Vb^* and bVa. It also follows that a perturbative treatment will always involve arbitrarily many particles. But, as stressed in Section 1, it is another matter to actually implement this dynamics. In a certain sense, the complex structure to the Hilbert space itself must partake of the dynamical evolution.

The Segal approach is essentially a version of "algebraic" quantum field theory. It therefore distinguishes between an abstract algebra $\mathcal{A}$ of physical quantities (a C*-algebra), the basic object of the theory, and Hilbert-space representations of this self-same object. The Hilbert space is not something given *a priori*: in the generic approach (the GNS construction), it appears as a quotient space of $\mathcal{A}$, canonically defined in terms of a choice of linear functional (state) on $\mathcal{A}$. The complex numbers associated with the Hilbert space are inherited from the algebra $\mathcal{A}$ (see Saunders (1988) for a heuristic review). In the Segal approach they derive rather from a real-linear transformation on the classical solution manifold of the wave equations. That one can relate the classical solution manifold to an abstract C*-algebra is itself an insight basic to Weyl's formulation of the canonical commutation relations (the CCRs).

Our starting point can then be succinctly stated thus: we consider a linear Weyl system ($\mathcal{A}$,W,M,ω), where W is a map from M into $\mathcal{A}$, $\mathcal{A}$ is a C*-algebra, M is a real-linear vector space, and ω is a bilinear non-degenerate antisymmetric form on M. W satisfies the algebraic relationships (the Weyl relations):

$$W(\mathbf{a})W(\mathbf{b}) = \exp(-i\omega(\mathbf{a},\mathbf{b})/2)W(\mathbf{b})W(\mathbf{a}).$$

In the framework of geometric quantization, M is the classical phase space with symplectic form ω. If one has a Hilbert space representation of $\mathcal{A}$ in which $W(\lambda\mathbf{a})$, $\lambda\in\mathbb{R}$, is continuous, then by Stone's theorem there exists a self-adjoint operator $\psi(\mathbf{a})$ which generates this group. $\psi(.)$ is called the Segal field over M; as a consequence of the Weyl relations, it satisfies a version of the CCRs:

$$[\Psi(\mathbf{a}),\Psi(\mathbf{b})] = i\omega(\mathbf{a},\mathbf{b}). \tag{1}$$

Since a solution to the KG equation is fully specified by the value of the field and its first derivative on a spacelike hypersurface, we may identify the phase space of the field with its solution manifold. The Segal field can thus be considered a map from the latter to an abstract C*-algebra.

3. Non-Relativistic Quantum Field Theory

A number of the basic ideas so far reviewed, as well as certain others that we shall need, are beautifully illustrated by NRQFT. I follow the formulation of Cook (1953). We suppose we have a 1-particle system with Hilbert space $\mathcal{h} = L^2(V,d\mu)$, with inner product $<\varphi,\varphi'> = \int \overline{\varphi(v)}\varphi'(v)d\mu$; the (symmetric) Fock space is denoted $\mathcal{F}(\mathcal{h})$, on which are defined creation and annihilation operators a*,a as maps from $\mathcal{h}$ into linear operators on $\mathcal{F}(\mathcal{h})$. Given $\psi \in \mathcal{F}(\mathcal{h})$, denote by ψ^n the restriction of ψ to the n-particle subspace of $\mathcal{F}(\mathcal{h})$; for $\varphi \in \mathcal{h}$ the creation and annihilation operators have the explicit action:

$$(a*(\varphi)\psi)^n(v_1,v_2,\ldots v_n) = \frac{1}{\sqrt{n}}\sum_{i=1}^{n}\varphi(v_i)\psi^{n-1}(v_1,..,\overset{0}{v_i},..,v_n)$$

$$(a(\varphi)\psi)^n(v_1,v_2,\ldots v_n) = \sqrt{n+1}\int\overline{\varphi(v)}\psi^{n+1}(v,v_1,\ldots,v_n)d\mu$$

(where $\overset{0}{v}$ means this entry is deleted). The important linearity properties follow by inspection:

$$a*(i\varphi) = ia*(\varphi)$$
$$a(i\varphi) = -ia(\varphi) \tag{2}$$

i.e., the creation (annihilation) operator is linear (antilinear) respectively. The CCRs also follow:

$$[a(\varphi),a*(\varphi')] = <\varphi,\varphi'> \tag{3}$$

in terms of which the (anti)linearity properties flow from those of the inner product <.,.>. Defining the Segal field by:

$$\Psi(\varphi) = 1/\sqrt{2}(a*(\varphi)+a(\varphi)) \tag{4}$$

we find that:

$$[\Psi(\varphi),\Psi(\varphi')] = i\ \ \mathrm{Im}(<\varphi,\varphi'>). \tag{5}$$

Indeed $\mathrm{Im}(<\varphi,\varphi'>)$ is the symplectic form for the Schrödinger field, regarded as a classical field theory (cf. (1)). Explicitly, when V is $\mathbb{R}^3$ and $d\mu$ is d^3x:

$$\omega(\varphi,\varphi') = -\frac{i}{2}\int(\overline{\varphi(x)}\varphi'(x) - \varphi(x)\overline{\varphi'(x)})d^3x.$$

We note that the inner product <.,.> can be directly related to ω:

$$<\varphi,\varphi'> = \omega(\varphi,i\varphi') + i\omega(\varphi,\varphi'). \tag{6}$$

(5) is now a consequence of (1) and (6). Note also that, by virtue of the (anti)linearity properties (2), it follows from (4) that $\psi(i\varphi) = i/\sqrt{2}(a*(\varphi)-a(\varphi))$; inverting we obtain:

$$a*(\varphi) = 1/\sqrt{2}(\Psi(\varphi) - i\Psi(i\varphi))$$
$$a(\varphi) = 1/\sqrt{2}(\Psi(\varphi) + i\Psi(i\varphi)). \tag{7}$$

Evidently if we *define* the creation and annihilation operators by these equations, they will automatically satisfy (3) by virtue of (1) and (6). If, further, Ψ is linear with respect to multiplication of its argument by i, the relations (2) follow from (7). This definition, together with (1) and (6), now characterize the theory abstractly. This is the hallmark of the Segal approach.

The second quantization of 1-particle operators *O*, where *O* is a polynomial in functions on V and differential operators on V, is defined as:

$$d\Gamma(O) = \mathbb{I} \oplus O \oplus (O\otimes\mathbb{I} \oplus \mathbb{I}\otimes O) \oplus (O\otimes\mathbb{I}\otimes\mathbb{I} \oplus \mathbb{I}\otimes O\otimes\mathbb{I} \oplus \mathbb{I}\otimes\mathbb{I}\otimes O) \oplus \ldots$$

(ignoring questions of domain). If $\{\varphi_i\}$ is any orthonormal basis in $\mathcal{h}$ this expression can be written as:

$$d\Gamma(O) = \sum_{i,j} a^*(\varphi_i) < \varphi_i, O\,\varphi j > a(\varphi_j). \tag{8}$$

In particular, using the formal methods of delta-function normalization, we may choose the φ's as "position eigenfunctions" ($\varphi_i(u) \leftrightarrow \delta^3(\mathbf{x}-u)$), so that $<\varphi_i, O\varphi_j>$ goes over to $O\delta^3(\mathbf{x}-\mathbf{x}')$, the sum goes over to an integral over $\mathbf{x}$ and $\mathbf{x}'$, and we obtain $d\Gamma(O) = \int a^*(\mathbf{x})Oa(\mathbf{x})d^3x$; correspondingly, $a^*(\mathbf{x})Oa(\mathbf{x})$ is identified as a "local observable".

We have now assembled all the essential ideas that we shall need. The only difference, in the relativistic case, is that we do not suppose the imaginary unit given *a priori*; we suppose instead that we must find a (real-) linear canonical transformation J on the solution manifold M, which satisfies $J^2=-1$. Such a transformation is called a *compatible complex structure* (that it is canonical implies $\omega(J\omega,J\omega) = \omega(\varphi,\varphi)$, as in elementary mechanics; note that multiplication by i is a canonical transformation, with respect to the non-relativistic symplectic form). Equipping M with this complex structure, we may regard it as a complex-linear space, the relativistic analog of $\mathcal{h}$ (the 1-particle subspace of the Fock space). The *classical* field solutions then automatically correspond to the quantum 1-particle states—what Ashtekar and Magnon (1975) have called the fundamental expression of wave-particle duality. The essential difference between relativistic and non-relativistic theory is then simply stated: in the former case, there are two complex structures available, which effectively coincide in the non-relativistic limit.

4. Locality in Quantum Field Theory

Whilst there is no causal structure present in the Galilean-covariant theory, there is nevertheless an expression of locality. In particular, one has the formal equal-time relationships:

$$[a(\mathbf{x}),a^*(\mathbf{x}')] = \delta^3(\mathbf{x}-\mathbf{x}') \tag{9}$$

obtained from Eq.(3) by the same formal methods we used to define local observables. The number density operator is a local function in the fields ($\rho(\mathbf{x}) = a^*(\mathbf{x})a(\mathbf{x})$), and hence commutes at spatially separated points. The interpretation of the number density operator as the q-number version of the (configuration space) probability density is further born out by the formal identification:

$$\rho(\mathbf{x}) = d\Gamma(\delta_{\mathbf{x}})$$

in which we regard $\delta_{\mathbf{x}} = \delta^3(\mathbf{x}-u)$ as a (1-particle) multiplicative operator[1] on $\mathcal{h}$. Note that its expectation value in the state φ is $\overline{\varphi(\mathbf{x})}\varphi(\mathbf{x})$; correspondingly, the expectation value of $d\Gamma(\delta_{\mathbf{x}})$ in an arbitrary state of $\mathcal{F}(\mathcal{h})$ yields a sum over analogous quantities, for each 1-particle state occurring in the Fock space state, weighted according to the integral of other states entering into this state. Except for the fact that all permutations of 1-particle states figure, this quantity is identical in all respects to that proposed by Schrödinger (1927) as the "matter field density" for a many-particle system.

The CR (9) for the fields a,a* is necessary and sufficient for the simultaneous "diagonalization" of the (uncountable infinity of!) "point" densities $\rho(\mathbf{x})$; this CR follows from the local form of the CR (3), "local" in the sense (i) *in configuration space* the supports of the functions φ,φ' can have zero intersection (ii) their inner product then vanishes. In the case of the Segal field $\Psi(\mathbf{x})$, it is the locality of the imaginary part which is relevant (i.e., the symplectic form). (Note that when creation and annihilation operators are defined by (7), the commutators $[a(\mathbf{x}),a(\mathbf{x}')]$ and $[a^*(\mathbf{x}),a^*(\mathbf{x}')]$ automatically vanish, for any compatible complex structure, whether or not the symplectic form is local.)

This notion of locality evidently applies when we stick to a Hilbert space setting and make no appeal to delta-function normalization. Although the "point" fields are unavoidable in the non-linear case, they are at best bilinear forms on $\mathcal{F}(h)\times\mathcal{F}(h)$ (the local densities are in general not even definable in this way). Their use will be avoided in what follows.

We see that the locality condition appears necessary if the associated c-number distributions are to be considered as arbitrarily specifiable initial data for the system in question. Related, as it is, to the possibility of obtaining data from measurements at spatially separated regions "without disturbance", the locality condition has a common origin with the requirement of microcausality in RQFT, despite its presence in a theory which admits no upper bound on the propagation of causal influences. The precise relationship between microcausality and the locality condition formulated above demands a more elaborate discussion of the relationship between quantum fields, as maps from a space of distributions on $\mathbb{R}^4$ (the "smeared fields"), and quantum fields as maps from c-number field configurations (or equivalently 1-particle states); in essence, the latter correspond to the quantization of the constant vector fields on the solution manifold, the former to the quantization of the observables on this manifold which generate these vector fields (*cf.* the dual role of observables as vectors in phase space, and as generators of vector fields on phase space, familiar in classical mechanics). For a clear and illuminating discussion, see Segal (1967). In the present framework, we gain a certain mathematical simplicity at the expense of a manifestly 4-dimensional setting: all commutation relationships are defined on a spacelike hypersurface (when the coordinate system is adapted to these, they are therefore "equal-time" CRs). However, it is apparent that commutators which are local, in the above sense, and which moreover have a covariant expression, will as a consequence satisfy microcausality.

5. The Connection Between Microcausality, Locality, and Antimatter

Recall that, when φ is a (complex) solution of the free KG equation, the form

$$(\varphi,\varphi') = i\int(\overline{\varphi}\frac{\partial}{\partial t}\varphi' - (\frac{\partial}{\partial t}\overline{\varphi})\varphi')d^3x \tag{10}$$

is Lorentz-invariant. Restricting to positive frequency solutions it is also positive definite, so one may use it to define an inner product on the space of such solutions, and thence a Hilbert space M^+ Similarly we can define the Hilbert space of negative-frequency solutions M^-(using the negative of (10)). Such constraints do not appear in NRQM, but M^+ can be cast in the familiar form of the latter theory by passing over to momentum space. One has the integral representation:

$$\varphi(x) = (2\pi)^{-3/2}\int e^{-ip\cdot x}\delta(p^2 - m^2)\Theta(p_0)\hat{\varphi}(p)\ \sqrt{2}d^4p$$

(where $\Theta(x) = 1$ for $x > 0$ and zero otherwise). Integrating over p_0 we obtain:

$$\varphi(x) = (2\pi)^{-3/2} \int_{P^3} e^{-ip \cdot x} f(\mathbf{p}) d^3p / \sqrt{2p_0} \qquad (11)$$

(where $f(\mathbf{p}) = \varphi(\sqrt{\mathbf{p}^2 + m^2}, \mathbf{p};\ p_0 = \sqrt{\mathbf{p}^2 + m^2})$.

Using this transform (what is sometimes called the "covariant Fourier transform") one can verify directly that the inner product takes the familiar form:

$$(f,g) = \int_{P^3} \overline{f(\mathbf{p})} g(\mathbf{p}) d^3p / p_0 \qquad (12)$$

Conversely, any function f on $\mathbb{R}^3$, square integrable according to (12), defines a positive frequency solution of the KG equation via (11).

The similarity with the NRQT is more apparent than real, however. The localization problem is normally viewed as a consequence of the fact that the triple $i\partial/\partial p_i$ is not self-adjoint with respect to the inner product (12). There is, however, a more fundamental phrasing of the problem (cf. Streater 1988): the positive-frequency solutions cannot have any straightforward relationship to physical data susceptible to laboratory control. Such solutions must be boundary values of functions everywhere analytic on the lower half-plane (that is, with *t* negative imaginary). It follows that they cannot vanish on any open set in $\mathbb{R}^3$. It seems, therefore, that whatever the difficulty of defining position operators, quantities such as $\varphi(\mathbf{x},0)$ can never be zero on any open set in $\mathbb{R}^3$, so they cannot possibly represent anything *subject to local laboratory operations parameterized by these coordinates.* We must conclude that such data is globally determined—that it is non-local, with respect to laboratory operations parameterized by the coordinates **x**.

This fact is perfectly well-known to physicists, but its general heuristic significance deserves, I humbly submit, wider appreciation. To this end, it is worth going into the argument in a little more detail. Suppose that f vanishes for negative p_0

$$\text{(i.e.,}\quad f(-\sqrt{\mathbf{p}^2 + m^2}, \mathbf{p}) = 0).$$

We then have, from (11):

$$\varphi(\mathbf{x}) = (2\pi)^{-3/2} \int_{P^3} e^{-ip_0 t} e^{i\mathbf{p}\cdot\mathbf{x}} f(\mathbf{p})\ d^3p / \sqrt{2p_0}.$$

It follows that as a function of *t*, φ is the boundary value of a function analytic everywhere on the lower half plane, since by assumption p_0 is positive definite. Therefore, by the "edge of the wedge" theorem, it cannot vanish on an open set in time (Streater and Wightman 1964, Th.2.17). So much is also true of the non-relativistic theory (for this reason it makes no sense to speak of localization in *time*). But here there is an important difference between the Galilean and Lorentz theories: the KG equation is hyperbolic, the Schrödinger equation parabolic; in the first case, but not the second, the value of a solution at a point $(\mathbf{x},t)$ is determined by its value (and first derivative) on the intersection of a spacelike hypersurface $\mathcal{Y}$ with the past light cone $\mathcal{Y}^-$ from the point $(\mathbf{x},t)$. That is, $\varphi(\mathbf{x}',0)$ and

$$\frac{\partial}{\partial t}\varphi(\mathbf{x}',t)\ |_{t=0}\ = \dot{\varphi}(\mathbf{x}',0)$$

with $\mathbf{x}' \in \mathcal{Y}$, determine the value of φ everywhere within the 4-volume $\mathcal{T}$ bounded by the 3-surfaces $\mathcal{Y}$ and $\mathcal{Y}^+$ If, then, φ and $\dot{\varphi}$ vanish on $\mathcal{Y}$ they vanish everywhere in $\mathcal{T}$; in particular φ vanishes on an open set in time; therefore it vanishes at all times.

This observation also helps us to understand the relationship between locality and microcausality. For suppose we take M^+, with the inner product (10), and define a representation of the creation and annihilation operators on $\mathcal{F}(M^+)$ to obtain the CR:

$$[a(\varphi), a*(\varphi')] = (\varphi, \varphi') = i\int (\overline{\varphi}\frac{\partial}{\partial t}\varphi' - (\frac{\partial}{\partial t}\overline{\varphi})\varphi')d^3x. \tag{13}$$

The RHS vanishes when the supports of $(\varphi, \dot{\varphi})$ and $(\varphi', \dot{\varphi}')$ do not intersect—and since the formulation is covariant, we would expect that in this way we establish microcausality (for the creation and annihilation operators). But we know that these operator fields do *not* satisfy microcausality (we must combine creation and annihilation operators to obtain causal fields). So it seems that locality together with covariance does *not* ensure microcausality. But, referring to our criteria for locality, the reader will see that (i) is not in fact satisfied: the supports of $(\varphi, \dot{\varphi})$ and $(\varphi', \dot{\varphi}')$ cannot have zero intersection, for neither can vanish on any open set of $\mathbb{R}^3$. To obtain locality, and therefore microcausality, we have to give an account of negative energies, or equivalently, of antimatter, at the level of the c-number solutions of the wave equation.

The heuristic insight which I have in mind is that the hyperbolic character of the wave equation in special relativity requires that the positive frequency c-number solutions be non-local, and that it is only when they are combined with the negative frequency parts, which must then be physically interpreted at the level of the state-space, that we can obtain locality, and hence causality for the associated operator fields. But how do we formulate a physical interpretation for the negative-frequency c-number solutions? For this we must turn to the Segal theory.

6. Segal Quantization of the Scalar Field

For the real field we have:

$$\varphi(x) = (2\pi)^{-3/2}\int (e^{-p\cdot x}f(p) + e^{-ip\cdot x}f(p))d^3p / 2p_0. \tag{14}$$

The amplitude of the second term must equal the complex conjugate of the first, if φ is to be real. In view of this correspondence, we can identify (complex!) 1-particle states f with *real* field solutions. We recall Dirac's concern with the fact that one must specify both φ, and its first derivative, to determine a "wave function" (unlike the usual situation in NRQM); since the states of NRQM are complex, and thus determine two real fields, we see that it is the pair $(\varphi, \dot{\varphi})$ which is the analog of the non-relativistic state.[2]

The symplectic form is (see Woodhouse 1980):

$$\omega(\varphi, \varphi') = \int (\varphi\frac{\partial}{\partial t}\varphi' - (\frac{\partial}{\partial t}\varphi)\varphi')d^3x = \int (\varphi\dot{\varphi} - \dot{\varphi}\varphi')d^3x.$$

The real functions φ, φ' are unconstrained, so the symplectic form is local. Since it is also covariant, the Segal field Φ, obeying the CCR of Eq.(1), is causal.

What is the analog of multiplication by i, applied to the complex momentum-space states, as a transformation on the real field solutions? Obviously it is not multiplication by i! To work it out, we simply appeal to (14); from this we have:

$$\varphi(\mathbf{x},0) = (2\pi)^{-3/2}\int (e^{i\mathbf{p}\cdot\mathbf{x}}f(\mathbf{p}) + e^{-i\mathbf{p}\cdot\mathbf{x}}\overline{f(\mathbf{p})})d^3p / \sqrt{2p_0}$$

$$\dot{\varphi}(\mathbf{x},0) = -i(2\pi)^{-3/2}\int (e^{i\mathbf{p}\cdot\mathbf{x}}f(\mathbf{p}) - e^{-i\mathbf{p}\cdot\mathbf{x}}\overline{f(\mathbf{p})})p_0 d^3p / \sqrt{2p_0}.$$

We rewrite the second of these using the (self-adjoint) operator

$$R = +\sqrt{-\Delta + m^2}$$

to obtain:

$$\dot{\varphi}(\mathbf{x},0) = -i(2\pi)^{-3/2} R \int (e^{i\mathbf{p}\cdot\mathbf{x}} f(\mathbf{p}) - e^{-i\mathbf{p}\cdot\mathbf{x}} \overline{f(\mathbf{p})}) d^3p / \sqrt{2p_0}.$$

It is now clear that under $f \to if$, the pair $(\varphi', \dot{\varphi})$ must transform as:

$$(\varphi, \dot{\varphi}) \to (-R^{-1}\dot{\varphi}, R\varphi) \tag{15}$$

that is, the new Cauchy data $(\varphi', \dot{\varphi}')$ is given by $\varphi = -R^{-1}\varphi$, $\dot{\varphi}' = R\varphi$ (this transformation is canonical; the associated complex structure is therefore compatible). But, as we shall see, the new Cauchy data has completely different support from the old; multiplying f by i is *not* a local operation in terms of the field configurations.

One might think that for complex fields this transformation will not be required; that in terms of the correspondence between states (as functions of momentum space) and classical field configurations, multiplication by i will be the same in the two cases. But this is to ignore the negative energy difficulty; this correspondence must also associate configuration space solutions with antiparticle states, in such a way as to ensure that the energy is positive. To see what is involved, we write $\varphi = \varphi^+ + \varphi^-$, with $\varphi^- = \overline{\varphi}^+$ (so that φ is real), and obtain (14) by appeal to (11). Multiplication of f by i corresponds to multiplication of φ^+ by i, but then it follows that φ^- must be multiplied by *minus* i to obtain a real solution. With this the action of the imaginary unit on φ^-, a plane wave φ^- will have positive energy. In the complex case, where there is no relationship between positive and negative frequency parts, we must still use this complex structure (given by (15), in terms of the Cauchy data), in order to ensure positive energy for the negative frequency solutions. This complex structure I shall call the *particle* complex structure (denote J_P); (ordinary) multiplication by i, the *charge* complex structure (denote J_C). The negative frequency solutions, using the particle complex structure, have positive energy; they are the antiparticle states.[3]

The generic form of the 1-particle inner product, in terms of a compatible complex structure J, is (*cf.* (6)):

$$<\varphi, \varphi'>_J = \omega(\varphi, J\varphi') + i\omega(\varphi, \varphi'). \tag{16}$$

This quantity is automatically anti-linear in its first entry, and linear in the second (taking J as the imaginary unit), as the reader can easily verify. If now we define creation and annihilation operators in accordance with (7) (with J replacing i with respect to its action on the φs):

$$a_J^*(\varphi) = 1/\sqrt{2}(\Psi(\varphi) - i\Psi(J\varphi))$$
$$a_J(\varphi) = 1/\sqrt{2}(\Psi(\varphi) - i\Psi(J\varphi)) \tag{17}$$

it follows from the CCR for the Segal field that (3) is satisfied with $<.,.>_J$ appearing on the RHS. As already remarked, from the (real) linearity of the Segal field, and using $J^2 = -1$, it also follows that $a(\varphi)$ is antilinear, and $a^*(\varphi)$ linear, as required. With their canonical action on $\mathcal{F}(M_J)$, we construct a concrete representation of Ψ, and thus of the Weyl algebra. The further requirement, that with respect to this complex structure the Hamiltonian be positive, is satisfied by the particle complex structure, but not by the charge one. But it is the creation and annihilation operators defined by the charge complex structure which are the physical fields, which add and subtract charge. Correspondingly, interactions gauge invariant with respect to the J_C unit preserve charge, whilst those invariant with respect to J_P preserve particle number. Only in kinematics can both be preserved (the non-relativistic limit is "kinematic" from the perspective of RQT). The relationship between the two kinds of operator fields follows from (17) (Bongaarts 1972). It is, as promised, of the form given by the plane wave expansion:

$$a_C(\varphi) = a_P(\varphi^+) + a^*_P(\varphi^-)$$
$$a^*_C(\varphi) = a_P(\varphi^-) + a^*_P(\varphi^+) \tag{18}$$

The CRs for these various fields are all determined by (3), where the RHS is given by (16). Evidently their locality and causal properties depend critically on the complex structure. In the case of the real field, we find (writing $\varphi(\mathbf{x},0) = f$, $\dot{\varphi}(\mathbf{x},0) = g$):

$$\omega(\varphi, J_p\varphi) = \omega((f,g),(-R^{-1}g, Rf)) = \int (fRf + gR^{-1}g)d^3x$$

(which is positive definite, since R, and hence R^{-1}, is positive). The hermitian inner product on M_p is:

$$<\varphi,\varphi'>_p = \omega(\varphi, J_p\varphi') + i\omega(\varphi,\varphi') = \int [(gR^{-1}g' + fRf') + i(fg' - gf')]d^3x. \tag{19}$$

It is an easy calculation to recover (12); clearly the energy is positive. But we see the explicit occurrence of R and its inverse in the real part of (19). This operator has been subjected to a systematic analysis by Goodman and Segal (1965), who showed that R is not only non-local, but *anti-local*, in the sense that for non-zero f in the domain of R, $(\text{supp}.f)^c \cap (\text{supp}.Rf)^c = \emptyset$ (here c indicates the set theoretic complement); that is, f and Rf cannot simultaneously vanish in any region unless f is zero (this result is only true for odd spatial dimensions). It follows that for any $E \subseteq (\text{supp}.f)^c$ that $(Rf)(x) \neq 0$ for all $x \in E$, and that as a consequence the real part of $<\varphi,\varphi'>_p$ is non-local; hence the particle creation and annihilation operators are not causal, as we have already seen in Section 4. (Since there is no other complex structure for the real field, there is no more to be said.) Exactly the same holds good for the complex field. The symplectic form is:

$$\omega(\varphi,\varphi') = 1/2\int (\bar{\varphi}\frac{\partial}{\partial t}\varphi' - (\frac{\partial}{\partial t}\bar{\varphi})\varphi')d^3x \ + \text{complex conjugate},$$

or equivalently, in terms of the (now complex) Cauchy data (f,g):

$$\omega((f,g),(f',g')) = 1/2\int (f\bar{g}' - g\bar{f}' + \bar{f}g' - \bar{g}f')d^3x.$$

and we find that the real part of the particle inner product is again non-local:

$$\omega((f,g), Jp(f',g')) = 1/2\int (gR^{-1}\bar{g}' + fR\bar{f}' + \bar{g}R^{-1}g' + \bar{f}Rf')d^3x$$

hence too the particle operators. The locality of the imaginary part guarantees that the Segal field is causal. But now we also have available the charge complex structure, $J_C:\varphi \rightarrow i\varphi$, or $(f,g)\rightarrow(if,ig)$; it too is compatible by inspection. The sesquilinear form defined by J_c is

$$<\varphi,\varphi'>_C = \omega(\varphi, J_C\varphi') + i\omega(\varphi,\varphi') = i\int (\bar{f}g' - \bar{g}f')d^3x. \tag{20}$$

That is, we have obtained the familiar "inner product" (10). We see that $<\varphi,\varphi>_C$ is real but indefinite (the complexified solution manifold M_C cannot then be made into a Hilbert space in the usual way[4]); on the other hand, (20) is patently local. Correspondingly, the charge creation and annihilation operators are causal. Since, strictly speaking, we do not have a Fock space, to define them properly we must express them in terms of the particle operators, with the result already given.

7. What is Local?

There is a remarkable simplicity and beauty to this formulation of RQT: *all* the curious features that appear in the kinematic structure of RQT can be introduced within the *canonical* non-relativistic framework at a single stroke: by the recognition that we must deal with two, quite distinct, kinds of imaginary unit, and that the one which attaches to the particle interpretation of the fields is non-local.

The connection with the Newton-Wigner representation is simply made. Consider again the non-local action of the particle complex structure $J_p:(f,g)\rightarrow(-R^{-1}g,Rf)$. Since R is positive and Hermitian we may take its square root S. Following Segal (1964), (1967), consider the new initial data $(\tilde{f},\tilde{g}) = (Sf,S^{-1}g)$. We now have the *local* action $J_p:(\tilde{f},\tilde{g})\rightarrow(-\tilde{g},\tilde{f})$. In the case of the real field we define:

$$\chi = \tilde{f}+i\tilde{g} = Sf+iS^{-1}g.$$

in terms of which $J_p\chi=i\chi$. The inner product may now be written as:

$$<\varphi,\varphi'>_p=\int \overline{\chi(x)}\chi(x)d^3x.$$

The KG equation, which in terms of the Cauchy data reads:

$$\frac{\partial}{\partial t}f = g, \ \frac{\partial}{\partial t}g = -R^2 f$$

becomes:

$$i\frac{\partial}{\partial t}\chi = iS\frac{\partial}{\partial t}f - S^{-1}\frac{\partial}{\partial t}g = iSg + S^{-1}R^2 f = R\chi$$

that is, we obtain a Schrödinger equation with Hamiltonian R. This form of the KG equation is familiar; it is the analog of the Foldy-Wouthuysen representation in the scalar case (we shall call it the Foldy- representation, after Foldy 1956). The coincidence of this representation with that of Newton and Wigner is well-known.

For the complex field we define:

$$\chi = Sf + iS^{-1}g$$
$$\xi = Sf - iS^{-1}g$$

whereupon $J_p:(\chi,\xi)\rightarrow(i\chi,-i\xi)$. In terms of these functions we find:

$$<\varphi,\varphi'>_p=\int\overline{\chi(q)}\chi(q)'d^3q+\int\xi(q)\overline{\xi(q)}'\,d^3q. \tag{21}$$

The complex KG equation is now:

$$i\frac{\partial}{\partial t}\chi = R\chi$$
$$-i\frac{\partial}{\partial t}\xi = R\xi$$

which establishes that ξ is the negative frequency part of the solution φ with Cauchy data (f,g). Equivalently, we may write:

$$J_p \frac{\partial}{\partial t}\binom{\chi}{\xi} = R\binom{\chi}{\xi}.$$

Despite the similarities to the Foldy representation, in the latter formalism the sesquilinear form is indefinite; previous treatments (Foldy 1956, Feshbach and Villars 1958) used the *charge* inner product. In terms of the pair (χ,ξ) this may be written:

$$<\varphi,\varphi'>_C = \int \overline{\chi(\mathbf{q})}\chi(\mathbf{q})' d^3q - \int \overline{\xi(\mathbf{q})}\xi(\mathbf{q})' d^3q \qquad (22)$$

(note carefully the differences with (21)).

If the new functions χ,ξ are unconstrained, it follows that *both* the particle and the natural inner products are local; however it also follows that fields φ, with Cauchy data (f,g), can no longer be freely specified (they, along with their local couplings, are anti-localized with respect to the data (χ,ξ)). If, on the other hand, we take as Cauchy data the functions (f,g), then the pair (χ,ξ) (and the data $(\tilde{f},\tilde{g})$) is constrained, and the particle inner product becomes non-local (in that case it also follows that neither integrand on the RHS of (22) can vanish; their difference does).

We cannot have it both ways. If $\tilde{\varphi}$ is what is local with respect to laboratory operations, then φ is not, and *vice versa.* Do we ever know, *with certainty*, that a single particle is localized in the laboratory? If we do, if particle number is what is local with respect to laboratory operations, then with respect to such operations φ is anti-localized. There is no contradiction with the result of Section 4, because $\tilde{\varphi}$ obeys a parabolic rather than hyperbolic equation. The fact that Newton-Wigner localized states can propagate outside the light cone is essential to the consistency of this result. But of course it is φ that is the covariant field, that is therefore local in the Einsteinian sense,[5] and with respect to which $\tilde{\varphi}$ is anti-localized. In particular, since the interactions usually introduced are local in this sense, they are anti-localized with respect to $\tilde{\varphi}$.

One would like to infer that it is the *charge distribution* that is local with respect to laboratory operations. From an epistemological perspective, this would mesh well enough with the operational definition of spatio-temporal intervals—that we count charge distributions and not particle number or mass as the fundamental macroscopic objects with respect to which space-time measurements are actually performed. It is the charge-current distribution of the "rigid bodies" used as measuring rods, and the bodies with respect to which the "radar method" yields a determination of distance, which are coupled to the radiation field (which also figure in contact forces), and not the gravitational coupling to mass. But it must be emphasized that although such interactions are local with respect to the data φ, and the (causal) fields create or destroy data φ, the action of the latter are represented by (18), that is, on $\mathcal{F}(M_p)$. Since, on pain of violation of charge superselection, coherent superpositions of particle and antiparticle states cannot be realized (equivalently: the $\mathcal{F}(M_p)$-vacuum is cyclic), the states actually produced by the action of these fields are always given by data $\varphi^{\pm}$, hence can never vanish on an open set in space.

One can also pose the dilemma as follows. To develop a phenomenological interpretation of such couplings, one must eventually (if only in an asymptotic-time limit, as in standard perturbation theory) represent their contribution to the Hamiltonian in terms of operators acting on positive-energy particles (and antiparticles). The normal ordering, which effects this representation in kinematics, is a non-local operation, by means of which one passes from the (formal) representation using the charge complex structure to that of the particle complex structure (Saunders 1991) (and one must suppose that the same will be true in dynamics). Inevitably, it seems, we must express the phenomenology in terms of positive-energy states, and these can only vanish over open sets in space when expressed by the data $\tilde{\varphi}$. What is local with respect to laboratory operations is, it seems, forever antilocalized with respect to what is "really" local, the data φ, whether referred to particle number and mass, or to charge.

It seems we have reached an impasse. One might hope to appeal to the infinite mass limit, in which case the spacetime intervals determined with respect to the data φ and $\tilde{\varphi}$ coincide, but this is no better than the appeal to the correspondence principle. Effectively, one simply consigns the difficulty to an area over-burdened enough already, the problem of measurement. Evidently we are far from a satisfactory understanding of the relationship of relativity to quantum theory, as Wigner (1986), Bacri (1988), and Fleming (1988) have recently stressed. It is, it seems to me, another virtue of the Segal approach that the difficulty is stated in such stark and simple terms.

Notes

[1]Such operators (properly speaking, bilinear forms) have received remarkably little attention in NRQM. What understanding we have is due to Wan and his collaborators (see Wan 1984 and references therein). The identity that holds between local q-number densities, and (the second quantization of) 1-particle bilinear forms, as also the formal similarity between the former and c-number 1-particle probability densities (the configuration-space expectation-values of the latter), is a consequence of the identity of the classical field solutions with the 1-particle *states*. To formulate a comparable equivalence in the relativistic case one must interpret the negative-frequency solutions as states (related to this, one must interpret delta-functions in terms of the states). This is essentially what is provided by the Segal approach.

[2]It is fortunate that Dirac did not have this insight; he might then have simply insisted that we work with the real KG equation, rather than the complex equation that Gordon and others actually used -throwing away the negative frequency parts as "unphysical" -and thus never have discovered the Dirac equation.

[3]In the non-relativistic theory we do not, of course, consider that the energy equals *minus* the square of momentum (divided by 2m). But why do we not, of course? (In the present framework: because we do not take $J = -i$. Equivalently: because we do not consider the complex conjugate wave-functions). In NRQT (as opposed to RQT) the imaginary unit occurs explicitly in the symplectic form; the two roots of -1 lead to two (non-coupling) theories, each represented by that root which yields a positive symplectic form (yielding the particle and antiparticle theory respectively). There is every reason to use both (antimatter also has a non-relativistic limit). Antimatter does not appear as an intrinsically relativistic effect—its dynamical role does (contrast *e.g.* Teller (1990), Weingard (1993)).

[4]If one could make sense of M_C as a Hilbert space, one would have available a "1-system" (i.e. particle-antiparticle system) theory in which states are in general indeterminate with respect to charge, but otherwise admit a covariant notion of localization. Evidently the physical interpretation of the localization difficulty, that one may change the particle number through pair creation, does not apply to charge. On the other hand, there seems little hope or reconciling such a physical picture with charge superselection (cf. the concluding comments of Section 7).

[5]Evidently Fleming's notion of "hyperplane covariance", applicable to the fields $\tilde{\varphi}$, goes some way to redressing the balance between the two notions of localization. There are intimate relations between Fleming's work and the Segal approach; see, in particular, Fleming and Bennett (1989), in connection with the operator R. It may well be that such a generalization of the notion of covariance to spacelike hyperplanes is all but inevitable, in view of the difficulties here encountered.

References

Ashtekar, A., and Magnon, A. (1975), "Quantum Fields in Curved Space-Times", *Proceedings of the Royal Society of London* A346: 375-94.

Bacri, H. (1988), *Localizability and Space in Quantum Physics.* Lecture Notes in Physics, Volume 308. Heidelberg: Springer-Verlag.

Bongaarts, P. (1972), "Linear Fields According to I.E. Segal", in *The Mathematics of Contemporary Physics*, R. Streater, (ed.). New York: Academic Press.

Cook, J. (1953), "The Mathematics of Second Quantization", *Transactions of the American Mathematical Society* 74: 222-45.

Feshbach, H. and Villars,F. (1958), "Elementary Relativistic Wave Mechanics of Spin 0 and Spin 1/2 Particles", *Reviews of Modern Physics* 30: 24-45.

Fleming, G. (1988), "Hyperplane-Dependent Quantized Fields and Lorentz Invariance", in *The Philosophical Foundations of Quantum Field Theory,* H. Brown and R. Harré (eds.). Oxford: Clarendon Press.

_ _ _ _ _ _ _. (1989), "Lorentz Invariant State Reduction, and Localization", *Philosophy of Science Association*, 2: 112-26.

_ _ _ _ _ _ _. and H. Bennett, (1989), "Hyperplane Dependence in Relativistic Quantum Mechanics", *Foundations of Physics*, 19: 231.

Foldy, L. (1956), "Synthesis of Covariant Particle Equations", *Physical Review* 102: 568-81.

Fröhlich, J. (1982), "On the Triviality of $\lambda\varphi_d^4$ Theories and the Approach to the Critical Point in d≥4 Dimensions", *Nuclear Physics* B200: 281-96.

Goodman, R., and Segal, I.E. (1965), "Anti-locality of Certain Lorentz-Invariant Operators", *Journal of Mathematics and Mechanics* 14: 629-38.

Novozhilov, Y. (1975), *Introduction to Elementary Particle Theory.* Oxford: Pergamon Press.

Saunders, S.W. (1988), "The Algebraic Approach to Quantum Field Theory", in *The Philosophical Foundations of Quantum Field* Theory, H. Brown and R. Harré (eds.). Oxford: Clarendon Press.

_________. (1991), "The Negative Energy Sea", in *The Philosophy of Vacuum*, S.W. Saunders and H. Brown (eds.). Oxford: Clarendon Press.

Schrödinger, E. (1927), "*Der Energieimpulssatz der Materiewellen*", *Annalen der Physik* 82: 265-72. English translation in E. Schrödinger, *Collected Papers on Wave Mechanics*. London: Blackie and Son, 1928.

Segal, I.E. (1964), "Quantum Fields and Analysis in the Solution Manifolds of Differential Equations", in *Analysis in Function Space*, W. Martin and I.E. Segal, (eds.). Cambridge: MIT Press.

______. (1967), "Representations of the Canonical Commutation Relationships", in *Cargèse Lectures in Theoretical Physics*, F. Lurçat, (ed.). New York: Gordon and Breach.

Streater, R.F. (1988), "Why Should Anyone Want to Axiomatize Quantum Field Theory?", in *The Philosophy of Quantum Field Theory,* H. Brown, and R. Harré, (eds.). Oxford: Clarendon Press.

________. and Wightman, A.S. (1964), *PCT, Spin and Statistics, and All That.* New York: W.A. Benjamin.

Teller, P. (1990), "Prolegomenon to a Proper Interpretation of Quantum Field Theory", *Philosophy of Science* 57: 594-618.

Wan, K.K., and Jackson, T.D. (1984), "On Local Observables in Quantum Mechanics", *Physics Letters* 106A: 219.

Weinberg, S. (1964), "Feynman Rules For Any Spin", *Physical Review* D133: 1318-32.

Weingard, R. (1992), "On the Interpretation of Quantum Field Theory", P*hilosophy of Science*, forthcoming.

Wigner, E.P. (1986), "Some Problems of Our Natural Sciences", *International Journal of Theoretical Physics* 25: 467-76.

Wilson, K.G., and Kogut, J. (1974), "The Renormalization Group and the ε Expansion", *Physics Reports* 12: 75-200.

Woodhouse, N. (1980), *Geometric Quantization.* Oxford: Clarendon Press.

Reversibility and the Interpretation of Mixtures in Quantum Mechanics[1]

Osvaldo Pessoa, Jr.

Universidade Estadual de Campinas, Brazil

The term "experimental philosophy" has been used to refer to the solution of what were considered philosophical problems by means of laboratory experiments. A recent example of this was the experimental violation of the Bell inequality, which ruled out certain philosophically appealing "realist local" theories as alternatives to quantum mechanics (QM).

Following the spirit of such experimental philosophy, this paper proposes a feasible test between two different interpretations concerning the nature of "mixtures" in QM. The use of delayed coincidence techniques seems to show that the process of mixing beams of light in different polarizations is reversible, favoring a weak version of the so-called "ignorance interpretation" over the "instrumentalist" view.

The first two sections survey in a conceptual way the philosophical discussion about the interpretation of mixtures. This is followed by a review of the mathematical notation and of procedures for preparing and analyzing mixtures. The argument that differently prepared but equivalent mixtures may be distinguished by measuring particle fluctuations is then showed not to be valid, at least in the example considered. This leads the way for the experimental argument proposed in favor of the ignorance interpretation.

1. The problem of interpretation of mixtures

"Given a beam of unpolarized electrons, should one think of each electron as having a definite spin orientation?" With these words, U. Fano (1957, p. 74) posed the problem of the *interpretation of mixtures* in QM. A beam with definite spin polarization corresponds to a "pure state" in QM, being represented by a state vector, while a beam of unpolarized electrons is usually associated with a "mixed state", being represented by a density operator. A single density operator can be resolved into many different combinations of pure states. Thus, the problem is whether the mixed state representing a given system should be thought of as a specific but unknown combination of pure states, or as a unique state in its own right.

In optics the discussion goes back to the 30's, when Birge had pointed out that there was "no scientific reason for the assumption, so commonly made in texts, that

PSA 1992, Volume 1, pp. 381-392

unpolarized light consists of *plane*-polarized components, oriented in all azimuths". He therefore concluded that "no experiment can give us any information on the nature of unpolarized light", besides the fact that "*if* such light is split into components, in any given apparatus, no preferential polarization will be found" (Birge, 1935, pp. 180, 182). If that is so, then the acceptance of an "instrumentalist" point of view would lead to the conclusion that *any* beam of unpolarized light is in the *same* state of polarization. That was the philosophical step taken by Fano (1957, p.76), for whom the definition of state should only be concerned with predictions of measurement outcomes in future experiments.

The opposing view, according to which an unpolarized beam of light consists of photons in definite but unknown polarization states, has been called the "ignorance" interpretation (IgI) of mixtures (term due to Putnam 1965, p. 98). It was the traditional view in statistical QM (see Fano, p. 76, and Park 1968, pp. 215-6), which separated clearly between the "objective" probability of measurement outcomes associated with pure quantum states, and the "subjective" probability arising from our incomplete knowledge of the microscopic state of a many-particle system.

Within this interpretation, one can speak of *different but equivalent* mixtures. This is the case of two beams A and B which are prepared by mixing different pure beams, but which yield the same mean values for any observable measured on the beams. If the preparation procedure is unknown, the ignorance view still maintains that equivalent mixed beams may be different.

In spite of being considered "unrealistic" by Fano (p. 74), the IgI can be classified as a "realist" view, since it conceives that distinct but indistinguishable physical states may underlie a same observable phenomenon. The belief in such reality arises from reasons other than empirical adequacy, reasons such as simplicity or uniformity of the physical theory. A mixture considered within the IgI will be referred to as a "realist" (or "classical") mixture. The "realism" of the IgI is however of a different sort from what is usually referred to as "the realist interpretation of quantum mechanics". For the latter, the elements of reality which underlie the observed phenomenon are variables which assign, for instance, well defined position and momentum to particles. The IgI, in contrast, is closer to what can be called "realism of the wave function" (the "naive realism" of Pearle 1986, p. 442), a view that assigns some sort of reality to probability amplitudes in configuration space.

2. The philosophical debate

In the 70's the discussion about the nature of mixtures became rather intense in the philosophical literature. We will examine an important argument in favor of the IgI in section 5, concentrating here on two different debates that took place, both involving *correlated* quantum mechanical systems. The typical example of such a composite system is the Einstein, Podolsky & Rosen (EPR) setup. In Bohm's (1951, pp. 614-619) version of the EPR setup, the correlated systems are two particles "entangled" with opposite spins, described by a pure composite state which has cylindrical symmetry about the trajectory of the particles. Another much explored example arises in the formal theory of measurements in QM. The measuring apparatus is considered a quantum system which becomes correlated to an object system during the measurement interaction.

The first debate arose after an argument due to van Fraassen (1972, pp. 325-31) that the IgI leads to inconsistencies. As pointed out by Hooker (1972, pp. 97-106) and Grossman (1974), implicit in van Fraassen's critique was the acceptance of the so-called "reduction assumption". This assumption applies to correlated systems which are described by a pure composite state. If such a state is non-factorable (entangled), then one cannot assign a pure state to each individual subsystem. For measurements performed on only one of the subsystems, the best available description of the subsystem is considered to be a "reduced density operator" obtained by means of the mathematical operation of taking a "partial trace" (see for instance d'Espagnat 1976, pp. 58-61). The reduction assumption accepts such an "improper mixture" as the actual state of the subsystem. There are no compelling reasons to accept such an assumption, so that the IgI can be sustained. There have even been attempts to dissolve the inconsistencies without rejecting neither the IgI nor the reduction assumption, within the framework of quantum logic (Gibbins 1983).

The problem raised for the IgI by the study of correlated systems is not that of formal inconsistencies. The problem is that when describing the state of a beam, we usually do not know whether the particles of the beam are correlated to other unobserved systems. Thus, given a beam of unpolarized light, we might have a classical mixture of pure states, or we might have one subsystem of an entangled composite pure state. This possibility weakens the ignorance interpretation. Our initial characterization of this realist view, to be called the *strong Ig*I, would have to be conditioned on the requirement that the beam is not correlated to other systems in the environment, which is quite a stringent condition. On the other hand, the *weak IgI*, which considers the possibility of correlations with the environment, does not answer Fano's question in the affirmative. A beam of unpolarized electrons might *not* consist of electrons in well-defined spin states (before a measurement takes place), if such electrons are correlated to other particles.

The second debate begun in the early 70's involved A. Fine's (1970) "insolubility proof" to the "measurement problem". Such a proof applies to *unitary* measurement interactions, *i.e.* it only involves the Schrödinger equation and not the projection postulate. Fine was able to show that there is no set of apparatus "pointer states" such that for any initial object state, the final composite state is in a realist mixture involving such pointer states. The important point for us here is that Fine's proof made use of the IgI, and assumed a special rule for the evolution of mixtures, which H. Brown (1986) called "real unitary evolution" (RUE). Such a rule applies the usual unitary evolution to each pure subsystem composing the mixture, leading to a new realist mixture. RUE prohibits replacing such an evolved state by any other mixture which according to the instrumentalist interpretation is equivalent to it.

The whole project of formulating insolubility proofs was criticized by Park (1973), who attacked the IgI which underlies Fine's approach. Contrary to Park's contention, however, the acceptance of his instrumentalist critique does not undermine the several *insolubility* proofs proposed in the literature, but only a possible *positive* solution to the measurement problem which would involve apparatus mixtures (see Pessoa 1990, p. 101).

Park (1968, pp. 214-7) is probably the most vocal defender of the instrumentalist position, within the framework of the "statistical interpretation of QM". According to this widespread view, the state vector does not refer to an individual system, but only to an "ensemble" of identically prepared systems. Park therefore does not have to address the additional problem of whether an unentangled individual particle can be in a impure mixture. According to the IgI, a single uncorrelated particle is always in a pure state, although our lack of knowledge might allow us to describe such a system as a mixture.

3. Mathematical notation

Having surveyed the problem of interpretation of mixtures in a conceptual way, let us introduce the mathematical notation for describing pure states and mixtures, and look at the experimental procedures for obtaining light beams in such states.

A pure state can be represented by a normalized vector $|\phi\rangle$ in an appropriate Hilbert space $\mathcal{H}$. According to the standard approach, the measurement of an observable represented by an operator $\hat{Q}$ in $\mathcal{H}$ yields as possible outcomes the eigenvalues a_i associated with the eigenvectors ϕ_i of $\hat{Q}$. If the pure state is written as a superposition of such eigenvectors, $|\phi\rangle = \Sigma_i\, a_i \bullet |\phi_i\rangle$, then the probability for an outcome a_i is given by $|a_i|^2$.

A mixture is represented by a density operator $\hat{W} = \Sigma_j\, w_j \bullet \beta\,[\Psi_j]$ acting on $\mathcal{H}$. The coefficients w_j give the probability of obtaining the eigenvalue β_j as the outcome of the measurement of an observable $\hat{R}$ with eigenvectors $|\Psi_j\rangle$. The operator $\beta\,[\Psi_j]$ projects any state vector onto the one-dimensional subspace spanned by $|\Psi_j\rangle$, and can be written[2] as $\hat{P}[\Psi_j] = |\Psi_j\rangle\langle\Psi_j|$. A mixture for which $\hat{W}^2 = \hat{W}$ corresponds to a pure state.

Now let us suppose that we are going to measure the observable represented by $\hat{Q} = \Sigma_i\, \alpha_i \bullet \hat{P}[\phi_i]$ for the system represented by the density operator $\hat{W} = \Sigma_j\, w_j \bullet \hat{P}[\Psi_j]$. What are the probabilities for obtaining the different eigenvalues α_i of $\hat{Q}$? One way of calculating this is by transforming from the basis $\{\Psi_j\}$ to the basis $\{\phi_i\}$, using a set of equations $|\Psi_j\rangle = \Sigma_i\, c_{ji} \bullet |\phi_i\rangle$. We would obtain: $\hat{W} = \Sigma_i\, \Sigma_i \bullet v_{ii} \bullet\bullet |\phi_i\rangle\langle\phi_i|$, where $v_{ii} = \Sigma_{d\, wj\, cji}{}^*$. The matrix $[\, v_{ii}]$is the "density matrix" in the representation $\{\phi_i\}$. The diagonal elements v_{ii} furnish the probabilities for measuring the eigenvalue α_i, while the off-diagonal terms (v_{ii}' ($i \neq i'$) express the "coherence" of the state, the fact that the system cannot be represented as a classical mixture of pure states $|\phi_i\rangle$. The density matrix can be shown to be self-adjoint ($v_{ii}' = v_{i'i}{}^*$) and positive definite (v_{ii} is real and ≥ 0), with unit trace ($\Sigma_i\, v_{ii} = 1$) (Fano, 1957, p. 77). In the representation $\{\Psi_j\}$, the density matrix has diagonal elements w_j and null off-diagonal elements.

Any density matrix may be diagonalized in some orthogonal basis of representation. Such a basis is unique if none of the diagonal elements are equal, and is a candidate for being the set of pure states that constitute the mixture, according to the IgI. But this interpretation should also allow for a mixture of non-orthogonal pure states, so that there will always be "ignorance" unless the *method of preparation* is known.

4. Preparation procedures

Let us now survey the operational procedures for characterizing the polarization state of a quasi-monochromatic beam of light. Assuming that the photons are not correlated to other systems, then it is sufficient to consider a Hilbert space of dimension K=2 spanned for instance by the vectors $|\phi_0\rangle$ and $|\phi_{90}\rangle$, which correspond to linear polarization at 0° and at 90°, in relation to some reference axis. To measure the polarization state of the beam, the usual procedure is to determine the four "Stokes parameters", which requires measurements of transmittance behind four filters: an isotropic filter, a horizontal linear polarizer (0°), a linear polarizer at 45°, and a right-circularly polarized filter (Shurcliff 1962, 19-25). With these K^2 real numbers one is able to determine the normalized density matrix representing the polarization state of light, which in general will not correspond to a pure state.

To obtain a pure beam, one can simply pass the original mixed beam through a dichroic polarizer, oriented say at 0°. Another way of doing this is to pass the beam through a birefringent analyzer such as a Wollaston prism, which separates the beam into two orthogonal components, say $|\phi_0\rangle$ and $|\phi_{90}\rangle$ (see figure 1). If a detector is placed in the channel corresponding to $|\phi_{90}\rangle$, the superposition between these two beams will be destroyed with a collapse of the state vector, and one can assume that the undetected beam that has been selected is pure and unentangled. In the 2-dimensional case there is a simple way to check whether the beam is pure, by measuring if all of the beam is transmitted through another appropriately oriented polarizer.

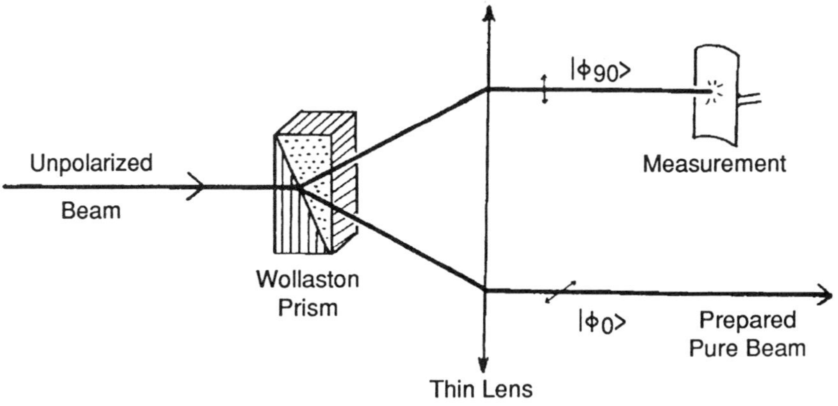

Figure 1: Preparation of a pure beam.

Once pure beams in different polarization states have been obtained in the laboratory, they can be "mixed" with each other, resulting in a beam which is a impure mixture. The simplest way for mixing two light beams is by using a beam splitter such as a half-silvered mirror (see figure 2a), but part of the beams is usually lost. In order to mix two beams with practically no losses, one can reverse the procedure for separating the orthogonal polarization components, as indicated in figure 2b.

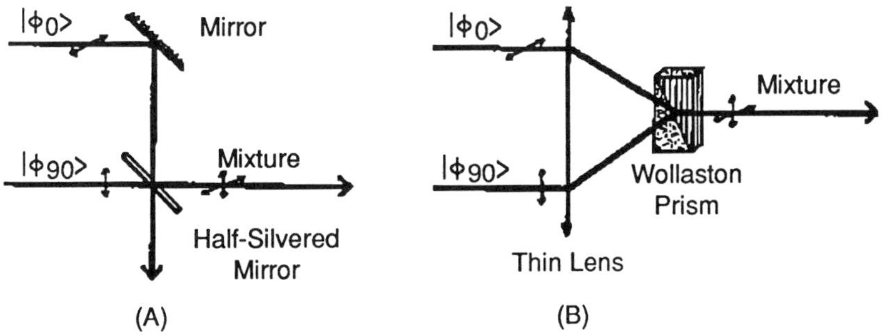

Figure 2: Procedures for mixing pure beams: (a) beam splitter; (b) reversed prism analyzer.

The use of *reversed analyzers* in polarization measurements dates back to Jamin in 1868, with the development of polarization spectroscopy (see Fran(on & Mallick 1971, pp. 55-63). The idea of using reversed analyzers to test the principles of QM was apparently introduced by Bohm (1951, p. 606), with the Stern-Gerlach apparatus, to show that the separation of a pure beam by an analyzer does not destroy the coherence between the beams (state collapse does not occur at the analyzers). The realization of this thought-experiment for single particles has only been achieved in the 80's, for neutron spin (Badurek et al. 1986, pp. 137-141) and for photon phase (Grangier *et al.* 1986, pp. 104-106).

Now that we are able to produce pure beams and to mix them, consider the two following mixtures (Park 1973, pp. 214-5). Mixture A is prepared by combining with equal intensities two pure beams linearly polarized at 0° and at 90°, the states of which are denoted by $|\phi_0\rangle$ and $|\phi_{90}\rangle$. The density operator is given by:

$$\hat{W}_A = (1/2)\cdot|\phi_0\rangle\langle\phi_0| + (1/2)\cdot|\phi_{90}\rangle\langle\phi_0| \quad (1)$$

Mixture B is prepared by combining equal amounts of pure beams linearly polarized at 45° and at 135°, in states denoted by $|\phi_{45}\rangle$ and $|\phi_{135}\rangle$:

$$\hat{W}_B = (1/2)\cdot|\phi_{45}\rangle\langle\phi_{45}| + (1/2)\cdot|\phi_{135}\rangle\langle\phi_{135}| \quad (2)$$

These latter states may be expressed as linear combinations of the states polarized at 0° and 90°:

$$\begin{cases} |\phi_{45}\rangle = (1/\sqrt{2})\cdot|\phi_{90}\rangle + (1/\sqrt{2})\cdot|\phi_{90}\rangle \\ |\phi 135\rangle = -(1/\sqrt{2})\cdot|\phi_0\rangle + (1/\sqrt{2})\cdot|\phi_{90}\rangle \end{cases} \quad (3)$$

Now when we represent $\hat{W}_B$ in terms of the basis $\{\phi_0, \phi_{90}\}$ we obtain the right-hand side of eq.(1). Mixtures A and B are therefore *equivalent*. They are both represented by the density matrix $1/2\cdot\hat{I}$, where $\hat{I}$ is the 2-dimensional identity matrix. This equivalence means that any attempt to distinguish the two mixed beams by means of a polarization analyzer fails. Whatever the orientation θ of the analyzer, the beam intensities measured in both channels would be the same.

Suppose that the analyzer prism is oriented at θ=0° so as to separate any beam into components polarized at 0° and 90° (figure 3). According to the IgI, the photons of mixture A that are in state $|\phi_0\rangle$ will all go through the same channel, falling on the same detector. Likewise for photons in state $|\phi_{90}\rangle$, which go through the other channel. In total, half of the beam will go through each channel.

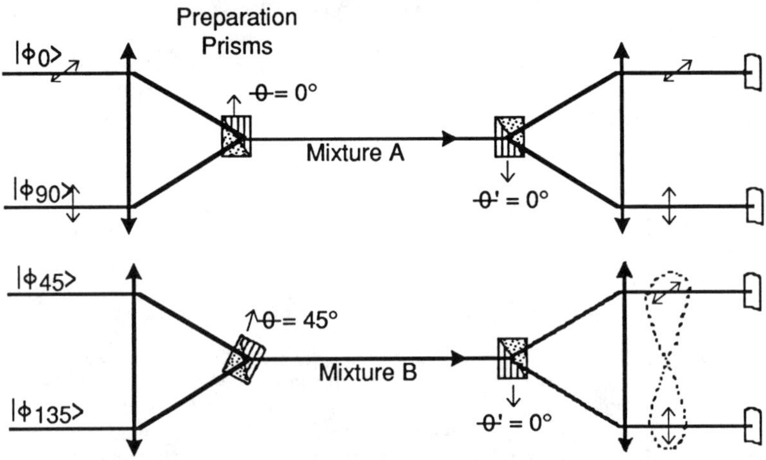

Figure 3: Measurement of polarizations at 0° and 90° on mixtures A and B.

In the case of mixture B, a photon in state $|\phi_{45}\rangle$ has a 50% chance of being counted in each of the two detectors. Likewise for a photon in state $|\phi_{135}\rangle$. In total, roughly half of the photons will be counted in each detector, which is the same as for mixture A. There is however a qualitative difference between the two cases. Before a detection occurs for mixture B, one cannot say that the photon went through one channel or through the other. The system is still in the superposition expressed by eq.(3), represented in figure 3 by dotted ribbons. One can say that the measurement of the polarization component oriented at 0° and 90°, on beam B, involves a *collapse of the state vector*, while on beam A it does not.[3]

5. Fluctuation argument for the ignorance interpretation

An instrumentalist tends to pay more attention to the practical limitations affecting operational procedures than a realist. Notwithstanding this tendency, two different arguments for distinguishing beams A and B have been given by defenders of the IgI, based on the existence of *particle fluctuations*. We will adapt such arguments, given originally for spin-1/2 particles, to the optical example presented in the previous section.

Consider the preparation of mixtures A and B. According to Grossman (1974, 333-338), one can never be sure that *exactly* half of the particles have been prepared in one of the two orthogonal polarization states constituting each mixture. The fluctuations in the preparation of the mixed beams would make them be described by slightly different density matrices, so that they could be distinguished.

The other argument is due to d'Espagnat (1976, pp. 100-102), for the case in which beams A and B have been prepared with *exactly* equal proportions of pure states. Consider again the measurement of linear polarization along 0° and 90° by means of an analyzer (figure 3). For beam A, since exactly half of the photons are in state $|\phi_0\rangle$, and half in state $|\phi_{90}\rangle$, then in the long run there will be no fluctuation in the number of counts obtained on each detector: $\sigma_A = 0$. For beam B, however, each photon has a 50% probability of being detected in each channel, and such random events are subject to fluctuations of the order of $\sigma_B = 1/2 \cdot \sqrt{N}$ (binomial distribution), where N is the mean number of photons per unit time in each beam. Mixtures A and B could therefore be distinguished by their particle fluctuations.

It is curious that these two argumments in favor of the IgI neutralize each other, like parallel coherent waves with opposite phases. D'Espagnat's fluctuations of *measurement outcomes* on the mixed beams is counterbalanced by the fluctuations arising in the *preparation* of the mixtures. In fact, if mixtures A and B are prepared by the procedure indicated in section 4 (figures 1, 2b and 3), and we consider that the initial light source has a Poissonian fluctuation of $\sqrt{N}$ (for a beam of N photons), then the measurements to distinguish beams A and B will yield the *same* fluctuation[4] of $\sigma_A = \sigma_B = (1/2 \cdot \sqrt{N}$.

The fluctuation argument for the IgI breaks down, at least for the preparation setup considered in this paper. The problem remains of whether such an argument can be made valid for some other experimental arrangement, or whether differently prepared but equivalent mixtures can never be distinguished by fluctuation measurements[5].

6. Reversibility as a test for the ignorance interpretation

The instrumentalist and the ignorance interpretations disagree on the issue of whether the process of mixing is *reversible*.

In QM, it is customary to assign an "entropy" to mixtures which is greater than the entropy of pure states (von Neumann 1932, pp. 379-90; Belinfante 1980, pp. 10-5). This accounts for the entropy increase accompaning measurements, since in general a measurement transforms a pure object system into a mixture (a process which involves state collapse). Such a definition of entropy fits in well with the instrumentalist view, for which the procedure of mixing different pure states is always *irreversible.*

For the IgI, however, if one knows how the pure beams were combined, then it is possible to reverse the mixing in a process of "unmixing" which yields the original pure beams again. Such a process would not have to involve any measurement, state collapse, dissipation of energy, or loss of part of the beams. If however one does not know how a mixture was prepared, then one would not know how to reverse the process.

The issue of the reversibility of mixtures can be used as a test in favor of the weak IgI. Consider a "photon cascade" in which pairs of photons correlated with orthogonal polarizations are emitted (see figure 4). The photon cascade used by Aspect *et al.* (1981) is obtained from a beam of calcium excited by tunable laser light, with the pair of correlated photons having frequencies 423 nm and 551 nm. Each pair may be emitted in any direction, but only those heading in opposite directions towards the detectors D_1 and D_2 are selected.

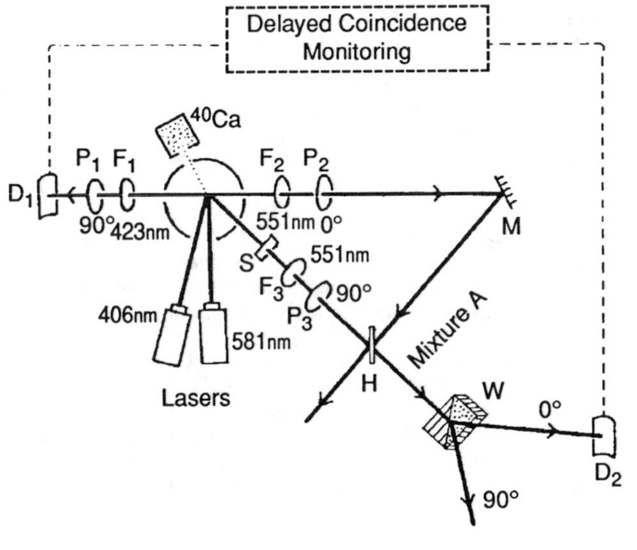

Figure 4: Experiment for reversing the process of mixing.

After appropriate filtration, pairs of photons may be detected in *delayed coincidence*. A photon detected at D_1 "triggers a temporal gate" for detection at D_2, and the probability of a count at D_2 becomes much greater than if no photon had been detected at D_1 (Grangier *et al.* 1986, pp. 101-102). This is a convenient way of "marking" or "individuating" an undetected photon, although in practice one cannot be sure that such a photon will fall upon the gated detector. While the shutter S is closed, the beam passing through analyzer W is pure and oriented at 0°, so that all of the beam falls on D_2 and one can measure a certain coincidence rate (in their setup, Aspect *et al.* (1981) measured 150 true coincidence counts per second, out of 10^5 individual counts per second at each detector).

Upon opening the shutter, an additional beam polarized at 90° can be mixed to the one polarized at 0° at the beam-splitter H, yielding mixture A defined in eq.(1). If this mixture is then separated by the analyzer into components at 0° and 90_, will the previous coincidence rate still be observed between counts in D_1 and D_2? We would expect so. The mere introduction of the additional beam, which does not modify the beam intensity detected at the final channel at 0°, should not affect the correlations, since different photons do not interact. *We expect the coincidence rate to be maintained* by the mixing and separation of the additional beam, and this would indicate that the *same photons* prepared at 0° from the cascade arrive at the detector after mixing and unmixing. We would have marked one of the two pure subsystems composing the mixture, and been able to retrieve the pure beam after unmixing.

Such a thought-experiment is relatively easy to perform, and if the result we expect turns out to be confirmed, then we have a good argument for claiming that the process of mixing is *reversible*. This lends support to the IgI, since such a view conceives that the pure subsystems composing the mixture maintain their individuality.

7. Conclusion

Two conclusions have been obtained. First, contrary to d'Espagnat's fluctuation argument, it seems that there is no simple way to distinguish between two "different but equivalent mixtures" if the preparation procedure is not known. We have arrived at this conclusion, however, only by looking at a particular type of experimental setup. The problem of whether this result is general or not thus remains open. In particular, it might be possible to distinguish beams described by the same density matrix by means of 2nd order coherence effects·5

The second conclusion is that there is an experimental argument for sustaining that the process of mixing light beams is reversible. In other words, if the preparation is known, we can devise a reversed setup so that we can be confident that the "same" particles constituting the original pure beams will constitute the corresponding unmixed pure beams.

As the photons are detected in D_2 they are not correlated to their pairs anymore, since these had to be previously measured in order to trigger the detection gate. We are therefore assuming that the particles constituting the mixture are not correlated to other systems. Thus, our argument can only lend support to what we have called the weak IgI.

We can now attempt to answer Fano's question. Given an unpolarized beam, we assert that each of the component particles is in a definite polarization state, as long as correlations with other systems in the environment can be neglected. There is however no simple operational means of distinguishing two unpolarized beams. Our adoption of the weak ignorance interpretation is based on those situations in which the procedure for preparing the mixture is known. Implicit in our answer, therefore, is the assumption that beams with known preparation have the same nature as beams the preparation of which we ignore.

Notes

[1]I wish to thank Linda Wessels and Stephen Kellert for discussions on the subject of this paper. Financial support was provided by the "Funda(No de Amparo " Pesquisa do Estado de SNo Paulo" (FAPESP).

[2]To see that such an operator effectively projects the vector $|\Psi\rangle = \Sigma_k b_k \cdot |\Psi_k\rangle$ onto $|\Psi_j\rangle$, yielding the vector $b_j \cdot |\Psi_j\rangle$, we just need to remember that the inner product $\langle\Psi_j|\Psi_k\rangle$ equals 1 if j=k, and 0 if $j \neq k$ (since the set of eigenvectors is an orthonormal basis of $\mathcal{H}$). We therefore have: $\hat{P}[\Psi_j]\,|\Psi\rangle = \Sigma_k b_k \cdot |\Psi_j\rangle\langle\Psi_j|\Psi_k\rangle = b_k \cdot |\Psi_j\rangle$.

[3]The photon is absorbed by an electron during detection, so it becomes ambiguous to speak of a "collapse" in one case but not the other. This ambiguity, however, does not arise for situations such as the detection of particles in a cloud chamber, where only part of the object's energy is absorbed by the detector. Another clear-cut example is that in which the object is an excited atom, and the emitted photon is taken to be the carrier of the interaction between object and measuring apparatus (see Pessoa 1990, pp. 81-91).

[4]Let us write out as $(n \pm \sigma)$ the number of particles and the standard deviation expressing the fluctuations (for reference, see Bevington 1969, pp. 33, 40, 60). For each of the four initial unpolarized beams subject to the Poisson distribution, we have:

$$(n_o \pm \sigma_o) = N \pm \sqrt{N} \tag{4}$$

The selection of a pure beam from each of these mixed beams (figure 1) leads to the beams $|\phi_o\rangle$, $|\phi_{90}\rangle$, $|\phi_{45}\rangle$, and $|\phi_{135}\rangle$ of figure 3, a process involving the binomial fluctuation:

$$(np \pm \sigma) = (1/2)\cdot(N \pm \sqrt{N}) \;\; (1/2)\cdot(N \pm \sqrt{N})^{1/2} \approx (1/2)\cdot N \pm (1/\sqrt{2})\cdot\sqrt{N} \tag{5}$$

For mixture A, all of the photons in the pure state $|\phi_o\rangle$ (and only these) go to the upper channel. At detector D_1 we therefore have:

$$(n_A \pm \sigma_A) \approx (1/2)\cdot N \pm (1/\sqrt{2})\cdot\sqrt{N} \tag{6}$$

For mixture B, we can first add the contributions from $|\phi_o\rangle$ and $|\phi_{90}\rangle$:

$$\begin{aligned}(n_m \pm \sigma_m) &= [(1/2)\cdot N + (1/2)\cdot N] \pm ([(1/\sqrt{2})\cdot\sqrt{N}]^2 + [(1/\sqrt{2})\cdot\sqrt{N}]^2)^{1/2} \\ &= N + \sqrt{N}\end{aligned} \tag{7}$$

This is the same fluctuation as in the source (eq. 4). Now since each photon in mixture B has a 50% chance of being detected in the upper channel, we apply the binomial fluctuation to the whole beam, as in eq.(5):

$$(n_B \pm \sigma_B) = (1/2)\cdot(N \pm \sqrt{N}) \pm (1/2)\cdot(N \pm \sqrt{N})^{1/2} \approx (1/2)\cdot N \;\; (1/\sqrt{2})\cdot\sqrt{N} \tag{8}$$

Equations (6) and (8) are the same, QED.

[5]It is straightforward to show that the result of endnote 4 can be extended to *any* two equivalent mixtures in *any* finite dimension. This follows as long as the initial unpolarized beams exhibit a Poissonian distribution for photon counts, which is typical of chaotic light sources as well as coherent ones (such as a laser above threshold). In principle, one could distinguish differently prepared but equivalent mixtures for more noisy sources.

The above result assumes that the beam intensities are constant, so that only 1st order coherence effects are present (particle fluctuations). For chaotic sources, this corresponds to sampling times T that are much larger than the "coherence time" τ_0 of the light source (typically 10^{-9} s). If however $T \ll \tau_0$, 2nd order coherence effects become important (wave fluctuations), and the fluctuations are given by $\sigma_0 = (N^2 + N)^{1/2}$ instead of $\sigma_0 = \sqrt{N}$. In this limit, different but equivalent mixtures *can* be distinguished by their fluctuations! An experiment can be readily performed by scattering laser light from plastic balls suspended in water, for which $\tau_0 \approx 10^{-1}$ s (Loudon 1973, pgs. 98-99, 214-221).

References

Aspect, A., Grangier, P., and Roger, G. (1981), "Experimental Tests of Realistic Local Theories via Bell's Theorem", *Physical Review Letters* 47(7): 460-463.

Badurek, G., Rauch, H., and Tuppinger, D. (1986), "Polarized Neutron Interferometry", in *New Techniques and Ideas in Quantum Measurement Theory,* D.M. Greenberger (ed.). *Annals of the New York Academy of Sciences* 480, pp. 133-146.

Belinfante, F.J. (1980), "Density Matrix Formulation of Quantum Theory and its Physical Interpretation", *International Journal of Quantum Chemistry* 17: 1-24.

Bevington, P.R. (1969), *Data Reduction and Error Analysis for the Physical Sciences.* New-York: McGraw-Hill.

Birge, R.T. (1935), "On the Nature of Unpolarized Light", *Journal of the Optical Society of America* 25: 179-182.

Bohm, D. (1951), *Quantum Theory.* Englewood Cliffs, NJ: Prentice-Hall.

Brown, H. (1986), "The Insolubility Proof of the Quantum Measurement Problem", *Foundations of Physics* 16: 857-870.

d'Espagnat, B. (1976), *Conceptual Foundations of Quantum Mechanics.* 2d ed. Reading, MA: Benjamin.

Fano, U. (1957), "Description of States in Quantum Mechanics by Density Matrix and Operator Techniques", *Reviews of Modern Physics* 29: 74-93.

Fine, A.I. (1970), "Insolubility of the Quantum Measurement Problem", *Physical Review* D 2(12): 2783-2787.

Françon, M. and Mallick, S. (1971), *Polarization Interferometers.* New York: Wiley & Sons.

Gibbins, P. (1983), "Quantum Logic and Ensembles", in *Space, Time and Causality,* R. Swinburne (ed.). Dordrecht, Holland: Reidel, pp. 191-205.

Grangier, P., Roger, G., and Aspect, A. (1986), "A New Light on Single-Photon Interferences", in *New Techniques and Ideas in Quantum Measurement*

Theory, D.M. Greenberger (ed.). *Annals of the New York Academy of Sciences* 480, pp. 98-107.

Grossman, N. (1974), "The Ignorance Interpretation Defended", *Philosophy of Science* 41: 333-344.

Hooker, C.A. (1972), "The Nature of Quantum Mechanical Reality: Einstein Versus Bohr", in *Paradigms and Paradoxes,* R.G. Colodny (ed.). Pittsburgh: Pittsburgh University Press, pp. 67-302.

Loudon, R. (1973), *The Quantum Theory of Light.* Oxford: Clarendon.

Park, J.L. (1968), "Quantum Theoretical Concepts of Measurement: Part I", *Philosophy of Science* 35: 205-231.

______. (1973), "The Self-Contradictory Foundations of Formalistic Quantum Measurement Theories", *International Journal of Theoretical Physics* 8: 211-218.

Pearle, P. (1986), "Suppose the State Vector is Real: the Description and Consequences of Dynamical Reduction", in *New Techniques and Ideas in Quantum Measurement Theory,* D.M. Greenberger (ed.). *Annals of the New York Academy of Sciences* 480, pp. 539-552.

Pessoa, O., Jr. (1990), *Measurement in Quantum Mechanics: Experimental and Formal Approaches*, Doctoral Dissertation, Bloomington: Indiana University.

Putnam, H. (1965), "A Philosopher looks at Quantum Mechanics", in *Beyond the Edge of Certainty,* R.G. Colodny (ed.). Englewood Cliffs, NJ: Prentice-Hall, pp. 75-101.

Shurcliff, W.A. (1962), *Polarized Light—Production and Use.* Cambridge, MA: Harvard University Press.

van Fraassen, B. (1972), "A Formal Approach to the Philosophy of Science", in *Paradigms and Paradoxes*, R.G. Colodny (ed.). Pittsburgh: Pittsburgh University Press, pp. 303-366.

von Neumann, J. ([1932] 1955), *Mathematical Foundations of Quantum Mechanics.* Translated by R. T. Beyer. (Originally published as Mathematische Grundlagen der Quantenmechanik. Berlin: Springer-Verlag.) Princeton, NJ: Princeton University Press.

Renormalization and the Effective Field Theory Programme

Don Robinson

University of Toronto

1. Introduction

Quantum field theory (QFT) poses the following well-known problem. Calculations of values for certain quantities begin with a first-order approximation to which are added higher order corrections. In many cases, the calculation of even the very first higher order correction term yields infinite values. These infinite values are due to divergences in the theory. These divergences can be traced to their source in the standard fundamental assumptions defining QFT. No matter which version of the formalism we adopt, we are trying to incorporate into the mathematics certain fundamental physical constraints on the nature of the vacuum state, on the states that can be occupied by systems, on the comeasurability of observables, and so on. The resulting formalism contains divergences.

Various calculational techniques known collectively as renormalization techniques were devised to cut off these divergent series at some finite stage and thereby keep the calculations from yielding infinities. Renormalized theories such as quantum electrodynamics (QED) have yielded remarkably precise and accurate predictions. Moreover, renormalizability has proven remarkably successful as a guiding principle and constraint on the construction of new theories. But how can this be? Given that the divergences arise in the formalism, and given that the formalism was designed to capture various fundamental assumptions, how can we take the theory, cut off the sums using renormalization techniques, and end up with such a successful theory? The main problem is that by renormalizing a theory we end up with a new theory containing the same fundamental assumptions as did the original but one in which the infinities derivable from those assumptions are removed. This is either a sign that the initial assumptions are somehow mutually incompatible or that the renormalization techniques are illegitimate, or perhaps both.

On the face of it, there would seem to be only two possible sorts of responses to this situation. Since the divergences can be shown to follow from standard fundamental QFT assumptions, one response is to give up the QFT framework entirely. This has been tried. The resulting theories are widely regarded as too phenomenological, too closely tied to experiment. On failing to find such a theory, one might hope that for

PSA 1992, Volume 1, pp. 393-403

every renormalizable QFT some new theory formulated along standard QFT lines will appear which is empirically equivalent to, or even more successful than, the renormalized one, but does not itself need to be renormalized. Research continues, for instance, into superstring theory. The trouble with this approach is the reverse of the trouble with the more phenomenological one. The resulting theories are widely perceived to be too far removed from any experimental data available in the foreseeable future and extremely difficult to calculate with. These two approaches would on the face of it seem to exhaust the possibilities. In the absence of either a satisfactory field-theoretic alternative or a satisfactory non-field-theoretic alternative to renormalizable QFT, it would seem as if the only thing left to do would be to face the problem and try to explain how renormalized theories can be so successful.

In 1980, Steven Weinberg suggested a third approach to the problem. This third approach is known as the effective field theory (EFT) approach to the problem of renormalization. On this approach, which I will explain more fully below, the standard field-theoretic apparatus is retained. Indeed, on the EFT approach, non-renormalizable quantities are included in theories formulated along standard quantum field-theoretic lines. They are effective theories in the sense that they are closely tied to particular stages of the development of experimental physics. On this approach theories are developed that are applicable only up to some threshold energy. Beyond each threshold new physics is needed. Mathematical models for EFT's were inspired by the description afforded by thermodynamics of the cutoffs between liquids, solids and gases (see Cao and Schweber (forthcoming) for an excellent historical survey of these developments and references to the relevant literature). What has thus far not been established is whether the EFT programme provides a satisfactory resolution of the renormalization issue.

Although EFT's have for the past fifteen years been a major topic for research in physics, there has been very little discussion of their philosophical implications. What little has been written about these implications reflects fundamental disagreement over them. At the heart of these disagreements lies a difference of opinion over the relation between EFT's and a hypothetical fundamental theory of everything, the hypothetical field-theoretic or non-field- theoretic replacements for renormalizable QFT. Some authors have claimed that EFT's are merely provisional theories on the way to a fundamental theory and will eventually be seen as approximations to that fundamental theory. Other authors have rejected that idea and suggested that there is no ultimate fundamental theory to be found in some limit of theory construction, that EFT's are the best we can hope for, that this tells us something new and exciting about the proper criteria for evaluating theories, and that the programme carries important philosophical implications. My aim in the present work is neither to solve the problem of renormalization nor to decisively settle the issue between the defenders of these two points of view. In part, I wish to point out that without further assumptions the EFT programme taken by itself does not carry any implications one way or the other. In other words, the EFT programme does not tell us whether or not there will eventually be some fundamental theory. The main point I wish to make is ontological. My point is simply that the implications of the EFT programme and their acceptability are bound up with certain ontological issues which turn in part on the question of entity, or experimental (as opposed to theoretical) realism. I argue that if some version of entity realism is tenable, then the EFT programme does not provide a resolution of the renormalization question. I now merely note in passing that I am well aware of the controversy concerning the ontology of QFT, specifically, whether the fundamental systems include particles, fields, or systems that have both particle and field aspects. I will use the language of particles and forces, rather than the more neutral but more honest language of quantum field systems, and trust that nothing I say below turns on the outcome of that debate. I will begin by describing EFT's.

2. Effective field theories

The best place to begin understanding the EFT programme is with a discussion of scale. The progress of particle physics is marked by the discovery of smaller and smaller particles found by probing smaller and smaller distances. To find out what is going on at smaller distances you need more elaborate equipment to produce beams of higher energy which produce more violent collisions. I will merely recite a few well-known facts to suggest the sizes and distances involved. The typical atom has a diameter of approximately 10^{-8} centimeters, the typical nucleus 10^{-12}, the proton 10^{-13}. Long-lived particles differ quite dramatically in mass. Photons, for instance, have zero mass. Electrons have mass approximately .5 (in MeV), protons approximately 938, and tauons approximately 1,784. The weak force, responsible for the decay of long-lived particles, has a very short effective range. It is negligible except at distances less than approximately 10^{-16} centimeters, or 1/10 of one percent of the diameter of the proton. And it is weaker than the strong force by a factor of approximately 10^{12}. The strong force is important for particles separated by a distance less than approximately 10^{-15} meters.

A fundamental theory of everything would provide a unified treatment of all these particles and forces and get them right at every distance and size. QED, though a renormalizable quantum field theory, is not such a theory. It is intended to describe electrons, photons, the interactions between them, and nothing more. It is one of the most highly successful physical theory we have ever had. It virtually ignores, however, whatever might be going on between much smaller or much more massive particles. How, then, can it be so successful? The answer is that what goes on at smaller and at larger scales can be, for reasons I will set out below, virtually ignored at the scale of electrons, photons and the range of the forces acting between them. Presumably, however, QED would at some point have to be fitted into a fundamental theory. Such a fundamental theory would describe whatever goes on at arbitrarily smaller and smaller distances. Ideally, it would have to incorporate the effects particles of every size have upon electrons and photons. QED is a renormalizable theory. QED provides us with an instance of the more general problem outlined at the outset, namely, how to explain the success of renormalizable theories.

The EFT approach takes seriously the differences between different distance and mass scales. At a distance scale of, say, 10^{-18} centimeters, certain particles and forces stand out. Others are too small to have any significant effect on what is going on. The effects of certain forces will appear only at much smaller or much larger distances. So we devise a theory of what is going on just at that particular scale. Next we focus on a smaller scale and, building on the previous theory, we devise a theory of what is going on at that and larger scales. Next, we go to a smaller scale yet and, building once again on the previous theory, we devise a theory of what is going on just up to and including that scale. And on we go. At each cutoff in scale we get a new EFT. Since the limit of moving to smaller and smaller distances is a point, we can understand this process of theory construction as yielding an infinite "tower" of effective field theories, each applicable up to some cut-off in scale, or distance, but not beyond. These cutoffs in scale are intimately related to renormalization. In renormalization we cut off certain divergent series before we have added infinitely many terms. Where we take these cutoffs in the sums mirrors the cutoffs between the effective field theories (for a very clear and helpful discussion of the relations between currently available apparatus, the measurement of renormalizable quantities, and where to incorporate the cutoffs, see Teller (1989), pp. 242-247).

Let us return to QED and see how the notion of an EFT applies in that case. Charged particles other than electrons are much heavier than the electron. At such relatively long distances there is not enough energy to produce these heavier particles, so for all intents and purposes we can practically ignore them. Also, there are much lighter particles but they interact relatively weakly and so we can effectively ignore them as well. Practically, but not entirely. Consider just the heavier particles. Their effects are actually incorporated into QED. When we build QED we incorporate the effects caused by particles that stand out at smaller distance scales. These effects appear in the theory as, "renormalisable (or, at least, less nonrenormalisable) interactions involving heavy particles" (Georgi 1989, p. 456). The strategy involved in building this tower of EFT's within the framework of QED is to incorporate effects from shorter distances into the theory and then when we build a theory for that shorter distance we explicitly include reference to the particles responsible for those effects. In this way, we can arrive at more and more accurate theories without mentioning the other particles in those theories. The constraint on the construction of effective theories is just that the nonrenormalizable interactions at one stage of construction must get replaced by particles corresponding to them at the next stage of construction. This is called a 'matching condition'.

Where does this process of construction lead? As I already indicated, one might think that there is a well-defined limit to this process, one in which we have described all particles and forces down to arbitrarily short distances. This is what workers in superstring theory are seeking. Against this suggestion, Cao and Schweber (forthcoming) claim we have good reasons to believe there is no well-defined limit to this tower of EFT's. They offer three reasons. The first is that we have no acceptable alternative to a fundamental renormalizable quantum field theory. As I mentioned above, the alternatives we have are widely regarded as either too phenomenological or too far removed from experimental data available in the foreseeable future. Their second reason is that the fundamental field theories we do have need to be renormalized but nobody has been able to provide an adequate explanation for the success of renormalized theories. Their third reason is that EFT's can, so to speak, stand on their own. The EFT programme, they claim, shows how we can do physics without having or even desiring a fundamental theory. This brings us to the question of whether EFT's can really stand on their own. This prompts a closer investigation into the relations between the various EFT's.

At each stage in the tower of EFT's we add new terms referring to particles that have not yet appeared in the tower. Someone who believes there might be an acceptable fundamental theory would regard the tower of EFT's as a tower of sub-theories of the fundamental theory. On the other hand, anyone who believes there can be no such fundamental theory of everything will say there is no well-defined limit to this process. On this view, it would not make sense to talk about the set of all particles and forces. There would only be particles and forces known at some particular time. Cao and Schweber suggest that there is no ultimate ontology, no fundamental theory of everything, nothing except the never-ending infinite tower of EFT's. Denying the existence of some ultimate ontology carries the implication that we must be content with some form of limitation on our ontology of quantum field systems. This limitation is reflected in what they call quasi-autonomous domains. They claim there is one such domain corresponding to each EFT. We must look at each EFT in isolation from every other EFT and not think of each one as a stage in the construction of some ultimate fundamental theory. They call themselves realists with respect to the cutoffs defining each of the EFT's and claim there is experimental evidence that the cutoffs are not imposed arbitrarily. Since there is, at least on this view, no limit theory to be renormalized, there is no question of having to explain the success of a renormalized fundamental theory. Each EFT can contain non-renormalizable interactions but this is not thought to be the problem it is in the fundamental theory because EFT's are not applicable at anything less than some finite distance. Thus, on this account of the EFT programme, we can avoid the question of renormalization if we are willing to accept the existence of quasi-autonomous domains and the consequences of that accceptance. These consequences amount to the rejection of several commonly used constraints on theory construction including, for example, logical consistency. I am focussing in the present work on the notion of quasi-autonomous domains because I believe it is what ultimately drives Cao and Schweber to these consequences.

Not all authors view things in this way. Georgi, for example, refers to the view I have just described as a, "peculiar scenario in which there is really no complete theory of physics—just a series of layers without end," and says that it is, "[m]ore likely, I think, [that] the series does terminate, either because we eventually come to the final renormalizable theory of the world, or (most plausible of all) the laws of relativistic quantum mechanics break down" (1989, p. 456). These two stopping places represent the field-theoretic and non-field- theoretic fundamental theories I alluded to above. Also, note that on Georgi's account the fact that the EFT's do not meet the commonly used constraints on theory construction is not as serious a problem as it is for Cao and Schweber. On Cao and Schweber's account, EFT's are all we will ever have. On Georgi's view, EFT's are approximations to the fundamental theory and so, arguably, the EFT's do not have to meet the same constraints imposed on the ultimate theory. In any case, the disagreement over whether there is or is not a stopping place in the construction of EFT's reflects a widespread tension in the philosophy of QFT. The tension is between, on the one hand, the desire to spell out the philosophical consequences of our current quantum field theories and, on the other, the common perception that current theory is perhaps too provisional to support any implications, at least any we ought to take seriously. Do we wait until we have a fundamental theory formulated along standard field-theoretic lines? Do we wait until we have a theory formulated along entirely new lines? One of Cao and Schweber's reasons from backing the EFT programme is that we have no alternative at the present time. If I can show that the idea of quasi-autonomous domains is not implied by the EFT programme, and if I can give reasons to believe that the EFT's are indeed approximations to some fundamental theory, then the mere observation that no satisfactory alternative has yet been found will not in itself provide any support for Cao and Schweber's interpretation of EFT's. Before I return to these issues I must first examine the idea which motivates the 'no stopping place' view, the idea that each EFT can be said to have its own quasi-autonomous domain.

3. Quasi-autonomous domains

Cao and Schweber claim that each EFT has its own quasi-autonomous domain of systems (fields, forces, particles) that can be said to exist. In Cao and Schweber's words, the physical world, "can be considered as layered into quasi-autonomous domains, each layer having its own ontology and associated 'fundamental' laws," with no, "possibility of deducing various entities from some basic ontology" (p. 34). EFT_1 describes just the particles and forces that are noticable or important up to the first cutoff. Other particles and forces, although they certainly will appear eventually in a theory with a higher cutoff, do not appear in EFT_1. When we try to read our realist commitments off EFT_1 we will come up with a list of particles and forces. These are just the particles and forces that are prominent up to the first cutoff or, as Georgi puts it, each EFT, "contains explicit reference only to those particles that are actually important at the energy being studied" (1989, p. 446). None of the particles or forces from above that cutoff appear in the formulation of EFT_1 and so, according to Cao and Schweber, we do not add them to our inventory. This returns us to the question of how an EFT can be so effective if it is built on the assumption that these other particles and forces do not exist. Well, things are not that simple. What we have actually done is "smuggled" the effects of these higher level particles and forces into EFT_1. They appear as non-renormalizable quantities, as I alluded to above in connection with QED. Thus, although these particles and forces do not appear as things referred to by terms in the theory itself, their effects are nevertheless incorporated into it. This is in part why Cao and Schweber refer to these domains as 'quasi-autonomous' rather than strictly autonomous.

What do these ontologies contain? According to Cao and Schweber each contains systems they refer to as, "structureless quasi-points" (see section 4A). They are structureless because if they had structure they would contain other systems. The theory would then have to describe them. Those systems do not come into the picture, however, until we are working on the next EFT_1 up the tower. They are quasi-points and not points because they are extended. Part of the internal structure of the systems of EFT_1 gets filled in by the systems of EFT_2, for EFT_2 describes what is going at distances of the scale of the interior of the systems of EFT_1. Part of the internal structure of the systems of EFT_2 gets filled in by the systems of EFT_3 , and so on. The structure would get filled in completely only if we had a fundamental theory of everything. In the absence of a fundamental theory, at least according to Cao and Schweber, there is no ultimate ontology. Also, what we say about theEFT_1systems is not quite the same as what we say about their counterparts in EFT_2. On Cao and Schweber's view, then, each EFT has its own quasi-independent ontology of extended structureless quasi-point systems.

Cao and Schweber appeal to these quasi-autonomous domains in support of their contention that each EFT can, so to speak, stand on its own. The autonomy of the EFT's is then appealed to in support of their contention that the EFT's need not be seen as approximations to some hypothetical fundamental theory. The EFT programme interpreted in this way shows, they claim, that empirical adequacy, rather than traditionally prominent criteria such as logical consistency, is the single most important criterion governing theory choice, that there is no need to have one fundamental theory even within physics, that the problems raised by renormalization are surmountable, and several other claims. If I can show that there is really nothing to this idea of quasi-autonomous domains, I will have taken away the central claim driving the whole program of interpreting EFT's in this way. That is why I have focussed on this particular ontological issue.

To get these quasi-autonomous domains of structureless quasi-point systems, Cao and Schweber rely on the following strategy. First they take a cutoff. Then they build a theory good up to the cutoff. Next, when they wish to know what exists according to individuals who accept that theory, they restrict themselves to asking what exists according to the theory they have up to that point. This strategy depends crucially on the assumption that realist commitments can simply be read off of a theory. If the theory contains the term 'photon' and we acccept the theory, then we add photons to our ontology. If the theory does not contain the term 'blue quark' then we do not add blue quarks to the list of things that exist. Reading ontological commitments off the theory is Cao and Schweber's route to quasi- autonomous domains. This strategy would break down, however, if there were some way to arrive at our ontological commitments other than simply reading them off our theory. If there were such a way, we might have to include entities from above the first cutoff in the ontology of the first EFT. Acceptance of an EFT would commit one to a belief in the existence of not only the entities referred to by the terms of that EFT but belief in the existence of entities not even referred to by terms of that EFT. This would mean, however, that the EFT programme would not carry the implications claimed by Cao and Schweber. This, in turn, would make EFT's look less like self-standing theories and more like approximations to some fundamental theory. This is where entity realism comes in. If we have theory-independent grounds for belief in, say, blue quarks, then even if EFT_1 does not contain the term 'blue quark' we would still have to include them in our list of things that can be said to exist given our acceptance of EFT_1.

Realism based on what can be manipulated experimentally, or at least what can be discovered by experimenters, rather than what can be inferred from theory, has been offered as a challenge to the idea that we can be realists only with respect to the referents of terms of a scientific theory. To be sure, typical formulations of scientific realism describe it either as the belief that the terms of a mature scientific theory typically refer, the belief that theories are approximately true, or the belief that the aim of science is to formulate theories describing some mind-independent structure to which we have substantial epistemic access. Hacking (1983) challenged these sorts of definitions of realism and offered a realism about experimentally manipulable entities and not just about things referred to by the terms of some theory. This move to entity realism allows one, arguably, to be a scientific realist without having resolved all the outstanding problems facing a theory of reference.

Cao and Schweber address Hacking's version of entity realism. Indeed, his is the only version they discuss, but they dismiss it as probably not relevant to EFT's. As I will explain more carefully in the next section, it is important for them to find fault with entity, or experimental realism, for if entity realism could be justified the EFT programme would not in any straightforward sense have any of the implications claimed by Cao and Schweber and so would not in any straightforward sense allow us to bypass the problem of renormalization. In the next section I will discuss this connection between entity realism and the EFT programme in more detail and respond to Cao and Schweber's criticisms of entity realism.

4. Entity realism and the effective field theory programme

Hacking has observed that when we look at how physicists build up their ontologies we find that experimenters come to believe in the existence of certain entities even though they either have no theory at all, at least not one that allows them to predict the existence of new entities, or they have several competing theories on hand. Hacking believes we should believe in the existence of entities when we understand their low-level causal properties well enough to be able to manipulate them to investigate the structure of other entities. This form of realism poses the following problem for Cao and Schweber's approach to EFT's. If we can in fact determine through extra-theoretical means that certain systems exist, systems other than those referred to by the terms of EFT_1, we will have to include them in the ontology of EFT_1 even though they are not explicitly referred to by the terms of EFT_1. Our ontology, in other words, will consist of all the particles and forces believed to exist period, whether we have a fundamental theory, an EFT, or no theory of them at all. This means we have one ontology, not several quasi-autonomous domains. Unless we have independent reasons to believe there are limits to the distances we can probe with experiment, this one ontology is not limited to what exists according to our most recently devised EFT.

Cao and Schweber claim that Hacking's version of realism is probably inappropriate in the case of EFT's. As I suggested above, if Hacking's version of realism is tenable, then it does not make sense to consider the ontologies of EFT's context dependent. They claim that Hacking's version of realism does not work for such entities as the Fayddeev-Popov ghost fields in gauge theory. Presumably, they are offering this as an example of a theoretical entity that was not discovered through experiment but was predicted on the basis of a certain derivation from gauge field theory. Weingard (1988) has argued that these ghost fields are merely artefects of the notation because they can be completely transformed away under gauge transformations. Be that as it may, I would suggest that even if Weingard is right about these ghost fields there are probably other theoretical entities in QFT that could be pointed out as having been inferred not from experiment but solely from theory. If this were true, Cao and Schweber would have another, even more plausible example. In the end, however, it will not make any difference, for the suggestion is that Hacking's version of realism is not appropriate to quantum field theory simply because some entities have been inferred on the basis of theory alone and have not been used at all in experiments. This will not do, however, as an argument against Hacking, for nowhere does he suggest either that such inferences are impossible or that experiment—manipulating entities—is the only way to arrive at belief. He only claims experiment is one way. He writes:

> Perhaps there are some entities which in theory we can know about only through theory (black holes)....Perhaps there are entities which we shall only measure and never use. The experimental argument for realism does not say that only experimenter's objects exist. (1983, p. 275)

Clearly, according to Hacking we can infer the existence of entities either on the basis of having found certain solutions to equations or on the basis of experimentation and manipulation (even if we have no theory from which we can derive predictions about new entities).

Cao and Schweber offer a second criticism of Hacking's version of entity realism. They claim it leads to belief in certain entities no one actually believes in. At this point I wish to shift attention away from Hacking's version of entity realism toward the experimental basis for belief in entities more generally. I do so not because I believe that issue is settled, but because the outcome of that particular debate over the applicability of Hacking's version is probably not relevant here. Even if his version needs revision, something much like it appears to be called for by the facts of QFT history. Thus, even if Hacking's version of entity realism needs revision, all I need for present purposes is that some version of entity realism will work for QFT. The sort of case I have in mind is Carl Anderson's 1932 discovery of the positron. Anderson built a cloud chamber and detected particles deflected by what was then the strongest magnetic field in existence. The particles deflected downward were believed to be electrons. Anderson found that many particles seemed to be deflected in the same sort of trajectory but upward rather than downward. He determined that their ionization did not match that of protons, the only other particle believed at the time to exist. He concluded that he was detecting a new kind of particle, the positron, comparable in mass to the electron but opposite in charge. As it turned out, the existence of these particles had been conjectured the year before, albeit somewhat reluctantly, by Dirac. Dirac had discovered two solutions to the equations of his theory, one positive in sign, the other negative. Anderson, when asked whether he was aware of Dirac's conjecture, replied that although he knew about Dirac's theory he, "was not familiar in detail with Dirac's work," because he was, "too busy operating this piece of equipment to have much time to read his papers," and that, "[t]he discovery of the positron was wholly accidental" (Pais 1986, p. 352). Dirac read his ontological commitments off his theory; Anderson got his commitments from experiment in the absence of predictions from a theory.

According to Hacking's version of realism, Anderson did not in 1932 have sufficient grounds for belief in positrons. Belief in their existence should have been withheld until positrons could be manipulated to investigate the structure or properties of other systems. There is, of course, a rapidly growing literature on the role of experiment in science. I only wish to say here that Anderson's discovery of the positron could be used to show that, at least according to some version of experimental realism, it is possible for an experimenter to find good reasons to believe in the existence of quantum field systems yet have neither a fundamental theory of everything nor even an EFT from which to predict the existence of such systems. And even on Hacking's version, if we came to believe in the existence of even one kind of field-theoretic entity as a result of experimental manipulation in the absence of an EFT predicting its existence, this would mean that more is involved in identifying our ontology than merely reading those commitments off the terminology of some EFT. Applying these ideas to EFT's, we see that if we wish to know what things exist according to EFT_1, we can either read our commitments off EFT_1 or we can determine them through experiment. If there are theory-independent grounds for belief in the existence of certain particles and fields, and there certainly seem to be such cases in the QFT literature, even by Cao and Schweber's own admission, then whether Hacking's particular version of entity realism or some other version is applicable to QFT, it makes no sense to talk of unique quasi-autonomous domains for each EFT. In other words, it does not make sense to talk about 'the ontology of EFT_1'as distinct from 'the ontology of EFT_1'. Provided we have experimenters who, like Anderson, add to the ontology of quantum field systems by experimenting or by manipulating without having derived predictions from some EFT, then it makes no sense to talk about quasi-autonomous domains as an implication of the EFT programme. Therefore, we only have to accept the fact that experimenters do add to the ontology of particle physics, even in the absence of any theoretical predictions, to have reasons to deny that EFT's have quasi- autonomous domains.

The idea that there is a single domain of quantum field systems suggests itself too when we consider how the effects of particles and forces from above, say, the first cutoff get incorporated into EFT_1. EFT_1 is built up in the following three-step process. First, we ignore the particles and forces above the first cutoff and devise a theory of particles and forces below it. Next, we take the particles and forces above the cutoff into account and calculate their effects on the particles below. Then we incorporate any significant effects from above into the theory we devised at the first stage, redefine the appropriate parameters, and ignore any negligible effects. Of course, when we look at the final theory we have constructed we will find no reference to the particles and forces from above the cutoff. This makes it seem as if they do not exist. But this is precisely because we either ignored them or incorporated their effects into parameters of the final theory. The process of theory construction in this case can be described as follows. First we pretend particles and forces above the cutoff do not exist. Then we admit they do exist and incorporate their effects into our theory. In the end, all direct reference to them drops out. According to Cao and Schweber, we can in effect pretend again that they do not exist. This is how they arrive at quasi-autonomous domains for EFT's. Furthermore, the structureless quasi-point systems of EFT_1 get some of their structure 'filled in' at the level of EFT_2. Some of the structure of the systems of EFT_2 gets filled in at the level of EFT_3, and so on. This picture suggests quite strongly that this sequence of sets of structureless quasi-point systems, in which each new set of systems fills in some of the structure of the previous set, converges on some well defined limit in which all the structure gets filled in. If anything looks like convergence, this does. This brings us back to the one domain. And this brings us back to the need for a fundamental theory of that one domain. And, unless we have independent grounds for believing that there is some limit in principle to the distances we can probe, or that the laws of relativistic quantum field theory break down at some distance, this brings us back to the need for an explanation of renormalization.

5. Conclusion

I began by describing the conditions under which the EFT programme arose. We tried to build into the formalism of QFT various physical assumptions about the vacuum, the states available to systems, the comeasurability of observables, and so on. These assumptions yielded infinite values for certain calculations. Renormalization techniques were developed to cut off the divergences producing these infinite values. These techniques seem to contradict the fundamental assumptions from which the infinite values can be derived, thereby producing a new theory which seems to be inconsistent. Alternatives to this theory, however, have been widely regarded as either too closely tied to experiment data available in the foreseeable future or as too far removed from experiment. In the absence of such alternatives, EFT's were developed. This raised the question of whether EFT's are, on the one hand, provisional theories approximating or leading to some entirely satisfactory fundamental theory or, on the other, theories which have to stand on their own and not lead to any fundamental theory.

According to Cao and Schweber, one implication of the EFT programme is that each EFT has its own quasi-autonomous domain. This claim lends credibility to the idea that EFT's can stand on their own and not be seen as approximations to some fundamental theory. I have argued that the idea of quasi-autonomous domains makes sense on the assumption that we can build up our ontological commitments in QFT only by the method of first identifying the referring terms of the theory we accept. But to accept this form of realis—theoretical realis—is to deny that we can build up our ontological commitments through experiment in the absence of theory. Anderson's discovery of the positron suggests that belief in field theoretic systems has been based on experiment in the absence of theory. But the existence of such cases in the history of QFT take away any motivation for adopting the idea that each EFT has its own quasi-autonomous domain of structureless quasi-points. And without this idea, it looks like there really is one domain for QFT. Moreover, the description provided by each successive EFT in the tower of EFT's seems to converge on a description of that one domain. Whether we regard this hypothetical fundamental theory as 'the one true theory' or as merely our best approximation is another matter. Thus, the EFT programme itself does not tell us that EFT's can stand on their own. Only if we reject entity realism does it even appear to do so. Now, it may be the case that we will at some point have good reasons to think that there is some limit even in principle to the distances at which we can probe at higher and higher energies. Or, it may be the case that we will find that the laws of relativistic quantum field theory no longer hold within some tiny region. But that is not something the EFT programme itself tells us. It is not a philosophical implication of EFT's. EFT's may very well serve as the best physics we can do in the absence of a fundamental theory but the EFT programme does not allow us to bypass the problem of renormalization. Where does that leave us? In the absence of an acceptable replacement for QFT, all we can do is try to identify the inconsistencies among the fundamental assumptions.

Notes

[1]I would like to thank Ian Hacking, James Robert Brown, and Jacqueline Brunning for their helpful questions and comments on this work. I would especially like to thank T. Y. Cao and S. Schweber for graciously allowing me to see a pre-publication copy of their forthcoming paper on renormalization.

References

Cao, T.Y. and Schweber, S. (forthcoming), "The Conceptual Foundations and Philosophical Aspects of Renormalization Theory".

Georgi, H. (1989), "Effective quantum field theories", in *The New Physics*, P. Davies (ed.). Cambridge: Cambridge University Press, pp. 446-457.

Hacking, I. (1983), *Representing and Intervening: Introductory topics in the philosophy of natural science*. Cambridge: Cambridge University Press.

Pais, A. (1986), *Inward Bound: of matter and forces in the physical world*. New York: Oxford University Press.

Teller, P. (1989), "Infinite Renormalization", *Philosophy of Science* 56: 238-257.

Weingard, R. (1988), "Virtual Particles and the Interpretation of Quantum Field Theory", in *Philosophical Foundations of Quantum Field Theory*, H. Brown and R. Harre (eds.). Oxford: Clarendon Press, pp. 43-58.

Bell's Inequality, Information Transmission, and Prism Models[1]

Tim Maudlin

Rutgers University

1. The Problem

Suppose we set about creating pairs of photons in the singlet state, allow them to separate to some considerable distance, and then send each photon into a polarizer oriented at some randomly chosen angle. Quantum theory predicts that in the long run we will observe the following behavior. On each wing of the experiment we will see perfectly random behavior. Roughly half of the photons will pass their polarizers and half will be absorbed. There will be no correlation between passage or absorption in one experiment and passage or absorption in the succeeding experiments. There will be no correlation between the passage or absorption and the angle at which the polarizer is set. But when we examine the correlations between the two wings of the experiment, something remarkable emerges.

Let us say that the two photons agree if they both pass their polarizers or both fail to pass, and disagree otherwise. In the long run, quantum theory predicts that agreement between the photons depends quite strikingly on the degree of misalignment between the two polarizers. When the polarizers are misaligned by an angle θ the photons agree $\cos^2\theta$ of the time (Figure 1).

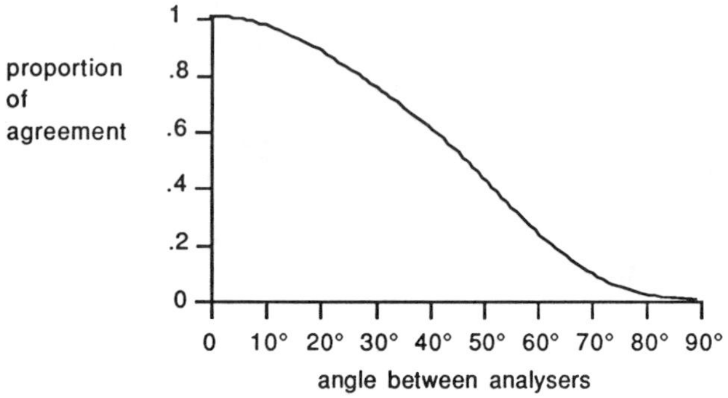

Figure 1: The Quantum Correlations

PSA 1992, Volume 1, pp. 404-417

This behavior is the stuff which violations of Bell's inequality are made on. If, for example, one restricts the polarizer on the right so that it can only be set at 0° or 30° and the polarizer on the left so it can be set only at 30° and 60° then one can generate correlations from the four possible experimental set-ups which violate Bell's inequality. I take for granted that this audience is already sufficiently aware of how these sorts of statistics violate Bell's factorization assumption.

Bell concluded that violations of the inequality demonstrate that the world is not locally causal, i.e., that these phenomena cannot be reproduced by any theory which postulates only locally defined physical states which cannot influence states at space-like separation (Bell 1976). Philosophers of physics have been wont to question this conclusion, especially since Jarrett's analysis of the Bell factorization condition into the so-called "localization" and "completeness" conditions (Jarrett 1984). Bell was, however, quite correct in his analysis. Statistics such as those displayed by the photons cannot be reliably reproduced by any system in which the response of each particle is unaffected by the nature of the measurement carried out on its distant twin. The photons remain "in communication" no matter how great the spatial separation between them. Instead of trying to deny these non-local (i.e., superluminal) influences, we should begin to study the role such influences must play in generating the phenomena.

As a heuristic aide, let us imagine a game which mimics the condition of the photons. You and a friend are to play the following game many times in succession. You begin each game together in a room, where you are permitted to devise any strategy you like. You leave the room by separate doors, and after some period of time you are each asked a question, which consists in a real number between 0 and 180 (corresponding to the setting of a polarization analyzer). You are each to answer either "passed" or "absorbed" to the question. Your object is to replicate the correlations depicted in Figure 1: in the long run, when the questions asked of you differ by θ, your answers agree $\cos^2\theta$ of the time.

We will begin our game under the assumption that each of you has no information about the question asked your partner. (The questions may be asked, or even decided upon, at spacelike separation, so in any local theory no such information could be available.) We will see that the quantum correlations cannot be recovered under this assumption. We will then see how the situation changes as we allow ever more information to flow between the sides. Trading in information allows us to operate at a high level of abstraction, ignoring the exact details of the communication channel. We can sensibly ask how much information must flow between the wings of our experiment if the quantum statistics are to be generated, leaving aside questions about how the communication is brought about. In the sequel we will take advantage of this studied agnosticism concerning mechanism and simply pursue the question of how much, at a minimum, one photon needs to know about the other.

2. Limits for Uncommunicative Partners

Let's start with two facts. First, the answers must agree whenever the same question is asked on both sides, no matter what the question may be. Second, when the questions differ by 90° the answers must differ.

It has long been noted that the perfect correlation in response when the polarizers are aligned requires any local theory to be deterministic (indeed, this was the argument of Einstein, Podolsky and Rosen). Employing stochastic mechanisms for answering questions only makes the situation *worse* for the two players since more information (i.e., the result of the stochastic process) must be communicated between the two sides. Without this information the players cannot possibly be sure to agree when the same question is asked. Since we are trying to minimize the amount of communication needed, we will confine ourselves for the moment to deterministic strategies.

Partners unable to communicate can satisfy the first constraint, perfect correlation when the polarizers are aligned, only by adopting identical strategies for answering questions. The partners must therefore be provided with identical instructions for answering all possible questions. We can represent the instruction set as a semi-circle with some segments colored black. If the polarizer is set at an angle whose corresponding point is colored black then the photon will be absorbed, otherwise it passes. Of course the instruction sets may be varied from run to run.

To satisfy the second criterion, perfect disagreement if the polarizers are misaligned by 90°, every black segment on the semi-circle must have a corresponding white segment displaced by 90°. Hence we need only specify the coloring of the region between 0° and 90°: the area between 90° and 180° will just be its inverse image (Figure 2).

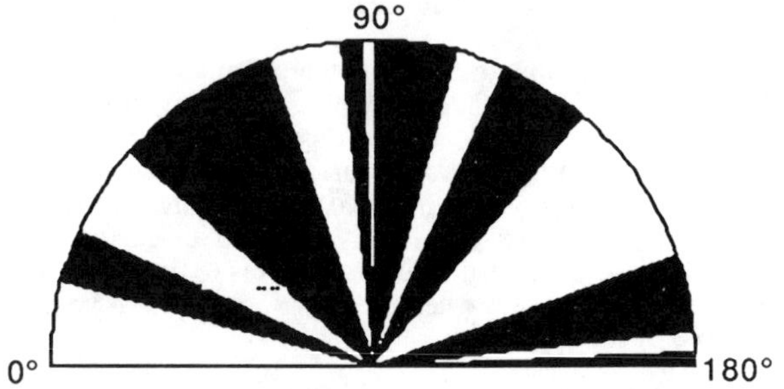

Figure 2: Instruction Set Satisfying the First Two Criteria

It follows that half of the semicircle will be colored white, half black. Only instruction sets constructed in this way can satisfy the first two requirements.

Having taken care of the cases of perfect alignment and perfect misalignment of the polarizers, we must now consider all of the intermediate cases. There are several way of proceeding, and we will follow one of the most obvious. Let's consider the behavior of the photons when the polarizers are only slightly misaligned, when θ is near zero. As Figure 1 shows, there remains a very high degree of correlation in these cases, so nearby regions on the instruction set should generally agree in color. The region surrounding a point colored black should itself tend to be colored black, the region near a white point white, so that small differences in angle will not usually lead to differing responses.

Given that half of the semicircle must be black and half white, the best one can do in this regard is to have all of the similarly shaded regions contiguous, a solid quarter-circle of white and a solid quarter-circle of black. Two examples of this maximal strategy are shown in Figure 3.

In the absence of communication the very best strategy you and your partner can adopt is the following. When you are still together in the room, randomly choose an angle α between 0° and 180°. Paint the region between α and $\alpha + 90°$ (mod 180°) on your instruction sheets black, that is, agree to answer "absorbed" if the question asked falls within that region and "passed" otherwise. Using this strategy each of you will answer "passed" approximately half of the time in the long run. There will be no correlation between the angle of either polarizer and passage or absorption at that polarizer. When the same question is asked on both sides you will always agree. When the questions differ by 90° you will always disagree. And when the questions are close to each other, e.g. when "50?" is asked on one side and "51?" on the other, you will give the same answer as often as you possibly could (absent communication).

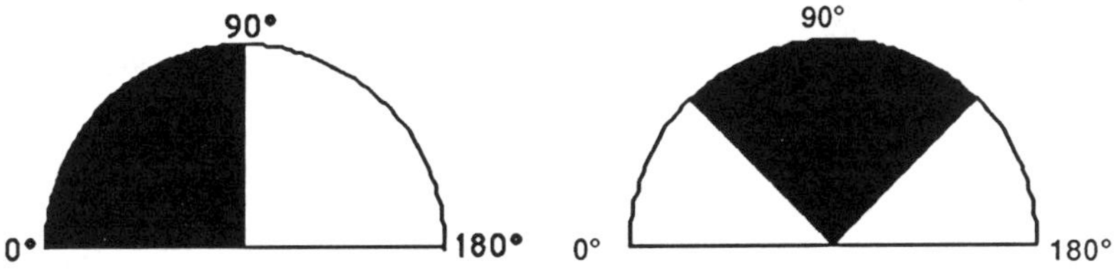

Figure 3: Two Maximally Correlated Instruction Sets

But just how often will you agree in this last case? The probability that you will *disagree* is just the chance that a black/white border in the instruction set lies between 50° and 51°. More generally, the probability that the questions "ϕ?" and "ψ?" get different answers is just the chance that such a border lies between ϕ and ψ. But this last probability is just proportional to $\phi - \psi$. If the slice between ϕ and ψ is twice as large then the likelihood of catching the border in the region is twice as great. (This holds as long as $\phi - \psi \leq 90°$, after which point one must worry about catching two borders.)

So thc probability of disagreement scales linearly with the angle of misalignment from 0% disagreement with no misalignment to 100% disagreement for 90° misalignment. This linear relation is the very best anyone can do if there is no communication between the two sides of the experiment. As has been often noted, one hallmark of the entangled systems is that as the polarizers become misaligned the particles hold on to the correlations more strongly than any pair of causally isolated particles could. They show more correlation at small angles than is possible with no communication given 100% agreement for no misalignment and 100% disagreement when the polarizers are at right angles.

3. How Much Does a Particle Need to Know?

In order to improve upon the linear approximation, either you or your partner must know something about the other's apparatus. The vital bit of information is the angle at which the distant polarizer is set. Just how much information about the setting on the distant wing is needed to be able to match the target curve? Let's start by asking how much your situation would improve if your partner could send one bit of information after you have been separated.

In terms of our game, imagine that after having left the room and before you have to answer the question posed to you, you can receive a message from your partner containing either a 1 or a 0. Which message is sent may depend on the question which has been asked of your partner. How could you best make use of this new resource?

The way to take advantage of this new source of information is to have two instruction sets rather than one. To maximize the difference between using one of the set rather than the other the two sets ought to be as different as possible (with respect to the placement of their black/white borders). The most extreme difference obtains between pairs whose black and white regions are displaced by 45°, such as the pair depicted in Figure 3. Let's call the instruction set on the left *Plan 0* and the instruction set on the right *Plan 1*. Your partner has the option of using either Plan 0 or Plan 1 and, by informing you of the choice, assuring that you use the same scheme. This provides a means of overcoming the problem of small angles.

The linear correlation fails to match the target in part because it requires more disagreement at small angles of misalignment than the photons display. This happens because when the polarizer settings ϕ and ψ are close to each other the probability that a black/white border will fall between ϕ and ψ is too high. Disagreement is likely to occur when the boundary in the instruction set happens to fall near to the direction of both polarizers. But using the one bit of information your partner is now able to avoid this danger. If her polarizer is set very near a black/white boundary for Plan 0 she can opt instead for Plan 1, thereby moving the boundary approximately 45° away from the polarizer setting. Suppose, for example, that your partner's polarizer is set at 91°. If you both follow Plan 0 your partner will answer "passed". If your own polarizer happens to be set at 89° you will answer "absorbed", giving a disagreement at small angle of misalignment (2°). By choosing Plan 1, though, this hazard disappears. Your partner answers "absorbed" at 91°, and Plan 1 assures that for any small angle of misalignment, indeed any misalignment up to 44°, you will agree.

Consider, then, the following strategy. You and your partner, while still in the room, randomly choose a direction between 0° and 180°. Plan 0 puts the black/white border at the chosen direction, Plan 1 offsets the boundary from this direction by 45°. When your partner is asked to respond she chooses the plan whose borders are *farthest* from her polarizer setting and communicates that choice to you using one bit of information. You adopt the corresponding plan when answering your question. What will be the result?

Since the boundaries in Plan 0 and Plan 1 are offset by 45°, the chosen plan will have borders at least 22.5° from the direction of your partner's polarizer. So if you and your partner always use this strategy you will *always* agree when the two polarizers are misaligned by less than 22.5°. Beyond 22.5° the rate of agreement drops linearly to perfect disagreement at 67.5°. If the angle of misalignment is between 67.5° and 90° you are certain to disagree with your partner. The overall pattern of correlation is shown in Figure 4, along with the target curve.

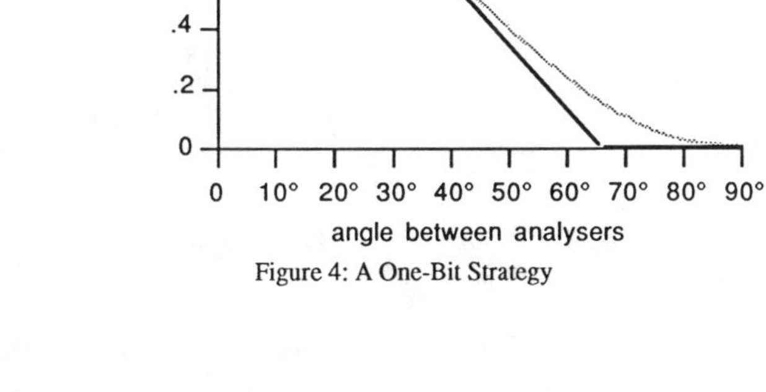

Figure 4: A One-Bit Strategy

So using only one bit of information we have overcome, even overcompensated for, the problem of small angles. We have devised a strategy which shows even higher degrees of correlation at small angles than do the photons, and more disagreement when the polarizers are misaligned by nearly 90°.

The single bit conveys information about the setting of the distant analyzer, locating it within a 90° region. The message "0" will be sent if the distant polarizer lies either between 22.5° and 67.5° or between 112.5° and 157.5°. Call the union of these areas "region 0". If the analyzer lies anywhere else, in what we may call "region 1", the message "1" will be sent.

Our one-bit strategy does not match the target curve exactly, so we still have some work to do. Perhaps some combination of the one-bit and zero-bit (linear) strategies can give an improved approximation. But instead of working with so impoverished a palette, we should first note that the one-bit strategy described above is but one of an infinite family of one-bit strategies available to us, each of which yields a unique correlation pattern.

When designing our first one-bit strategy we maximized the difference between the placement of the boundaries in Plan 0 and Plan 1. If we adopt less extreme measures we get less extreme results. Suppose, for example, that the black/white boundary of Plan 1 is shifted only 20° from the orientation in Plan 0 rather than 45°. In that case the plan whose boundary is farthest from the polarizer angle might be only 10° away. Perfect correlation would only be guaranteed for small angles from 0° to 10°, with the probability of agreement falling linearly from perfect correlation at 10° to perfect disagreement at 80°. (Alternatively, one could still use the original plans of Figure 3 but only switch to Plan 1 if the distant polarizer is set within 10° of the boundary in Plan 0.) If one used this new one-bit strategy over a long series of runs the resulting correlation pattern would be like that depicted in Figure 4 except that the flat portion runs from 0° to 10° and from 80° to 90°.

In general, any one-bit strategy with an offset angle of θ between Plan 0 and Plan 1 will give rise to a graph which is flat from 0° to $\theta/2$ and drops off linearly until reaching 0 at 90°- $\theta/2$. The zero-bit strategy is the limit as θ approaches 0: the simple linear relation. In the sequel we will refer to any strategy which produces a graph flat from 0° to θ and linearly dropping off to 0 at 90° - θ a *θ-strategy*. Having the ability to communicate one bit of information makes available to us all of the strategies from the linear 0°-strategy to the maximal 22.5°-strategy.

How nearly can we match the target curve by using a mix of one-bit strategies? Surprisingly well: we can exactly match the target curve in the region from 0° to 22.5° (and, by symmetry, from 67.5° to 90°). This perfect match results from using the probability density $\rho(\theta) = (\pi - 4\theta)(1 - \sin^2\theta) = (\pi - 4\theta)\cos 2\theta$ for choosing among the strategies between 0 and 22.5 (θ in the probability density has been expressed in radians rather than degrees).[2] The probability of choosing a strategy with $0° \leq \theta <$ 22.5° is

$$\int_0^{\pi/8} (\pi - 4\theta)\cos 2\theta \, \partial\theta \cong .848$$

The remaining 15.2% of the time the strategy 22.5° is chosen.

The result of this mix of strategies is an exact match to the target curve between 0° and 22.5° and between 67.5° and 90° with a linear regime from 22.5° to 67.5°. Linearity cannot be avoided in this region if one uses only one-bit strategies since *all* of those strategies are linear between 22.5° and 67.5°. So our mix of one-bit strategies, although an excellent approximation, fails to give a perfect match to the target.

In order to extend the region of perfect match further more information must be sent. If your partner can send two bits of information then she can choose between four possible plans of action, such as those illustrated in Figure 5.

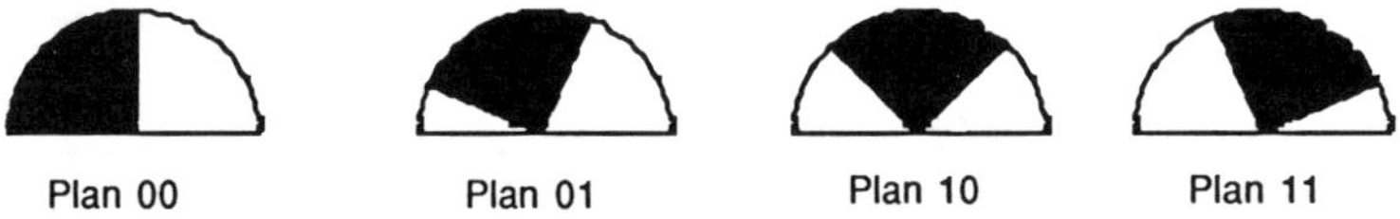

Figure 5: Plans Used in a Two-Bit Strategy

One of the plans can always be chosen so that the black/white border lies no nearer than 33.75° to the orientation of your partner's polarizer, thereby extending the flat part of the response curve to 33.75°. Using the same weighting function we can now obtain a perfect match with the target curve from 0° to 33.75° using one-bit strategies 84.8% of the time, two bit strategies (between 22.5° and 33.75°) 13.2% of the time, and the 33.75° strategy the remaining 2% of the time. The result of this mix is a curve which is linear only between 33.75° and 56.25°. Two-bit strategies send enough information to localize the distant polarizer setting to within a region which occupies a quarter of the original semi-circle.[3]

Further improvements can be made in the same way. Absolutely perfect match to the target requires using three-bit strategies 1.74% of the time, four-bit strategies .22%, five-bit strategies .028% and so on, without end. To *exactly* match the target, your partner must in principle be capable of sending messages of unbounded informational content, but will be required to send complex messages only rarely. In the long term, to achieve the behavior that photons display requires that your partner transmit an *average* of 1.174 bits of information per experiment. (A slightly more interesting number (as we will see) is the average value of 2^N, where N is the number of bits sent in a run. This represents the average for the number of states of the distant polarizer that one can distinguish between on the basis of the message sent, and comes out to 2.408.) You and your partner must begin by choosing a particular strategy, with the probability for each strategy determined by $\rho(\theta)$. Most of the time you will choose a strategy which can be implemented using one-bit messages, but occasionally much longer messages will be needed.

4. Evaluation of Results

The conclusion of our investigation into informational content is frustratingly equivocal. Given that the vital unknown, viz. the setting of the distant polarizer, can only be precisely conveyed using messages of infinite informational content, one cannot but be pleasantly surprised at much can be achieved using only one-bit messages. Even with an unlimited choice of polarizer settings available on each wing, the one-bit strategies can do an admirable job.

Admirable, but not perfect. If the predictions of quantum mechanics are correct, if the real long-term proportion of agreement between photons tends to $\cos^2(\phi - \psi)$ when the polarizers are set at ϕ and ψ, then one-bit methods will not do. Winning the game in this case requires that your partner at least have the capacity to send messages of unbounded length, messages which locate the orientation of the distant polarizer to any desired degree of precision. And if the capacity to send such messages exists, if in at least some instances the reaction produced on one side is so exquisitely sensitive to the setting on the other, then our miserly economies in the marketplace of information seem pointless. If the capacity for such precise sensitivity exists, why not use it all the time, one partner always informing the other of the exact position of her analyzer?

In fact, this is precisely what happens in the standard (non-relativistic) quantum mechanical picture of wave collapse. Each photon starts out in a perfectly indefinite state of polarization. When the first photon is measured the wave-function changes so that the second photon becomes definitely polarized *either parallel or perpendicular to the angle at which the first was measured.* So wave collapse transmits an infinite amount of information. If we could see the state of the second photon after the collapse, determine exactly what its state of polarization is, we could infer the orientation of the distant polarizer (or, more accurately, we could infer that either the distant polarizer is set at a particular angle or perpendicular to it). Of course, the exact polarization state of the incoming photon is not observable, so we cannot use wave collapse to send messages.

If our task is to perfectly match the target curve, the calculated average value of 1.174 bits of information per experiment appears as an insignificant number. The low average masks the fact that unbounded capacities for information transmission must be available, capacities just like those built into wave collapse. But if we relax the conditions of our game in another way this number, or similar averages, assumes real significance.

5. Simulators

One of the rules of the game we constructed was that each player, when asked a question, *must respond.* If the game is played 1000 times then we collect 1000 sets of responses, from which the proportion of agreement can be calculated. This corresponds to the demand that every photon which enters a measuring device must either be recorded as having passed or having been absorbed by the polarizer. (In other experiments this corresponds to every electron being found to have spin up or spin down.)

In experimental practice, though, this ideal is never realized. Actual detectors have only limited efficiency. Many, sometimes even most, of the particles simply fail to be detected at all. No usable data can be gleaned from a run of the experiment in which one of the particles avoids detection so such runs are discarded.

Translating these circumstances back to our game, imagine that each player is accorded a third option: instead or answering either "passed" or "absorbed", one can opt to refuse to answer at all. Data from an experiment in which at least one player remains silent are ignored; only runs in which both respond are counted.

In his 1982 paper "Some Local Models for Correlation Experiments", Arthur Fine devised a clever way to use this third, silent, response to eliminate any need for communication between the two wings of the experiment. Fine called these schemes "prism models", but I prefer the term "simulators" for reasons which will soon become apparent. The basic idea is very simple. In our one-bit strategies, one partner depends on information from the other, information as to whether the distant polarizer is set in region 1 or region 0. Suppose no communication can be provided, so even a one-bit message can't be sent. What can the players do?

One thing they can do is simply *guess* into which region the polarizer setting falls. For example, they might guess that the left-hand polarizer is set in region 0, and so agree that the right-hand player will use plan 0. If the left-hand polarizer *is* set in region 0, all is well: both players follow plan 0 and their behavior is just as if the left-hand setting had been communicated to the right. But if the left-hand polarizer is not set in region 0 then the left hand player is to *refuse to answer the question at all.* The experiment is deleted from the data on grounds of failure of detection, *and only the cases in which the guess was correct get into the data.*

In this way two players who cannot communicate at all can create the impression that local responses directly depend on the setting of the distant device. Partners who adopt the strategy outlined in the previous paragraph will produce the correlations of Figure 4 for the runs in which they both answer. The price for this conjuring trick is paid in the coin of efficiency: to simulate sending a one-bit message, 50% of the experiments must be disqualified due to one partner refusing to answer. The "detectors" must miss one "particle" in half the runs.

As the informational content of the message increases the efficiency of the corresponding simulation decreases. To simulate communication in the two-bit strategy the partners must guess in which of four equal regions one of the detectors is set, with that partner refusing to answer if the setting is outside the chosen region. Both partners answer, on average, in only one of four experiments and the other three runs are discarded. Simulation of the three-bit strategy requires efficiency to fall to one out of eight, and so on.

We can now calculate a preliminary approximation for the overall efficiency of a mix of strategies which can, without any communication, simulate the quantum correlations. Suppose we want, after a certain number of experiments, to have 1,000 *recorded* experiments which display the correct correlations. As we have seen, when communication is possible, approximately 848 of the 1,000 experiments will use a one-bit strategy. At an efficiency of 50% the non-communicating simulators must make twice as many guesses, i.e., they must use 1,696 experiments (on average) to get the 848 correct one-bit responses. To simulate the needed 132 successful two-bit strategies, 528 attempts (on average) are required, the approximately 17 three-bit strategies need 8 x 17 = 136 tries, and so on. Taking the sum over all of the strategies, weighted by their respective efficiencies, we find that the simulators need an average of 2,408 experiments to succeed in their illusion. Of these, 1,408 will not be recorded since one of the particles will not be detected. The remaining 1,000 successful experiments will, however, display the $\cos^2(\theta)$ pattern.

Successful simulation of an N-bit strategy requires, on average, 2^N attempts. The significance of our calculation of the average value of 2^N per experiment, 2.408, can now be made clear. On average, a successful simulation of the selected strategy will require 2.408 attempts, and we recover the same result: simulators need, on average, 2,408 tries to achieve 1,000 recorded results.

The simulators can improve their efficiency somewhat by adjustments in their method. We will develop the most efficient scheme in two steps. First, consider the effect of adopting any given θ-strategy. The graph of agreement has the form of perfect agreement when the angle of misalignment between the polarizer is less than the critical angle θ, dropping off linearly to perfect disagreement when the angle lies between 90° - θ and 90°. The difficulty for the simulators is establishing the perfect correlation between 0° and θ. Following the pure zero-bit strategy, the partners will disagree whenever the black/white boundary falls between the two polarizer orientation angles. The likelihood of this occurring is simply proportional to the angle of misalignment.

The simulators circumvent this problem essentially by assigning a *forbidden zone* to one of the particles. The forbidden zone surrounds the black/white borders, and if the polarizer setting for that particle falls within the forbidden zone, the particle refuses to be detected. If we want to assure that the particles never disagree when the angle of misalignment is less than θ, we need to set up a forbidden zone which extends θ on each side of the borders. If the polarizer setting on the chosen side falls outside the forbidden zone, any polarizer setting within θ of it on the opposite side will elicit a matching result.

The method of improving efficiency of our crude one-bit simulators is now obvious. Most of the crude one-bit simulators gratuitously extend the forbidden zone on one side so that the entire zone covers 90° of the possible 180°. More efficient simulators would restrict the zone to cover only 4θ out of the 180°. The more efficient scheme is illustrated in Figure 6 for θ = 10°.

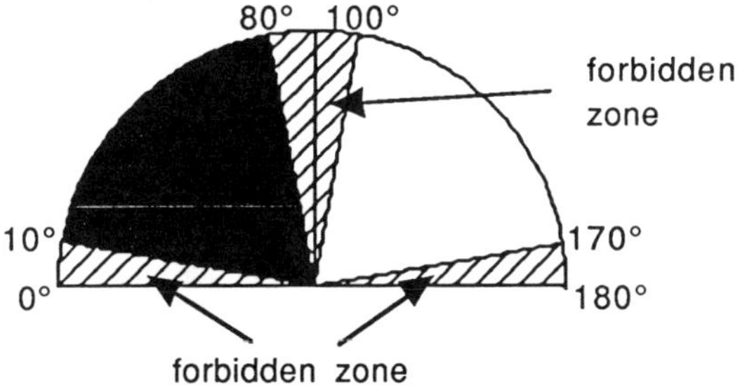

Figure 6: Efficient Simulator for θ = 10°

In the regime for $0 < \theta \leq 22.5°$ the original simulation scheme had a constant efficiency of 50%, but for the new method the efficiency is $(\pi - 4\theta)/\pi$ (θ expressed in radians).

What distribution of strategies should these more efficient simulators use? The *observed* frequency of a strategy (i.e., the frequency with which the strategy give usable data since both partners answer) is the *actual* frequency of the strategy multiplied by its efficiency. We want the observed frequency of every strategy to be proportional to $\rho(\theta) = (\pi - 4\theta)(\cos 2\theta)$. So the actual distribution must be $k(\pi/(\pi - 4\theta))((\pi - 4\theta)(\cos 2\theta) = k\pi\cos(2\theta)$, where k is some proportionality constant. Integrating,

$$\int_0^{\pi/4} k\pi\cos(2\theta)\delta\theta = k\frac{\pi}{2} = 1$$

we find that $k = 2/\pi \cong .6366$. The overall efficiency for this more efficient simulation is about 63.66%: on average, 1,571 experiments will yield 1,000 usable results and the results will show the proper pattern of correlation.

In complete detail, then, one way to achieve the quantum correlations without any communication is as follows. While still in the room, you and your partner randomly choose an angle θ between 0° and 180°. You construct an instruction set by coloring the region from θ to $\theta + 90°$ (mod 180) black. Next, an angle ϕ between 0° and 45° is randomly chosen using the probability density $2\cos(2\phi)$. The regions which extend ϕ on either side of the black/white boundary on one of the instruction sets is designated the forbidden zone. If the partner with that instruction set is asked a question that lies within the forbidden zone (if the polarizer is oriented at one of the forbidden angles), she refuses to answer. In the long run you will both answer 63.66% of the time, and in the long run the statistics displayed for the cases in which you both answer will match the target curve: you will seem to be doing what the photons do. Let us call this plan of action *Scheme A.*

Scheme A can be improved on slightly by dividing the forbidden zone evenly between the two particles rather than concentrating it in one of the instruction sets. Again, we want the θ-strategy to guarantee agreement between recorded pairs of particles so long as the angle between the polarizers is less than θ. Disagreement between responses only occurs when the polarizer settings fall on opposite side of a black/white border in the instruction set. For this to occur when the angle between the polarizers is θ requires that *one* of the polarizer settings falls within $\theta/2$ of the border. So by adding to each instruction set a forbidden zone which extends $\theta/2$ from each boundary the θ-strategy can be simulated. Call this method of simulation *Scheme B.*

Scheme B poses a slightly greater analytical problem than Scheme A. Complications arise because the efficiency of a Scheme B strategy varies with the angle of misalignment between the polarizers. When the polarizers are perfectly aligned the probability of getting both particles to respond is $(\pi - 2\theta)/\pi$, the probability that the chosen orientation of the polarizers falls within the common forbidden zone around one of the two boundaries. This efficiency decreases linearly as the angle of misalignment grows, reaching $(\pi - 4\theta)/\pi$ when the angle is θ. The efficiency remains at this level up to the maximum angle of misalignment, $\pi/2$.

To spare the reader tedious calculational detail, we will simply quote the main results.[4] In order for the recorded pairs to show the desired $\cos^2\theta$ correlations, Scheme B strategies must be chosen in accord with the probability density

$$\rho(\theta) = \frac{1}{\cos^4(\theta)}\left[\left(\frac{\pi}{2}+1 - 2\theta\right)(\cos 2\theta) - 4\sin\theta\cos\theta\right].$$

Using this weighting, Scheme B achieves the desired results with an efficiency of $(\pi + 2 - 4\phi)/2\pi\cos^2\phi$, where π is the angle of misalignment measured in radians. When the polarizers are perfectly aligned, both particles will respond about 81.8% of the time. Efficiency decreases until, at the misalignment of $\pi/4$, it reaches a minimum of 63.33%, equal to Scheme A.

There is no obvious way to improve on the efficiency of Scheme B. The total area of the forbidden zones cannot be decreased and Scheme B constitutes the best distribution of the forbidden zones between the two instruction sets. I conjecture that Scheme B is the best one can do in the absence of any information-transmitting connection between the two wings of the experiment. One can get the right correlations between the particles in those cases when both particles are detected, but joint detection of the particles must fail from 18.2% to 36.3% of the time.

6. Comparison with Fine

Schemes A and Scheme B demonstrate how close local realistic theories can come to matching the predictions of quantum theory. With inefficient detectors, one can exactly match the quantum correlations. They do not, of course, match all the predictions of quantum mechanics since it implies no comparable limits on the efficiency of ideal detectors. The schemes also refute Arthur Fine's analysis of the import of the prism models.

Fine thought that the models could be understood by reference to the algebra of Hermitean operators in Hilbert space. Based on that analysis, he claimed to construct a "maximal model", a model of maximal efficiency for the case when each wing can be set to only two positions This model has two salient features. First, the efficiency of joint detection in the maximal model is 50% (Fine 1982, p. 287). Second, Fine believed that on principle, the hidden states could not respond to more than three possible analyzer settings (p. 286-7). Taking Fine at his word, W. D. Sharp and N. Shanks (1985) argue that prism models will be reduced to unacceptably low efficiencies when all analyzer settings are allowed, namely 0% joint detection rates (p. 555). Schemes A and B obliterate these supposed constraints, achieving higher efficiency than 50% for experimental conditions with an infinity of settings available, and using states which will respond to an infinity of different settings.

Abner Shimony (1989, footnote 8, p. 31) challenges Fine to demonstrate that the prism models can be expanded to cover all possible polarizer settings; Schemes A and B answer this challenge. In terms of Fine's later (1989) paper, Schemes A and B solve the $\infty \times \infty$ problem (i.e., and infinite set of possible observations can be made on each side), and obviate the need for any "rigging" of the models to the experimental set-up. These models beat the efficiency constraints announced in that paper since the probability of detecting neither particle is not the square of the probability of detecting a single particle.

The key to understanding violations of Bell's inequality is not operator algebras but information transmission. And from this point of view, Jarrett's distinction between theories which violate parameter independence and those that violate outcome independence are misleading. Deterministic theories, such as our information transmission scheme, violate parameter independence. (They cannot be manipulated to send messages if the hidden states cannot be prepared.) Indeterministic theories, such as standard quantum mechanics with wave collapse, violate outcome independence. But the indeterminism just makes matters *worse* from the point of view of information transmission. If your partner uses an indeterministic method to settle on her response then *not only the parameter setting but also the response* must be transmitted to you if you are to match the quantum statistics. For example, you can't possibly guarantee matching outcomes on identical analyzer settings if you don't know how your partner responded. And if your partner uses a stochastic mechanism to decide on the response, you can not know how she has responded without information being sent. Classical wave collapse performs this function of information transmission (again, in a way that cannot be manipulated from the outside to send signals). When the first particle is detected the wave function for the partner changes to an eigenstate which reflects *both* the exact parameter setting *and* the outcome on the distant wing. An infinite quantity of information is sent in that transformation. Furthermore, the information transmission provided by wave-collapse *cannot* be simulated even by inefficient local models: that is why our simulator schemes are all deterministic.

If detector efficiencies can be raised sufficiently, our local simulator models can be ruled out empirically, showing that information *must* be transmitted between wings of the experiment. Standard quantum theory with wave collapse postulates the instantaneous transmission of infinite quantities of information in every experiment. Non-local deterministic hidden variables theories can reduce this demand on information transmission between the wings, but not eliminate it. And even in the best cases, the *potential* for transmission of unbounded amounts of information must exist if the quantum predictions are to be recovered.

Notes

[1]I would like to thank an anonymous referee for very helpful suggestions.

[2]Any target probability function $f(\theta)$ defined for $0 \leq \theta \leq \pi/2$ such that

i) $f(0) = 1$

ii) $\frac{\partial^2}{\partial\theta^2} f(\theta) \leq 0$ for $0 \leq \theta \leq p/4$

iii) $f(\theta) = 1 - f(\pi/2 - \theta)$

can be matched by this method. The appropriate probability density is

$$\rho(\theta) = \frac{1}{2}(\pi - 4\,\theta)\,\frac{\partial^2}{\partial\theta^2}\,f(\theta).$$

If all strategies from 0 to $\pi/4$ are available, the match will be exact. If only strategies from 0 to $\phi < \pi/4$ can be used then one uses the probability density above for the available strategies $< \phi$ and concentrates the remaining weight on strategy ϕ. The resultant curve will match the target exactly from 0 to ϕ with a linear connecting piece from ϕ to $\pi/2 - \phi$.

[3]The two-bit strategy we have chosen allows the recipient of the message to place the distant setting somewhere within a disconnected set of regions which together cover one fourth of the total possibilities. Other strategies are possible, strategies which allow one to locate the distant setting within a more connected region. Some two-bit strategies have features which allow a perfect match to the target curve from 0° to 45° with deviations in the region from 45° to 67.5°. No exhaustive analysis of all possible strategies exists, to my knowledge, but I see no reason to think that some other scheme could do significantly better than this.

[4]In general, any function $f(\phi)$ which satisfies the three criteria listed in footnote 1 can be simulated by Scheme B. The appropriate probability density is the solution of

$$\rho(\theta) = \frac{\partial^2}{\partial\theta^2}\left[\frac{\pi \int^{\pi/4} \phi\rho(\phi)d\phi - \theta}{f(\theta)}\right].$$

References

Bell, J.S. (1976), “The Theory of Local Beables”, *Epistemological Letters* : March 1976, 11-24, reprinted as chapter 7 of Bell’s 1987 *Speakable and Unspeakable in Quantum Mechanics*, Cambridge: Cambridge University Press.

Fine, A. (1982), "Some Local Models for Correlation Experiments", *Synthese* 50: 279-294.

_____. (1989), "Correlations and Efficiency: Testing the Bell Inequalities", *Foundations of Physics* 19: 231-478

Jarrett, J. (1984), "On the Physical Significance of the Locality Conditions in the Bell Arguments", *Nous* 18: 569-589.

Sharp, W.D. and Shanks, N. (1985), "Fine's Prism Models for Quantum Correlation Experiments", *Philosophy of Science* 52: 538-564

Shimony A. (1989), "Our Worldview and Microphysics" , in *Philosophical Consequences of Quantum Theory: Reflections on Bell's Theorem*, J.T. Cushing and E. McMullin (eds.). Notre Dame: University of Notre Dame Press

Part XI

REALISM, METHODOLOGY AND UNDERDETERMINATION

Convergent Realism and Approximate Truth[1]

David B. Resnik

University of Wyoming

1. Introduction

Convergent realists typically maintain the following theses:[2]

T1: Some scientific theories, literally interpreted, are at least approximately true.

T2: Some scientific theories, literally interpreted, genuinely refer.

T3: The history of mature sciences is a progressive approximation to the truth.

T4: The world (reality) described by scientific theories is independent of our thoughts or theoretical commitments.

Convergent realists usually offer the infamous "success of science" argument as the strongest defense of their view. Scientific realism, according to convergent realists, is the only view that does not portray the success of science as miraculous (i.e., unexplainable). Various versions of convergent realism have been defended by writers such as Richard Boyd (1973, 1980, 1984), W.H. Newton-Smith (1978), Hilary Putnam (1975, 1978), J.J.C. Smart (1963, 1985, 1989), and Russell Hardin and Jay Rosenberg (1982). Critics of convergent realism include Larry Laudan (1981), Arthur Fine (1984), and Bas van Fraassen (1980).

Although convergent realism has suffered numerous attacks during the last decade, it is not my aim in this paper to rehash those criticisms or offer a general synopsis of this position.[3] Instead, my goals will be more focused. I will examine the role that approximate truth plays in arguments for convergent realism and I will diagnose some difficulties that face attempts to defend realism by employing this slippery concept. Approximate truth plays two important roles in the defense of convergent realism: it functions as a truth surrogate and it helps explain the success of science. I shall argue that the concept of approximate truth cannot perform both of these roles. If approximate truth adequately fulfills its role as a truth surrogate, then it cannot explain the success of science. If approximate truth can adequately explain the success of science, then it cannot function as a truth surrogate.

PSA 1992, Volume 1, pp. 421-434

2. A Surrogate for Truth

The concept of approximate truth usually has two important functions in scientific realism: 1) it performs the role of a truth surrogate; and 2) it is vital to explaining the success of science. In addition to these two roles, it is often maintained that approximate truth is important in defining realism.[4]

There are three main reasons why realists rely on a truth surrogate, such as approximate truth, truthlikeness, or verisimilitude. Each of these reasons for using a surrogate for truth stem from a desire to fend off potentially damaging objections to realism. The first reason arises from the need to reconcile realist metaphysics with our epistemological predicament (Niiniluoto, 1987). Most scientific realists maintain that science is a truth-seeking activity; the aim of science is to produce true descriptions (theories) of the world or at least to make progress toward the truth.[5] But since theories contain laws and generalizations, they have an infinite number of observational consequences; and since we can never check all of these observational consequences, we can never know that a theory is true; it could always be refuted by an observation. We might be able to know that a theory is highly confirmed, has survived numerous severe tests, and so on, but we cannot know that a theory is (completely) true or that we are even making progress toward the truth.

However, if we can never know that we are achieving the goals of science, then science would appear to be an irrational or at least futile activity, given a realist account of these goals. Although we might stumble upon a true theory, we would not know it; although we might be making progress (i.e., getting closer to the truth), we would not know it. All we could know is that some of our theories have not yet been refuted. In order to avoid this unpalatable consequence, scientific realists have proposed that scientists can seek a more epistemically accessible goal, such as verisimilitude (Popper 1963, 1968), approximate truth (Boyd 1984; Newton-Smith 1981) or truthlikeness (Niiniluoto 1986). While true theories are epistemically inaccessible—we can never know that a given theory is true—approximately true theories might be more epistemically accessible.[6] Thus, it might be possible to know that some of our theories are approximately true or that the history of science indicates a progressive approximation toward the truth.

The second reason why realists rely on approximate truth comes from an awareness of the role of approximations and idealizations in science. Ronald Laymon (1980, 1985) and Nancy Cartwright (1983) both convincingly argue that even the most highly confirmed theoretical laws in the physical sciences only provide approximately correct descriptions of the world. Physical theories typically make a wide variety of idealizing assumptions, e.g., ignoring friction, treating molecules as points, etc. .... Yet a true theory should not make these idealizing assumptions; it should provide an exactly correct description of the world. In order to circumvent this objection, realists have maintained that scientific theories need not be true, but only approximately true (Niiniluoto, 1987).

A third, more hypocritical, reason why realists may rely on a truth surrogate is the need to avoid having their doctrine refuted by false scientific theories or evidence from the history of science (M. Resnik, 1990). If realists hold that theories of the mature sciences are true, then realism will be refuted as soon as one theory of a mature science is proven false. In order to avoid this consequence, realists can propose that the best scientific theories are approximately true, though strictly speaking, false. Theories can be approximately true even though they have false observational consequences. This view portrays realists as hypocrites since many realists (e.g., Boyd

1984, Putnam, 1978) hold that realism itself is an empirical hypothesis capable of being tested or refuted. Yet the use of approximate truth would appear to make realism a non-empirical, untestable hypothesis, since it would allow one to maintain that scientific realism is true no matter how many scientific theories are shown to be false.

Thus, scientific realists appeal to approximate truth (or some related notion) because it is a less demanding and more accommodating concept than truth; it is a surrogate for truth. In order for the concept of approximate truth to perform this function, it should satisfy the following conditions:

(C1): Approximately true theories must be more epistemically accessible than true theories.

(C2): Approximately true theories need only provide approximately correct descriptions of the world.

(C3): Approximately true theories may have false observational consequences. These conditions are not entirely unrelated. Indeed, one might argue that C2 entails C3 and C1 or that C3 entails C2 and C1. Although each condition is important, I suspect that C1 is both the most important and the most difficult to satisfy. Unnecessary polemics aside, C1-C3 are conditions that realists who appeal to approximate truth should accept. After all, if the concept of approximate truth does not satisfy these conditions, then why bother developing such a concept in the first place? We might as well just stick to the less problematic concept of truth.[7]

3. Explaining the Success of Science

A second, crucial role that approximate truth plays in arguments for realism is that it is supposed to help explain the success of science, and hence, provide support for a particular kind of realism: convergent realism. Versions of the success of science argument (Boyd 1973, 1980, 1984; Putnam 1978; Smart 1963, 1985, 1989) typically make use of approximate truth. It is alleged that scientific realism is the only position that does not make the success of science seem to be a miracle. The version of this argument which has received the most attention is due to Boyd (1973, 1980, and 1984). Boyd's abductive argument goes as follows:

1) Scientific methods are instrumentally reliable (i.e., successful). Scientific methods yields theories that are highly confirmed, predictively and explanatorily successful, and so on.

2) The only plausible explanation of the instrumental reliability of scientific methods is that some theories in the mature sciences are at least approximately true and referentially successful.

3) Hence, some theories in the mature sciences are at least approximately true and referentially successful (i.e., a version of scientific realism is true). (Boyd 1984, pp.58-61).

Could approximate truth explain the success of science? This depends in large part on what we mean by "approximate truth" and "success of science." Unfortunately, writers who defend the success of science argument, including Boyd, do not offer an explicit account or definition of "approximate truth." Indeed, Laudan has remarked that until convergent realists specify what they mean by "approximate truth", their central theses can be regarded "as so much mumbo jumbo (Laudan 1981, p.32)". While I will not go so far as Laudan goes in bashing convergent realism, I agree that it is incumbent on convergent realists to provide a suitable analysis of approximate truth.[8]

Although there have been many attempts to analyze the concept of approximate truth,[9] not all of them will provide convergent realists with a concept that can help explain the success of science. Since convergent realists use approximate truth to help explain the success of science, they must use an analysis of approximate truth that gives this concept explanatory power. How might a concept have (or fail to have) explanatory power? One way that a concept might lack explanatory power would be if it is defined in terms of the very phenomena it is invoked to explain. For instance, if one defines genetic 'fitness' in terms of actual reproductive success, then the concept of fitness will lack explanatory power in evolutionary biology, since explanations of reproductive success that appealed to fitness would be circular. However, if one defines 'fitness' in terms of expected reproductive success, then this concept will have explanatory power, since expected reproductive success is independent of actual success.[10] Similarly, in order for the concept of approximate truth to have explanatory power it must not be defined in terms of the very phenomena it is invoked to explain. In other words, it must satisfy the following condition:

(C4): Approximate truth must not be defined in terms of the very phenomena it is invoked to explain (e.g., the success of science).

Although one might argue that the concept of approximate truth must meet other requirements in order to have explanatory power, such as causal efficacy, statistical relevance, and so on, these requirements might presuppose controversial theories of explanation. Presumably any adequate account of explanation will hold that circular explanations do not explain.

What shall we mean by "success of science"? Although most writers would agree that science is remarkably successful, there is much disagreement over what counts as success. For the purposes of discussion, I shall describe five characteristics of a successful science. These characteristics are not necessary and sufficient conditions, but sciences that we would call successful will exhibit some or all of these characteristics. They are as follows:

a) Predictive Success: Scientific theories yield accurate predictions.

b) Explanatory Success: Scientific theories provide good explanations.

c) Confirmational Success: Scientific theories are highly confirmed.

d) Methodological Success: Scientific methods produce theories which exhibit predictive, explanatory, and confirmational success.

e) Practical/Technological Success: Scientific theories provide us with information and technology that enables us to achieve practical goals and purposes, such as human well-being, power, health, and happiness.

It should be noted that I have excluded something like "referential success" or "descriptive success" from the characteristics of a successful science. This oversight is deliberate. Although Putnam (1978) is a big fan of referential success, it might be argued that allowing science to exhibit this kind of success begs-the-question in favor of realism. If we assume that theories in mature sciences genuinely refer, and we subscribe to a causal theory of reference, then it can be argued that the causal structure (i.e., relations, properties, and entities) described by these theories exists. With good reason, anti-realists, such as Laudan (1981) are not willing to admit that theories in mature sciences genuinely refer, since granting this assumption is tantamount to adopting a version of scientific realism. "Descriptive success" would also seem to beg-the-question in favor of realism, if we mean by "X is descriptively successful" something like "X gives a correct (or true) description of the world (reality)."

In sum, if the success of science argument for realism is to have any force at all, convergent realists must not assume the very premise (i.e., realism) that they are trying to prove. If truth and reference are to play a crucial role in explaining the success of science, then these two concepts should not enter into our understanding of the success of science. Thus, I shall interpret success in epistemological/practical terms, since this interpretation of success does not smuggle in any unjustified realist assumptions.

4. Three Approaches to Approximate Truth

Having set the stage for my main thesis, I shall now argue that no concept of approximate truth can fulfill both conditions 1-3 and condition 4 described above. In order to defend these claims, I shall distinguish between three different ways of analyzing approximate truth. These different approaches to approximate truth are derived from four common theories of truth, the correspondence theory of truth, the semantic theory, the coherence theory, and the pragmatic theory. I shall argue that none of these kinds of analyses allow approximate truth to function as *both* a truth surrogate and an explainer of the success of science.

According to metaphysical analyses of approximate truth, approximate truth should be understood in metaphysical terms: a theory is approximately true if and only if it gives us an approximately true description of reality. Truth is understood metaphysically via a correspondence theory of truth. Thus, approximately true theories, like true theories, correspond to reality, and the difference between the two is that true theories have a better correspondence. Science makes progress by adopting theories which have an increasing degree of approximate truth (or correspondence to reality). In his earlier work on verisimilitude, Popper (1963) gave a metaphysical analysis of approximate truth.[11]

This analysis of approximate truth has some definite advantages. First of all, it should allow realists to use approximate truth to explain the success of science. If a theory is approximately true, then we should expect that it will make accurate predictions and yield good explanations, since the theory will be an approximately correct description of reality. The circularity problem would be avoided because approximate truth would be a metaphysical notion, while success would be an epistemological/practical one.

However, there are some definite problems with using a correspondence theory of approximate truth. If approximate truth is analyzed in terms of truth, and truth is understood as correspondence to reality, then approximate truth will be not be any more epistemically accessible than truth. Approximate truth will be no more more epistemically accessible than truth because both concepts will ultimately be understood in terms of correspondence to reality. In order to determine whether our theory corresponds to reality, we would need to step outside of our theory and theoretical commitments and verify this correspondence, and, following Kant, most philosophers would maintain that this is an impossible task. The move to approximate truth does not improve our predicament, since we will face the same problem in determining whether our theory approximately corresponds to reality.[12]

Convergent realists who wish to understand approximate truth in terms of a correspondence theory of truth might do better to employ Hartry Field's version of this theory. Field (1972, 1977, 1986) proposes a correspondence theory based on Alfred Tarski's (1944) recursive definition of truth. According to Tarski, 'truth' can defined recursively in a given object language, L, as a concept that entails biconditionals of the form T:

(T) X is true if and only if p

where "X" is the name of a sentence in the object language and "p" is the sentence itself. The classic example of such a biconditional is:

"Snow is white" is true if and only if snow is white.

Each of these biconditionals constitute a partial-definition of 'truth' in a language L. While Tarski saw himself as giving an analysis of truth that captures the intuitions engendered by the correspondence theory of truth (Tarski 1944, p.342), and others have interpreted his theory as providing a version of the correspondence theory of truth (Niiniluoto 1987, p.138), his analysis need not be understood this way. In its barest form, Tarski's theory of truth can be understood simply as a device for disquotation or what Quine calls "semantic ascent" in a given language (1970, pp. 10-13).

Although Field employs Tarski's definition of truth, he supplements Tarski's biconditionals with a causal theory of reference. The reason he supplements Tarski's theory is that he, like many realists, would like to give truth a causal/explanatory role, and Tarski's theory of truth, in its barest form, cannot perform this task (Grover 1990). According to Field's account of truth, expressions of the form "p is true" are to be eliminated and replaced by "p denotes" or "p applies", and these expressions, in turn, can be eliminated with the aid of a causal theory of reference (Grover 1990). In order for a theory to be approximately true, on this view, its terms must refer to genuine features of the world—it "carves the world at its joints"—and it must be highly confirmed. This view, though not explicitly defended by anyone whom I am aware of, may capture that spirit of Boyd's (1984) use of the phrase "approximate truth."[13]

The main advantage of employing a causal theory of reference in an analysis of approximate truth is that it can give this concept explanatory power. If a theory does in fact "carve the world at its joints", then we would have good reasons to believe that the theory will be highly successful, that it will give good explanations, accurate predictions, and so on. Explanations of the success of science will not be circular, on this view, because approximate truth is understood in metaphysical terms, while success is understood in epistemological/practical terms.

While it is tempting to invoke a causal theory of reference in order to solve some problems with approximate truth, the causal theory of reference is not without its own difficulties.[14] Moreover, an analysis of approximate truth based on a causal theory of reference does not provide realists with a satisfactory truth surrogate because it makes approximate truth no more epistemically accessible than truth. In order to know whether a theory is approximately true, according to this analysis, we must know that its terms refer to genuine features of the world and that it is highly confirmed. While it is not difficult to know that a theory is highly confirmed, it is exceedingly difficult to know that its terms refer to genuine features of the world. In order to know that any given theory refers to genuine features of the world, we must know 1) that these features (e.g., entities, properties, relations, etc.), actually exist; and 2) that the theory provides us with an adequate description of these features.

I submit that it is no easier to know that a theory "carves the world at its joints" than it is to know that it is true. If we assume that the ontological claims of a theory can only be tested via their observational consequences,[15] and that theories have an infinite number of observational consequences, then it is impossible to know that a theory's ontological claims are true. Even if the theory's ontological claims are highly confirmed, it will always be possible that some recalcitrant observation will disconfirm them. As Laudan (1981) reminds us, history is replete with examples of highly confirmed theories that made incorrect ontological claims (i.e., did not refer), such as the theory of the optical ether, the phlogiston theory, and the caloric theory of heat. I do not mean to imply that it is impossible for a theory to "carve the world at its joints"—we could always stumble on the correct ontology for a given domain—but if a theory does accomplish this task, we will not know it. Thus, this analysis of approximate truth will also not satisfy convergent realism's desire for a truth surrogate, since it also has the consequence that we cannot know that our theories are approximately true or that we are making progress toward the truth.

An alternative analysis of approximate truth, which I shall dub the semantic analysis, is also based on Tarski's (1944) semantic analysis of truth. But the semantic analysis of approximate truth, unlike Field's analysis of truth, does not supplement Tarski's biconditionals with a theory of reference; it employs a disquotational theory of approximate truth. Ilkka Niiniluoto (1987) has developed a semantic theory of truthlikeness based on Tarski's theory of truth. Niiniluoto's theory is not a theory of approximate truth *per se*, since he distinguishes the concept of truthlikeness from several related concepts, including probability, partial truth, degree of truth, and approximate truth. Nevertheless, his theory can be evaluated in terms of its ability to provide a surrogate for truth. Since approximate truth is also generally regarded as a surrogate for truth, we can view Niiniluoto's theory as an alternative approach to approximate truth.

Niiniluoto bases his analysis of the truthlikeness of theories on a prior analysis of the truthlikeness of statements. He attacks this problem in a very systematic, technical fashion, applying his notion of truthlikeness to singular statements, monadic generalizations, and polyadic generalizations (theories). Since his entire system is based on an account of the truthlikeness of statements, I will focus my discussion on this part of his theory. If problems arise in this essential part of his account, it is likely that they will infect his entire analysis.

Niiniluoto defends what he calls the similarity approach to truthlikeness (Niiniluoto, 1987, p. 198). The general idea is that the truthlikeness of a statement, S1, depends on the similarity between the state of affairs, A1, described by S1 and the state of affairs, A2, described by the true statement, S2. Niiniluoto's formal definition of truthlikeness is as follows:

> If g, is a potential partial answer to problem B, its degree of truthlikeness depends on the similarity between the states of affairs allowed by g and the true state expressed by the true element h* of B. (Niiniluto 1987, p.198).

In the above quote, B is a cognitive problem, i.e., a question or request for information. Corresponding to the problem is a non-empty set of complete, partial answers (or elements), one of which, h*, is the most informative true statement (or target statement). The question answered by h* is "Which element of B is true?" (Niiniluoto 1987, p. 129). Truthlikeness is a measure of how close a given element in B is to the target statement, h*.

Since the similarity approach to truthlikeness requires that we be able to determine how similar a state of affairs described by a particular statement is to the true state of affairs, Niiniluoto takes great care to define a similarity function that provides a measurement of the similarity between different states of affairs, relative to a cognitive problem. Following Hilpinen (1976) and Tichy (1976), Niiniluoto analyzes similarity in terms of possible world semantics. In order to avoid difficulties that arise from comparing one whole world to another, Niiniluoto suggests that possible worlds need only be compared at a specific descriptive depth, which again, is relative to a cognitive problem. Thus, Niiniluoto modifies his definition of truthlikeness:

> The degree of truthlikeness of a statement g does not measure its distance from the actual world in all of its variety, but rather *from the most informative true description of the world* (Niiniluoto 1987, p. 206).

By "most informative true description of the world" Niiniluoto means something like "the description that provides the most complete answer to a given cognitive problem."

I find Niiniluoto's analysis of truthlikeness extremely impressive, and I have no doubt that it is perhaps the best current attempt to provide an explication of a truth surrogate, and it is not my aim in this essay to attack Niiniluoto's analysis *per se*.[16] Instead, I will argue that this approach to approximate truth fails to provide the convergent realist with a satisfactory truth surrogate and it cannot give adequate explanations of the success of science.

The truthlikeness approach does not provide a satisfactory truth surrogate because it makes approximately true theories (truthlike theories) just as epistemically inaccessible as true theories. Niiniluoto has not provided an adequate solution to what he calls the "epistemic problem of truthlikeness" (1987, p. 263). The epistemic problem of truthlikeness results from that fact that the target statement of any interesting cognitive problem, h*, is unknown.[17] Since we don't know the true statement, how can we possibly know which statement is closer to the true statement (more truthlike)? Niiniluoto tries to get around this problem by arguing that the epistemic probability of a statement is a reliable indicator of a statement's degree of truthlikeness. While he admits that this approach presupposes that we have an epistemic probability distribution for any given cognitive problem at our disposal, he cheerfully maintains that the problem of estimating the degree of truthlikeness "is neither more nor less difficult than the traditional problem of induction" (Niiniluoto 1987, p. 263).

It would appear that Niiniluoto has offered a reductio of his own position, since the problem of induction is not an easy one to solve. Indeed, if we had an adequate solution to this problem, then there would be no need to develop a surrogate for truth, since we could use inductive methods to determine whether a given statement or theory is true.

The truthlikeness approach does not provide good explanations of the success of science because the concept must be relativized to a particular language and a particular cognitive problem within that language. But to adequately explain the success of science we need a concept of approximate truth that transcends any particular language and particular cognitive problem. Scientists from different cultures and language communities can understand, test, and apply scientific theories. One often mentioned example of scientific success, the quantum theory of matter, was tested, developed, and applied by scientists speaking different natural languages, including English, German, and Danish (Cline 1987). Niiniluoto's theory cannot adequately explain how these scientists developed the quantum theory because it is restricted to a particular language.

By relativizing truthlikeness to particular cognitive problems, Niiniluoto's theory also has difficulties explaining the success of science. Different scientists might use the same theory for a variety of different cognitive problems; theories are not accepted or rejected simply because they solve or fail to solve one cognitive problem. Darwin's theory of natural selection, for example, offered solutions to several different problems, including the origin of species, the variation of life, and the adaptedness of organisms to their environment (Ruse 1979). Moreover, it offered an excellent (highly truthlike) solution to the problem of adaptation but a poor solution to the problem of variation.[18] If all theories solved only one cognitive problem, then Niiniluoto's theory might offer inadequate explanations of the success of science, but the history of science indicates that this is seldom the case.

Epistemic analyses of approximate truth are analogous to epistemic analyses of truth, such as Nicholas Rescher's (1973) coherence theory of truth or C.S. Peirce's theory of truth as "the opinion which is fated to be ultimately agreed to by all who investigate" (Peirce 1955, p.38). Epistemic analyses of approximate truth analyze this concept in epistemic terms, such as confirmation, probability, rational belief, or acceptability.[19] Although epistemic analyses of approximate truth are not as common as other analyses, they have defenders. Ronald Laymon has suggested that approximate truth be interpreted in terms of a "confirmational history" (Laymon 1985, p. 159). An approximately true theory, according to this suggestion, is one that is good enough for the purposes of confirmation or disconfirmation. Laymon's analysis of approximate truth is epistemic because 'confirmation' is an epistemic term.[20]

When Putnam (1975, 1978) lapses into talk about approximate truth, he may also have something like an epistemic conception of approximate truth in mind. In his *Reason, Truth and History* (1981), Putnam presents us with a concept of truth as idealized rational acceptability:

> The claim that science seeks to discover the truth can mean no more than that science seeks to construct a world picture, which, in the ideal limit, satisfies certain criteria of rational acceptability (Putnam 1981, p. 130).

While one might argue that Putnam's philosophical views changed between 1975 and 1981, the concept of truth Putnam defends here fits in well with his views on truth and realism defended in his earlier works. The concept of truth that Putnam defends in this passage is inherently epistemic, since "rational acceptability" is an epistemic term. It should follow that approximate truth, according to Putnam, is also an epistemic concept, since approximate truth would be an approximation to idealized rational acceptability.

The main virtue possessed by epistemic analyses of approximate truth is that they allow this concept to function as a truth surrogate. Since approximate truth is interpreted in epistemic terms, it is epistemically accessible. It is inconceivable, on this view, for there to be true or approximately true theories that we cannot know. It is also possible for approximately true theories to have some false consequences and to be merely approximately correct descriptions of the world. Approximate truth is a very undemanding and accommodating concept if it is interpreted in epistemic terms.

However, epistemic analyses of approximate truth will not allow this concept to explain the success of science. If "approximate truth" and "success" are both understood in epistemic terms, then explanations of success will be circular. In order to explain the success of science, realists need a concept of approximate truth that is, as Putnam says, "radically non-epistemic" (Putnam 1978, p.125).

5. Conclusion: a Conflict of Roles

I have argued that none of these analyses of approximate will enable this concept to both serve as a truth surrogate and as an explainer of the success of science. Metaphysical analyses of approximate truth enable this concept to explain the success of science, but they do not provide realists with a useful truth surrogate; epistemic analyses of approximate truth allow this concept to serve as a truth surrogate but they do not allow it to explain the success of science; and semantic analyses of approximate truth do not allow this concept to perform either role.

None of these approaches to approximate truth allow this concept to perform its two primary roles in convergent realism because these roles pull in opposite directions. In order for approximate truth to have explanatory power, it must not be understood in terms of the phenomena it is invoked to explain, i.e., the success of science. Since most realist "success of science" arguments interpret success in epistemic terms, i.e., in terms of confirmation, predictive success, and so on, then approximate truth cannot be analyzed in epistemic terms if it is to have explanatory power. But if approximate truth is not analyzed in epistemic terms, then it cannot perform its role as a truth surrogate. If approximate truth is radically non-epistemic, then it will be no more epistemically accessible than truth. In short, convergent realists cannot have their cake and eat it too.

One might object that I have only shown that several theories of approximate truth do not allow this concept to serve both as a truth surrogate and as an explainer of the success of science, but that other theories might allow it to perform both of these roles. While I admit that I have not examined every theory of approximate truth, the objections I raise against the theories I have examined apply to other theories as well. The problems inherent in these particular theories are indicative of difficulties that plague several different ways of characterizing approximate truth. In logical parlance, my argument against convergent realism's use of approximate truth takes the form of a universal generalization, where the theories selected are arbitrarily chosen individuals.

Although I have not given a general argument against realism or approximate truth, my conclusions do undermine attempts to defend scientific realism that offer the approximate truth of scientific theories as an explanation of the success of science. While one could give such arguments, they would have persuasive force only if they use a concept of approximate truth that is not an adequate truth surrogate. But why should realists bother developing a truth surrogate that is not really a truth surrogate? This methodological move would seem to be as pointless as developing a "diet" Coke that has just as many calories as the real thing.

Notes

[1]I am grateful to James Forrester, Allan Franklin, James Martin, and Michael Resnik for helpful comments and criticism.

[2]Not all convergent realists hold these claims. For example, Putnam (1978) does not assert thesis 4.

[3]Most anti-realist replies to convergent realism have focused on the second premise of the argument, while other replies have questioned the whole argument form (inference to the best explanation) itself. For more on this topic, see Almeder (1989).

[4]Boyd (1984) maintains that approximate truth is important in defining realism. However, I will not offer an all-encompassing definition of "scientific realism" since I am skeptical about attempts to do so.

[5]I do not mean to imply that theories have truth values or can be identified with a set of sentences (statements or assertions), although those who adopt the syntactic (or received) view of theories (Carnap 1966, Hempel 1965) take this position. Those who defend the semantic view of theories, however, (Suppe 1989, van Fraassen 1981) regard a theory as a set of models.

[6]By "epistemically accessible" I mean simply "can be known by us".

[7]Of course the concept of truth is also problematic, but presumably it is less problematic than approximate truth.

[8]Recently, Niiniluoto (1987) has provided a powerful analysis of truthlikeness, but his analysis has had little impact on arguments for convergent realism. Although Niiniluoto defends a version of realism—he calls it critical realism—he does not offer the explanation of the success of science as the main argument for his position. See Niiniluoto (1984).

[9]See, for example, Popper (1963, 1968, 1972), Reichenbach (1949), and Miller (1974), Tichy (1974), Hilpinen (1976), Niiniluoto (1987).

[10]For further discussion of this point see D.Resnik (1988).

[11]In his later works, Popper (1968, 1972) abandoned the correspondence theory of truth and adopted Tarski's theory of truth.

[12]Many writers have attacked the correspondence theory of truth. See, for example, Putnam (1981).

[13]See Boyd (1984, pp.62-63) for evidence of his commitment to a causal theory of reference and its possible role in an analysis of approximate truth.

[14]For an account of some difficulties with the causal theory of reference, see Putnam (1981). For skeptical comments concerning reference see Quine (1960). For criticism of reference's role in convergent realism, see Laudan (1981).

[15]This argument does not apply to theories whose ontological claims can be tested directly via our sensory apparatus or through highly reliable instruments.

[16]For those who would like to attack Niiniluoto's analysis per se, the concept of similarity itself is a good place to start. Goodman (1972) has attacked this concept as hopelessly vague and indeterminate, and I'm not convinced that Niiniluoto has overcome Goodman's objections.

[17]There is no problem in estimating the degree of truthlikeness of a statement where h* is known. But as Niiniluoto himself acknowledges, there is little use in saying that a statement is like the truth if we already know which statement is true.

[18]Darwin's theory did not give a satisfactory solution to the problem of variation because he lacked a theory of inheritance that explained the origin of variations.

[19]By the phrase 'epistemic term' I mean a term that refers to knowledge, evidence, or justified belief.

[20]It should be noted that Laymon does not offer his suggestion as a way to argue for realism. Indeed, he bases his analysis of approximation on some prior account of what it means for an approximation to be realistic.

References

Almeder, R. (1989), "Scientific Realism and Explanation", *American Philosophical Quarterly* 26, 173-186.

Boyd, R. (1973), "Realism, Underdetermination, and a Causal Theory of Evidence", *Nous* 7, 1-12.

______. (1980), "Scientific Realism and Naturalistic Epistemology", *PSA 1980,* volume2, P.D. Asquith and R. Giere (eds.). East Lansing: Philosophy of Science Association, pp. 613-662.

______. (1984), "The Current Status of Scientific Realism", in *Scientific Realism,* J. Leplin (ed.), Berkeley and Los Angeles: University of California Press.

Cartwright, N. (1983), *How the Laws of Physics Lie*. Oxford: Clarendon Press.

Cline, B. (1987), *Men Who Made a New Physics*. Chicago: University of Chicago Press.

Field, H. (1972), "Tarski's Theory of Truth", *Journal of Philosophy* 69: 347-75.

______. (1977), "Logic, Meaning, and Conceptual Role", *Journal of Philosophy* 74: 379-409.

______. (1986), "The Deflationary Concept of Truth", in *Fact, Science, and Morality: Essays on A.J. Ayer's Language Truth and Logic*, G. Macdonald and C. Wright (eds.). Oxford: Basil Blackwell.

Fine, A. (1984), "The Natural Ontological Attitude", in *Scientific Realism.*

Goodman, N. (1972), *Problems and Projects*. Indianapolis: The Bobbs-Merrill Company.

Grover, D. (1990), "Truth and Language-World Connections", *Journal of Philosophy* 87: 671-87.

Hardin, R. and Rosenberg, J. (1982), "In Defense of Convergent Realism." *Philosophy of Science* 49: 604-615.

Hilpinen, R. (1976), "Approximate Truth and Truthlikeness", in *Formal Methods in the Methodology of Empirical Sciences*, M. Przlecki, K. Szaniawski, R. Wojicki (eds.), Dordrecht: Reidel.

Laudan, L. (1981), "A Confutation of Convergent Realism", *Philosophy of Science* 48: 19-49.

Laymon, R. (1980), "Idealization, Explanation, and Confirmation", *PSA 1980*, volume 1, P. Asquith and R. Giere, (eds.). East Lansing: Philosophy of Science Association, pp. 336-352.

_ _ _ _ _ _ . (1985), "Idealizations and the Testing of Theories by Experimentation", in *Observation, Experiment, and Hypothesis in Modern Physical Science*, P. Achtinstein and O. Hannaway, (eds.). Cambridge: The MIT Press,.

Miller, D. (1974), "Popper's Qualitative Theory of Verisimilitude", *British Journal for the Philosophy of Science* 25: 178-88.

Niiniluoto, I. (1984), *Is Science Progressive?* Dordrecht: Reidel.

_ _ _ _ _ _ _. (1987), *Truthlikeness*. Dordrecht: Reidel.

Newton-Smith, W. (1978), "The Underdetermination of Theories by Data", *Proceedings of the Aristotelian Society*, pp. 71- 91.

_ _ _ _ _ _ _ _ _ _ _. (1981), *The Rationality of Science*, London: Routledge and Kegan Paul.

Peirce, C.S. (1955), "How to Make Our Ideas Clear", *Philosophical Writings of Peirce*, J. Buchler, (ed.). New York: Dover.

Popper, K. (1963), *Conjectures and Refutations*. London: Routledge and Kegan Paul.

_ _ _ _ _ _. (1968), *The Logic of Scientific Discovery*. London: Hutchinson.

_ _ _ _ _ _. (1972), *Objective Knowledge*, London: Oxford University Press. Putnam, H. (1975), *Mathematics, Matter, and Method*, volume1. Cambridge: Cambridge University Press.

_ _ _ _ _ _. (1972), *Objective Knowledge*. Routledge and Kegan Paul.

_ _ _ _ _ _ . (1981), *Reason, Truth, and History*. Cambridge: Cambridge University Press. Quine, W.V. (1960), Word and Object. Cambridge: The MIT Press.

_ _ _ _ _ _ . (1970), *Philosophy of Logic*. Englewood Cliffs.: Prentice-Hall.

Reichenbach, H. (1949), *The Theory of Probability*. Berkeley: University of California Press.

Rescher, N. (1973), *The Coherence Theory of Truth*. Oxford: Oxford University Press.

Resnik, D. (1988), "Survival of the Fittest: Law of Evolution or Law of Probability?" *Biology and Philosophy* 3: 349-62.

Resnik, M. (1990), "Immanent Truth", *Mind* 99, 407-24.

Ruse, M. (1979), *The Darwinian Revolution*. Chicago: University of Chicago Press.

Smart, J.J.C. (1963), *Philosophy and Scientific Realism.* London: Routledge and Kegan Paul.

_ _ _ _ _ _ _ _. (1985), "Laws of Nature and Cosmic Coincidences", *The Philosophical Quarterly* 35: 272-80.

_ _ _ _ _ _ _ _. (1989), *Our Place in the Universe.* Oxford: Basil Blackwell.

Suppe, F. (1989), *The Semantic Conception of Theories and Scientific Realism.* Urbana and Chicago: University of Illinois Press.

Tarski, A. (1944), "The Semantic Conception of Truth and the Foundations of Theoretical Semantics", *Philosophy and Phenomenological Research* 4: 341-76.

Tichy, P. (1976), "Verisimilitude Redefined", *The British Journal for the Philosophy of Science* 27: 25-42.

van Fraassen, B. (1980), *The Scientific Image.* Oxford: Clarendon Press.

Realism and Methodological Change

Jarrett Leplin

University of North Carolina, Greensboro

1. The Relation of Truth and Utility

An effective way to challenge scientific realism these days is to ask whether evidence for a theory's truth can ever outstrip evidence for its utility. Granted that we might have strong evidence of a theory's explanatory, predictive, heuristic, and technological utility, might we have *in addition* evidence of its truth? A negative answer claims that all the evidence in support of a theory is, immediately, evidence of pragmatic virtues, and only indirectly, if at all, evidence of truth. Realism then reduces to a willingness to infer truth from utility. Evidently, none of the evidence by which theories are tested will pertain to this inference, so the only recourse for sustaining it is to a metalevel at which truth and utility are compared in the abstract. And there enormous problems loom. That utility provides the only purchase on truth prevents establishing independently, even in a single case, that a useful theory is true, and so blocks any meta-inductive, realist argument. At the same time, a record of rejected but useful theories offers the anti-realist a meta-inductive argument.

This strategy against realism is common to many influential empiricist epistemologies. For example, Bas van Fraassen's instrumentalist, or constructive empiricist, anti-realism argues that utility is always a weaker thesis than truth, and deploys the epistemological maxim that one should stop at the weakest hypothesis that the evidence supports.[1] This, in fact, is van Fraassen's *only* argument against realism; the rest is refutation of bad realist arguments. Arthur Fine (1984) claims that to any realist explanation of empirical facts there corresponds an equally good pragmatic explanation, and argues that the realist explanation cannot in principle be differentially supported. Ian Hacking's (1983) opposition to theoretical realism depicts theory as subservient to experiment. Theory has only the pragmatic role of posing and helping to solve experimental problems, and the most that can be said for any theory is that it does this well. Nancy Cartwright (1983) takes the subject matter of theoretical laws to be conceptual artifacts that relate to purportedly real entities only as idealizations relate to the entities they deliberately misrepresent for pragmatic purposes. The question of truth, then, does not properly arise for theoretical laws. Instead, laws are valued for their pragmatic functions of unification and explanation, functions they perform only to the extent that they misrepresent entities whose causal role in producing

PSA 1992, Volume 1, pp. 435-445

empirical phenomena earns them ontic status. Truthlikeness and pragmatic success are inversely related.

One realist response is to bite the bullet, and maintain that utility just *is* evidence of truth, as though the connection were written into scientific method and to question it would amount to no more than the classical, skeptical demand to justify induction.[2] But a normative, naturalized epistemology denies that anything is "written in" to scientific method; a method or standard for warranting scientific claims itself requires warrant. From this perspective, nothing is evidence for anything unless found regularly to accompany that thing in practice.

Another realist response is to persist in seeking differential support for realism at the meta-level, arguing, for example, that the hypothesis of truth has an explanatory power unequaled by that of utility (Cf. Alan Musgrave 1988). The question then arises as to how such additional explanatory resources, supposing they could be made out, connect with truth. How, in other words, is inference to the best explanation to be warranted?

An interesting, recent alternative, due to Paul Horwich (1991), identifies utility, not with evidence of truth, but with truth itself. Horwich does not quite put it this way, for it is not a pragmatism reducing truth to practice that he favors. Rather, he identifies the *belief* that a theory is useful with the *belief* that it is true. There is nothing more to believing a theory than satisfying conditions for attributing belief in its utility: a commitment to explaining phenomena in terms of it, to developing it, to basing predictions upon it, and so forth. If, however, these beliefs are identified, it is difficult to see how the concepts of truth and usefulness are to be distinguished.[3] For Horwich, the ascent from utility to truth is no ascent at all; the inference under challenge is trivial because nonampliative.

What interests me here is that all parties to the debate accept the rhetorical thrust of the anti-realist's challenge; no one proposes to distinguish truth from utility evidentially. According to Horwich, who will not distinguish them at all, the mental state of believing a theory is the *same* as the mental state of being disposed to use the theory. Then any reason to believe a theory must be a reason to believe it useful. Anti-realists suppose that a reason to impute truth, if there were one, would have to consist in a demonstration of utility. It seems to be assumed all around that truth implies utility; only the converse is problematic.

I question this assumption. It seems to me perfectly coherent to suppose that the truth is utterly useless, and to suppose, furthermore, that we come into possession of good reason so to regard it. Horwich denies that we can fail to believe true a theory that we believe useful. I suggest that we can fail to believe useful a theory that we believe true, and for good reason. That is, bracketing the question of what makes us believe the theory true, that belief carries no commitment to its utility and may, to the contrary, commit us to its *nonutility*. I further suggest that there could be a rational basis for belief in a useless theory, that utility does not exhaust the range of considerations that legitimately bear on theory acceptance.

Of course, if I am even to raise these possibilities I must be allowed to circumscribe to some extent the range of features that count as a theory's utilities. If just any of a theory's credentials, or any interest, pragmatic or conceptual, that it occasions, is regarded as "use", then the inference I contest is true, but vacuously so. There can be no question of distinguishing truth from utility on epistemically probative grounds if the concept 'utility' is taken *in advance* to be so loose as to incorporate whatever might be said on behalf of a theory. By way of circumscription, I shall focus on explanations and predictions of empirical phenomena, the uses normally at issue in evaluating realism (without troubling over whether these include the more instrumental functions that so impress Hacking).

For example, Fine holds that the hypothesis that a theory is *reliable* when used to explain and predict observations explains anything that hypothesizing its truth would explain. He infers that the former hypothesis is always preferable to the latter because it is entailed by the latter and is therefore less metaphysically committal. On van Fraassen's version, it is therefore more probable. I claim that the entailment at issue here does not hold. The relevant sense of 'use' is not so attenuated that a reason to believe that a theory is not useful need itself constitute a use and so trivially defeat this claim.

2. Empiricist Constraints on Realism

My thesis arises in connection with realism, but its significance is broader. It pertains to the standards and expectations for theories—criteria of theory choice or theory acceptance—whether or not one favors a realist interpretation of theories meeting those standards and expectations. A focus on utility as central to evaluation reflects the adage that action is the measure of conviction. Belief in a theory must make some practical difference. It reflects a positivist demand for the "cash value" of attributions of unobservable properties, a demand that leads us to interpret subjective probabilities as dispositions to place wagers. It reflects a neo-positivist preference for intervention over representation.[4]

I do not dispute the centrality to the standards of modern science of a focus on utility in evaluation. That attitude has taken us far, producing theories of astonishing technological power, but is it inevitable? We can trace substantial, historical changes in scientific method—from deduction from phenomena to postulation of unobservables; from disallowing to allowing fundamental, statistical laws; from fixing fundamental constants of nature empirically to requiring their deducibility from general principles; from observation to detection. Could standards of theory choice come to countenance theories that fail empiricist criteria of warrant in a more general way? More to the point, could such a development enjoy the kind of rationale that underwrote the methodological changes we have seen already?

My formulation of the problem facing realism as an inferential gap in the ascent from utility to truth focuses on theories. But consider, by way of motivation, theoretical entities. What if there are things in the world that we cannot possibly use; black holes or other galaxies, for instance? Must it be impossible to know that there are such things, or justifiably to award them ontological standing, because use is preordained as the route to knowledge? Hacking, whose pragmatic realism forces him to deny that we can know of such things, accepts the challenge in this question, but his answer begs it: it happens, just conveniently, that extra-galactic inquiry is limited to modeling (Hacking 1989).[5] What if we had unifying theory in cosmology rather than a plethora of incompatible models; what would then become of Hacking's restrictions on knowledge? It is *incidental* that theoretical entities exceeding those restrictions are also to be faulted on other grounds.

Standards of evaluation change under pressure of new information. As we no longer tie observation to the eye when less direct means of detection prove both feasible and unsurpassable, so we can weaken the standard of laboratory control if a well understood process proves unproducible. We needed to produce the W and the Z to confirm electroweak theory partly because *they were producible*; failure to identify them, once the appropriate modifications were in place at CERN, would have been an anomaly for the theory. No similar requirement to produce X particles can reasonably be supposed to hold Grand Unified Theories (GUTs) hostage, for according to GUTs, X particles have properties that preclude their production. An epistemology anchored in utilitarian concepts of production and intervention could be a transitory luxury, destined to be a relic of primitive theorizing.

There is nothing remarkable in this. In pressing technology to greater levels of energy production, it is surely foreseeable that there be limits beyond which we cannot reasonably hope to bring theoretical entities within the compass of technological access. Such limits, if only recently discerned in high energy physics, have long been evident in astronomy. The admission that utility gives the only possible purchase on truth ties realism to a methodology of evaluation that is admittedly vibrant but potentially mortal.

3. Theories that Imply their own Nonutility

That the ontic commitments distinguishing GUTs from theories achieving a lower level of unification—from electroweak theory, for example—ought not, according to GUTs themselves, to be empirically confirmable does not, of course, constitute a reason for belief in GUTs alternative to empirical confirmation. But let me postpone, for the moment, the question of what is the right epistemic attitude to take toward a theory that dictates its own unconfirmability (whether or not this proves generally true of GUTs in particular[6]). I first wish to show that recent developments in theoretical physics do raise this question, that we now have—for the first time—a body of theory that by its own lights is not assessable by the utilitarian standards of theory choice to which physics has properly been held. To do this, I shall briefly review the prospects for empirical testing in quantum cosmology.

There appears to be only one possible account of the early history of the universe, only one way, at least in major outline, that the universe could have evolved and exist at all. That uniqueness is persuasive, despite the fact that no empirical discoveries are based on it, and the empirical predictions that it suggests carry little confidence. Physicists have had to face the possibility that the ultimate unified theory of the fundamental forces of nature would not reveal its accuracy at the evidential level at which current norms of rationality direct that theories be judged. There is a clear respect in which current ideas in quantum cosmology imply and explain their own nonutility, why such ideas, if correct, should not be expected to manifest that correctness in pragmatic terms.

Ten years ago Stephen Weinberg (1981) predicted the detection of 300 instances of proton decay per year in ultra-pure water filling a salt mine in Ohio. This may have been the last confident prediction of empirical evidence for unification of forces beyond electroweak theory. After one year without a single clear case of proton decay, physicists began adjusting their estimates of the proton's lifespan. Move it up an order of magnitude or two, and detection is not to be expected. But whether protons last, on average, 10^{31}, 10^{32}, or 10^{33} years is not crucial to grand unification as such; one simply adjusts one's estimates of the masses of the posited decay products.[7] What is crucial is that ordinary matter be ultimately unstable, and physicists seem pretty confident of that although they no longer think this instability detectable. In fact, the Ohio experiment was conceived not so much as a test of theory as an attempt to determine the proton's lifespan.[8]

The cases of magnetic monopoles and gravitons are similar. Although their detection would dramatically confirm current theoretical ideas for the unification of forces, the failure to detect them is not interpretable as a failure of theory. Gravitons, representing the quantization of the gravitational field, are widely believed, following Einstein, to yield no differential effects to distinguish quantum from classical gravity. Magnetic currents do not appear in natural sources, and magnetic monopoles are too massive to produce. More importantly, GUTs are strictly consistent with a Maxwellian asymmetry of electric and magnetic field interactions. Monopoles need not exist at all; or, at least, they may be of zero density in space. If we could produce energies sufficient to create the massive particles that serve as quanta for the force fields that maintain local gauge symmetries in current theories unifying the remaining fundamental forces, then we would indeed have a test. But while reflecting on the extent to which past technological leaps have exceeded past dreams may give hope, there is really no good reason to expect this ever to be possible.

We might think to back off somewhat from the standard of laboratory test, but maintain the thrust of empiricism, by seeking evidence from observational astronomy. But according to inflationary theory, the conditions that produced the phenomena that quantum cosmology purports to explain—a single, unified force; the observed uniformity in the global distribution of matter; the expansion of the universe; the strength of the gravitational force—are confined to an inflationary stage that antedates the observable universe. And the conditions under which a single, unifying force existed at the end of the inflationary phase were obliterated by the ensuing process of spontaneous symmetry breaking. The mere fact of our own existence is incompatible with the realization of events and structures that might corroborate theoretical convictions about the origin of the universe and the unification of forces. This is basically because unification depends on local gauge symmetries that prevail only under unstable conditions; whereas life requires stability.

The basis for confidence in unification is not a basis for expectations of being proved right. It is possible that convictions about the ultimate structure of the world do not admit of the sort of independent, empirical corroboration that authorizes belief by current standards of scientific objectivity. There is no *a priori* reason that they should, that their correctness should happen to have empirical manifestations accessible to us. There is now reason *a posteriori* to conclude otherwise. There is even reason to expect such convictions to be systematically belied by the sort of evidence that *is* accessible. Or, to put it another way, there is no reason to expect the products of fundamental, physical processes to reflect the underlying symmetry or simplicity of the mathematical equations governing those processes. What we can observe is more suggestive of a world diverse in fundamental laws and types of substance than of the sort of world that actually carries conviction for the theorist.

4. The Epistemic Situation

Empiricist philosophy traditionally makes two assumptions not carefully distinguished: that truths about the natural world are bound to have consequences attestable in experience, and that only via observable consequences could truths about the natural world be recognized or learned. The first is plainly contingent. But as the second tends to be regarded as *a priori* and pragmatically necessitates the first, that contingency is overlooked. As the first assumption is not even plausible, it would be fortunate if the second proved false. So let me turn to the question of whether such theoretical ideas as those at the forefront of quantum cosmology might be warrantable by other than empiricist criteria of evidential support.

What grounds these ideas is not what would normally be recognized as independent evidence, but rather a pattern of reasoning that is essentially abductive and explanationist. The epistemic situation looks roughly as follows. We have a theory T that is well confirmed on empiricist grounds. (To sustain our example, we can take T to be the general program of quantum electrodynamics carried out through the unification of electromagnetism and the weak force.) T leaves many phenomena, $\{p\}$, unexplained. (These include, for example, the fact that there are *three* (remaining) forces; the relative strengths of the forces; the fact that elementary particles come in certain distinct families, and that some particles are distinguished only by their mass; the value of the unit of electric charge; where the W and Z particles get their mass; as well as facts about the structure and origin of the universe to which T is not so relevant, but which a fully quantized theory of forces would have to address.)

The success of T supports a direction to pursue in explaining $\{p\}$. (Unify electroweak theory with quantum chromodynamics; develop a renormalizable quantum theory of gravity, in which gravity establishes a new local gauge symmetry.) A new, more comprehensive theory T' (a GUT or, more comprehensively, Supergravity) is developed along the lines suggested by T. (It will strengthen the example to appeal as well to certain explanatory achievements due to superstring theory; although superstrings are not really a development of T, they can be accommodated consistently within T'.) T' explains $\{p\}$ and also exhibits virtues not specifically anticipated but of a kind always valued in physics, virtues found, to a lessor extent, in historically successful theories. These virtues have the general nature of *internal coherence*: T' reveals striking connections among previously known phenomena and among diverse physical principles, including principles that, while important and successful in physics, were not directly relevant to T and were not part of the rationale T provided for the development of T'. (For example, T' connects the constants of nature and reveals new applications for the second law of thermodynamics, the principle of least action, Mach's principle, and Heisenberg uncertainty. Moreover, T' avoids in a natural way—especially in superstring theory—the troublesome infinities that had to be finessed in T by suspicious mathematical means; T' does not *need* to be renormalized retroactively.) T' also exhibits a certain *inevitability*, connected with its claim to uniqueness. It can be shown that requirements implicit in the program initiated by T must be violated by any alternative to T'. (This inevitability is particularly striking in superstring theory, where the twin requirements of manifest covariance and quantization restrict the equations for the strings to the point that no alternatives are left open.)

So impressive are these virtues that if anything anywhere in physics were different, it seems the whole structure would collapse. Indeed, unlike T, T' seems to leave no major explanatory lacuna. (Admittedly, this judgment is a matter of emphasis; some explanations are speculative, and their adequacy controversial: the development of structure in the universe through symmetry-breaking phase transitions; the stability, despite quantum fluctuations, of the suppressed, or compactified, dimensions of spacetime. Then too, there are profound disagreements as to whether certain things *need* explaining, and whether the failure to explain them is a liability; for example, the fact that empty spacetime is flat.) T' is optimal not merely with respect to $\{p\}$, but also with respect to explanatory challenges unassociated with the program that produced it.

However, while T' predicts unknown phenomena (new particles, for example), it predicts no new *observable* phenomena; it admits of no empirical test comparable to that confirming T. All its explanations are retroactive. We may express this disadvantage by saying that although T' is empirically supported by results that confirm T, T' is not *differentially* supported with respect to T. Differential support—whether analyzed in terms of independence of experiments, novelty of predictions, or just temporal newness of empirical results—is clearly a condition for acceptance of theories under prevailing standards. T''s utilities do not meet these standards.

A methodology that warrants T' must count explanationist virtues sufficient, in the absence of differential support, for theory acceptance. However, the situation is not so bleak as this admission might suggest. The sort of explanationism needed to support T' is not the rather weak and discredited rule of "inference to the best explanation". It is not merely that T' offers the best available explanation of $\{p\}$. It offers the *only* available explanation. The explanation it offers is *good*, which the best explanation, in general, need not be. And it implies that no alternative explanation could match its internal coherence and inevitability. We cannot rule out an alternative predicting the same observable phenomena (that is, an alternative to T'; T is not such an alternative, as it does not explain what it was the point of developing T' to explain); but unless T' is false, no alternative will forge comparable connections or so reinforce our confidence in basic principles.

This uniqueness is unusual. Usually, in excluding alternatives, a theory does not exclude the possibility of an alternative's being equally virtuous by explanationist criteria. Philosophical theories of theory evaluation typically give criteria of theory *choice*, and it is assumed, often explicitly, that *all* evaluation is comparative.[9] But not only has T' no known rivals; T' itself gives reason to deny the possibility of rivals. T' is not simply a theory about the world; it is also a theory about theories about the world (to the effect that only theories exhibiting certain formal symmetries are admissible in the unification of forces). The distinction between theories about the world and theories about theories would seem to prevent any theory's so co-opting its own domain. But this distinction, so entrenched in Positivism's circumscription of the scopes of science and philosophy, has really been untenable since Einstein introduced his conception of "theories of principle".

The alternative to T' seems to be that $\{p\}$ have *no* explanation; they are just initial conditions of the universe.[10] There is some precedent for this uniqueness in the preclusion of hidden variables in quantum mechanics. And naturally such a built-in pre-emption of rivals could just be wrong. But then if there were rivals to T', we would have a different epistemic situation, one in which T' would not appear so compelling and the question of warranting it would be of less moment.

This is a condition for which comparative theories of evaluation are unprepared. Influenced perhaps by historical cases to which our example does not conform, perhaps by a positivist legacy of formal algorithms for generating rivals to a given theory,[11] such philosophical theories take for granted a context of theory competition. They lack the means to credit a theory unopposed, whereas under prevailing standards of science want of opposition itself redounds to a theory's credit.[12]

5. Prospects for Methodological Change

What would warrant the explanationist methodology needed to warrant T'? One thing that can warrant a methodology is that it is supported by a theory that is independently warranted.[13] This means that by *prevailing* methodological standards, a theory is accepted which *then* has the effect of supplanting those standards. This seems to pose a problem of consistency. Can a theory fail to support a standard by which it excels, instead supporting a different one? If a methodology picks out a certain theory as good or best, does it not follow that the theory will possess attributes that this methodology selects for? If so, how could a different methodology, which selects for different attributes, be one that acceptance of the theory would incline us to prefer in our subsequent evaluations of theories? Once we have an established theory sufficiently impressive to serve as a model of what a good theory should look like, we will naturally adopt its features as criteria of adequacy or acceptance in the methods or standards to be applied in considering further theories (modulo some general area of inquiry or domain of problems). (More precisely: once we have identified features that covary with the successfulness of theories at advancing our epistemic ends, we will prefer theories that present those features in our further evaluations.) But how did a theory thus influential get established in the first place, if the operative methodology was a different one counseling a preference for different features? Do we have coherent dynamics for warranting methodological *change*?

To make sense of the process, we must understand 'theory' and 'methodology' in a certain way. We must view a theory as a network of hypotheses that can be revised and supplemented without detriment to the theory as a whole. That is, theories are not sets of propositions defined by their members; changes *in* theory need not be changes *of* theory. Theory individuation may be tied to the fate of certain key elements, but theory identity affords latitude for development. Secondly, we must view methodology in terms of criteria of acceptability for theories, not in terms of an abstract logical relation of inductive warrant that definitively is or is not sustained independently of the scientific standards contingently in force. So far as anyone knows, there is no such abstract logical relation, and if there were its being sustained could not be recognized independently of scientific standards anyway. John Worrall (1988) regards it as logically fallacious to suppose a theory established under one methodology to establish in turn a contrary methodology. That would indeed be fallacious if it meant that a theory proves incorrect the very methodology that proves it correct. For what is incorrect proves nothing. But the relation between theory and methodology is not one of proof. A Methodology *authorizes* inference on the basis of evidence; it does not *compel* belief. Nor can a theory *force* a change in methodology; at most it *warrants* a change, by disputing assumptions on which the appropriateness of the prevailing methodology depends, or revealing previously unrecognized bases for belief.

With these understandings, a theory accepted in virtue of certain features recommended under one methodology can also contain or come to contain other features that support a different methodology.[14] The methodological change that this process permits does not, Worrall to the contrary, oblige one to recant the judgment that the replaced methodology really did support, and was not just thought or reputed to support, the theory that proved its downfall. The components originally recommended must, however, continue to fare well by the successor methodology if the problem of consistency is not to recur at a finer level of analysis.

For example, a methodology of inductivism supports Newtonian Mechanics, which in its full development incorporates a universal force of gravitation. The success of this construct leads to a new toleration of theoretical entities, eventually to countenancing the postulation of a universal ether to support a wave theory in optics. Newtonian Mechanics continues, however, to be a viable theory, judged by the standard of predictive and explanatory power that accompanies the licensing of theoretical posits that transcend the resources of inductivist methodology.

For this process to work for T' would require that explanationism be warranted by T (or other theories preceding T'). We would have to come to value explanationist virtues, extant or yet to emerge, of T *over* its differential empirical confirmation, much in the way that we came to value the predictive and explanatory power of theoretical entities over inductive connections to phenomena in evolving a methodology of hypothetico-deductivism. (For example, the identification of new symmetries and reduction in the number of forces would have to be valued above the laboratory production of new particles.)

While this development is certainly conceivable, it is not at all clear that it *ought* to occur, apart from the influence of T' which it would be circular to allow. One might argue historically that it *ought not* to occur, because a methodology of experimental control evolved precisely in order to constrain unsuccessful speculations fueled by undisciplined desire for explanatory power. It is only fair to point out, however, that the situation we now face is *new*. We have not before had a body of theory, recommended on explanationist grounds, that implied and explained its own failure to be differentially testable. The upshot, so far as I can see, is that while skepticism remains a warranted posture, it need not be the *only* warranted posture; rational acceptance of T' is a yet a possibility. So long as methodology is understood in terms of criteria of warrant rather than formal relations of inductive support, there is no inconsistency in supposing conflicting epistemic stances to be warranted.

Such ambiguity is in fact characteristic of revolutionary change in science. Revolution poses a problem of rationality if evaluative standards are relativized to methodologies that revolution separates. But the problem is not that there is too little rationality; rather, there is too much. A judgment in favor of change can be definitive only in retrospect.[15] Only in retrospect can we track the contributions to our epistemic ends of alternative methodologies of assessment, and a question about whether change *ought* to occur is basically a question about how our ends are best achieved. Resistance to an explanationist methodology favoring T' could lose legitimacy if T' remains unopposed while accruing ever greater and unprecedented explanatory virtues. In the meantime, the strongest conclusion we can reach is that acceptance of T' is not in principle irrational.

6. Conclusion

Ultimately conviction responds to evidence; there must be evidence to support the explanation for the failure of evidence to support deeper theorizing. At least that minimal an empiricism is firmly established, and our example respects it. There is plenty of evidence, compelling by prevailing empiricist standards, of the undetectability of X particles, for example. But empiricism at this level does not establish the sort of connection between belief and prediction or control on which the presumption that beliefs must be actionable depends. One is free in this situation to remain skeptical, to prefer a permanent suspension of credence with respect to matters that, in the nature of the case, transcend differential evidential support. But that might not prove the *only* rational stance. One might instead transcend the standard of differential support, changing methodologies for reaching and testing conclusions, on the strength of new information as to what kinds of support are potentially available. This option is defensible. Whether it can ever be preferable is speculative.

Notes

[1]*The Scientific Image* (1980) remains the fullest statement of van Fraassen's anti-realism. He attempts to dissociate his position from instrumentalism, most recently in *Laws and Symmetry* (1989, pp. 191 ff.), by conceding the statement-making function of theories. But as he allows theories to be appraised only for their utility, never as vehicles for advancing knowledge, the label fits in spirit.

[2]Larry Laudan (1981) has had a field day with such *aprioristic* attitudes in philosophy of science. The seminal statement of the anti-*apriorist* line that permeates Laudan's recent work is his (1986).

[3]This consequence poses an interesting problem for Horwich, who defends a redundancy theory of truth and would reject a pragmatist reduction of truth as anti-realist.

[4]In *Representing and Intervening* (1983), Hacking describes his position as "neo-positivist".

[5]Of course Hacking would not deny that there is "evidence" of black holes or other galaxies. Rather he contends that the sort of evidence there is, as it does not (and, so far as we can foresee, cannot) include "intervention", fails to carry credence.

[6]By treating unification as a process that proceeds through separate stages, a recent GUT, based on the gauge group SU(15), introduces new gauge forces "at energies accessible to accelerator experiments". See P. Frampton and Bum-Hoon Lee (1990). At the stage of lowest energy, particles ("leptoquarks") light enough to detect are produced.

[7]It is crucial to particular versions; the simplest GUT based on the gauge group SU(5) is ruled out by higher orders of lifetime.

[8]Experiments to detect proton decay are a good example of the view taken of experiments in J. Leplin (1986), and are also a congenial example for Hacking.

[9]"All evaluations of research traditions and theories must be made *within a comparative context*", declares Laudan (1977, pg. 120; italics in original). According to Richard Miller (1987, pg. 177), scientists' own theoretical goals presuppose a comparative context: "... the approximate truth which scientists seek has, itself, an intrinsically comparative dimension." On such positions, a theory without rivals cannot be warranted.

[10]In *Theories of Everything* (1991), John Barrow argues that it is not possible to dispense entirely with initial conditions in explanation, although it may be possible to arrange for their values to have no measurable effect on the observable universe.

[11]For a rejection of such algorithms, see L. Laudan and J. Leplin (1991, pp. 456-57).

[12]John Worrall (1988) maintains *both* that there is a unique, invariant methodology of theory construction (for want of which epistemology collapses into relativism), *and* that (according to this historically transcendent methodology) theories must be tested against rivals. If this is correct, the methodology of theory evaluation is itself untestable and without warrant. Of what epistemic moment, then, are its pronouncements as to the merits of theories? How is the uniqueness or historical invariance of a methodology supposed to answer relativism's challenge to the rationality of theory preferences? Surely that challenge requires that methodology, whether one or many, be itself warranted.

[13]This is the basic strategy of Laudan's (1984).

[14]In his (otherwise cogent) reply to Worrall, Laudan (1989) never answers Worrall's charge of logical fallaciousness. I do not know whether he would accept my solution.

[15]Lakatos held a similar view at the level of theory change, though with different rationale and without allowing for definitiveness.

References

Barrow, J. (1991), *Theories of Everything*. Oxford: Oxford University Press.

Cartwright, N. (1983), *How the Laws of Physics Lie*. Oxford: Oxford University Press.

Fine, A. (1986), "Unnatural Attitudes: Realist and Instrumentalist Attachments to Science", *Mind* 95: 149-179.

Frampton, P. and Bum-Hoon, L. (1990), "SU(15) Grand Unification", *Physical Review Letters*, 64: 619-621.

Hacking, I. (1983), *Representing and Intervening*. Cambridge: Cambridge University Press.

______. (1989), "Extragalactic Reality: The Case of Gravitational Lensing", *Philosophy of Science* 56: 555-582.

Horwich, P. (1991), "On the Nature and Norms of Theoretical Commitment", *Philosophy of Science* 58: 1-15.

Laudan L. (1981), "A Confutation of Convergent Realism", *Philosophy of Science*, 48: 19-49.

_____. (1986), "Methodology's Prospects", in *PSA 1986*, volume 2, A. Fine and P. Machamer (eds.). East Lansing: Philosophy of Science Association, pp.347-355.

_____. (1977), *Progress and its Problems*. Berkeley: University of California Press.

_____. (1984), *Science and Values*. Berkeley: University of California Press.

_____. (1989), "If it Ain't Broke, Don't Fix it", *British Journal for the Philosophy of Science* 40: 369-376.

Laudan, L. and Leplin, J. (1991), "Empirical equivalence and Underdetermination", *Journal of Philosophy*, 88: 449-473.

Leplin, J. (1986), "Methodological Realism and Scientific Rationality", *Philosophy of Science* 53: 31-53.

Miller, R. (1987), *Fact and Method*. Princeton: Princeton University Press

Musgrave, A. (1988), "The Ultimate Argument for Scientific Realism", in *Relativism and Realism in Science*, Nola, R. (ed.). Dordrecht: Kluwer Academic Publishers, pp. 229-253.

van Fraassen, B. (1980), *The Scientific Image*. Oxford: Oxford University Press.

________. (1989), *Laws and Symmetry*. Oxford: Oxford University Press.

Weinberg, S. (1981), "The Decay of the Proton", *Scientific American*, 6/81. pp. 74 ff.

Worrall, J. (1988), "The Value of a Fixed Methodology", *British Journal for the Philosophy of Science*, 39: 263-275.

Historical Contingency and Theory Selection in Science[1]

James T. Cushing

University of Notre Dame

1. Introduction

The central theme of this paper is that historical contingency plays an essential and ineliminable role in the construction and selection of a successful scientific theory from among its observationally equivalent and unrefuted competitors. Let me make a few clarifying remarks at the outset. It is *not* my claim that in *all* cases of theory selection there is in practice this radical type of underdetermination.[2] Nor am I concerned with the rather trivial and philosophically uninteresting historical contingency of who did what when. For example, in response to Napoléon's fishing for a favorable comparison between himself and Newton, Lagrange is said to have lamented (Gillispie 1960, p. 117; Moritz 1914, p. 167)[3]: "Newton was the greatest genius that ever existed, and the most fortunate, for we cannot find more than once a system of the world established." That is, Newton had *discovered* the fundamental laws of mechanics and of gravitation and there was nothing of comparable significance left for anyone else to accomplish in that arena. So, good show for Newton and hard luck for Lagrange. Of course, someone else *might* have discovered these *same* laws. Then, the laws would bear another's name, but they would be the same laws. This is surely historical contingency, but of a benign and philosophically uninteresting variety. Neither shall I consider any possible contingency in the necessary conditions for human understanding (Cushing 1991). Nor is 'contingency' in the sense of (other) possible worlds having a fundamentally different (objective) structure of interest to me here. Rather, I shall discuss the acceptance and rejection of observationally equivalent, alternative and, indeed, *incompatible* descriptions or theories of our *actual* world. For this purpose, I need not enter into the argument about the existence of *necessary* laws, as opposed to (mere) descriptions of (contingent) regularities (van Fraassen, 1989). After all, there *are* laws and theories that are accepted by the scientific community and I shall focus on how successful scientific theories come to be accepted. There is no space here to give all of the necessary historical background for my argument. Only a brief summary is presented, since the details are readily available (even if scattered throughout the literature) and relatively uncontroversial (*e.g.*, Beller 1983a, 1983b, 1992; Cassidy 1992; Cushing 1990, 1992a, 1992b, 1992c).

PSA 1992, Volume 1, pp. 446-457

For the purposes of this paper, I make a simple distinction between two components of a scientific theory: its formalism and its interpretation. The claim is that these are conceptually separable, even if they are often entangled in practice. To simplify matters, I also restrict my remarks to the case of quantum mechanics, since that will be sufficient. Very simply, what I mean by a formalism is a set of equations and a set of calculational rules for making predictions that can be compared with experiment (i.e., "getting the numbers right"). The (physical) interpretation refers to what the theory tells us about the underlying structure of these phenomena (i.e., the corresponding picture story about the furniture of the world—an ontology). Hence, *one* formalism with *two* different interpretations counts as *two* different theories. This is essentially the type of distinction that Jammer (1974, pp. 2-17) also makes in his book on *The Philosophy of Quantum Mechanics*.

2. A Specific Case

To argue for my claims, I shall trace the origins and eventual fate of the so-called causal quantum theory program. In the fall of 1926, Madelung suggested a hydrodynamical interpretation of quantum mechanics. He accepted Schrödinger's wave equation as his starting point and, by means of a mathematical transformation, recast it into the form of the equations describing the flow of an ideal fluid. All of the mathematical details aside, Madelung suggested interpreting the Schrödinger equation as representing a physical "fluid" of identical particles. One difficulty of this interpretation was that these equations had a term, for the force acting on a given particle, that appeared to depend in an unphysical way on the locations and velocities of all of the other particles in the fluid. The physical meaning or significance of this term will be a recurrent theme of, or actually a problem for, various causal interpretations I discuss. It would later be named the quantum-force or quantum-potential term. It is clear that Madelung was attempting to provide a largely classical picture of or explanation for quantum phenomena.

This formal equivalence between the Schrödinger equation and the classical hydrodynamical equations, as well as the analogy of the Hamilton-Jacobi formulation of classical mechanics to classical wave optics, led de Broglie to his own attempts at a largely classical formulation of quantum mechanics. Before I sketch de Broglie's attempts, let me recall that, until the mid-1950's, de Broglie was very much in the Copenhagen camp, having been converted by 1930 after his bitter experience at the 1927 Solvay Congress. David Bohm's 1952 papers were to have a profound "reconversion" effect on de Broglie and he acknowledged that Bohm had successfully overcome the original objections to a pilot-wave model. I return to this part of the story later. The main goal that de Broglie had was to unify the wave and particle dualism into a *single* coherent picture or model. His hope was to treat the "particle" as a mathematical singularity in the center of an extended wave. This *extended*, continuous part of the wave would "sense" the environment (*e.g.*, obstacles, slits, *etc.*) and thus vary the motion of the singularity accordingly. He wanted two related wave solutions, one for the singularity and the other for the continuous wave. This latter would account for the statistical behavior of a collection of particles. A classical conception of actually-existing entities in a continuous space-time background was to underlie his theory or world view. Thus, de Broglie's solution to the wave-particle duality was a synthesis of wave *and* particle, versus the wave *or* particle of the eventual Copenhagen interpretation. However, the existence and nature of the singular solution to the wave equation for the general case of motion in the presence of a force field proved quite complex mathematically. In the face of these severe mathematical difficulties, de Broglie decided, as an interim measure, to accept, or simply *postulate*, the existence of a particle accompanied by its phase wave. This was his "pilot-wave" theory—one very similar to

Madelung's—that he presented to the 1927 Solvay Congress. Conceptually, it was a mixture of waves and particles and that did not incline Schrödinger kindly toward it, since, at this time, Schrödinger wanted an interpretation based *wholly* upon the wave concept as being fundamental. However, one might have expected Einstein to be partial toward de Broglie's theory, since it allowed one to picture actually-existing physical particles following well-defined trajectories—the existence of an observer-independent, objective physical reality. Einstein was at that time, and remained thereafter, quite partial toward a belief in such a reality. Why Einstein was distrustful of such a model is related to his own abortive attempt at a hidden-variables type of theory for quantum mechanics.

A fascinating unpublished manuscript in the Einstein Archives is titled "Does Schrödinger's Wave Mechanics Determine the Motion of a System Completely or Only in the Sense of Statistics?".[4] In it, Einstein tells us that to each solution of the wave equation there corresponds the motion of an *individual* system that is determined unambiguously and uniquely! His basic idea was that the time-independent Schrödinger equation can be used to find the kinetic energy for any given wave function. Einstein was able to give an expression for *uniquely* assigning the velocity of a particle in terms of a given wave function. Conceptually, this is very much in the spirit of Madelung's hydrodynamical model. However, in an addendum to this manuscript, Einstein mentioned that Walther Bothe had pointed out a difficulty with this scheme by means of an example of coupled systems. The expression for the kinetic energy of a system of coupled resonators consisted of more than just the two-body terms expected classically. Einstein expressed the hope that it might be possible to overcome this difficulty, but nothing specific followed. The fact that this paper, originally presented to the Prussian Academy in early 1927, was never published indicates that this "entanglement" problem remained grounds, for him, to reject this particular "classical" attempt at interpreting quantum mechanics. The general, at least conceptual, similarities among this attempt by Einstein, the Madelung hydrodynamical model and the de Broglie pilot-wave theory probably account for Einstein's lack of interest in de Broglie's 1927 Solvay Congress presentation, to which I turn next.

Under a time constraint to present a paper on the interpretation of quantum mechanics at the 1927 Solvay Congress that October, de Broglie contented himself with a presentation of the pilot-wave point of view. His ideas were not well received, while the purely probabilistic interpretation of Bohr, Born, Heisenberg and Dirac was defended strongly. Furthermore, Wolfgang Pauli offered what appeared to be a telling counterexample that showed de Broglie's theory to be inconsistent with empirical evidence. De Broglie attempted a response, but it was not convincing. It was only years later that David Bohm showed Pauli's argument to be specious. But, the indifference, and negative reaction, to his pilot-wave theory at the Congress, coupled with Heisenberg's uncertainty paper and the generally favorable attitude toward and acceptance of the "Copenhagen" view, all took their toll on de Broglie. By 1928 he had himself been converted to the "Copenhagen" dogma.

So, by 1927-1928, the issue of the "Copenhagen" vs. a causal interpretation had been *effectively* settled. As I have indicated elsewhere (1992a), there were "external" pressures that lent urgency to the need to find *the* correct interpretation of the formalism of quantum mechanics and rational, but as I have also argued, not wholly compelling reasons—even though many *took* them to be so—for settling on "Copenhagen". Even by September of 1927, *prior* to the October Solvay Congress, Niels Bohr's Como lecture had to a large extent solidified the matrix of what would become the Copenhagen interpretation. For my purposes here, the important point is that the general acceptance of the Copenhagen interpretation—even if still somewhat ill defined then and later—precluded any consideration of causal interpretations. The possibility of a causal interpretation was generally *believed* to be and *accepted* as a dead issue by 1928. With the appearance in 1932 of John von Neumann's *Mathematical Foundations of Quantum Mechanics* there then appeared to be a logically irrefutable proof that it was *impossible* to have any type of hidden-variables theory that gave *all* of the same predictions as standard quantum mechanics. It was many years before John Bell established conclusively that von Neumann's "proof" was irrelevant to the hidden-variables theories I have discussed.

Thus, there had been, prior to 1952, several attempts at causal interpretations of quantum mechanics. But, all had been found objectionable because of the nonlocal nature of the quantum potential and also, for some, because, given von Neumann's theorem, it remained unclear that a causal theory could actually be *completely* equivalent to standard quantum theory in its observational consequences. Finally, in 1952 David Bohm gave a brilliant and detailed exposition of a hidden-variables interpretation of the *same* formalism as standard quantum mechanics. Using the same type of decomposition of the wave function as Madelung had (but, apparently unaware of Madelung's work, as well as of de Broglie's pilot-wave theory), Bohm arrived at the equivalent of the quantum-force form of Newton's second law. He was able to show that his theory was *absolutely* observationally equivalent to the standard "Copenhagen" interpretation, that the notorious quantum-mechanical measurement problem (or, the collapse of the wave function) did *not* exist in his theory and that the Heisenberg uncertainty relations reflected a practical, but not an in-principle, limitation on the accuracy of observations. That is, David Bohm's 1952 paper presented a logically consistent and empirically adequate causal interpretation of the formalism of quantum mechanics. He exhibited explicitly a causal interpretation of quantum mechanics, something forbidden by "Copenhagen", and did so in terms of hidden variables, something *believed* to have been forbidden by von Neumann's "proof". In a sense, Bohm's 1952 work can be seen as an exercise in logic—proving that "Copenhagen" dogma was not the only logical possibility compatible with the facts. True, Bohm's theory does have some classically unexpected features, such as the highly nonlocal quantum potential. It does, though, provide us with an ontology of actual particles moving along continuous, even if at times irregular, trajectories in space-time. Such an ontology is not nearly as radical a departure from that of classical physics as is that associated with the Copenhagen interpretation. While the nonlocality of Bohm's theory may appear unpalatable to some, it is worth pointing out that the Copenhagen interpretation has the same nonseparable structure and other bizarre features as well. That being the case, nonlocality itself gives one little reason to choose "Copenhagen" over "Bohm". One can then turn to other criteria, such as intelligibility, simplicity, fertility and the like.

3. Contingency in the Choice

What was the reaction to Bohm's 1952 paper? Initially, de Broglie was against Bohm's ideas (which were similar to his own pilot-wave theory of 1927) and he raised the same objections against Bohm's theory that had been raised against his own. Interestingly enough, when David Bohm sent Pauli a copy of the paper in which Bohm showed Pauli's objections to the causal interpretation to be specious, Pauli never responded. Furthermore, Pauli's views on the nature of science and its relation to his conception of God (basically, a cosmic bookkeeper who enforced statistical causality) made it inconceivable to him that anything like a return to a "classical" world with causality and picturable, continuous processes in space-time was either possible or anything less than a disgusting loss of nerve and a return to darkness. Bohm did, as we have seen, produce a causal version of quantum mechanics—one capable of a *realistic* interpretation with a largely classical (micro) ontology. Bohm's work of the early

1950's reconverted de Broglie to his former ideas. For de Broglie the issue at stake was not (classical) determinism, but rather the possibility of a precise space-time representation for a clear picture of microprocesses. In this, his expectations were similar to Einstein's. With the concept of the quantum potential, one could provide a *model* for fundamental processes.

Subsequent to Bohm's papers, Edward Nelson in 1966 showed that a single particle subject to Brownian motion, with a diffusion coefficient ($h/2m$) and no friction, and responding to imposed forces in accord with Newton's second law, $\boldsymbol{F} = \mathrm{m}\,\boldsymbol{a}$, obeys *(exactly)* the Schrödinger equation. (See also, Goldstein 1987; Dürr *et al.* 1990.) Although there is randomness, a radical departure from classical physics is unnecessary so that the resulting theory is probabilistic in a *classical* way. Nelson's work has been generalized to the relativistic case. This alternative program has thus shown a great deal of fertility for generalization within its own resources, not just as *ad hoc* moves. That is, *if*, say in 1927, the fate of the causal interpretation had taken a very different turn and been accepted (over the "Copenhagen" one), it would have had the resources to cope with the generalizations essential for a broad-based empirical adequacy. We could today have arrived at a *very different* world view of micro-phenomena. If someone were then to present the (merely) empirically equally as adequate Copenhagen version, with all of its own counterintuitive and mind-boggling aspects, who would listen! That is, a highly "reconstructed" but entirely plausible bit of history could run as follows (all around 1925-1927). Heisenberg's matrix mechanics and Schrödinger's wave mechanics are formulated and shown to be mathematically equivalent. Hence, the Dirac transformation theory and an operator formalism are available as a *convenience* for further development of the formalism to provide algorithms for calculation. Study of a classical particle subject to Brownian motion (about which Einstein surely knew something!) leads to a "classical" understanding of the already discovered "Schrödinger" equation, which is then given a "de Broglie-Bohm" realistic interpretation. A "Nelson" model underpins this interpretation with a visualizable account of microphenomena. "Bell's" theorem is proven and taken as convincing evidence that there is a type of nonlocality present in quantum phenomena.

Since it is central to my line of argument here, let me consider in some detail what impact a Bell type of theorem might have had in 1927. I believe that many of us have wondered why someone hadn't come up with a Bell-type of inequality or theorem long before 1964. I recall that Anton Zeilinger, during informal conversation, made just such a remark at a quantum theory conference in Joensuu, Finland, in August of 1990. Of course, it is one of the hallmarks of a profound insight that it so changes our own way of thinking about a problem that the solution provided by that insight soon comes to appear inevitable and almost "obvious". Nevertheless, all of the conceptual tools and background existed in 1927 to establish, along the lines of a Bell-style argument, that we cannot have, in the actual world in which we do exist, any empirically adequate theory that preserves *both* determinism *and* locality. If that had happened, then Einstein, who did not dispute the empirical adequacy of quantum mechanics, and the rest of the quantum physicists would have perceived in sharp relief the choice between determinism and locality in *any* theory. That is, nonlocality may well turn out to be a feature of the *world*, not just a property of some deviant theory or other.

The crucial issue here is how, in an observer-independent reality, Einstein would have evaluated or weighted causality (or determinism) versus nonlocality (or nonseparability), given that one *had* to go. Arthur Fine, in his *The Shaky Game*, has analyzed extensively Einstein's position on realism and the quantum theory.[5] Fine tells us that the concept of determinism (or causality) was of central importance to Einstein and that he rejected the notion that probability in quantum theory should be accepted as an in-principle and fundamental feature of nature, as opposed to being an interim necessity in the present state of development of the theory (Fine 1986, pp. 103-104). It is true that nonlocality appears to present a *prima facie* case for a conflict with the first-signal principle of special relativity, but we know that the nonlocality of quantum theory cannot be used for signaling. Furthermore, even though we today look upon relativity as a theory of principle, it remains the case that the *empirical* basis for this principle is a lack of *observational* conflict with its basic hypotheses. That is, relativistic invariance could turn out to hold only at the observational level, not necessarily at the level of abstract space-time as usually envisioned in special relativity. Still, if push had come to shove, it is likely that Einstein would have opted for causality over space-time and, perhaps, to compromise on some of his *desiderata* for space-time.[6]

Since Einstein was committed to a causal (or deterministic) fundamental theory, the type of proof of a "Bell" inequality that I have in mind for my purposes here would assume both determinism and a version of locality to underpin the arguments used by Stapp (1971, 1985) and, in an even simpler form, by Peres (1978). I could continue this counterfactual line of argument by pointing out that von Neumann's celebrated impossibility "proof" would never have come to be an obstacle to hidden-variables theories, by allowing an early version of a Bohm interpretation on the scene to make a virtue of the quantum potential. We have already seen that, early on, the *equations* of quantum mechanics had been recast into a form that was conducive to a Bohmian interpretation, but that their nonlocal feature remained a sticking point. This would have provided an explicit counterexample to the "Copenhagen" interpretation's claim to completeness and finality and would have been an antidote to that dogma's establishing its hegemony. And, Don Howard (1990, p. 69) has documented Einstein's early inclination, already in 1924, to interpret quantum interference effects in terms of *actual* physical interactions among microentities. That is, a "causal" theory might then have appeared as a preferable, although certainly not a perfect, alternative. While it is true that Einstein in 1952 thought Bohm's solution "too cheap" (i.e., too simple) (Fine, 1986, p. 57), this was still *before* the world or he had been confronted with the stark choice forced by Bell's theorem.

A "no-signalling" theorem for quantum-mechanical correlations could have been established and put to rest Einstein's objections to the nonseparability of quantum mechanics. This important point is the following. If one considers a system S consisting of two subsystems S_1 and S_2 which are spatially separated at some time, then Einstein felt that ".. the real factual situation of the system S_2 is independent of what is done with the system S_1... ." (Fine, 1986, p. 103) Einstein worried what it would even mean to do science if such were not the case. But, a no-signalling theorem would have shown that relativity could be respected at the practical or observational level and that the nonlocality present in nature was of a "benign" variety. This could reasonably have been enough to overcome his objections to the nonlocal or nonseparable nature of a "de Broglie-Bohm" interpretation of the formalism of quantum mechanics. Bohm's interpretation would certainly have been possible in 1927. These models and theories could have been generalized to include relativity and spin. The program is off and running! Finally, quantum statistics follow naturally in a causal stochastic interpretation and this causal interpretation can be extended to quantum fields.

It is essential to appreciate that this "story" is neither *ad hoc* (in the sense of these causal models having as their sole justification an origin in successful results of a rival program) nor mere fancy, since all of these developments exist in the physics literature. However, "Copenhagen" got to the top of the hill first and, to most practicing scientists, there seems to be no point in dislodging it.

4. Two Problems for the Realist

In her recent article, Yemima Ben-Menahem (1990) has nicely summarized the problem that the Duhem-Quine underdetermination thesis poses for the scientific realist. Succinctly put, if there are two equally empirically adequate successful scientific theories (agreeing on all *possible* empirical tests and, therefore, being observationally indistinguishable) that support radically different (and, in fact, incompatible) ontologies, then such a situation must frustrate the scientific realist in his (or her) search for *the* correct scientific theory that gives a true picture of the world (even with limits set by reasonable caveats). As long as the two theories under consideration differ in relatively minor respects as regards their ontologies about the furniture of the world, one can simply decide to bracket these as inessential matters. The case I have considered involves one theory that represents the fundamental physical processes in the world as being inherently and *irreducibly* indeterministic and another theory that is based on an *absolutely* deterministic behavior of the physical universe. This would not appear to be a minor or irrelevant difference.

It is perhaps worthwhile here to enter a disclaimer or two. I am *not* centrally concerned with the question of the *refutation* of scientific theories. That is, in terms of the title of a well-known collection of essays on the Duhem-Quine thesis, *Can Theories Be Refuted?* (Harding 1976), I would be willing to give away a "yes" answer in the following sense. Even if one grants that there are many theories that can reasonably be rejected on evidential grounds (those are the *easy* cases!), there still remain other cases, I claim, in which viable, fertile theories have been rejected. My position is not a radical one claiming that practical underdetermination always exists in *all* cases,[7] but rather that there are some cases (at least *one* important instance) in which genuine underdetermination does exist. The case discussed in this paper does not concern just the (mere) compatibility of two essentially different theories with the *presently* available data, but involves a much more deep-seated indistinguishability. While a choice can be, and has been, made on the basis of non-evidential criteria, the question must then be faced of the basis for such criteria and of the role historically contingent factors have played in fashioning them. One must resist an urge to seek resolution of the underdetermination problem in terms of future developments that *may* take place in science. That alone would be more a declaration of belief than an argument. It is also true, though, that, even if one grants validity to the case presented in this paper, it would remain just *one* example from theoretical physics and may be peculiar to that area, having little relevance to the philosophy of science in general.

The basic issue here is a belief in the (at least effective) uniqueness of a correct scientific theory, with the selection process being "objective" and not involving in any ineliminable fashion "subjective" criteria such as coherence ("beauty"), simplicity or minimum mutilation (Ben-Menahem 1990, p. 267). By coherence I most specifically do *not* mean just lack of logical contradiction, since both theories I have discussed in the preceding sections are logically consistent and neither is pejoratively *ad hoc* in nature. Scientists typically take for granted the practical uniqueness (in any given era) of successful scientific theories. Thus, in an address delivered before the Berlin Physical Society in 1918, Einstein (1954, pp. 221-222) allows the *theoretical* (i.e., logical) possibility of more than one empirically adequate theory, but then goes on to make the (rather startling) declaration of faith that this is not a *practical* problem since, at any given time, the "world of phenomena" (which sounds pretty objective) uniquely determines one theory as superior to all others. Similarly, Heisenberg (1971, p. 76), in his retrospective reconstruction of what was going on in Copenhagen in 1926, states his opinion that a correct formalism in practice essentially uniquely determines *the* interpretation. It is just this (even practical or effective, as opposed to merely conventional) uniqueness that I have refuted earlier with a specific example.

Now this situation represents, in a sense, a double threat to the scientific realist. To begin with, the almost universally accepted Copenhagen interpretation has traditionally been a serious (perhaps, even arguably, an insurmountable) challenge to a realistic construal of quantum mechanics. The core of the difficulty is the measurement problem, one entailment of which is that a physical system cannot (*in principle*) possess definite (but merely unknown to us) values of all physically observable attributes (such as position and velocity). Taking such a theory seriously as an actual representation of the physical world (at the level of *individual* microentities) requires that we accept a rather bizarre ontology, one that may not even be conceptually coherent. The measurement problem has been around now for sixty-five years and has resolutely defied any successful, generally-accepted solution. This has provided effective ammunition for the antirealist who begins at the level of microphenomena, accepts ("Copenhagen") quantum mechanics as *the* fundamental and exact theory of all physical processes, and then throws down the challenge to the realist to construct a coherent (*realistic*) ontology consistent with the demands of quantum theory (van Fraassen 1980). That is, such an antirealist begins his (or her) argument in the micro-realm, extrapolates to the macro-realm of everyday experience and leaves the ensuing conundrum at the doorstep of the realist. In fairness, though, I must point out that the realist has relatively easy going in the domain of macro-phenomena (e.g., everyday objects, bacteria, dinosaurs, *etc.*) (McMullin 1984), but then encounters difficulties in carrying these explanatory resources down to the domain of micro-phenomena.

At first sight, the Bohm interpretation of quantum mechanics would seem to offer consolation and a potentially powerful means of rebuttal (of antirealism) to the realist. That is, this interpretation, which represents a microentity as a wave *and* a particle (not as a wave *or* a particle as does "Copenhagen"), lends itself readily to a realist construal of even fundamental physical processes that develop completely deterministically in a continuous space-time background. (True, there are some highly nonlocal, nonclassical effects present, but this is equally true for the Copenhagen interpretation.) However, while it is true that this Bohm interpretation is consonant with (and even conducive to) a realist position, it is empirically indistinguishable from an ontologically incompatible interpretation ("Copenhagen"). So, the realist has no grounds for requiring a realistic interpretation other than predilection or fiat. Realism is in double jeopardy here: "Copenhagen" is anathema to realism, while "Bohm", which provides a consistent realistic interpretation, presents an underdetermination dilemma and thus blocks the realist from achieving the desired goal. That is, *if* one can erect mutually incompatible ontologies on a given formalism, then that does (or may) pose a genuine problem for the realist since a proponent of realism would have to prove, or argue strongly for, a claim that, once genuinely different ontologies are proposed, it will be possible to extend the formalism along different lines, *because distinct ontologies involve distinct physical magnitudes.* This *may* happen, but *need* it? That seems more a declaration of belief than an argument for a position.

Of course, there is one obvious move still open to the realist at this point. He can claim that any two theories that are empirically indistinguishable can (by *definition*) differ only in inessentials. That is, the scientific realist *could* write off the difference between indeterminism and determinism in the ontology of the world as an insignificance, but that would indeed be strange for one concerned with a reliable and meaningfully complete picture of the world. This radical conceptual difference between inherent indeterminism and absolute determinism also makes it virtually impossible to conceive of a "dictionary" that would map the language of one of these theories onto the other (i.e., to map a concept onto its negation). It is not absolute endgame, though, for the realist who still wants to have *a* theory that will yield *the* correct and actual ontology of our world. Since empirical adequacy and logical consistency together do not alone provide sufficient criteria to choose between the two theories presented above, one can enlarge these criteria to include factors such as fertility, beauty, coherence, naturalness and the like. Certainly, it is the case that the scientific community quite early on did make a decisive selection in favor of the Copenhagen interpretation. That is, *one* theory was in fact chosen. Could the actual criteria used have been objective, or at least atemporal in the sense that they are not, in an essential way, (unstable) products of historical contingency or accident? I have indicated above that either of these theories passes equally well a test for fertility, in the sense of possessing the internal resources to cope with anomaly and new empirical developments that have actually occurred, as well as for suggesting new avenues for research and generalization. And, I have argued (1992a), an examination of the actual historical record shows that key motivating factors, for certain crucial assumptions about the features that an "acceptable" theory must have, were based upon the philosophical predilections of the creators of the Copenhagen version of quantum mechanics and upon highly contingent historical circumstances that could easily have been otherwise.

So, what is the relevance of this claim for the scientific realist? Well, if the historical record indicates that a rearrangement of highly contingent factors could plausibly have led to a radically different scientific theory and world view being accepted as correctly and uniquely representing the physical world, then one can reasonably question the value of a philosopher's rational reconstruction to pass judgement on a scientific theory. Philosophers typically study the successful theory already accepted by the scientific community. They doubtless have the ability to reconstruct rationally, and hence to legitimate, *any* theory that has already survived the scrutiny of the scientific community. But for historical contingency, though, they might find themselves doing just as well reconstructing and justifying an *essentially* different, equally successful and widely accepted theory. So, what is the value of the exercise, except as a check for noncontradiction? Each of these reconstructions could be equally *rational*, but there would not necessarily be anything rational to choose between them.

5. Conclusions

It has *not* been my purpose here to argue in favor of Bohm's interpretation over Bohr's, but rather to question any compelling necessity demanding "Copenhagen". So, let me reiterate the main unsettling concern that I have raised: but for a plausible temporal reordering of certain historically contingent events, our world view of fundamental microprocesses might well be one of determinism rather than indeterminism—a world view requiring a less radical departure from those classical principles already ensconced in the early part of this century. This is a particular instance of a larger issue. I claim that asking what might have happened at certain critical junctures of theory construction and selection, and *why* it did not, is more than just idle speculation fit only for a free Saturday afternoon. This raises serious epistemological and general philosophical issues about the even effective uniqueness of our most successful scientific theories and about the reliability of the knowledge science gives us concerning the structure of our world at the most fundamental level. Similar concerns have been raised by Feyerabend (1989), by Pickering (1984) and by the social constructivists in general. However, the usual ploy is to argue that logic and scientific criteria *alone* did not uniquely constrain theory selection so that, therefore, other sociological factors were operative, thus intimating that *anything* could be made to go in science. That is, they stir up the pot and then go home. I have not just alluded to some type of in-principle underdetermination, but have indicated a plausible alternative historical sequence that would have led to a radically different theory choice. While not just anything can be made to go, historical contingency does produce a considerable and essential underdetermination of theories.

Notes

[1]Partial support for this work was provided by the National Science Foundation under Grant No. DIR89-08497.

[2]Thus, I am not attempting to "refute" the claim made by Laudan and Leplin (1991, pp. 449-450) that "there is no general guarantee of the possibility of empirically equivalent rivals to a given theory" (my emphasis). Rather, I offer one important example of such underdetermination.

[3]I thank Michael Crowe for the specifics of this reference.

[4]I thank Arthur Fine for bringing the existence of this manuscript to my attention and Jürgen Renn and Robert Schulmann of the Einstein Archives for providing me with a copy of this handwritten document.

[5]I do not mean to imply that Fine necessarily supports the following interpretation.

[6]Don Howard (1991) has taken some exception to Fine's reading of the "young" vs. the "old" Einstein on the causality-separability issue.

[7]It is this general and universal type of underdetermination that Laudan and Leplin (1991) argue against.

References

Beller, M. (1983a), *The Genesis of Interpretations of Quantum Physics*, 1925-1927. Unpublished Ph.D. dissertation, University of Maryland.

_ _ _ _ _ . (1983b), "Matrix Theory Before Schrödinger", *Isis* 74, 469-491.

_ _ _ _ _ . (forthcoming), "The Birth of Bohr's Complementarity—The Context and The Dialogues", *Studies in History and Philosophy of Science* (1992).

Ben-Menahem, Y. (1990), "Equivalent Descriptions", *British Journal for the Philosophy of Science* 41, 261-279.

Bohm, D. (1952), "A Suggested Interpretation of the Quantum Theory in Terms of 'Hidden' Variables, I and II", *Physical Review* 85, 166-193.

Cassidy, D. (1992), *Uncertainty: The Life and Science of Werner Heisenberg*. New York: W.H. Freeman and Company.

Cushing, J.T. (1990), *Theory Construction and Selection in Modern Physics: The S Matrix*. Cambridge: Cambridge University Press.

_ _ _ _ _ _ _. (1991), "Quantum Theory and Explanatory Discourse: Endgame for Understanding?", *Philosophy of Science* 58, 337-358.

_ _ _ _ _ _ _. (forthcoming), "Causal Quantum Theory: Why a Nonstarter?", in *The Wave-Particle Duality* (1992), F. Selleri (ed.). London: Plenum Publishing Co.

_ _ _ _ _ _ _. (forthcoming), "Underdetermination, Conventionalism and Realism: The 'Copenhagen' vs. the Bohm Interpretation of Quantum Mechanics", in *Correspondence, Invariance and Heuristics: A Festschrift for Heinz Post* (1992b), Steven French and Harmke Kamminga (eds.). Dordrecht: Kluwer Academic Press.

_ _ _ _ _ _ _. (forthcoming), "What if Bell Had Come Before 'Copenhagen'?", in *Proceedings of the International Conference on Bell's Theorem (Cesena, Italy),* F. Selleri et al. (eds.). Singapore: World Scientific Publishing Co.

Dürr, D., Goldstein, S. and Zanghi, N. (1990), "On a Realistic Theory for Quantum Physics", in *Stochastic Processes, Geometry and Physics (*1992c) *, Albevario* et al. (eds.). Singapore: World Scientific Publishing Co., pp. 374-391.

Einstein, A. (1954), "Principles of Research" in Ideas and Opinions. New York: Dell Publishing Co.

Feyerabend, P. (1989), "Realism and the Historicity of Knowledge", *The Journal of Philosophy* LXXXVI (8), 393-406.

Fine, A. (1986), *The Shaky Game: Einstein, Realism and the Quantum Theory.* Chicago: The University of Chicago Press.

Gillispie, C.C. (1960), *The Edge of Objectivity.* Princeton: Princeton University Press.

Goldstein, S. (1987), "Stochastic Mechanics and Quantum Theory", *Journal of Statistical Physics* 47, 645-667.

Harding, S.G. (ed.) (1976), *Can Theories be Refuted?* Dordrecht: D. Reidel Publishing Co.

Heisenberg, W. (1971), *Physics and Beyond.* New York: Harper & Row.

Howard, D. (1990), " 'Nicht Sein Kann Nicht Sein Darf', or the Prehistory of EPR, 1909–1935: Einstein's Early Worries About the Quantum Mechanics of Composite Systems", in *Sixty-Two Years of Uncertainty,* A.L. Miller (ed.). New York: Plenum Press, pp. 61-111.

_ _ _ _ _ _. (1991), "Review Essay: Arthur Fine. *The Shaky Game: Einstein, Realism, and the Quantum Theory*", *Synthese* 86, 123-141.

Jammer, M. (1974), *The Philosophy of Quantum Mechanics.* New York: John Wiley & Sons.

Laudan, L. and Leplin, J. (1991), "Empirical Equivalence and Underdetermination", *The Journal of Philosophy* LXXXVIII (9), 449-472.

McMullin, E. (1984), "A Case for Scientific Realism", in *Scientific Realism,* J. Leplin (ed.). Berkeley: University of California press, pp. 8-40.

Moritz, R. E. (1914), *Memorabilia Mathematica.* New York: The Macmillan Company.

Nelson, E. (1966), "Derivation of the Schrödinger Equation from Newtonian Mechanics", *Physical Review* 150, 1079-1085.

Peres, A. (1978), "Unperformed Experiments Have No Results", *American Journal of Physics* 46, 745-747.

Pickering, A. (1984), *Constructing Quarks: A Sociological History of Particle Physics*. Chicago: University of Chicago Press.

Stapp, H.P. (1971), "S-Matrix Interpretation of Quantum Theory", *Physical Review* D3, 1303-1320.

______. (1985), "Bell's Theorem and the Foundations of Quantum Physics", *American Journal of Physics* 53, 306-317.

van Fraassen, B.C. (1980), *The Scientific Image*. Oxford: Oxford University Press.

___________. (1989), *Laws and Symmetry*. Oxford: Oxford University Press.

Part XII

ISSUES IN THE PHILOSOPHY OF PSYCHOLOGY

What Price Neurophilosophy?[1]

Eric Saidel

University of Wisconsin-Madison

If the appropriate cognitive model of the mind is a connectionist model, then what does the future of folk psychology (FP) look like? "Bleak" say the eliminative materialists. In several recent articles and books Paul and Patricia Churchland have been arguing by demonstration that there are models on which to explain all behavior—not merely cognitive behavior—models which are more fruitful than FP. These models, which are connectionist in nature, are supposed to describe a better theory of behavior than FP, and so will eventually replace it. In this paper I look at several arguments the Churchlands have advanced for the thesis that FP will be replaced in a mature cognitive science, i.e., that connectionism will supplant FP. What these arguments have in common is a vision of the direction cognitive science is taking, and a conviction that FP is incompatible with that future. I generally share the eliminativists' vision of the future of cognitive science, however, I don't think FP will be eliminated in that future.

This paper is divided into two main sections. In the first I look at two connectionist-based arguments the Churchlands have offered for the conclusion that we should abandon folk psychology. These arguments both presume that FP, as it now stands, will fail to reduce to neuroscience. In the second section I argue that the issue of reduction is orthogonal to the question of the future of folk psychology.

1. The Evidence of Connectionist-Type Models

The eliminative materialists have turned from their general anti-FP writings of the early eighties to more specific claims in the latter part of the decade. The new connectionist approach to studies of cognition has not only enlivened cognitive science, it has also given the eliminative materialists something to point to when arguing against FP. Ramsey, Stich, and Garon (1991), for example, have used connectionism to explicitly argue against FP; they claim that the two are incompatible, and that if connectionism is the correct theory of the mind, FP must be jettisoned.[2] The Churchlands have argued more by example: they have tried to demonstrate how neuroscience and models which are connectionist in spirit are able to solve cognitive problems. The suggestion is that FP is no longer the only game in town, and that this new theory will be able to supplant it.

Patricia Churchland's (1986) book has "a primary objective [of] show[ing] that neuroscience matters to philosophy" (p.482). Her point is that we should not limit

PSA 1992, Volume 1, pp. 461-468

ourselves to top-down analysis in order to understand how the brain/mind works. I endorse this conclusion wholeheartedly; we can only be assisted in our efforts to understand how the mind works by also trying to understand how the brain works. Insofar as this project is concerned Churchland is able to remain neutral on the question of the place of FP in our completed understanding of the mind/brain. However, she is not optimistic about the future of FP; "I think it most unlikely that a theoretical unification of neuroscience and psychology will involve a retentive reduction of mental states as currently understood with neural states as currently understood" (p.286). As it stands this claim is not threatening to an adherent of FP.[3] The threat is in what Churchland thinks the reduction will be like.

> Once it is recognized that folk psychology is not immune to scientific improvement, this reveals the possibility that what will eventually reduce to neuroscience are generalizations of scientific psychology that have evolved a long way from the home "truths" of extant folk psychology. ... What may eventually transpire, therefore, is a reduction of the evolved psychological theory, and this evolved theory may end up looking radically different from folk psychology - different even in its *categorial* profile. In other words, the psychological generalizations that are eventually considered ripe for reduction may be both richer and substantially revised relative to current generalizations in folk psychology. If that is the direction taken by the co-evolution of psychology and neuroscience, future historians of science will see folk psychology as having been largely displaced rather than smoothly reduced to neurobiology. (p.312)

Despite her careful equanimity in this passage it is clear that Churchland thinks that this is the direction that will be taken by the co-evolution of psychology and neuroscience. Even to one who endorses this co-evolution, it is not obvious why it is that the theory which reduces to neuroscience should have evolved to be of a different species than FP. Churchland apparently thinks there are two kinds of evidence for this conclusion. One is that FP is supposedly inadequate. A discussion here of her reasons for believing this would take us too far afield. She also points to the overwhelming success of neuroscience as a reason to be dubious about the future of FP. It is to this argument that I now turn.

Both Churchlands have spent a great deal of effort in press lately to demonstrate the ability of neuroscience to explain various cognitive tasks. These arguments appear to have two points. One is mostly a cheerleading goal, that neuroscience has a great deal to offer us in our study of the mind. The other is that FP has very little to offer in comparison. As this latter point is implicit rather than explicit in their discussions, what the argument is for this is not clear. Apparently it is this. Their discussions show not only that neuroscience is teaching us a great deal about the way the brain works, but that models built on neuroscientific principles, namely connectionist-type models, are able to resolve various problems facing cognitive science. For example, they illustrate how beliefs might be related to the world. One of the examples they discuss is that of a crab simulated in a network of simple units (P.M. Churchland 1986, P.S. Churchland 1986).

Two sets of information are relevant to the crab. One is the pair of angles at which its eyes must turn from straight ahead in order to focus on some food. The other is the pair of angles it must extend its upper arm and forearm in order to grasp that food. Given the first pair of angles, the crab is then able to calculate the proper angles of extension of its arm so that its claw would move precisely to the locus of the object. This is done without any explicit representation of a belief that the object is in such and such a place. The point of the model is that this can be done without reference to beliefs. If this kind of research proves fruitful, the Churchlands seem to be saying, this would provide a unified whole to scientific explanations, including explanations of behavior. We would then have reduced psychology to neuroscience, and would have eliminated FP.[4]

Suppose the Churchlands are correct; suppose this study of neuroscience will yield explanations of behavior in terms of neuroanatomy, even so that this supports the eliminativist program is far from clear. In order to support elimination it must be the case either that the explanations from neuroanatomy are incompatible with FP, or that FP, in its co-evolution with neuroscience has altered so dramatically that we would no longer recognize it. Patricia Churchland clearly thinks the latter is likely. However, I see no reason to think so. In fact, given that I take the identifying commitment of FP to be that behavior is to be explained in terms of beliefs and desires, I think that as long as what goes on is a co-evolution of FP and neuroscience, we have no reason to think that FP will evolve out of recognition. A closer look at the crab story will make this point clearer.

What the crab does is map a representation of its visual environment onto a representation of its tactile environment. Once this mapping has been completed it is able to obtain food. The implicit suggestion in holding this up as a model of cognition is that what enables us to successfully navigate our environment is many, much more complex, similar mappings. Suppose this is correct, what would this mean for the evolution of FP? Clearly our conception of FP would change somewhat as we learned more about the different mappings, and our explanations of behavior would become somewhat more sophisticated. However, those explanations of behavior would still intrinsically refer to the representations of the environment. That is, beliefs would have a primary role in the explanations of behavior, and these beliefs would still be fundamentally understood as representations of the environment. Furthermore, one of the ways in which the crab is over-simplified is that it does not have desires, all that it represents is food, and it always "wants" food. If behavior is to be understood on the model of such mappings, these mappings would have to include in some way a representation of the needs of the organism.[5] If it failed to do so, it would not be able to explain why it is that someone reached for food, rather than for drink. But to include such a representation is to include desires in the explanation of behavior. Again, I'm not claiming that our understanding of desires would not evolve in this situation, only that we would still be explaining an agent's behavior by reference to her representations of her environment and her needs. That is, we would still explain behavior by reference to beliefs and desires.

A *co*-evolution of FP and neuroscience involves neuroscience putting pressure on the evolution of FP, but it also involves FP exerting pressure on the evolution of neuroscience. There is as much reason to think FP will evolve out of recognition as there is to think neuroscience will evolve out of recognition.

A similar, and clearly related focus of interest is Paul Churchland's "Prototype-Activation" model of explanation (1986, 1989a, and 1989b Chapter 10). Briefly, he looks to connectionist networks as a model for a new theory of explanation. Connectionist models are exemplary feature detectors; the training of such a model is arguably equivalent to teaching the model to group similar objects together in some sort of feature-space. Churchland uses this aspect of connectionism as a model for explanation. Specifically, he argues that explanation is a matter of activating the correct prototypes in exactly the same way that a connectionist network can be said to recognize an object because of its match to a stored (i.e., learned) prototype. Churchland admits this to be, at best, an unexplored area, but thinks that it holds great promise:

> [T]he model brings a welcome and revealing unity into a stubborn diversity of explanation types, and the model is itself an integral part of a highly unified back-

ground conception of cognitive activity, one that encompasses with some success the general structure and activity of biological brains, and the structure and cognitive behavior of a new class of artificial computing systems. (1989b, p.230)

This model of explanation would replace models which refer to laws. Most analyses of explanation conclude that an event is explained by showing how it fits with certain laws about events of that type. For example, the ball's breaking the window is explained by reference to laws about the fragility of objects, and the kinetic energy of the moving ball. The explanation characterizes the event with a description that makes sense of the result given certain laws and regularities. One cannot deduce from the fact that Mary took birth control pills that she did not become pregnant, but because birth control pills, when working properly, prevent pregnancy, we can explain why Mary is not pregnant by mentioning that she takes birth control pills.

At least one problem with the traditional models of explanation, Churchland points out, is that they make extraordinary demands on our explanations of behavior. We would explain why it is that John ran screaming from the building by reference to the fire consuming the building. Presumably, what supports this explanation is the law that if a building is aflame the human inhabitants run screaming from the building, if they can. Then why is Carol running into the building? Because she's a firefighter. And there is an appropriate law supporting that explanation. Churchland points to all the thousands upon thousands of laws to which we must have ready access in order to explain behavior, if this view is correct. He suggests that it is more likely that explanation works by fitting a prototype to a situation. Certain aspects of a situation will key a prototype, and other aspects are explained by showing how they match this prototype. Our prototype of a building on fire is that certain people run from the building as best they can, and there are others trying to put the fire out, and assisting people who are having trouble leaving the building. Thus we can explain both John's behavior and Carol's without having to refer to several laws to do so.

The problem this raises for FP is that FP gains its explanatory power by the subsumption of an event to a law. That is, FP is by its nature an adherent to the received view of explanation. John and Carol's behavior is to be explained, according to FP, by showing how their behavior is to be expected given their beliefs and desires, and given the relevant laws. In the prototype-activation model there is no room for this kind of explanation.

Suppose Churchland is correct. Suppose the prototype-activation model is the correct way to think about explanation. Even so, I think this provides less challenge to FP than it does support. A weakness of FP is, as Churchland points out, that it implies that each one of us has access to uncounted laws, something which seems unlikely at best. If the prototype-activation model provides us a way to avoid that problem, it need not do so at the cost of FP. In fact, the prototype-activation model can only explain certain facets of an event if it includes beliefs and desires. And the explanation of other facets of that event will come into much sharper focus if beliefs and desires are part of the prototype. We can explain why it is that someone ran from the flaming building, and why someone ran into it by activating the fire prototype. But we cannot explain why it was that John ran out and Carol ran in unless we point to John's desire to escape from the fire and Carol's to do her duty. And we might be able to explain why it was that Carol tripped by pointing to the prototype of rushing behavior and showing how in that prototype accidents are often occurring. However, if we look at our fire prototype again, and include mental states, we might see that often the firefighter who has to enter the fire has a strong desire not to, and as a result subconsciously attempts to interfere with the success of that act. While the prototype-activation model might work as a better model of explanation, what it would achieve in the realm of explaining behavior is not a replacement of folk psychology, but a shift in our understanding of folk psychological explanations. The result would not be an elimination of FP, but a strengthening of a weak aspect of FP.

2. The Relevance of Reduction

The two arguments discussed in the previous section share an assumption. Implicit in each is the premise that folk psychology will fail to reduce to neuroscience, and that this failure is to the detriment of folk psychology. As I remarked above, I think the issue of whether or not folk psychology smoothly reduces to neuroscience to be a red herring in discussions about the future of FP. However, I think closer attention to this issue will prove rewarding.

There are two basic possible futures for folk psychology vis à vis reduction. One is that further investigations into the nature of the mind and the brain show us that the concepts of folk psychology can be derived from those of neuroscience. In this case FP will be said to have been reduced to neuroscience.[6] This would not bode ill for FP, though; even were we to eliminate the terminology of mental states from our discussions about behavior, these discussions would only sound different. The neuroscientific terms used to explain behavior would be those which were similar in extension to beliefs and desires. Our science of behavior would retain beliefs and desires, although perhaps under different names. (Of course, if we find that we cannot explain behavior by reference to beliefs and desires, then the successful reduction of FP to neuroscience would mean very little. However, the important factor in this case would be the failure of folk psychology to explain phenomena within its domain, not its reduction to neuroscience.)

The other alternative is that FP might fail to reduce to neuroscience. However, this in itself would not spell disaster for FP. To see this it is helpful to consider the failure of transmission, or classical, genetics to reduce to molecular genetics. My discussion of this example follows Kitcher's (1984) analysis.

Presumably, folk psychology would fail to reduce to neuroscience because the explanations given by folk psychology would not be derivable from neuroscientific principles. One of the ways in which classical genetics fails to reduce to molecular genetics is that "a derivation of general principles about the transmission of genes from principles of molecular biology" fails to explain "why the laws of gene transmission hold (to the extent that they do)" (p.339). These two cases should be seen as analogous. We are supposing that inasmuch as the laws of FP work, they cannot be explained by reference to neuroscience. And inasmuch as the laws of transmission genetics work, they cannot be explained by molecular genetics.

Why does molecular genetics fail to explain the laws of transmission genetics? Kitcher discusses a particular example of this failure. Suppose we ask why nonhomologous chromosomes assort independently. The answer has to do with how chromosomes pair up during meiosis. (Chromosomes line up with their homologues and produce recombinant pairs; these pairs then divide, each member going to a different gamete. "The assignment of one member of one pair to a gamete is probabilistically independent of the assignment of a member of another pair to that gamete". While genes near each other on the same chromosome are likely to be transmitted together, genes on nonhomologous chromosomes will assort independently.) This answer raises other questions, for example, "why are nonhomologous chromosomes distributed independently at meiosis?" These questions lead to a discussion of the formation of the spindle (a structure formed in the nucleus during meiosis), and eventually to the explanation that an important factor is that "chromosomes are not selectively oriented toward the poles of the spindle". This factor is missed in a molecular discussion of the phenomena. It is missed because the relevant natural kinds from the perspective of answering why nonhomologous chromosomes assort independently are not available from the molecular perspective. (Notice that the complexity of the molecular discussion has nothing to do with why it misses the relevant factor.)

The original explanation gains its force by identifying meiosis as a particular kind of process, and then showing how certain regularities are to be expected in this kind of process. However, this process is not identifiable from the molecular perspective. From the molecular perspective we see that, for example, the molecules that make up the chromosomes are separated because of gravity in some cases, electromagnetic forces in others, or nuclear forces in still others. The molecular account focuses too much on these details, and as such "cannot bring out that feature of the situation which is highlighted in the cytological story". Molecular genetics fails to explain the laws of transmission genetics because the natural kinds needed to explain the laws of transmission genetics are not available from the molecular perspective (pp.347-51).

However, this does not mean that molecular genetics has no role to play in relation to transmission genetics. Certain fundamental questions of transmission genetics can only be explained by reference to molecular genetics (for example, transmission genetics assumes that genes replicate, but cannot explain how, such an explanation is easily found on the molecular story). Importantly, though, certain questions relevant to transmission genetics are not answerable when approached from molecular genetics.

Despite its inability to reduce to molecular genetics transmission genetics is safe from elimination because there are explanations and predictions that it makes that molecular genetics is unable to make. That is, it works well as a scientific theory within its domain. There are interesting, and useful, generalizations that it makes that molecular genetics misses. Two principles are derivable from this: 1) the inability of a science to reduce to a more basic science need not lead to its elimination; and 2) what is important with respect to elimination is the success that a science has within its own domain. These principles are immediately applicable to folk psychology. FP might fail to reduce to neuroscience, but this would not spell its doom. If there are types of phenomena that FP can explain that neuroscience cannot, then FP will be retained despite its inability to reduce. If there is nothing that FP buys us above neuroscience, then its failure to reduce would be terminal. However, the resulting elimination of FP would be caused not by its failure to reduce, but instead by the arrival of a competing theory that better explains the phenomena.

There is a third principle to be gleaned from the example of classical genetics and molecular genetics. Suppose we look at a particular example of meiosis and ask questions about the behavior of the chromosomes in that example. The explanation of how the structures divide will have to do with the bonds that held them together. Such an explanation will be completely within the domain of molecular genetics. Molecular genetics will be easily able to explain an individual token of the phenomenon of chromosome division even though it is unable to give an explanation of the type of phenomenon. That is, the type of phenomenon identified from the molecular perspective is wholly different from that identified from the perspective of transmission genetics. The appropriate place to look for an explanation of the type of phenomenon is in a theory in which the phenomenon is a recognizable type. In this case, this is within transmission genetics, not molecular genetics.

This lesson can be directly applied to the relation between folk psychology and neuroscience. We might be able to explain why it is that Carol rushed into the burning building by reference to certain neuroscientific facts about Carol. (I think that this is not unlikely.) However, this, in itself, doesn't guarantee that we are able to give a satisfactory neuroscientific explanation of why it is that several different firefighters rushed into several different buildings. Neuroscience may look at these phenomena as being of several different types. However, the behavior of the firefighters is of one type when viewed from the folk psychological perspective. Even if it fails to reduce to neuroscience, folk psychology may well be the appropriate place to look for an explanans for this explanandum.

3. Concluding Remarks

Paul Churchland (1981) draws an analogy between folk psychology and alchemy. I think that this analogy can be useful in evaluating the prospects for folk psychology. It is useful not because the ontology of folk psychology is, like that of alchemy, made up of mere chimera, but because alchemy provides us with a model of what must happen for a science to be abandoned. The elimination of the ontology of alchemy had to do with its failure to make any explanatory or predictive difference, not with its inability to reduce to chemistry.

It may be that folk psychology will be replaced; I don't think it will, but nothing I've said here belies that possibility. Certainly, I agree with the eliminativists, folk psychology is a theory, it is logically possible that it be replaced. And any candidate for a new theory of behavior is likely to come from the neurosciences, as the eliminativists think. My point here has been that, while we should continue to study the brain in order to learn about the mind, we have yet to be given reason to think that these studies will tell us that our behavior cannot be accurately explained by reference to beliefs and desires. What would lead to the rejection of folk psychology as an adequate theory of behavior is not the successes of connectionism, nor the inability of folk psychology to reduce to neuroscience, but the explanatory and predictive failures of folk psychology. Until folk psychology is demonstrated to be a flawed theory we need not overly worry about its future.

Notes

[1]I am indebted to Berent Enç and Malcolm Forster for comments and suggestions based on earlier drafts of this paper, and to Elliott Sober for helpful discussions and suggestions.

[2]See Forster and Saidel (ms) for a discussion of Ramsey, Stich, and Garon's arguments for the incompatibility of FP and connectionism. Forster and Saidel argue that Ramsey, Stich, and Garon have not given sufficient reason to believe that FP and connectionism are incompatible.

[3]As I've said, I think the study of the brain can only help our efforts to understand the mind. While this is meant as a materialist position (a dualist could also endorse brain studies as a way of finding out how the mind and brain communicate), I mean it to be understood as neutral with respect to the question of reduction. It may turn out that the psychology will reduce to neuroscience, in the way water reduces to H_2O; it may be that psychology will fail to reduce, because, for example, of the multiple instantiality of mental states; it may even turn out that mental states are eliminated. All I am arguing in this section is that, given that neuroscience is relevant, that in itself

does not give us reason to think mental states will be eliminated from our ontology. For a discussion of the relevance of reduction see section 2 below.

[4]Paul Churchland (1986) discusses extensively the use of what he calls "state-space sandwiches", i.e., mappings from one representation of an n-dimensional space (such as the 2-D representation of the crab's visual space) to another (such as the 2-D representation of the crab's arm-extension space). His discussion is of a much greater depth than the crab example I look at here. While it is very helpful for understanding the power of these mappings, I don't think it undermines what I say here.

[5]I do not wish to take any position here on what desires are. Perhaps these mappings would be able to include desires without including representations of the needs of the organism. Instead they might have some way of potentiating the food beliefs when the organism needs food, and the water beliefs when the organism needs water. My point here is only that there must be some way in which the organism's needs have an effect on its actions.

[6]I am attempting to avoid issues related to the difficulty of specifying conditions for reduction here. While we might count the reduction of Classical Mechanics to the Special Theory of Relativity as a paradigm case of reduction, the laws of the former are not deducible from the laws of the latter. However, there is a relation between the two, something like deducibility, which would also hold between FP and neuroscience, if FP reduces to neuroscience. It is this relation I am referring to when I claim that the concepts of FP can be derived from the concepts of neuroscience.

References

Churchland, P.M. (1981), "Eliminative Materialism and the Propositional Attitudes", *The Journal of Philosophy* 78: 67-90.

Churchland, P.M. (1986), "Some Reductive Strategies in Cognitive Neurobiology", *Mind* 95: 279-309.

Churchland, P.M. (1989a), "Folk Psychology and the Explanation of Human Behavior", in *Philosophical Perspectives 3, Philosophy of Mind and Action Theory,* J.Tomberlin (ed.). Atascadero: Ridgeview Publishing Company, pp. 225-240.

Churchland, P.M. (1989b), *A Neurocomputational Perspective.* Cambridge: MIT Press.

Churchland, P.S. (1986), *Neurophilosophy: Toward a Unified Science of the Mind-Brain.* Cambridge: MIT Press.

Forster, M. and Saidel, E. (forthcoming), "Connectionism and the Fate of Folk Psychology: A Reply to Ramsey, Stich, and Garon" in *Philosophical Psychology.*

Kitcher, P. (1984), "1953 And All That. A Tale of Two Sciences", *The Philosophical Review* 93: 335-375.

Ramsey, W., Stich, S., and Garon, J. (1991), "Connectionism, Eliminativism, and the Future of Folk Psychology", in *Philosophy and Connectionist Theory,* W.Ramsey, S.Stich, and D.Rumelhart (eds.). Hilldale: LEA, pp. 199-228.

Darwin and Disjunction: Foraging Theory and Univocal Assignments of Content[1]

Lawrence A. Shapiro

University of Pennsylvania

According to Jerry Fodor, "[h]uffing and puffing and piling on the teleology just doesn't help with the disjunction problem; it doesn't lead to univocal assignments of intentional content" (1990, p. 72). That Fodor separates these two claims with only a semicolon is unfortunate, for each merits individual scrutiny. First, there is the claim that appeals to the theory of evolution by natural selection will not help with the disjunction problem. I will argue that the truth of this claim depends upon how we characterize the relation between a representational state and that which it is about. Fodor assumes this relation must be a causal one, and this assumption leads inevitably to the disjunction problem. But there is another plausible account of representation available that is not so burdened. If the disjunction problem is an artifact of causal approaches to representation, and if Fodor is right that evolutionary theory won't lend to its solution, then, we might decide, so much the worse for causal approaches to representation. Second is Fodor's claim that evolutionary theory does not ground univocal assignments of intentional content. Fodor's argument for this claim threatens the tenability of the functional role account of content I present as an alternative to the causal approach. Hence, it is essential that I show how evolutionary considerations do fix assignments of intentional content. This will take up most of my time, but first there are the preliminaries to cover.

1. The Disjunction Problem

The disjunction problem confronts many recent attempts to explain naturalistically how a cognitive state comes to be about some state of the world. Common to these attempts is the identification of the meaning of a cognitive state with those properties of the world the instances of which are nomically sufficient for a tokening of the state. So, to take Fodor's (1990) example, the mentalese predicate 'dog' has in its extension dogs because the property of being a dog is lawfully related to the property of being a tokening of 'dog'. 'Dog' is about dogs because dogs cause its tokening. This analysis of meaning is appealing because it is free of any intentional or semantic vocabulary: it offers a way out of Brentano's circle.

But absent from this account of meaning is a feature that has been hailed as one of the marks of the mental—the capacity to misrepresent. If the mentalese predicate 'dog' is truly to mean *dog*, we want to leave room for the occasions when we think

PSA 1992, Volume 1, pp. 469-480

dog but no dog is present. It is here we encounter the disjunction problem. Suppose, to accommodate the need for misrepresentation, we concede that not only do dogs cause tokenings of 'dog', but so also would cats on dark nights. Hence, the story goes, a tokening of 'dog' caused by a cat on a dark night is a misrepresentation. But, if a cognitive state expresses those properties whose instances would cause its tokening, then, since an instance of the property of being a cat on a dark night is sufficient to cause a tokening of 'dog', 'dog' must express not simply the property of being a dog, but rather the property of being a dog *or* a cat on a dark night. Therefore, a tokening of 'dog' caused by a cat on a dark night is not, after all, a misrepresentation.

2. Enter Darwin

Now, how might Darwin help us to solve this problem? The idea, which Fodor himself considers (1987, pp. 104-106) and which Dretske later hones (1988), is that we can preserve misrepresentation by distinguishing normal circumstances, where state C expresses whatever properties the instances of which cause it, from abnormal circumstances, where C will be caused by instances of properties that it, C, does not express. In the latter case, C misrepresents its cause as something else—as whatever would be its cause if circumstances were normal. We then use appeals to the theory of natural selection to define which circumstances, relative to C, are normal; and we do this by citing in which circumstances tokenings of C are adaptive, i.e., confer on the possessor of C a selective advantage.

So, to shift examples, a frog's fly detector[2] is normally situated in an environment where what it causes the frog to snap at are flies, because, presumably, it is the function of the fly detector to provide its possessor with nourishing flies. Frogs in a laboratory can be made to snap at bee-bees. But here, we can say, the fly detector is being tricked. The laboratory is not a normal environment for the fly detector because it is not in the laboratory setting that the fly detector evolved as an adaptation. Hence, in the laboratory, the frog misrepresents bee-bees as flies. Even though both being a fly and being a bee-bee are properties whose instances are nomically sufficient to trigger the firing of the fly detector, the firing of the fly detector means *fly*, not *fly or bee-bee*, because, selection considerations show us, fly detectors have the function to detect flies and not bee-bees.

3. Darwin and the Disjunction Problem

However, Fodor doesn't believe the theory of natural selection will help to circumvent the disjunction problem. His argument is this. While it is true that a detector of flies confers a selective advantage upon frogs, so too is it true that a detector of little ambient black things confers on the frog a selective advantage. The latter is true because in the frog's normal environment the detector of black things leads to the capture of flies. The moral is, Fodor tells us, "Darwin doesn't care how you describe the intentional objects of frog snaps. All that matters for selection is how many flies the frog manages to ingest in consequence of its snapping" (1990, pp. 72-3). Because frogs equipped with fly detectors and frogs equipped with black dot detectors will be equally successful in a frog's normal environment, appeals to Darwin won't decide between descriptions of the intentional object of the fly detector as *fly* or as *black dot*. "So," Fodor concludes, "it's no use looking to Darwin to get you out of the disjunction problem" (1990, p. 73).

Concisely stated, Fodor has argued for the claim that "the context: *was selected for representing things as F* is transparent to the substitution of predicates reliably coextensive with *F*" (1990, p. 73). In other words, the theory of natural selection, according to Fodor, is transparent regarding descriptions of the sorts of things that satisfy selection pressures. A frog needs to consume flies to survive, but whether it represents flies *as* flies, or as black dots, or as things of the order *Diptera*, or as flies-or-bee-bees (where all flies-or-bee-bees happen to be flies) is irrelevant to its success as a consumer of flies. Hence, we can assign to the fly detector the function to detect flies, black dots, things of the order *Diptera*, etc. Detectors detect *de re*, as it were; and this is why, despite Darwin, the content of fly detector firings will be disjunctive. So, appealing to the theory of natural selection to solve the disjunction problem won't work.

4. Functional Role Accounts of Representation

I will now argue that fly detectors have been selected for their ability to represent things as food (or, perhaps, prey) and *not* merely as anything named by other predicates reliably coextensive with 'food' (or 'prey'). On my view, the theory of natural selection is not transparent. My argument for this claim will come in two stages. First, in this section, I will sketch an alternative account of representational content—one that makes the content of a state depend upon its functional role within a system rather than simply its causal covariation with instances of properties in the world. Then, in the next section, I will defend the claim that the fly detector is part of a feeding system, and hence it has states that ought to be assigned the content *food.*

Functional role accounts of content define the content of a state by appeal to the state's role in the functioning of the system of which it is a part.[3] For instance, Hatfield (1988 1991) argues that certain states of the early visual system represent edges of distal surfaces because, adopting Marr's (1982) analysis of early vision, the early visual system is *for* the representation of surfaces in the viewer's environment; and the function of "edge" detectors within this system is to represent the presence of edges. To say that edge detectors have the function to represent edges, for Hatfield, is simply to say that within the early visual system states of a particular sort *stand for* edges of surfaces. Standing for edges is the role these states play in the functioning of the visual system.

Significantly for my purposes, the functional role account of content resists the disjunction problem. The disjunction problem surfaces against the assumption that states represent in virtue of their causal relations to properties in the environment. Content becomes disjunctive because causal relations are cheap. However, states of the edge detector do not represent edges merely because of their correlation with surface edges in the world—indeed, there are surely edges that will not cause a firing of the edge detector (edges that are equally illuminated from both sides) and there are surely non-edges that will (sharp changes in luminance caused by, perhaps, sharp color changes on a flat surface). Rather, states of an edge detector represent edges because, a) they are states within a sensory system that is for the representation of distal surfaces, and b) in the context of this system it is the state's regular correlation *with edges* that contributes to the visual system's function to represent surfaces. States of the edge detector represent edges because it is their job to represent edges in a system that has the job to represent the layout of surfaces in the environment.

5. Constraints on Biological Function Ascriptions

In the last section I sketched an approach to representation that avoids the disjunction problem. However, hard work has yet to be done before appeals to Darwin can be made to supplement usefully this approach. For, assignments of content to the states of a system can be only as fine-grained as the function of the system to which the states contribute is determinate. And, according to Fodor, the theory of natural selection does not license determinate function ascriptions. Fly detectors are adaptive if they make for fitter frogs—but, Fodor argues, fly detectors are adaptive whether we take them to have the function to detect flies, black dots, things of the order *Diptera*, or flies-or-bee-bees (in environments where all flies-or-bee-bees are flies). If Fodor is right about this, a functional role approach to the content of the fly detector's states will be no better off than the causal analysis with its associated disjunction problem.

Let's now consider what makes the fly detector a food detector and not a black dot detector. First, we should note that Fodor never denies that a fly detector is a detector; and it is worth pondering why Fodor would not deny this. Or to put a slightly different twist on the issue, it would be instructive to imagine the sort of defense Fodor would mount against the reductionist's charge that what we have been calling a detector is nothing more than a bunch of frog neurons.

It is not difficult to construct a Fodor-like response to this charge, for Fodor more than anyone else has been a champion of the scientific legitimacy of functional talk (for especially clear instances of his championing see Fodor (1974) and the introduction to Fodor (1981)). So, here is what Fodor would say (minus the characteristic wit) in defense of calling the fly detector a detector. First, Fodor would affirm that the detector in the frog is a bunch of neurons, but he would deny that the property of being a detector is the property of being neurons x, y, and z in a frog's brain. Fodor would point out that the "level of abstraction" at which a bunch of frog neurons becomes a detector "collapses across the differences between physically quite different kinds of systems" (1981, p.8). This is why we cannot identify the properties of being a detector and being frog neurons x, y, and z.

This talk of levels of abstraction, moreover, is legitimate because there are generalizations—empirically testable ones—that we can make about detectors but that we lose if we identify the property of being a detector with the property of being a bunch of frog neurons. This is because some special sciences, e.g., neuroethology, comparative and cognitive psychology, might like to generalize about detectors in animals other than frogs or in cognitive systems other than the fly detector. For these sciences to generalize over detectors is for these sciences to make an ontological commitment: it is to claim that within these sciences detectors are a natural kind. But, natural kinds in special sciences will not, usually, be natural kinds in sciences "below" them—detectors are not type identical to neurological kinds, and neurological kinds are not type identical to chemical kinds—and so the generalizations of a special science cannot be recast in a more basic science. Since these generalizations are valuable, since they provide explanatory and predictive successes, we are justified in talking about detectors. To paraphrase Fodor, the best argument for the existence of detectors is that we need them to do our science (1981, p. 29).

That, I believe, is how Fodor would respond to the reductionist who questions the existence of detectors. And, I think, Fodor would be perfectly correct in so responding. But, we can make the same sort of response to Fodor's claim about the transparency of the theory of natural selection. If I can make plausible the claim that our science has reason to distinguish food detectors from black dot detectors then Fodor should accept that frogs might have one sort of detector without having the other. If there are generalizations that are lost when we downgrade the grain of functional ascription to the point where fly detectors become black dot detectors then Fodor ought to be satisfied that the context *was selected for representing things as F* is not transparent to the substitution of predicates coextensive with *F*.

I shall now argue that it is only under fairly determinate function assignments that the fly detector is covered by the sorts of generalizations made by those sciences in which Darwin is a prominent figure. More generally, biological sciences (and I place in this group cognitive ethology, animal learning, behavioral ecology, and branches of cognitive psychology including perception) must impose specific intentional contexts upon the representational systems they examine if they are to generalize over these representational systems. In particular, I will argue that unless we take the fly detector of the frog to have the function to represent food (or prey)—and so have states with the representational content *food* (or *prey*)—we will miss two important and related sorts of generalizations: (1) the fly detector is an adaptation; (2) as a food detector the fly detector is likely to obey certain general principles of foraging theory. Characterizations of the fly detector as a black dot detector or a bee-bee detector will prevent application of these sorts of generalizations to the fly detector. Because generalizations like these are central to contemporary biological sciences, Darwin, I claim, would care a great deal how we specify the intentional objects of fly snaps.

Regarding the first generalization, the claim that the fly detector is an adaptation, we should note that adaptation plays a crucial role in Darwin's theory of evolution, for it serves as the mechanism by which variation in heritable traits results in differential reproductive success. Without the notion of adaptation, the theory of natural selection would collapse. Differential reproductive success, according to this theory, is a consequence of better and worse adaptedness to an environment. And, given the importance of the concept of adaptation to the theory of evolution, it is essential that we have a method by which to identify which traits of an organism are adaptive.

Lewontin, though not entirely sanguine about the methods currently in use, is clear about what the concept of adaptation entails: "[t]he concept of adaptation implies a preexisting world that poses a problem to which an adaptation is the solution" (1978, p.213). The biologist Pittendrigh (1958) effectively uses this manner of thinking in his classic study of the fruit fly's adaptive response to the problem of desiccation. Pittendrigh discovered that the fruit fly has evolved a variety of transducers, transducers that encode information about light, darkness, and temperature, which, despite the less than reliable (though reliable enough) correlation of these conditions with humidity, all adapt the fly to its dry environment. Anticipating Lewontin's characterization of adaptations as solutions, Pittendrigh claims that a "given problem—a given pattern of selection—is met by a multiplicity of solutions in different (and even in the same) organism (*sic*) (1958, p.400). More generally, for an adaptation to be an adaptation, it must be a solution to some selection problem. Thus, to talk about types of adaptation—a practice too deeply rooted in evolutionary biology to excise—one must also talk about problems to which adaptations are solutions. Typical categories of problems evolutionary biologists propose include predator avoidance, feeding, reproduction, etc. When an evolutionary biologist talks about the adaptedness of a trait, she is talking about its adaptedness relative to this set of problems.[4]

Now, one might question what the proper sort of attitude to this taxonomy of problems ought to be—i.e., one might want to resist taking a realist attitude toward selection problems.[5] However, we need not address this concern to draw the following point: adaptations can only be defined relative to some taxonomy of selection pressures, and so functions will be adaptations only under some descriptions. Given the taxonomy of selection pressures biologists, ethologists, ecologists and psychologists currently use (and, I would add, use to such success that one should alter it only under conditions of extreme duress) the proper characterization of the fly detector is as an adaptation to the need for food. A fly detector, that is, is the frog's solution to the need for food. As such, its function is to detect food and so the state it enters when it fires represents the presence of food. To deny this, or to claim that the fly detector has the function to detect, say, black dots, is to place the fly detector outside the reach of the only scheme available for individuating kinds of adaptations. There is no biological need for black dots. The detection of black dots *qua* black dots does not answer any kind of selection problem the theory of natural selection names. Only when we take the function of the fly detector to be the representation of food does its adaptive value become apparent.

To give up this point about the relativity of the notion of adaptation to a taxonomy of selection problems, we should note, would be to forsake the sort of generalization Pittendrigh draws about the fruit fly's various transducers. Pittendrigh shows that many different kinds of transducers all serve the same (adaptive) function: the representation of humidity. The adaptively significant variable in the fruit fly's environment is not darkness, or temperature, or time of day; it is moisture. It is for the conservation of water that these transducers were selected. Accordingly, the assignment of more proximate contents to these transducers (i.e., *darkness, temperature, time of day*) leads to two grievous consequences. First, the generalization that these things are adaptations is lost. It is not that darkness transducers indicate darkness that explains why these transducers are adaptive (or, more exactly, that darkness transducers indicate darkness can at best be only part of the explanation), it is the fact that darkness, in the fruit fly's environment, is well enough correlated with moisture for the indication of darkness to be adaptive. In other words, it is *qua* solution to the problem of desiccation that nature selects darkness transducers. Assigning to the darkness transducer the content *darkness*, or assigning to the fly detector the content *black dot*, veils facts about to which selection problems these mechanisms are solutions, and so obscures the fact that these mechanisms are adaptations. Secondly, a more proximate attribution of content to the darkness transducer leads one to miss the important fact that the darkness transducer, the temperature transducer, the time of day transducer, etc., are all adaptations to the *same* problem—all serve the same adaptive function. Any ascription of content more proximate than *humidity* or *moisture* will distinguish the functions of these transducers. The generalization that they have all been selected to solve the same problem will be lost.

I take the argument in the preceding paragraphs to show that, despite what Fodor claims, Darwin would indeed care about how we characterize the function of the fly detector. The context *was selected for representing things as F* is in fact opaque to the substitution of predicates reliably coextensive with *F* because the context *representing things as F provides a solution to selection problem P* is opaque to the substitution of predicates reliably coextensive with *F*. This in turn is because the theory of natural selection subscribes to a taxonomy of selection problems, and, consequently, to count as a solution to one of these problems an adaptation must be suitably described. Only under certain descriptions will we appreciate, for instance, that the distinct transducer mechanisms of different species—or, as with the fruit fly, that distinct transducer mechanisms *within* a species—are all adaptations to a single kind of selection pressure. And so, the context *was selected for representing things as F* cannot, after all, be transparent.

Let's turn now to the second important reason to assign to the fly detector the function to represent food. The reason is this: frogs *treat* black dots like food, so, plausibly, they *represent* black dots as food. This claim takes shape when we turn from Fodor's misleadingly simplistic picture of the frog's feeding behavior to the more sophisticated accounts of feeding behavior we find grouped under the heading 'foraging theory'. As best as I can determine, ecologists have not yet constructed foraging models for the frog, but we can benefit from examining a foraging study that Jaeger and Barnard (1981) performed upon the red-backed salamander, *Plethodon cinereus*. The salamander also feeds upon flies, and does so with the help of a visual detector. Moreover, the behavior of the feeding salamander is *prima facie* similar to that of the frog's: sometimes the salamander will pursue prey and snap at it with its tongue; other times it will wait in ambush for prey. Given the similarities between the frog and the salamander, I do not think I am "sneaking" into the following account of salamander foraging anything that should cause Fodor to charge me with unfairness to his point about fly detectors—I am merely examining the actual feeding behavior of an organism that employs equipment very similar to the frog's. I claim that this examination will show that salamanders, and by parity of example, frogs, represent those things that fly around in their environment as food.

First a word about foraging theory. Foraging theory is a species of optimization theory.[6] Optimization theory analyzes the behaviors of organisms in terms of the costs and benefits associated with the behavior. An optimization model will make hypotheses about the currency in which these costs and benefits are accrued, and about the constraints under which the organism must perform when attempting to maximize its holding of currency. An optimization theorist assumes that the organism under study will behave in such a way as to maximize a particular currency within the constraints defined by the model. Foraging theory is the branch of optimization theory concerned with feeding strategies. So, for instance, suppose we were to ask whether crabs tend to eat mussels of a size that, under specified constraints, will maximize the crab's rate of energy return, i.e., will crabs select mussels of a size that provide them with the most calories per amount of time they spend to find and open them? This question suggests a hypothesis about the *currency* of the costs and benefits facing crab behavior: crabs have been selected for maximizing their rate of energy return.[7] Note that this hypothesis is not the only one available to describe the forces that have shaped the crab's feeding strategy. There might have been selection upon crabs not to maximize their rate of energy return, but to maximize their energy efficiency. Assume that each mussel provides 1 Kcal of energy. A crab that in one hour burns 50 Kcal to eat 100 mussels is as efficient as a crab that in the same amount of time burns 100 Kcal to eat 200 mussels, or 1 Kcal to eat 2 mussels. However, if crabs require a net gain of 100 Kcal per hour to survive (assume a crab requires this much energy to keep its body running), only the crab with the highest rate of energy return will survive. So, the selection problem is the maximization of rate of energy return—not energy efficiency.

In addition to this hypothesis, there will be others about the constraints under which the crab feeds. Given the tools with which the crab is equipped—its claws, or *chelae*—the crab will have more difficulty opening larger, more energy-rich, mussels than smaller, less meaty, mussels. In foraging theory, time a predator spends in manipulating prey is called *handling time*. Handling time is one of the constraints a crab must face in its efforts to maximize its rate of energy return. A second constraint is searching time: the number of mussels in a given feeding patch, and the sizes of mussels in a given patch, will vary. Given that crabs have been selected to maximize their rate of energy return, and that they must spend time handling and searching for prey, we can compute which sizes of mussels crabs ought to favor in any given patch (Elner and Hughes 1978). This computation is possible once we have established, through empirical investigation, how much time it takes the crab to open mussels of various sizes, how much energy mussels of different sizes furnish the crab, and the distribution of differently sized mussels in the patch. Without worrying about the mathematics involved, the following prediction emerges. Assuming there are available to the crab mussels of $n > 1$ different sizes (with different sizes providing different rates of energy return), the crab's tendency to select a mussel of a particular size will be independent of the abundance of mussels of that size and will depend instead upon the densities of mussel types of greater profitability (Pyke, Pulliam, and Charnov 1977, p.141). As the abundance of more profitable mussels increases the crab will become more specialized, tending to select mussels of higher profitability and ignoring mussels of lesser profitability, *regardless* of the abundance of the latter. Correlatively, as the abundance of more profitable mussels decreases, the number of kinds of mussels in the crab's diet will increase: the crab will become less selective.

Krebs and Davies (1987, pp.69-70) list three virtues of optimization models like the foraging model for the crab. First, such models provide quantitative predictions and so are easily tested. If the crab does not perform as predicted, we must revise one or more of the assumptions: one or more of the hypotheses about the selection problem the crab faces and the constraints under which it behaves is wrong. Perhaps the crab has been selected for energy efficiency rather than energy maximization. Perhaps handling time and searching time are not the only constraints on the crab's capacity to feed. Also constraining its ability to maximize its energy gain may be the presence of predators that feed on crabs and so limit the time the crab can spend foraging. Given recalcitrant data, it may not always be possible to decide which of the hypotheses in the model is wrong, but this is a problem that, as Duhem has taught us, any science must face. Second, assumptions about currency and constraints are made explicit in the model. For the crab, the selection pressure is the need to maximize rate of energy return; the constraints are the number and kinds of mussels available, and the time it takes the crab to locate and open a mussel. These assumptions about selection pressures and constraints can be quantified, thus enabling the precise predictions I mentioned as the first benefit of optimization models. Last, optimization models are quite general. The model describing the foraging behavior of crabs also fits the foraging behavior of animals that feed under the same broadly defined conditions as the crab—animals like the sunfish (Werner and Hall 1974) and the mantid (Charnov 1976).[8]

This last point about the generality of foraging theory brings us back to talk of salamanders and frogs. In their studies of the salamander, Jaeger and Barnard (1981) found that a salamander in an environment dense with flies of two different sizes indeed specialized on the more profitable, larger, fly. The salamander, that is, adopts the same foraging strategy as the crab, which, we have seen, will tend to select more profitable mussels when these are available. Moreover, only when larger prey were scarce would the salamander strike at smaller flies. This is the analogue to the prediction that the crab will increase the number of kinds of mussels in its diet when more profitable mussels become scarcer.

In an apparently anomalous result, Jaeger and Barnard found that if they increased the population of small flies so that it was double the size of the population of large flies, the salamander would increase its number of strikes at small flies. Because the abundance of large flies did not change, foraging theory predicts that the salamander should not have shifted its foraging strategy to include more frequent attacks upon the smaller prey. However, Jaeger and Barnard explain, given the high density of small flies swarming about, the chance that the salamander misrepresents a small fly as a large one increases: "salamanders often made 'mistakes' in striking at small flies at high densities" (1981, p. 646). Supporting the claim that the salamander mistook a small fly for a large one is the fact that salamanders do not always hit a target with their first attack; but they would persist in their attacks against a large fly much more vigorously than they would for a small fly. Jaeger and Barnard report that salamanders that went after a small fly when large ones were readily available "then corrected those mistakes by aborting their attacks" (1981, p. 646).

The conclusion I draw from this brief discussion of foraging theory is that the best reason to take crabs to represent the slick hard objects that lie on the ocean bed as food is that they *treat* them as food. Likewise, the salamander represents those things buzzing around its head as food because it *treats* them as food. Foraging theory tells us exactly how predators like the crab or the salamander (or, I would speculate, the frog) interact with the food in their respective environments, and so the objects in their respective environments to which the generalizations of foraging theory relate them are, for that animal, food. What better reason to say that the salamander represents black dots as food than the fact that, by all behavioral indications, the salamander treats the black dots as if they were food?

On the other hand, what should we expect of a salamander that represents the flying black dots in its environment as flying black dots? Given that black dots are not any sort of currency that wants optimizing, that there is no selection pressure for black dots (if there were, we could presumably build an optimization model predicting the behavior of organisms that seek to maximize their intake of black dots—a model that would be as general across species as the foraging theory model we have discussed), it stands to reason that the function of the salamander's fly detector is not to represent black dots. On the contrary, the function of the salamander's fly detector is the same as the function of the crab's mussel detector, which is the same as the function of the starling's worm detector and the crow's whelk detector, and, finally, the frog's fly detector: to represent food.

In closing this section, we should note that generalizations like those we find in foraging theory depend upon generalizations of the first sort I discussed, i.e., those concerning the notion of adaptation. Foraging theory assumes that organisms have evolved feeding strategies that are well adapted to the selection problems in their environments. We cannot begin to understand the feeding strategies of a crab or a salamander until we decide that they are, first, *solutions*, and second, that they are solutions *to the need for food.* In focusing upon proximate descriptions of representational content, Fodor misjudges the generality of Darwin's conception of selection problems and adaptations and so fails to see that the context: *was selected for representing things as F* is opaque. We can only substitute for F predicates that make the representation of something as *F* adaptive. Hence, we can only substitute predicates that make plain the sense in which the representation of F is a token of the abstract (in the sense that it generalizes over species kinds) type *solution to the need for food.* Representing things as food fits the bill. Representing things as black dots does not. It is not surprising, then, that the generalizations of foraging theory apply to a variety of organisms that search for food, but not to organisms that search for black dots.

6. Conclusion

I have argued first that there is an alternative approach to representation that does not seem susceptible to the disjunction problem. As I understand it, the disjunction problem is especially virulent in those accounts of representation that make central causal relations between representing states and represented conditions. As I suggested, this in itself may be sufficient reason to favor views of representation that do not emphasize the role of causation in content ascription—approaches like the functional role account Hatfield and others have pushed.

Second, I argued that the context *was selected for representing things as F* is not transparent to the substitution of predicates reliably coextensive with *F*. Ironically, my argument rests on appeals to the practices in those special sciences that make use of such predicates—a strategy for which Fodor ought to have a lot of sympathy. Traits that are selected for what they represent are selected because they better adapt their possessor to its environment. But adaptation is adaptation to something, and our taxonomy of adaptations will thus be a reflection of the taxonomy of problems to which they are solutions. This is why evolutionary theory cannot be transparent: only under particular descriptions will adaptations appear as solutions to the pressures an uncaring environment imposes upon its inhabitants.

My project in this paper has been quite limited. I have not made claims about what contributions evolutionary theory might make to a study of higher level cognitive capacities, especially those involving conceptual thought. How far up the ladder of cognitive abilities evolutionary considerations will lend an explanatory role is an open question. I suspect they will not make it to the top of the ladder, where beliefs and desires are about the sorts of things that cannot possibly be viewed as solutions to selection pressures. On the other hand, Darwin probably has a lot to tell us about the representational content of the states in our own perceptual systems. Certainly, if my arguments have been on track, appeals to evolutionary theory ground ascriptions of function, and hence content, to cognitive mechanisms as primitive as the fly detector in the frog.

Notes

[1]In writing this paper I have benefitted from many useful discussions with Gary Hatfield. Thanks also go to Gary Ebbs.

[2]Here and throughout I use the label 'fly detector' neutrally to name the mechanism in the frog that fires at moving black specks—discussion of how to characterize the function of this mechanism comes in section five.

[3]Haugeland (1978) suggests such an account; but for mature statements other than those I cite in the text see Dennett (1987) and especially Kosslyn and Hatfield (1984) and Matthen (1988).

[4]For more on this taxonomy of selection pressures see Tinbergen (1951, p. 157), Williams (1966, p. 263), and Rozin and Schull (1988, pp. 522-523).

[5]As the line I am taking would suggest, however, I am comfortable with a realistic attitude toward selection problems. If we let our sciences settle our ontology, as Fodor does, and we agree that biological sciences cannot do without posits of selection pressures, we have good reason to treat these pressures realistically.

[6]For descriptions of optimization theory see Maynard Smith (1978) and Krebs and Davies (1987, chapter 3). For descriptions of foraging theory in particular see Pyke, *et al.* (1977); Stephens and Krebs (1986). Stephens and Krebs (1986, chapter 10) contains a number of convincing responses to Gould and Lewontin's (1978) criticisms of optimization theory.

[7]Maynard Smith (1978) construes optimization theory as a means by which to test which selection pressures have acted upon an organism. This view of optimization theory is consistent with the "currency" approach I discuss above when we assume that there is selection for the maximization of some kind of currency. Selection pressures, that is, can be identified with needs for particular kinds of currency. Animals that do a better job at maximizing this currency within certain constraints are fitter.

[8]The foraging strategy I mention will be optimal for a given forager when: (1) the predator encounters individual prey items; (2) the prey are handled one at a time dur-

ing which searching for additional prey is impossible; (3) recognition of prey type takes no time; (4) the prey cannot harm the predator. See Charnov (1976) for a more encompassing list. Also see Stephens and Krebs (1986) for modifications to the basic prey model that will accommodate changes in these basic assumptions. The claim foraging theorists make is that any animal to which the above conditions apply will respect the foraging strategy I discuss in the text.

References

Charnov, E. (1976), "Optimal Foraging: Attack Strategy of a Mantid", *The American Naturalist* 110: 141-151.

Dennett, D. (1987), "Styles of Mental Representation", in *The Intentional Stance,* D.Dennett (ed.). Cambridge: MIT Press, pp. 213-225.

Dretske, F. (1988), *Explaining Behavior: Reasons in a World of Causes.* Cambridge: MIT Press.

Elner, R.W. and Hughes, R.N. (1978), "Energy Maximization in the Diet of the Shore Crab, *Carcinus maenas*", *Journal of Animal Ecology* 47: 103-116.

Fodor, J. (1974), "Special Sciences", *Synthese* 28: 77-115.

_____. (1981), *Representations: Philosophical Essays on the Foundations of Cognitive Science.* Cambridge: MIT Press.

_____. (1987), *Psychosemantics: The Problem of Meaning in the Philosophy of Mind.* Cambridge: MIT Press.

_____. (1990), "A Theory of Content, I: The Problem", in *A Theory of Content and Other Essays,* J. Fodor (ed.). Cambridge: MIT Press, pp. 51-87.

Gould, S.J. and Lewontin R. (1978), "The Spandrels of San Marco and the Panglossian Paradigm: A Critique of the Adaptationist Programme", *Proceedings of the Royal Society* B205: 581-598.

Hatfield, G. (1988), "Representation and Content in Some (Actual) Theories of Perception", *Studies in History and Philosophy of Science* 19: 175-214.

______. (1991), "Representation in Perception and Cognition: Connectionist Affordances", in *Philosophy and Connectionist Theory,* W. Ramsey, S. Stich and D. Rummelhart (eds.). Hillsdale: Lawrence Erlbaum Associates, pp. 163-195.

Haugeland, J. (1978), "The Nature and Plausibility of Cognitivism", *The Behavioral and Brain Sciences* 1: 215-226.

Jaeger, R. and Barnard, D. (1981), "Foraging Tactics of a Terrestrial Salamander: Choice of Diet in Structurally Simple Environments", *The American Naturalist* 117: 639-664.

Kosslyn, S. M. and Hatfield, G. (1984), "Representation Without Symbol Systems", *Social Research* 51: 1019-1045.

Krebs, J.R. and Davies, N.B. (1987), *An Introduction to Behavioural Ecology*, 2nd ed. Oxford: Blackwell Scientific Publications.

Lewontin, R. (1978), "Adaptation", *Scientific American* 293: 213-230.

Marr, D. (1982), *Vision*. San Francisco: W.H. Freeman and Co.

Matthen, M. (1988), "Biological Functions and Perceptual Content", *The Journal of Philosophy* 85: 5-27.

Maynard Smith, J. (1978), "Optimization Theory in Evolution", *Annual Review of Ecology and Systematics* 9: 31-56.

Pittendrigh, C. (1958), "Adaptation, Natural Selection, and Behavior", in *Behavior and Evolution*, A. Roe and G. Simpson (eds.). New Haven: Yale University Press, pp. 390-416.

Pyke, G., Pulliam, H. and Charnov, E. (1977), "Optimal Foraging: A Selective Review of Theory and Tests", *The Quarterly Review of Biology* 52: 137-154.

Rozin, P. and Schull, J. (1988), "The Adaptive-Evolutionary Point of View in Experimental Psychology", in *Stevens' Handbook of Experimental Psychology*, 2nd ed., Volume 1, R.C. Atkinson, R.J. Herrnstein, G. Lindzey and R.D. Luce (eds.). Toronto: John Wiley and Sons, Inc., pp. 503-546.

Stephens, D.W. and Krebs, J.R. (1986), *Foraging Theory*. Princeton: Princeton University Press.

Tinbergen, N. (1951), *The Study of Instinct*. New York: Oxford University Press.

Werner, E.E. and Hall, D.J. (1974), "Optimal Foraging and Size Selection of Prey by the Bluegill Sunfish (*Lepomis mochrochirus*)", *Ecology* 55: 1042-1052.

Williams, G. (1966), *Adaptation and Natural Selection*. Princeton: Princeton University Press.

Thought and Syntax[1]

William Seager

University of Toronto

Psychological Externalism (PE) is the view that the contents of intentional psychological states are determined by factors external to the subject of those states. For example, PE holds that the content of a belief will be established at least in part by features of the environment, broadly construed, around and more or less detached from the believer. The range of such external factors is quite large, and various forms of PE emerge depending on which sorts of factors are endorsed. Generally speaking, these factors are those that establish the reference of the terms used to specify the contents of psychological states. To convey the flavour of this doctrine, let me briefly review some of the candidate external factors.

1. Demonstrative reference

One's beliefs can be about something picked out demonstratively independent of one's beliefs about the object. Even if one would deny that one's beliefs are about a certain object when described in some particular way they nonetheless may well be about that object. For example, at a cocktail party one might use the phrase 'the man with the martini' to refer to someone across the room via some demonstrative mechanism—just glancing at him might suffice (the example is from the classic Donnellan (1966)). It could well be that the man in question has no martini but only water in his martini glass, and even that "he" is no man. The creature that one's beliefs are about is not determined by the descriptive content of one's beliefs. However, with its retention of the internally determined descriptive content, this stands as a rather weak form of externalism.

2. Causal determination of reference of natural kind terms

Putnam (1975) tells us that natural kind terms, like 'water' or 'gold', get their reference via a causal chain that starts with an archetypal act of dubbing in which at least the implicit intention of the dubbers is to link a term to the essence of that kind, an instance of which stands before them awaiting naming. The use of the term through successive generations causally links the archetypal dubbers to current language users. Although the exact nature of this causal relation is hard to spell out, the theory has some plausibility. It clearly leads to a kind of PE. Since belief contents are specified, in part, via natural kind terms, as in the belief that water is wet, the stuff that this belief is about is externally determined. On Putnam's infamous Twin Earth (a physical duplicate of our

PSA 1992, Volume 1, pp. 481-491

earth save that there the substance that is phenomenologically identical to water and plays the same role as water is made of XYZ instead of H_2O) my physical duplicate has no beliefs about water (i.e., H_2O) but only about XYZ. (Actually, he is my physical duplicate only in relevant respects. We are not true duplicates since I am mostly water and he is mostly XYZ, but presumably this difference is psychologically irrelevant.)

3. Determination of reference of terms via community standards of usage

Tyler Burge (1979) asks us to imagine a counterfactual situation in which the word 'arthritis' applies to 'various rheumatoid ailments' not restricted solely to inflammation of the joints. In that counterfactual community one cannot (or at least not very easily) have beliefs about arthritis. For example, whereas one of us could mistakenly believe that arthritis can occur in the thigh, a physical duplicate from the counterfactual community would not have this belief (the counterfactual belief, with its attendant counterfactual content, would indeed be true). Thus, again, the content of belief is not determined by the internal state of the subject but rather is, at least in part, determined by the standards of linguistic usage of the social community which provide the resources for the ascription of intentional psychological states.

4. Determination of content by rationality constraints

According to Donald Davidson (many places, but see e.g., (1970),(1987)) the intentional states properly ascribable to a subject answer to a complex set of constraints on interpretation which demand maximization of rationality, truth of belief and "goodness" of desire. Although I do not think Davidson has ever presented such a crude argument as the following, a form of PE emerges from this thesis if we suppose that two temporarily physically indistinguishable persons could be immersed in distinct matrices of events each of which favours a different interpretation of its subject's intentional life[2]. It does not seem hard to imagine such cases. The dispositional resources of the brain are great, and given suitably distinct inputs the states of two initially identical brains will diverge radically under their differential pressures. The behaviour which these brains generate will also diverge and with it the most reasonable psychological interpretation of this behaviour. Such a conclusion is reinforced when we note that interpretation is retroactive in the sense that the overall rationality of the subject may be best preserved by modification of earlier assignments of intentional states, thus our two physically identical subjects will properly be said to have different beliefs, desires, etc. I believe that the possibility of such cases is, and should be, one ground for Davidson's so-called anomalous monism.

The foregoing are some of the more prominent candidates for external factors which individuate, at least in part, intentional psychological states independently of the subject's intrinsic state and, especially, the subject's physical state (for a somewhat different, as well as more detailed taxonomy of externalism, see McGinn (1989), especially the initial part of chapter one). They all agree in their denial of what may be called the *local physical supervenience* of intentional psychological states. This is, roughly speaking, the doctrine that any two physically identical persons must also be identical with respect of their intentional psychological states.

Now, one might wonder whether PE was true; many philosophers think that it has great plausibility, although there is room for disagreement about the credibility of any of the particular routes to PE sketched above (perhaps the Davidsonian route sketched in 4 is the most radical, and correspondingly least plausible).

One might also wonder whether the truth of PE would have any consequences for the science of psychology. Scientific psychology is presumably in the business of explaining what intentional mental states *are* and how they causally interact with each other, other psychological states, and the world to produce perception and action. Yet, if PE is true, there is no scientific psychology *of the subject*, since the subject of intentional psychological states is not the actual and complete locus of these states. The intentional states sprawl disconcertingly about the world in various undisciplined ways, essentially involving apparently quite non-psychological aspects of the world. It is not a psychological fact, nor one that psychologists are competent to judge about or, one would think, should even be interested in, whether the oceans are full of H_2O or XYZ. Could it be that even though PE is true it is irrelevant to scientific psychology? Jerry Fodor is one who thinks so, though, while he concedes the plausibility of PE, he might still prefer to rephrase the last remark as even if PE is true, it is irrelevant to psychology. He has an argument for this hopeful position as well, which I want to examine here (the argument appears in chapter two of Fodor (1987)).

Here's the argument. PE is irrelevant to psychology. Why? Because it individuates mental states in non-causal or causally irrelevant ways. That is, PE makes distinctions among mental states which mark no causal difference. The slogan or, speaking with more dignity, the principle that no distinction without a causal difference is one which any science interested in causal explanation ought to abide by. Thus Fodor's reproving example of the difference between H and T particles, which physics must ignore just because it is a clear case of a distinction with no corresponding causal difference (1987, pp. 33ff.). The example is worth spelling out, since it is supposed to be an absurd parody of PE, and we are urged to transfer its patent absurdity to the apparently less patent case of PE. Fodor has a coin and a definition: for any particle, x, x is an H-particle if Fodor's coin is heads up and x is a T-particle if the coin is tails up. Although there is nothing wrong with this definition (it is coherent) and although it certainly marks out a distinction which the universe has no difficulty exemplifying, the distinction it marks brings with it no causal difference whatsoever. Obviously, the causal powers of a given particle are not affected in the slightest by its being an H rather than a T particle. Similarly, although PE may well be true, and mark a real distinction in some sense, it too is a distinction that measures no difference in causal power. Whether my mental state is one about water (H_2O) or t-water (XYZ) has no bearing on its causal powers. Since psychology is a *science* and is thus interested in causal explanation and/or the delineation of the causal powers of its proper objects it has correspondingly no interest in PE, no matter whether PE is true or not. Of course, since PE tells us what beliefs, and intentional psychological states in general, *are*, this indifference generates a problem about what the proper objects of an authentically scientific psychology could be. Here enter Fodor's own views about narrow content and the syntactic theory of mind about which much more later. For now, remain focused on the nature of causal powers.

What is it to say that two events or things are "causally equivalent" or have the *same* causal powers? Fodor's definition is this:

> (CE) X and Y are causally equivalent if and only if any effect which X can bring about in a context, c, Y can bring about in c; and vice versa.

To speak negatively and equivalently, X and Y are not causally equivalent iff in some context there is an effect that X (or Y) can bring about which Y (or X) cannot. Example: The first and second printed copy of today's newspaper are causally equivalent (assuming that they are, for all intents and purposes, physically indistinguishable) since they could be replaced in any context with no ensuing difference in effect. Today's and yesterday's papers are not causally equivalent (at the very least the dates at the top of each page will be different and thus could lead to different effects).

Naturally, it is important to know how contexts are to be specified. For example, are we permitted to include in our description of a test context *epistemic* facts like 'it is known (by Jerry) whether a given copy of a newspaper was the first or second printed copy'? If so, then the two copies are obviously not causally equivalent in that context, for the one may cause the belief, in Jerry, that he is holding the first printing; the other that he is holding the second, and these are different beliefs, different effects. Further, these are quite literally and very robustly different effects—there is, presumably, a difference in Jerry's brain attendant upon acquiring one of these beliefs rather than the other.

If permissible, such contexts are obviously of some importance in the externalism debate. Suppose Jerry knows that Oscar is from Earth and Twin-oscar is from Twin Earth. Then even if Oscar and Twin-oscar are physically identical if the one utters 'I want water' Jerry ascribes a different desire than he would have to the other upon his making the same (sounding) utterance. But how could Jerry know the difference? Not by any intrinsic *mark* for that would destroy their hypothesized physical indistinguishability. Suppose we include in the context the demand that Jerry have tracked how Oscar or Twin-oscar came to appear before him. They must arrive via different "paths"—sufficiently different to ground Jerry's knowledge of their identities but still similar enough to ground the intrinsic physical indistinguishability of Oscar and Twin-oscar. Actually it's not hard to imagine contexts like these. We only need to find one such differentiating context to ground a difference in causal power according to (CE).

We might call such contexts externally constrained and it is manifest that locally physically identical items can produce different effects in them. In an externally constrained context one cannot replace Oscar by Twin-oscar without allowing for an external source of distinguishability between them. Presumably, Fodor would say that this procedure is illegitimate. But I get to define my contexts the way I want to and if I want to impose a condition of external distinguishability as part of the identity conditions of certain contexts I don't see what will stop me. It's not incoherent to define contexts with external constraints.

Of course, externally constrained contexts individuated in terms of epistemic facts that count as *the same* will not necessarily be physically identical, nor will different contexts necessarily be physically distinct; externally constrained contexts will not abide by any local physical supervenience principle. But so what? To demand that contexts supervene on local physical state comes close, I think, to begging the question here. And while this oddity no doubt rankles those who already favour Fodor's views on causal powers, this by itself does not show that the notion of externally constrained contexts is incoherent.

Still, Fodor would object (and I think it would be almost irresistibly tempting to agree with him) that from the point of view of one trying to do science such contexts are an abomination. This is because, as good physicalists, we are independently sure that, unlike externally constrained contexts, causal relations do supervene on the physical state of the world. To allow externally constrained contexts would be to allow that H particles are causally distinct from otherwise identical T particles. This is surely ridiculous or at least hopelessly counter-productive with respect to formulating a reasonably clear scientific picture of particle interactions. How can we formulate a principle of causal equivalence which rules out the ridiculous or counter-productive? Taking the bull by the horns, we can just say that where two items are physically indistinguishable intrinsically then only contexts that make them indistinguishable tout court are allowable. Differential histories and differential environments that serve to externally mark out the difference between otherwise physically identical subjects will be ruled out. (One may imagine that the contexts we are interested in are ones in which the two, perhaps intrinsically physically indistinguishable, subjects are switched instantaneously and invisibly, say by the hand of God.) In all such contexts, intrinsic physical duplicates will have the same causal powers. In such contexts, causal powers will supervene locally on physical state. To the extent that a science is concerned only with such contexts, that science will maintain local supervenience of causal powers. Perhaps science is concerned only with these. After all, externally constrained contexts are incredibly messy and perhaps only shift the site of interesting causal efficacy from the physical duplicates to the attendant patterns of distinguishability. It is clear that it is only because there is a physical difference surrounding, as it were, our duplicates that they can produce distinct effects. This is not scientifically mysterious, but it would be perverse to expect a science to include all possible cases of externally explicable differential effects as aspects of the causal powers of its target objects. (Ask yourself whether you think it is part, and a strange part, of the causal powers of alpha particles to produce surprise in Rutherford as opposed to no surprise in some modern student wearily repeating his scattering experiment.)

Let us accept this and see where it leads. I think it will lead to trouble for Fodor's own favoured view of the *science* of psychology: the so-called syntactic theory of mental representation and cognition. Roughly speaking, this is the theory that mental representations are realized *in* and borrow their causal powers *from* a purely physical model of a syntactic system which mirrors the syntactic properties of the original representations. For example, on this theory, the mental representation of P & Q is quite literally formed from sub-representations of P and Q joined by some neural simulacrum of the logical operation of conjunction (for Fodor's explication, see Fodor (1978)). The trouble I think the above characterization of causal powers will lead to is that *syntactic form* will turn out to be just as irrelevant to psychological theorizing as broad mental content. This is because syntactic form no less than semantic content is subject to a kind of externalism: syntactic form does not supervene on local physical state.

In a rather crude and preliminary form, this is not hard to see. Begin with a text, like this page. Why does it exemplify syntactic properties, such as "having occurrences of *words* (or *sentences*, etc.) on it"? Would any physical duplicate of this page exhibit the same syntactic structures? Obviously not. Suppose we duplicate this page in a linguistic context in which 'page' is a radically different part of speech. The language at issue looks like English (there may be lots of differences between it and English but they don't turn up on this page) but is in fact radically different (e.g., '.' is a word, ' ' (the blank space) is a character and 'x' represents a word divider). Although this kind of syntactic externalism seems pretty clear cut, a closer look is both interesting and informative.

As an sobering illustration of the difficulty of isolating syntactic structure on the basis of local physical structure, consider the problems we had with Egyptian hieroglyphics (for an exhaustive presentation of hieroglyphic grammar as well as a tutorial, see Gardiner (1950)). Only the completely fortuitous discovery of the famous Rosetta stone in 1799 by Napoleon's army led to an understanding of their meaning and it did so by literally providing a hieroglyphic text with an accompanying Greek translation. This case provides several points of interest. First, even with the parallel texts it took some 20 years before a reasonable key to the hieroglyphics was found by Jean François Champollion (aided, incidentally, by the work of the physicist Thomas Young) and it was many years after that before scholars could confidently translate arbitrary hieroglyphic texts. Even with a parallel text, it was not a trivial task to decode hieroglyphic writing. And while scholarly interest naturally focused on the *meaning* of hieroglyphic texts, one of the stumbling blocks remained the ignorance of their syntax, which is simply not apparent in an isolated sample. Medieval speculations, usually of the most fantastical sort, ignored the crucial distinction between the ideographic component of hieroglyphics and that of the phonetic component (which indeed are interspersed in a complex way).

Second, the discovery of the syntactic structure of hieroglyphic writing was dependent upon prior knowledge of the meaning of a sample text. If the parallel text of the Rosetta Stone had been from another unknown language rather than Greek it would not have been possible to decipher the hieroglyphics. Syntax is never apparent from the physical structure of a text. I think we often overlook this point because we think of the syntax of the natural languages we know and are masters of or the syntax of the artificial languages of logic or computer programming which we have constructed so as to make syntactic structure unambiguous (at least from the point of view of those who already know the function of such quasi-linguistic systems). Such basic syntactic categories as "word" and "sentence" seem to be obviously present in the physical form of our words and sentences. Words are separated by blank spaces, sentences delimited by a period and a capital letter. But of course we know that such niceties are not necessary nor have they always been present in every language. The ancient Greeks did not separate words by spaces (nor did the Egyptians) for example and wrote only in capital letters. In hieroglyphics, symbols sometimes serve to mark concepts directly and at other times signify phonetic values. Whether a text is to be read from right to left or the reverse is signified by the direction that asymmetrical figures face (e.g., a picture of a man in profile has a "natural" direction to it—that of the direction of gaze; the same goes for a snake symbol). But of course there is nothing necessary about this interpretation of direction. This is nicely illustrated in hieroglyphics. If the symbols are "looking" to the right then the text is read from right to left, that is, in the way opposite to the direction of gaze. In Hieroglyphics, the identity of a word—surely one of the rock bottom syntactic features of a language—can be expressed in many different ways: purely ideographically, purely phonetically, by a partial phonetic representation, or by a mixture of phonetic and ideographic clues.

It seems that knowledge of syntactic structure is formed by noting the joint consequences of a hypothesis about the meaning of the text coupled to a hypothesis about the syntactic structure. This hypothesis pair will face the data of further texts together and, if success is near, will be mutually corrective through a kind of feedback loop. It is easy to see from this that syntax must be externally determined in the sense that two physically indistinguishable texts could exemplify distinct syntactic structures. This is because a particular text will not carry enough information to rule out all but one of the possible alternative hypotheses about its syntactic form. This information will be found external to the text both in further text samples and knowledge, or hypotheses, about meanings of the hypothesized syntactic elements of the target text. It is important to see that I am not simply relying on the general doctrine of underdetermination of theory by evidence here (though indeed this might be sufficient to make the case). Underdetermination is a genuine practical issue—not just a somewhat artificial philosophical worry—where evidence is scanty and any particular text is scanty evidence. Taking external evidence into account, the syntactic form of the target may be known beyond reasonable doubt, but since the context of external evidence can be varied without any difference to the target text, the syntactic form of this target can vary across these contexts with no physical variation in the target.

I think that this quite obvious point, coupled with Fodor's views about causal roles and the scope of science, will have serious consequences for Fodor's theory of the mind. Fodor is one who espouses the Representational Theory of the Mind (RTM). In Fodor's version of RTM, the brain is seen as a syntactic engine. In some sense, words and sentences (of a language of thought) are really, literally, written in the brain. This explains, supposedly, both the semantic evaluability of states of mind and their causal powers. In virtue of being syntactic they can sustain semantic interpretation; and since syntactic items must be physically realized, they can enter into causal relationships with other physical items, like behaviour.

But we have seen that whether a particular physical item is a particular syntactic item is not supervenient upon local physical state. To be a syntactic element of the language of thought the item needs the requisite history of causal interaction with other syntactic items. Two distinct histories of the right sort might produce two sample texts that were locally physically indistinguishable. These items would nonetheless differ in their syntactic form.

It might be objected that there is some kind of confusion between ontology and epistemology in the above. Our *knowledge* of syntactic form depends on external factors but this does not entail that syntactic form *itself* is determined by external factors. Such an objection appears to have some force for examples like that of hieroglyphics. It is possible to suppose that Champollion's deciphering of the hieroglyphics is actually radically mistaken with respect to both the meaning and the syntax of hieroglyphic writing. Although the data we possess comports very well with Champollion's hypotheses it could nonetheless be wrong. And, it could be urged, the truth of the matter ultimately depends upon the intentions about linguistic usage of the ancient Egyptians who originally inscribed the hieroglyphics. This is even more plausible in cases where we have less information about the source of the text in question. Suppose we come across a cave, a wall of which is covered by small scratches of various shapes and sizes. This might be syntax and it might not be. There is no local way to tell—no intrinsic determinant of whether these scratches are part of a syntactic system or not, any more than there is some intrinsic way to tell what they might mean even granting that they are part of a syntactic system with meaning. But we might argue that they nevertheless possess a determinate syntactic structure simply in virtue of being produced with the intention of conforming to some syntactic system.

So it might be that we could reduce, so to speak, the true syntactic form of a text to the beliefs, desires and intentions of those who produced the text. After all, a "text" that appears magically out of nothing is both meaningless and devoid of syntax (it contains no sentences, words, nouns or verbs, nor even letters of the alphabet). But this point is not helpful. For since Fodor's approach assumes that the brain itself has a syntax which is to account for intentional psychological states there is no room to suppose that this syntax is dependent upon earlier beliefs, desires and intentions. (Or do we have here a most peculiar argument for the existence of God, whose intentions are necessary to make our very thoughts determinate in their narrow content.) To have a *desire that p* just *is* to have a syntactic structure within one's brain that, within a determinate context, means p.

Well, Fodor tells us that psychology is interested in psychological explanation as causal explanation. Broad content is irrelevant since it does not map onto causal powers. This is because causal powers supervene locally whereas broad content doesn't. So broad content talk allows for distinctions without any attendant causal differences. We can now clearly see that the same ought to go for syntactic talk. Syntax also does not supervene locally. So it shouldn't map onto causal powers either. To show this we need only exhibit a syntactic distinction with no attendant causal difference. But this is easy since physically identical inscriptions can be syntactically distinguished in virtue of external factors.

But let's consider an example in the vein of a language of thought theory of mentality. I say it is possible to imagine a set of inscriptions that are systematically ambiguous as to their syntactic form. Here's a sample such inscription: 'Jerry loves Jenny'. You don't know whether to read this right-to-left or left-to-right unless you already have assumed that it has normal English syntax. Now consider the language of thought. Do the brain states of someone at an instant, or for a suitable short space of time, which make up the inscriptions of this language necessarily admit of only one syntactic interpretation? Given the above, there is no more reason to think so than to think that the physical form of a discrete message eliminates all but one syntactic hypothesis. Could a particular brain be syntactically ambiguous and have the ambiguity settled by evidence taken from the brains of others? I don't see why not.

It might now be objected that the case of textual syntax is misleading. In the case of *texts* the fact that syntactic structure does not supervene on local physical structure is clear and uncontroversial. But texts are, in an important sense, static. Perhaps the fact that the syntax of the brain is in some sense active or dynamic will somehow restore local supervenience. The argument for this would go something like this. In contrast to static inscriptions those *causal processes* of the brain which involve and express syntactic relations will serve to individuate the syntactic elements of the internal code. For example, whereas the form: P v Q & R, is syntactically ambiguous as it stands but might be determinate given an external convention about the precedence of connectives, in an active context the form will be disambiguated by the system itself transforming P v Q & R into some other form, say, P v Q. If the system makes this transformation then the syntactic form of the original was (P v Q) & R. Two brains embodying distinct syntaxes might be inscriptionally identical with respect to this form but would nonetheless differ in dispositional properties about its "use". These different dispositions would have to be grounded in a physical difference, and thus the brains could not be completely identical. Speaking abstractly, it is such dynamic differences that would reveal the syntax of the brain.

This argument can be recast as one about ordinary inscriptions. Although the syntax of a given inscription does not locally supervene on the physical structure of the inscription, perhaps the syntax would become apparent if we could watch it being produced. Thus the production system plus produced inscription would determine syntax on the basis of local physical structure. But such an example makes it clear that syntax would *not* be apparent from observation of production alone. It is surely a syntactic feature of an inscription that it be read from right-to-left rather than left-to-right, but this is not a feature that need be duplicated in the production of a text (although of course it usually is in most of the syntactic forms we are used to—but not reverse Polish Notation in logic for example).

The argument must appeal to the dispositions of the system to transform inscriptions in certain ways and to refrain from other transformations. The mere patterns of production of inscriptions will not settle questions about syntactic form in every case. In principle at least, there is no problem about this however, since presumably every disposition is grounded in physical states and different dispositions will supervene on different physical states. The suggestion is then, that syntax supervenes on inscriptional form plus dispositions to transform these forms in particular ways. Both of these factors supervene locally so brain syntax is locally supervenient on the physical.

Still, we must recall what is meant by calling certain states of the brain, brain *syntax*. It is not just that there are states of the brain with causal powers—all brain states possess all sorts of these and most brain states are not syntactic states. To call them syntactic is to demand both that they be causally efficacious in virtue of their form but also to demand that they be systematically semantically evaluable.

It is true that sometimes when philosophers speak of the "syntax" of the brain they have this very much weaker notion in mind. For example, consider this passage from Dennett: "But the brain ... is just a syntactic engine; all it can do is discriminate its inputs by their structural, temporal, and physical features and let its entirely mechanical activities be governed by these "syntactic" features of its inputs" (1987, p. 61). Dennett's use of scare-quotes reveals his awareness that it strains the sense of the word "syntax" to call any set of causally efficacious states of the brain syntactic simply in virtue of their efficacy. Fodor's view is distinguished by the stronger claim and a more standard use of "syntactic".

The semantic evaluation of putative syntax may vary from external context to context (as in the twin earth cases, etc.) but the syntactic identity of the brain states at issue is dependent upon semantic evaluability as well as causal efficacy. States will be syntactic states if they both interact and *would* interact so as to preserve semantic interpretation (within a given context). Now, as we have seen in the case of simple texts, the possibility of non-supervenience of syntax on inscriptional form is dependent upon multiple semantic interpretations of the inscription as it is successively hypothesized to contain distinct syntactic structures (it is a different but obviously still important question whether distinct semantic interpretations of a *given* syntactic form are possible, as the standard forms of PE would claim). If there is only one sensible semantic interpretation for all possible syntactic renditions then the syntactic rendition which underlies that semantic interpretation is correct. Samples of texts will in fact never possess this feature. This is clear if we stop to think that syntactic form is not a *given* but rather depends upon semantic interpretability for its discovery.

Consider, as a rather amusing example, the so-called Voynich manuscript. Brought to America in 1912 by the rare book dealer Wilford Voynich, this is a known to be a centuries old text, although its exact age is disputed (it has been attributed to Roger Bacon but many think it postdates Columbus' trips to the New World) which now resides in Yale University's Beinecke Rare Book and Manuscript Library. It is written in an unknown language that nonetheless appears to present a relatively unproblematic syntax. That is, the "characters" are quite clearly separated and "words" are broken by blanks, etc. Still, no one has ever deciphered the manuscript and its subject matter is completely mysterious. It is possible that it is utter nonsense and thus embodies no syntax whatsoever. (In fact, this is unlikely seeing as it presents reasonable statistics of character occurrence.) In the face of indecipherability, alternative syntactic hypotheses might be attractive. In 1921, Professor William Romaine Newbold, in an address to the American Philosophical Association (which I very much doubt would be even marginally acceptable today) put forth the hypothesis that the actual syntax of the Voynich language was in the almost microscopic strokes that made up the apparent characters of the text. These micro-strokes formed a Latin shorthand which Roger Bacon (Newbold was a champion of the Baconian hypothesis) used to encode some truly remarkable, if not outright fantastic, reports of very early scientific work that, if they could be verified, would completely overturn our understanding of the history of science. For example, Bacon is claimed to have discovered both the reflecting telescope and the compound microscope by which he observed the spiral structure of the Andromeda nebula and spermatozoa respectively (for details about the Voynich manuscript see Newbold (1928), Brumbaugh (1978) and Poundstone (1988)). Newbold's deciphering hypothesis won the approbation of such distinguished philosophers as Étienne Gilson (who never claimed to *understand* Newbold's baroque system of decoding practices). But in truth, although Newbold was a respected scholar, this was a crackpot hypothesis. Still, if it had actually made the text sensibly interpretable it would have commanded respect and could have revealed the true syntax. True syntax, as opposed to mere physical structure, is a matter of a formal interpretation of physical structure that ensures semantic interpretability.

Will the brain's active structure present a set of physical tokens that uniquely provide for semantic interpretability? The question reduces to the question of whether the dispositions of the brain which govern the transformations of putative syntactic elements suffice to limit semantic interpretability to a single winner from all possible contenders. These dispositions are what distinguishes the active syntax of brains from the static syntax of inscriptions. I can't see any reason to believe that a particular brain will uniquely specify a definite syntax. Rather, the syntax manifested by a brain, if any, will depend upon the semantical hypotheses which will be made by those investigating neural structures. These investigators will bring standards of evidence and indeed evidence (perhaps from other brains) that might decisively settle the question of syntax. Such external factors being crucial to the successful prosecution of their scientific task, no brain by itself will suffice to uniquely determine syntactic structure.

Nor can we appeal here to some deep realist metaphysics, in which the world itself embodies a correct syntax within the brains of animals. Realism about syntax can rest on realism about intentional attitudes towards syntactic structures (as in the earlier example of the cave markings) but the syntax at issue here is that which ostensibly underlies these intentional attitudes themselves. Such "deep syntax" cannot be divorced from the strategies of interpretation which, bringing to bear all evidence that might be relevant whether external to the brain or not, might serve to break the brain's code.

Suppose that on a planet somewhat like Twin Earth (PSSTTW) the psychologists believe in the RTM and the language of thought. Suppose they have worked out the brain writing system and can read brains like a book. They do this by taking certain brain states to form a syntactic system to which they extend a semantic interpretation. Naturally, they believe that the causal powers of the brain reside in its physical structure but since syntax is embedded in this structure they feel happy to say that causal powers are also mapped out by this syntax. Now consider someone from a future Earth, where the language of thought has also been decoded. But due to various historical reasons and accidents the syntax of the brain is taken to be quite different than that worked out on PSSTTW. The reason is straightforward. There is vastly different evidence on the two planets, each favouring the appropriate language of thought syntactic hypothesis. Nonetheless, Oscar on future Earth and Twin-oscar on PSSTTW are physically indistinguishable. If Oscar raises his arm then so does Twin-oscar and both events are caused by physically indistinguishable processes in their respective brains. But those processes need not be *parsed* in the same way on PSSTTW and Future Earth. How they will be parsed is a matter of interpretation of syntax which answers to the strategic considerations sketched above. Is it thus *indeterminate* which syntax they embody? No, it is determined by the syntax that the rest of the inhabitants of the respective planets embody and for which there is very good evidence.

The lesson we were supposed to learn from externalism is that states that do not map out causal powers are irrelevant to psychology. So broad content intentional states which are individuated non-locally and defy local supervenience are irrelevant to psychology. Now this argument can be paralleled in the case of the syntax of the language of thought. Syntactic states do not supervene locally and thus do not map out causal powers. It is possible to imagine cases where the same causal powers (in the sense intended by Fodor) exist "beneath" distinct syntactic states. Thus these syntactic states are irrelevant to a psychology which is interested in the causal explanation of behaviour.

I take it that Fodor would not like this conclusion. Since I think it would be hard to argue that syntax is locally supervenient, how can the conclusion be avoided? By dropping the overly stringent demands on the individuation of causal powers in favour of one sensitive to the explanatory objectives of the science at issue? But then broad content might end up relevant to psychology again after all. Perhaps we should rejoice in this conclusion.

Notes

[1]I would like to thank several people for their help with this paper: Deborah Brown, Tori McGeer, Bob Murray, Bill Munroe, Janice Porteous, Ted Snider and the philosophers of Dalhousie University.

[2]While I do not think Davidson has ever put forth this argument for PE, he does explicitly accept a form of PE (see Davidson (1987)) which seems to be closely akin to that described in 2 above.

References

Brumbaugh, R.S. (1978), *The Most Mysterious Manuscript.* Champaign: Southern Illinois University Press.

Burge, T. (1979), "Individualism and the Mental", in *Midwest Studies in Philosophy*, 4: 73-121.

Davidson, D. (1970), "Mental Events", in *Davidson"s Essays on Actions and Events*, NY: Oxford University Press, 1980.

________. (1987), "Knowing One's Own Mind", in *Proceedings and Addresses of the APA* volume 60, 3: 441-58.

Donnellan, K. (1966), "Reference and Definite Descriptions", in *Philosophical Review* 75: 281-304.

Fodor, J. (1978), "Propositional Attitudes", in *The Monist*, volume 61, no. 4. Reprinted in *Fodor"s RePresentations*, Cambridge: MIT Press, 1983.

____. (1987), *Psychosemantics*. Cambridge: MIT Press.

Gardiner, A.H. (1950), *Egyptian Grammar*, Second Edition, Oxford: Oxford University Press.

McGinn, C. (1989), *Mental Content*. Oxford: Basil Blackwell.

Newbold, W.R. (1928), *The Cipher of Roger Bacon*, R. Kent (ed.). Pennsylvania: University of Pennsylvania Press.

Poundstone, W. (1988), *Labyrinths of Reason: Paradox, Puzzles and the Frailty of Knowledge*. New York: Doubleday.

Putnam, H. (1975), "The Meaning of 'Meaning'", in *Putnam"s Mind, Language and Reality, Philosophical Papers* volume 2. Cambridge: Cambridge University Press.

Color Perception and Neural Encoding: Does Metameric Matching Entail a Loss of Information?[1]

Gary Hatfield

University of Pennsylvania

It seems intuitively obvious that metameric matching of color samples entails a loss of information. Metamers are distributions of wavelength and intensity (or "spectral energy distributions") that perceivers cannot discriminate. Consider two color samples that are presented under ordinary white light and that appear to normal observers to be of the same color. It is well-established that such color samples can have quite different surface-reflective properties; e.g., two samples that appear to be the same shade of green may in fact reflect strikingly different patterns of wavelengths within the visible spectrum (see Figure 1). In this case, sameness of appearance under similar conditions of illumination does not entail sameness of surface reflectance. Spectrophotometrically diverse materials appear the same. It would seem then that information has been lost, that the visual system has failed in its task of chromatic discrimination.

This intuition implicitly relies on a conception of the function of color vision and on a related conception of how color samples should be individuated. It assumes that the function of color vision is to distinguish among spectral energy distributions, and that color samples should be individuated by their physical properties. I shall challenge these assumptions by articulating a different conception of the function of color vision, according to which color vision serves to partition visible objects into discrimination classes. From this perspective, objects are chromatically individuated by their membership in a particular equivalence class. Spectrophotometric diversity may in some cases (though not in all) be consistent with sameness of class membership. Metameric matching need not entail a loss of (pertinent) information.

My argument requires the articulation and adjudication of competing conceptions of the function of color vision. For my stalking horse I will examine the conception of color vision advanced by Barlow (1982c) and his followers (Buchsbaum & Gottschalk, 1983). I will contrast this conception with one derived from an approach to vision in the spirit of Marr's (1982), though it will turn out that it is not Marr's own analysis of color. My argument presents a function-based analysis of the content of color sensations (thereby extending Hatfield 1988, consistently with Matthen 1988); however, someone who accepted function-ascription in biology but was squeamish about content-ascription could reformulate the argument using only function-ascriptions.

PSA 1992, Volume 1, pp. 492-504

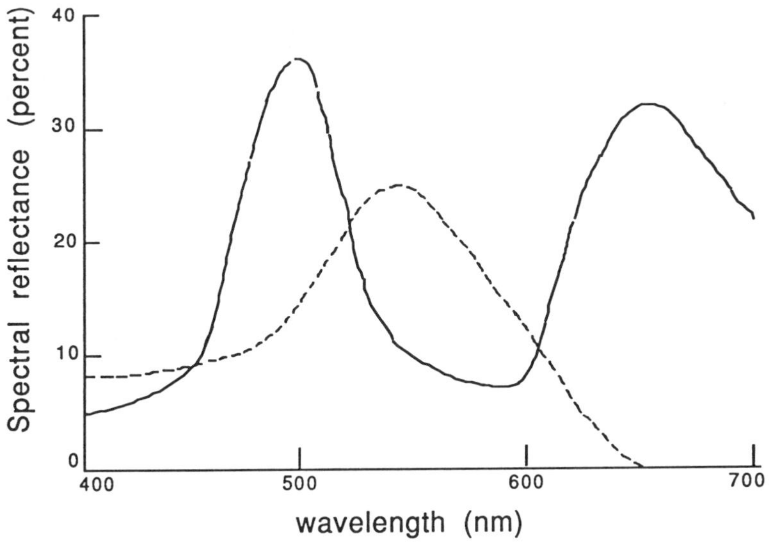

Figure 1. Two spectral reflectance distributions that produce matching greens in daylight for normal human observers. (After Hurvich 1981, 207.)

1. Task Analysis and Perceptual Content

In recent years there has been increasing attention to the role of task analysis in the investigation of psychological systems. A task analysis specifies what it is that a particular system does, where this "doing" cannot be captured through mere behavioral description; it is a specification of the function of a given system, or of its task within the economy of the organism. To take a non-psychological example, there are many things that the digestive system "does", including producing growling noises when inputs have been light. But its function, its contribution to the economy of the organism, is to break down nutrients for distribution in the body. In psychology, the task analysis approach seeks to determine the function of various psychological systems, or their contribution to the psychological or cognitive economy of the organism.

Teleology lingers in the background of task analyses (sometimes being brought into plain sight). Such teleology is underwritten by appeal to natural selection. In the popular Wrightian analysis of functions (and its descendants), a function is ascribed to a type of system through an etiological analysis: a system's function is whatever it does that explains (evolutionarily) its presence in organisms of a certain type (Wright 1973; Matthen 1988, Millikan 1984). To ascribe a function is to make a conjecture about the adaptive significance of a given structure, and hence about the characteristics of the structure that led to its fixation and subsequent maintenance in a (temporally persisting) species of organisms. Such conjectures are difficult to confirm or disconfirm, but they are not totally immune from empirical evidence (and science is in the business of venturing beyond the data). In any event, task analysis is central to physiology and psychology, and an appeal to adaptation and evolution currently offers the most promising means of legitimizing the latent teleology of such analyses.

When applied to representational systems such as the visual system, task analysis provides a means of ascribing representational content to states of the system.

Matthen makes this point in treating perceptual systems as systems that have the function of detecting the presence of certain environmental conditions; the "on" state of the detector mechanism is then ascribed representational content in accordance with the environmental condition it serves to detect. Thus, the "on" state of an edge detector has the content *edge* (Matthen 1988). More generally, various states of the visual system are assigned content in virtue of the environmental characteristics it is their function to represent: determinate shapes, sizes, motions, positions, and colors, to name a few (Hatfield 1988, 1991).

The notions of task analysis, etiological function ascription, and function-based content ascription have not received universal endorsement. But for present purposes I will take them as given. My aim is to show how competing task analyses have been given for the function of the visual system and in particular for the reception and post-retinal transmission of information about color. These competing analyses lead to different conceptions of the content of color perception, or of the function of color detecting mechanisms in the visual system.

2. Contrasting Task Analyses of the Visual System

Marr (1982) should be credited for drawing attention to the important role of task analysis in the investigation of perceptual systems. But he did not invent task analysis. Visual theorists have proposed (or presupposed) conceptions of the function of vision from earliest times (Aristotle 1984, 436b18- 437a9; Ptolemy 1989, bk. 2), and more recently Gibson (1966) made investigation of the natural functions of the senses central to his "perceptual systems" approach. Marr's own analysis, which assigns to human "early vision" the function of producing a representation of the spatial and chromatic properties of the distal scene, shares much with Gibson's (1950, 1966) analysis of the visual system. According to this conception, the perceptual apparatus is "distally focused", that is, it is tuned toward representing structures at a distance, and is not particularly tuned toward the representation of its own proximal states (e.g., the state of the retina).

Marr contrasts his approach with that of Barlow, and he criticizes the latter for merely describing the activity of single cells without addressing the question of the function of that activity (Marr 1982, 12-15, 19). In their work on space and color coding Barlow and his followers have emphasized the problem of the encoding and neural transmission of the physical characteristics of the retinal image, as opposed to the problem of representing distal scenes. This is not, as Marr charges, because Barlow failed to appreciate the importance of task analysis itself. Rather, it is because he (and his followers) gave a different task analysis than that favored by Marr. Examination of these contrasting task analyses will bring into relief the issues about metamers in color vision that I wish to address.

Far from failing to appreciate the notion of task analysis, Barlow explicitly proposes to analyze the functions of the sense organs. Thus, in an early paper he asserts that birds' wings are for flying, and asks, in effect, what sensory relays are for (1961b, 217). Indeed, his early work on fly detectors (Barlow 1953) led him to formulate the "password" hypothesis, according to which certain physical characteristics of stimuli act as "releasers" to initiate adaptively appropriate behavior, such as tongue-shooting (1961b, 219-220). But by and large, Barlow and his followers have emphasized the function of the senses as recorders of the physical properties of the image on the retina. Barlow described the function of the retina as follows: "The retina is a thin sheet of photoreceptors and nerve cells lining the back of the eye where the image is formed. It is obvious that its functional role is to encode the image falling on the retina as a pattern of nerve impulses in order to transmit the picture up the optic nerve to the brain"

(Barlow 1982b, 102). More recently, Sterling has written that "the retina's task is to convert th[e] optical image into a 'neural image' for transmission down the optic nerve to a multitude of centers for further analysis" (1990, 170). Although these authors are analyzing the function of the retina, not of the visual system as a whole, their conception of retinal functioning is part of a larger conception of how the visual system works. According to this conception, the retina records the physical image for efficient transmission to central visual centers (Barlow 1961a, Woodhouse and Barlow 1982), where subsequent processes "interpret" the neurally encoded image (Barlow 1982a, 31-32). Barlow in fact conceives of the senses as physical instruments designed to accurately encode the physical properties of the proximal stimulus; remarking on the familiar practice of labeling the external senses as "exteroceptors", he observes that "these so-called exteroceptors are really specialised interoceptors; they sense the outer world only by means of its physical and chemical influence on the special sense cells of the nose, the ears, or the eyes" (Barlow 1982a, 1). De facto, he has adopted what Gibson once labeled the conception of the senses as "channels of sensation" (1966, 3), by which he meant that one conceives the sense organs as recording their own state, rather than as components in a system that has the function of perceiving the distal environment. (Of course, all concerned agree that sensory transducers are affected by physical energy; the matter in question is the functional analysis of the detection and representational systems in which the transducers serve.)

Marr, by contrast, conceives vision—or at least early vision in primates—as having the function of generating a representation of the distal layout (1982, 41-42, 268-269). In connection with this conception of early vision, he considers the transduction process from the standpoint of the reception of stimulus characteristics that will allow recovery of distal properties. In considering the coding of spatial information on the retina, he focuses on the problem of detecting the aspects of the image that are informative of distal spatial structure. He thus looks for aspects of the image that "correspond to real physical changes on the viewed surface" (1982, 44). Hence, he emphasizes zero-crossings because of their correlation with physical edges in the world (pp. 49-50, 54). Similarly, in considering what encoding of the retinal image might best serve as input for stereopsis and detection of directional motion, he emphasizes that "by and large the primitives that the processes operate on should correspond to physical items that have identifiable physical properties and occupy a definite location on a surface in the world" (p. 105); again, Marr proposes that zero-crossings might be appropriate (p. 106). A Barlowist might object that detection of zero-crossings is not a good idea because it does not lead to an efficient encoding of the retinal image (Eckert, Derrico, and Buchsbaum, unpublished). But Marr would be nonplussed: he did not view the visual system as having the function of encoding the retinal image for transmission to higher centers. Rather, he saw the retina as detecting distally-relevant features of proximal stimulation. In Marr's analysis, the image is not conceived simply as a two-dimensional pattern that must be encoded with minimum loss of information about the image itself, as if the problem of retinal encoding were one in video engineering.

An important moral of this brief comparison is that depending on one's conception of the function of vision, one will have differing conceptions of the properties that are encoded and hence represented in the process of neural transmission. Barlow, reflecting the inspiration of his approach in engineering and communication theory, emphasizes the reliable coding of generic physical properties present at a sender (such as spatial frequencies on the retina) so that the physical properties could be reconstructed at the receiver (in the visual cortex). Ultimately, he suggests, a completely reversible code, in which one could fully specify the causes of sensation, would depend upon a completed scientific (read: physical) description of sensory energy (1961a, 354-359). Detection of physical properties is primary; information about such properties subsequently enters into inferential reasoning about distal causes, and only through such "interpretation" does it yield representations of the distal (behaviorally relevant) world. Marr, by contrast, reflecting the inspiration of his approach in biology and psychology, looks to the biological significance of sensory systems and hence conceives of them as systems for representing organismically significant properties of the distal environment. These properties are described in a biological vocabulary: Marr (1982, 32) describes the functional organization of spider vision to detect mates by detecting a pattern of light characteristic of a conspecific of opposite sex; in early human vision, he emphasizes the representation of distal spatial structures of human scale. These biological properties of course have a physical realization; indeed, the physical properties of light, distal objects, and optical media place constraints on the mechanisms of visual representation and detection. Nonetheless, in Marr's conception, as in Gibson's, the senses are taken to be detectors of biologically relevant properties. And the regularities upon which sensory systems depend for their functioning—Marr's "physical assumptions" (1982, 44-51) and Gibson's ecological regularities (1966, ch. 1)—are not laws of physics, but local (earthly, or even niche-specific) regularities in the relations between organism and environment. The difference between the "physical instruments" and "perceptual systems" approaches can be articulated more fully by returning to the case of color.

3. Competing Task Analyses of Color Vision[2]

What is the function of color vision? No single answer can be given for all organisms. Even if the question is narrowed to the function of color vision in vertebrates, it still admits a variety of answers, depending on how widely one construes "function". Here I intend to leave aside various functions that color vision has assumed in complex human societies, and to focus on conceptions of the function of color vision explicit or implicit in the literature associated with the Barlow and Gibson/Marr approaches. These competing conceptions of the function of color vision lead to differing task analyses and hence to differing conceptions of the content of color perception, and in particular of the properties in the world that perceived colors represent.

Barlow and his followers ascribe to the color receptors in the retina, and to color vision itself, the function of encoding accurately (and hence discriminating among) spectral energy distributions in the visible range of the spectrum. As Barlow puts it, "For colour vision, the task of the eye is to discriminate different distributions of energy over the spectrum" (1982c, 635). Buchsbaum and Gottschalk echo the same point: "The visual system is concerned with estimating the spectral functional shape of the incoming colour stimulus" (1983, 92). A spectral energy distribution is a well-defined physical magnitude mapping wavelength against intensity within a given sample of light, and thereby producing a particular "spectral functional shape". According to Barlow's conception, an optimal color system would encode each spectral energy distribution differently. Our trichromatic system is good, but not perfect, at encoding various distributions. It discriminates many distributions, but for some physically distinct combinations of wavelengths it gives the same response. This is the phenomenon of metamerism, illustrated in Figure 1.[3] According to Barlow (1982c) and Buchsbaum and Gottschalk (1983), metameric matching (same response to distinct energy distributions) is a failure of the visual system, a loss of information. It counts as a failure in the context of a specific conception of the function of the color system: to discriminate among spectral energy distributions. This approach to color vision is consonant with Barlow's conception of the visual system as a physical instrument. It treats the problem of color encoding much as a video engineer might treat the problem of building a good television camera: as the problem of accurately encoding the physical characteristics of a signal within given dimensions of variation. It may not be the happiest conception of the task of vertebrate color vision.

Marr's (1982, 250-264) brief treatment of color is an extension of his general program: he ascribes to the color system the function of determining the color of a distal surface, where "color" is defined as the spectral distribution of the surface reflectance. Although he differs from Barlow and his followers in emphasizing the distal focus of color vision, he nonetheless adopts the common attitude that color is to be understood as a physically-defined property of the distal stimulus, in this case, its surface-reflective characteristics. As far as can be told from Marr's analysis, he considered it the function of color vision to represent each physically distinct surface reflectance differently. Hence, he too might consider surface metamerism to reveal a deficiency in color vision.

In this case Marr has not been true to the spirit of his approach, which enjoins the investigator to reflect upon empirically given regularities of the visual environment (as extant during the evolution of the visual system) in seeking to understand the functioning of the visual system. That is, he does not provide a set of "physical assumptions"—ecological regularities pertaining to the earthly environment—for color vision, corresponding to those he provided for the recovery of the spatial structure of distal surfaces (1982, 44-51). His analysis of color vision does not analyze the characteristics of the distal stimulus as fully as it might; rather, it simply accepts the usual spectrophotometric description.

What would correspond, in the case of color, to the "physical assumptions" of Marr's analysis of spatial perception? Consider the spatial case more fully. Marr's analysis sought to discover regularities of reflective surfaces in earthly environments that a visual system might be built to exploit in reducing the informational equivocation of the retinal image. Marr conjectured that the visual system might have evolved to make use of certain surface regularities, such as evenness of grain, which allow it implicitly to restrict the domain of permissible perceptual outcomes, thereby allowing the system, in environments for which the regularities hold, to recover successfully surfaces that it otherwise could not. In the case of color, a similar restriction would occur if it were supposed that the function of the color system is not to discriminate all possible spectral energy distributions, or surface reflectance characteristics, within the range of the visible spectrum, but rather is to permit useful discriminations among determinate, environmentally given classes of surface reflectances. Adjudication of the functional adequacy of color discrimination would require consideration of the actual distribution of surface reflectances and of the organismic significance of differences among objects with differing surface reflectances for a token species kind in its characteristic environment; it could not be carried out by an abstract analysis of spectral resolving power alone. (This is not to deny the usefulness of analyzing the system's abilities in this regard, on which see section 4.)

Assume that one function of color vision is to enhance the discriminability of objects and surface features, and that a particular color system serves to promote the discrimination of healthy green plants from soil and rocks. Such a color system must be able to discriminate the surface reflectances of green plants from other reflectances. In evaluating the proficiency of the system, it would be of no consequence if there were physically possible but not actual (nonplant) metameric matches to green plants that the system could not discriminate. As long as such potentially equivocal stimuli were not extant in the environment, the fact that the color system could not discriminate them would not imply a functional deficiency. Similarly, for the purpose of enhancing the discriminability of foliage, it would be of no consequence if various types of soil and rocks possessed metameric surface reflectances. Indeed, it might well be an advantage if classes of surfaces that were biologically equivalent in relation to a given organism appeared to be of the same color to that organism, despite spectrophotometric variations in surface-reflective properties (an advantage consisting in fewer irrelevant differences among sensory representations). Under this analysis, an adaptively better color system would be one that allowed the organism to do a better job of discriminating environmentally-significant objects or surface characteristics than could a conspecific with less sensitive or no color vision.

An approach to color vision of the sort just canvassed has in fact been taken by investigators who adopt a comparative and evolutionary approach. Jacobs (1981) is one such investigator. He approaches color vision with the working hypothesis that its primary function is to enhance object discrimination, and ultimately, object recognition, by providing an additional source of information, beyond achromatic differences in surface reflectances, for discriminating objects with characteristic surface compositions. He reports that achromatic luminance discrimination is very good in vertebrates, but that performance is enhanced by as much as one third when color is added (1981, 168-169). He repeats the familiar conjecture, harking back to Polyak (1957), that one function of color vision might be to aid in the discrimination of ripened fruit (taking on a red, orange, or yellow color) from the surrounding green foliage (Jacobs 1981, 160, 179). He concludes that "it is hard to believe that color visual systems did not evolve in concert with the particular spectral energy distributions that are critical that each species be able to discriminate", though he adds that firm evidence for this view has yet to be found (p. 174). In any event, the implied conception of the function of color vision is clear: its function is to facilitate discrimination among biologically relevant, environmentally given classes of object surfaces.

Jacobs (1981, 160) distinguishes three uses of color perception in the cognition of objects: 1) object detection, by which he means the discrimination of separate objects (say, a red object from its green surroundings), 2) object recognition, or the recognition of an object as being of a certain kind (say, an apple), and 3) signal properties of color, or the use of color to discover further properties of a kind of object (that the apple is almost ripe). Each of these uses goes beyond the bare perceptual representation of color itself; each involves further representational or cognitive capacities, from the simple cognition of something as an object, to its cognition as an object of a certain kind, or as an object of that kind possessing one variable property rather than another. These are all cases in which the perception of color aids in a further cognitive achievement, involving additional representational content beyond that of mere color perception. I wish to ask about the representational content of perceived colors (or of color sensations) themselves.

The preceding analysis of the function of color vision suggests that the various color perceptions have as their content groups of surface reflectances. Focusing on the case of the perception of the colors of surfaces of objects, let us say that the content of our color perceptions is various object surface-colors. "Surface colors" themselves are determined in relation to a particular kind of visual system: to have the same surface color is to appear the same to normal observers under prevailing conditions of natural illumination. As metamerism reveals, surface color cannot be equated with a physical property such as spectral reflectance, because, under the present definition, two objects can have the same surface color but different reflectances. Color, as a property of the surfaces of objects, is that group of (often physically disjunctive) surface reflectances that form an equivalence class in relation to a given visual system in its characteristic photic environment.[4] It is a relational property, which must be defined in relation to a specific type of visual system, and it does not constitute a well-formed physical kind. Indeed, from the point of view of physics various equivalence classes are heterogeneous; they are only grouped together as color kinds because of their effects on a token kind of visual system. Color shares this relational aspect with other biologically constituted properties, such as *nutrient*.

On this conception, color sensations have their own representational content, prior to the subsequent cognitive categorization of objects into kinds. Color vision functions to enhance the discriminability of surfaces of objects of different kinds, but it falls to higher cognitive processes to recognize those objects for the kinds of objects they are, and hence to perceive their utility or lack thereof. For surface-perceiving visual systems such as those of primates, presumably this is so even if in a given environment the only reason to discriminate red from green is in order to be able to discern ripe fruit amidst foliage. Differences of surface color are represented in the processes of early vision, which are processes for representing the surface layout of the environment. The content of the representations produced by such processes is limited to surface properties. Color adds a new dimension of discriminability, or a new class of represented surface differences. But even if only ripe fruit is red, the bare perception of *red* doesn't mean *ripe fruit* unless the system it is part of is itself specialized for fruit detection. But presumably, if detection of ripe fruit was a selective pressure on the development of color vision, the animals in question could already discriminate fruit on the basis of other surface-reflective properties (shape, texture, achromatic luminance). Adding color gave them an added surface feature to use in discrimination. The content assigned to perceived colors in this case pertains to the surface feature, not to the more sophisticated cognitive achievement of recognizing an object as food of a certain sort.

As this sketch makes clear, content assignment to color sensations relies on a task analysis of the color system, and this analysis itself implicitly contains and is guided by evolutionary conjectures. The disparate tendencies of Barlow's physical instruments approach and of the perceptual systems approach as developed in this paper are rendered explicit in their respective analyses of the evolution of trichromacy.

4. Evolution, Optimization, and Trichromacy

The physical instruments and perceptual systems approaches adopt quite different analyses of the shift from dichromacy to trichromacy during the course of mammalian evolution. Barlow and his followers address this topic by asking how well trichromatic systems discriminate among all possible spectral energy distributions within the visible spectrum. Trichromatic systems do well, though not perfectly, as metamerism shows. On the ecological conception adopted by the perceptual systems approach, the relevant query is not how good trichromacy is at covering the spectrum; rather, one should ask what new (or improved) discriminations of environmentally-significant object surfaces trichromacy allows. Both approaches appeal to evolution, for Barlow couches his analysis of trichromacy in terms of optimality, and he assumes that evolving systems are driven toward optimal performance (within resource constraints). Comparing Barlow's appeal to evolution with other evolutionary accounts of color vision will allow us to see both the usefulness and the limitations of his physical instruments approach.

Barlow and his followers evaluate trichromatic systems for their efficiency in coding color information. They ask, in effect, what the optimal coding for discriminating among spectral energy distributions might be, and then they test various assortments of receptors—mainly, trichromatic and tetrachromatic—for their sensitivity, concluding that a trichromatic system does remarkably well (Barlow 1982c, Buchsbaum and Gottschalk, 1983). The analysis is rigorous and ingenious. Thus, Barlow considers various ways in which distinct sinusoidal functions of wavelength and intensity ("comb-filtered" spectral energy distributions) can be resolved by color systems with various receptor properties. He concludes that, given the broad receptivity of human cones, little or no advantage in resolving such functions would be gained by having four types of cone rather than three. He offers this finding as an explanation for why trichromacy might have evolved in mammals (1982c, 641).

Barlow's argument appeals to the controversial notion that evolution optimizes. Recent work cautiously endorses the claim that optimizing selection has played a role in evolution (Travis 1989). Careful statements of the optimization approach, such as that of Maynard Smith (1978), avoid the assumption that organisms are in some general sense optimally designed. Maynard Smith characterizes optimization theories as attempts to formulate concrete hypotheses about the selective forces at work in shaping the diversity of living things. He contends that, when properly formulated, such hypotheses make specific assumptions of three kinds: 1) about the kinds of phenotypes that are possible given present species characteristics, 2) about what is optimized, and 3) about the mode of inheritance of the trait in question. As Maynard Smith (1978, 33-34) stresses, point 2) is a conjecture about the selection forces that have been at work in fixing a trait. He argues that candidate optimizing explanations must include as part of the hypothesis under test a specification of the trait that is being optimized and of the selection forces that operate upon it.

Optimization arguments have been applied to the evolution of the visual system with apparent success. Thus, Woodhouse and Barlow (1982, 136) have found that the spacing of receptors in the fovea is very near the theoretical limit set by the physical optics of the eye; they offer evolutionary optimization as the explanation. Others have found that photoreceptors in deep-sea fish have absorption properties that maximize photic sensitivity in a light-starved environment (Lythgoe 1979, 82-83). It is important that optimization arguments be constrained by assumptions of type 1), pertaining to possible phenotypes. Barlow (1982c) explains the lack of optimal spacing among the three types of cones in mammalian trichromats by appealing to the tradeoff between optimizing color sensitivity and spatial resolution. For the purpose of sampling spectral energy distributions, even spacing among the peak sensitivities would be desirable. As it happens, the "red" and "green" cones cluster at 535 and 570 nm while the "blue" cone is at 440 nm. Barlow speculates that the close similarity between red and green cones allows them to be pooled for the purposes of spatial resolution, thereby effectively doubling the number of foveal receptors (1982c, 642). Goldsmith (1991), in an extensive review of the interplay between optimization and constraints on phenotypic possibilities, offers a quite different explanation. The distinction between red and green cones is relatively recent (65 million years), and presumably stems from a mutation in the gene for an ancestral green cone. In some dichromatic species of New World monkeys, a related gene for the green cone regularly produces variants with a spread of 30-35 nm. Goldsmith conjectures that molecular genetic constraints fix the possible red and green cone variation in the range of 535-570 nm, and that this variation set the phenotypic boundaries within which selection for trichromacy could act. He concludes that "the capricious course of mammalian evolutionary history, rather than adaptation by natural selection, is probably primarily responsible for the spectral positions of the long- and mid-wavelength cone pigments" (1991, 317). (The blue cone, which has long been fixed in the genome, is not present foveally and hence doesn't enter the argument.) While further work may be needed to determine the relative roles of genetic constraint and selection pressures in this case, it is clear that optimization arguments should seek to specify the domain of phenotypic possibilities.

As section 3 has shown, characterization of what is being optimized may be even more fundamental. The optimization arguments of Barlow and his followers suggest that in color vision the trait to be optimized is the power to resolve individual spectral energy distributions or physical surface reflectances. They present no argument that the adaptiveness of color vision depends upon this ability; indeed, they provide no argument that this ability would be biologically adaptive. Instead, they simply assume that spectrophotometric resolution is the appropriate measure of performance. By contrast, I have emphasized the adaptive feature of color vision suggested by Jacobs

(1981), namely, increased discriminability of object surfaces. The appropriate measure of the adaptiveness of trichromacy over dichromacy on this conception hinges on new or enhanced discriminability of surfaces in an animal's environment.

The extant studies of the relation between environment and evolution in color vision do not support the Barlow approach. The evolution of distinct cone pigments, a prerequisite for color vision, probably was not initially driven by a demand for spectral differentiation. Pigments with a range of sensitivity maxima would increase the range of optic sensitivity of the eye, thereby permitting increased discriminations among surface reflectances without necessarily permitting differential spectral sensitivity. (In order for multiple cone types to be exploited for color discrimination, the available neural machinery must be sensitive to differences in activity between or among cone types; a system that summed across cone types would enjoy enhanced optic range without color vision—see Goldsmith 1991, 301-304, Jacobs 1981, 178-179.) In an extensive study of the relation between environmental conditions and cone types, McFarland and Munz (1975) concluded that in certain tropical fishes, a system of two cone types evolved in order to enhance the contrast between objects and their background in spectrally restricted underwater photic environments. In such environments, dark objects can best be discriminated with a receptor whose sensitivity matches the peak spectral transmission of sea water (which, at a depth of 25 m, is nearly monochromatic). Light objects can best be discriminated with a pigment whose sensitivity is offset from the background light. On the basis of comparing several species with different feeding habits and inhabiting different photic environments, McFarland and Munz argued that "the evolutionary selection of multiple photopic systems, and of color vision itself, is probably related to to the maximization of contrast against monochromatic backgrounds" (1975, 1045). Although little work has been done measuring the environment in which primate trichromacy evolved, Jacobs reports that investigation of the environment of one South American primate supports the view that the principal color discriminations are "among subtle shades of green or between contrasting colors and green" (Snodderly 1979, as quoted in Jacobs 1981, 175). Here, ecological considerations suggest that the finest discriminations are needed within the greens, and otherwise between the greens and the entire red/yellow end of the spectrum.

The physical instruments approach of Barlow and his followers is not without its place. Optimization arguments can help to guide the formulation of functional and evolutionary hypotheses; maximization of sensitivity to various physical properties of stimulation is one form of optimization. Rigorous specification of the physical capacities of sensory systems can thus arm the investigator with candidate hypotheses about function. It would be a mistake, however, simply to assume that there has been evolutionary pressure to optimize sensitivity for the stimulus dimensions of greatest interest to physicists or to video engineers. Judgments of function must be tested by taking the animal/environment relation into account. Only by learning how sensory systems actually are used can we determine what they are for.

5. Conclusion

Assume that the function of early vision in primates is to provide representations of adaptively-significant features of the distal environment. The task of the system should then be described by denominating the adaptive significance of the distal properties. On this conception, the visual system is not a physical instrument for recording the values of the proximal stimulus as described in physical optics. Rather, it is a perceptual system with the function of representing surfaces as an aid to detecting food and other significant objects. Extended to the case of color vision, this approach suggests that metameric matching need not entail a loss of information. If color vision has the function of discriminating particular environmentally-given classes of object surfaces, the mere possibility of metamerism may be irrelevant to an assessment of its performance. Further, environmentally extant metamers need not entail a discriminatory deficiency if their discrimination would not yield a biologically significant partition of environmental surfaces. The representational content of color perception might best be conceived in terms of partitions of object surfaces into discrimination classes that are conjoined with adaptively-significant objects, and not in terms of a physical specification of spectral energy distributions.

Notes

[1]An earlier version of this paper was presented at the Cornell Cognitive Science Symposium and to the Departments of Psychology and Philosophy of Dalhousie University, both in June, 1991. I thank each audience for their stimulating discussion. Larry Shapiro has given me helpful comments and criticisms on a more recent draft.

[2]For the purposes of this section, as in the rest of the paper, "color" is used to mean what color scientists call "hue", or "chromatic color" (Boynton 1990, Hurvich 1981).

[3]Metamerism can be defined for samples of light received at the eye, or for surfaces illuminated by a given light source; my discussion takes surface metamers as its primary example.

[4]This conception of color as a property of objects is similar to Beck's definition of color as "the property of light by which two objects of the same size, shape, and texture can be distinguished" (1972, 181; his definition extends to achromatic color). It is opposed to the philosophical analyses of Hilbert (1987, 99), who equates colors in objects with individual physical surface reflectances, and of Hardin (1988, 111-112), who contends that, failing a reduction of color to a physical property such as surface reflectance, it should be categorized as an illusion.

References

Aristotle (1984), "Sense and Sensibilia", in his *Complete Works*, 2 volumes, J. Barnes (ed.). Princeton: Princeton University Press, 1:693-720.

Barlow, H.B. (1953), "Summation and Inhibition in the Frog's Retina", *Journal of Physiology*, 119: 69-88.

________. (1961a), "The Coding of Sensory Messages", in *Current Problems in Animal Behavior,* W.H. Thorpe and O.L. Zangwill (eds.). Cambridge: Cambridge University Press, pp. 331-360.

________. (1961b), "Possible Principles Underlying the Transformations of Sensory Messages", *Sensory Communication,* in W.A. Rosenblith (ed.). New York: Wiley, pp. 217-234.

________. (1982a), "General Principles: The Senses Considered as Physical Instruments", *The Senses,* in H.B. Barlow and J.D. Mollon (eds.). Cambridge: Cambridge University Press, pp. 1-33.

________. (1982b), "Physiology of the Retina", in *The Senses,* H.B. Barlow and J.D. Mollon (eds.). Cambridge: Cambridge University Press, pp. 102-113.

________. (1982c), "What Causes Trichromacy? A Theoretical Analysis Using Comb-Filtered Spectra", *Vision Research* 22: 635-644.

Beck, J. (1972), *Surface Color Perception.* Ithaca: Cornell University Press.

Boynton, R.M. (1990), "Human Color Perception", in *Science of Vision,* K.N. Leibovic (ed.). Berlin: Springer-Verlag, pp. 211-253.

Buchsbaum, G. and Gottschalk, A. (1983), "Trichromacy, Opponent Colours Coding and Optimum Colour Information Transmission in the Retina", *Proceedings of the Royal Society of London B*, 220: 89-113.

Eckert, M.P., Derrico, J.B., and Buchsbaum, G. (unpublished), "The Laplacian of Images Is a Special Case of Predictive Coding in the Retina", paper presented at the 1989 ARVO meeting.

Gibson., J.J. (1950), *Perception of the Visual World.* Boston: Houghton Mifflin.

________. (1966), *The Senses Considered as Perceptual Systems.* Boston: Houghton Mifflin.

Goldsmith, T.H. (1991), "Optimization, Constraint, and History in the Evolution of Eyes", *Quarterly Review of Biology* 65: 281-322.

Gould, S.J., and Lewontin, R.C. (1979), "The Spandrels of San Marco and the Panglossian Paradigm: A Critique of the Adaptationist Programme", *Proceedings of the Royal Society B* 205: 581-598.

Hardin, C.L. (1988), *Color for Philosophers: Unweaving the Rainbow.* Indianapolis: Hackett Publishing Company.

Hatfield, G. (1988), "Representation and Content in Some (Actual) Theories of Perception", *Studies in History and Philosophy of Science*, 19: 175-214.

_______. (1991), "Representation in Perception and Cognition: Connectionist Affordances", in *Philosophy and Connectionist Theory,* W. Ramsey, D. Rumelhart, and S. Stich (eds.). Hillsdale, NJ: Lawrence Erlbaum, pp. 163-195.

Hilbert, D.R. (1987), *Color and Color Perception.* Stanford, CA: Center for the Study of Language and Information.

Hurvich, L.M. (1981), *Color Vision.* Sunderland, MA: Sinauer.

Jacobs, G.H. (1981), *Comparative Color Vision.* New York: Academic Press.

Lythgoe, J.N. (1979), *The Ecology of Vision.* Oxford: Oxford University Press.

Marr, D. (1982), *Vision.* San Francisco: Freeman.

Matthen, M. (1988), "Biological Functions and Perceptual Content", *Journal of Philosophy* 85: 5-27.

Maynard Smith, J. (1978), "Optimization Theory in Evolution", *Annual Review of Ecology and Systematics* 9: 31-56.

Millikan, R.G. (1984), *Language, Thought, and Other Biological Categories: New Foundations for Realism.* Cambridge: MIT Press.

McFarland, W.N., and Munz, F.W. (1975), "The Evolution of Photopic Visual Pigments in Fishes", *Vision Research,* 15: 1071-1080.

Polyak, S. (1957), *The Vertebrate Visual System.* Chicago: University of Chicago Press.

Ptolemy, C. (1989), *L'optique*. Edited and translated by A. Lejeune. New York: Brill.

Snodderly, D.M. (1979), "Visual Discriminations Encountered in Food Foraging by a Neotropical Primate: Implications for the Evolution of Color Vision", in *Behavioral Significance of Color,* E.H. Burtt (ed.). New York: Garland Press, pp. 237-279.

Sterling, P. (1990), "Retina," in *The Synaptic Organization of the Brain,* 3d ed, G.M. Shepherd (ed.). New York: Oxford University Press, pp. 170-213.

Travis, J. (1991), "The Role of Optimizing Selection in Natural Populations", *Annual Review of Ecology and Systematics* 20: 279-296.

Woodhouse, J.M., and Barlow, H.B. (1982), "Spatial and Temporal Resolution and Analysis", in *The Senses,* H.B. Barlow and J.D. Mollon (eds.). Cambridge: Cambridge University Press, pp. 133-164.

Wright, L. (1973), "Functions," *Philosophical Review,* 82, 139-168.

Part XIII

SPACETIME

When is a Physical Theory Relativistic?[1]

Roland Sypel and Harvey R. Brown

Oxford University

1. Introduction

It may be thought that the question 'when is a physical theory relativistic?' has a fairly straightforward answer, namely: 'when it is Galilean invariant' (for classical mechanics) or 'Lorentz invariant' (for special relativity). In the context of the modern 'spacetime theory' approach to relativity physics, the answer is more elaborate, in the interests of greater precision. In this paper, we examine two recent discussions of the issue, or aspects of it, found in a 1990 paper of Arntzenius and the well-known 1983 study of Friedman.

Arntzenius constructs a 'sure-fire method' of constructing a Lorentz invariant spacetime theory from a non Lorentz invariant one. We argue that if the original theory is anything like a typical dynamical theory in physics, its 'completeness'—in a sense to be defined—will render the new theory underdetermined.

Friedman offers a general formulation of the relativity principle (or rather what it is for a spacetime theory to satisfy it) based on two conditions: an 'indistinguishability' requirement and one based on the notion of 'well-behavedness'. We argue that Friedman's technical formulation of the first condition is faulty: as it stands it implies that relativistic electrodynamics renders inertial frames linked by Galilean boosts (or any linear coordinate transformations) indistinguishable. We also question whether the second condition, which is violated in Newtonian mechanics with absolute space, is indeed a necessary condition for satisfaction of the relativity principle.

We finish with some general comments on the 'spacetime theory' approach and a brief analysis of Friedman's 1983 remarks on the outcome of applying the Galilean transformations to Maxwell's field equations in electrodynamics.

2. Theories, models and invariance

In a recent article (Arntzenius 1990) F. Arntzenius, in the course of a discussion of tachyon theories, introduces the following criterion of Lorentz invariance:

PSA 1992, Volume 1, pp. 507-514

> I shall call a theory Lorentz invariant if every Lorentz boost of every model of the theory is also a model of the theory. A model here is taken to be a differentiable manifold, a Minkowski metric and a set of geometric objects such as tensor fields. (Arntzenius 1990, p.231.)

(We take 'boost' to here have its usual meaning; it does not involve translations or rotations.) Arntzenius then goes on to claim that:

> A sure fire method of constructing a Lorentz invariant theory is to pick some non Lorentz invariant theory, and simply to add the rider that any Lorentz boost of a model of the non Lorentz invariant theory is a model of the new theory. This new theory will be Lorentz invariant by construction. (Arntzenius 1990, p.231.)

In this definition we must firstly be clear about what exactly is meant by a model. Wigner (Wigner 1956) introduced the notion of a 'complete description' of a physical system, which he defined as a "a full specification of the paths of all particles, together with a full description of all fields at all points of spacetime." Given such a description, one may then ask the question whether or not it is compatible with the laws of nature, or more specifically with the equations of a particular theory. This distinction between a description of a system and its compatibility with given laws is encapsulated in Earman's division of models into two categories, kinematically possible and dynamically possible (Earman 1974). For Earman, a model of a theory T is "a specification of the values of the 'state variables' of T for each instant of time." He goes on to describe a kinematically possible model as being one

> ...which takes no regard of the 'equations of motion' of T but which does take regard of the intended interpretations of the state variables, e.g., the values assigned to velocity variables must be real numbers. (Arntzenius 1990, p.273.)

Dynamically possible models form a subset of the kinematically possible ones, namely those which satisfy the laws of T. A more formalised notion of a model of a theory involves a four dimensional manifold endowed with an affine connection and an associated metric. In the case of special relativistic theories the affine connection is flat and the associated metric is the Minkowski metric. For Newtonian spacetime, the affine connection is flat but the four dimensional manifold does not have a metric. Instead there is a well defined notion of a temporal interval between any two points and thus a set of simultaneity planes, these being collections of points with zero temporal interval between them. Each simultaneity plane forms a Euclidean three-space. The manifold will eventually have defined upon it certain dynamical quantities characteristic of the particular theory under consideration. These quantities would typically be fields, charge distributions, particle trajectories and so on. The dynamical quantities (and in principle also possibly the metric and affine connection) also feature in the equations or laws of the theory.

A fundamental distinction between Wigner's and Earman's approaches to allowed models on the one hand, and Arntzenius' approach on the other, is that in the former cases the criterion by which models are dynamically allowed is given in terms of laws or equations of the theory. In Arntzenius' case there is no reference to laws or dynamical equations. There is simply a listing of allowed models. The theory is defined, as it were, extensionally rather than intensionally. (In the case of the original non Lorentz invariant theory, it is not in fact clear exactly how the allowed models are chosen, but in the case of the new Lorentz invariant theory it is certainly a question of allowing the boosted models by fiat rather than as a result of any laws.) Such an approach to theory formation is quite at odds with real scientific practice and is idiosyncratic to say the least.

Our criticism of this approach to the definition of a theory goes further, however, than a mere rejection on general methodological grounds. Let us suppose that we allow such law-independent listings of allowed models to count as theories in some purely abstract sense. There is a more specific problem for Arntzenius' particular formulation of Lorentz invariance. The problem is essentially one of underdetermination, since, as we shall show, under Arntzenius' formulation, a given set of initial conditions for a system does not give rise to a unique subsequent evolution of that system but, providing the theory is complete in a sense to be discussed shortly, allows more than one possible development.

Before showing this underdetermination we briefly consider exactly what is meant by a Lorentz boost. Every coordinate transformation has associated with it a point transformation of the manifold. For a coordinate transformation (x^1) to (y^J) the associated point transformation h is such that $x^1(p) = y^1(hp)$ for any point p. The boosted point hp bears the same relation to the new coordinate system as the original point p did to the original coordinate system. Such a point transformation is of course a purely mathematical notion which sets up a one to one correspondence between spacetime points. To translate this into the situation of boosting experiments into motion we need to also consider the question of the geometric objects (fields, masses, charges etc.) defined on the spacetime points. For every point transformation of the manifold there is a corresponding induced transformation of the geometric objects. This is such that any given geometric object G is transformed to a new object h*G such that $h^*G_y(hp) = G_x(p)$ (the values on the left hand side of this equation are referred to the new coordinate system, whilst those on the right are referred to the original system). This equation defines a new model in which the geometric objects bear the same relation to the new coordinate system as the original objects did to the old system. In other words, in the particular case when the coordinate transformation is to a relatively moving coordinate system, the new model represents the original experiment boosted into motion. (We avoid discussion here of the question whether we need a global boost of the whole manifold or just a local boost of some given region which is assumed isolated from all external influences.) If the coordinate transformation is a Lorentz boost, the new model will be a Lorentz boost of the original model. If the theory in question is Lorentz invariant, the boosted model will also belong to the set of allowed models of the theory. For a non Lorentz invariant theory the boosted model will not belong to the set of allowed models of the theory.

We now consider the question of the underdetermination introduced by Arntzenius' sure-fire method, that is, we show that, given the Arntzenius criterion, a given set of initial conditions can lead to more than one subsequent evolution. Suppose we are given a non Lorentz invariant theory and a set of initial conditions defined on any flat space-like hypersurface. These would include the values on that hypersurface of any fields, charges, or other dynamical entities included in the given theory. The boosted experiment will include a new set of initial conditions, defined on a new hypersurface H, such that the new initial conditions look the same in the boosted frame as did the original initial conditions in the original frame. This corresponds to starting the same experiment in a relatively moving frame. We need to assume at this stage that there is an allowed model of the original theory which has just these boosted values on the new hypersurface H. *This assumption ensures that the original theory is complete in the sense that it allows a description of the moving experiment.* Such completeness is normally ensured in a given theory by the fact that it contains general laws to which any set of initial conditions compatible with those laws can be applied. In Newtonian particle mechanics, for example, there is no restriction on the allowed initial values of positions and momenta. In Maxwell's equations the only restriction on initial values of fields is that those values be compatible with the equations themselves. A complete theory in this sense is one which allows a description in which the initial conditions can be boosted into motion. The Lorentz boosted model (including the initial conditions and the subsequent evolution) is, by hypothesis, not an allowed model of the original non Lorentz invariant theory. However, there will be some other model m of the original theory which coincides with the boosted model on the hypersurface H. In the new Lorentz invariant theory this model m is still allowed, but the Lorentz boosted model is also allowed. Since these two models coincide on the hypersurface H, we have underdetermination. Thus we see that Arntzenius' sure-fire method by enumeration of allowed models is not sufficiently stringent—some kind of dynamical constraint must be taken into consideration when listing allowed models in order to rule out underdetermination.

3 An alternative approach to the principle of relativity

We come now to the statement by Friedman (Friedman 1983) of the principle of relativity, or rather of a necessary condition for a 'spacetime' theory to satisfy the principle. This is itself based on earlier work by Earman (Earman 1974) to which we return in the next section. The formulation is motivated at least in part by a desire to avoid the problematic concept of 'form invariance'. Whether this concept is indeed problematic is open to question (Brown 1993) and we shall not pursue the issue here. In describing Friedman's alternative approach to the principle of relativity we first note a minor difference of terminology in the notion of coordinate transformations and their associated point transformations. Where above we defined the associated point transformation by x(p)=y(hp), defining a 'carry-along' transformation which associates a new point having the same relation to the new coordinate system as the original point had to the old system, Friedman uses the point transformation g defined by x(gp)=y(p). This reflects the inverse symmetry which exists between active and passive transformations, namely that to effect a given change in the relation between a point and a coordinate system we may either transform the coordinate system or perform the inverse transformation on the point. This differing definition of the associated point transformation will have little bearing on subsequent discussion. Clearly the two point transformations h and g are inverses of each other (i.e., $g=h^{-1}$) and if g belongs to some covariance group, so also will h (by the group property). The only difference we need note at present is that where above we defined the associated map h* by $h^*O_y(hp)=O_x(p)$ we must now define g* by $g^*O_x(gp)=O_y(p)$.

When considering the principle of relativity in model theoretic form Friedman claims that 'The state of motion of a frame is given by the components of the affine connection and, if present, the absolute space vector field V in that frame.' (p.151). He therefore concludes that the notion of boosting a frame into motion is related to the transformation of the affine connection D to g*D and the absolute velocity vector V to g*V. Two frames, connected by a coordinate transformation (x^i) to (y^j) are considered equivalent if, given an allowed model $<M,D,V,O_1...O_n>$, the boosted model $<M,g^*D,g^*V,O_1...O_n>$ is also an allowed model, where g is the point transformation associated with the coordinate transformation (x^i) to (y^j). Note that here we boost only D and V and not any of the other geometric objects O_i. The principle of relativity now requires that all inertial frames are equivalent in this sense. The classical principle relates inertial frames via a Galilean transformation whilst the special principle relates them via a Lorentz transformation.

This way of stating the principle of relativity in terms of the equivalence of frames is associated with the notion of the indistinguishability of frames: "Two reference frames are equivalent just in case no 'mechanical experiment' can distinguish them." (p.150) By a 'mechanical experiment' here is meant an experiment concerning particle trajectories, which are considered as uncontroversially observable. To see the difficulty with this account we consider Friedman's treatment of relativistic electrodynamics. A model of this theory takes the form <M,D,g,F,J,T> where M is the manifold, D is a flat affine connection compatible with the Minkowski metric g, F and J are respectively the Maxwell field tensor and four-current, and T is a set of tangent vectors defining the allowed particle trajectories. The boosted model is then <M, f*D,g,F,J,T> where f is a Lorentz boost. Friedman concludes that relativistic electrodynamics does not distinguish between inertial frames because the new boosted model is also allowed, being in fact identical to the original model. This identity depends on the fact that f*D = D i.e., that a Lorentz boost does not change the affine structure. However, the affine structure is in fact preserved under all linear transformations, including, for example, Galilean boosts. Hence, under Friedman's reasoning, relativistic electrodynamics would also fail to distinguish between Galilean frames. The fundamental problem here is that Friedman's method of arriving at a boosted model does not correspond to the intuitive notion of taking an experiment and boosting it into motion. This is underlined by the recognition that in the example of relativistic electrodynamics quoted here the boosted model is in fact identical to the original model and can hardly capture the idea of boosting an experiment into motion.

That Friedman seems to some extent to be aware of the fact that all is not well with his notion of how to boost a model is seen in a footnote which appears a few pages later (p.154) where he notes that equivalent frames of reference are assumed to have the same spatiotemporal geometry and must therefore preserve metrical structure as well as the affine connection and absolute velocity. This of course is at odds with the method offered in the main text for boosting models, where only the affine connection and, if present, the absolute velocity, are considered. Without this additional and, by his own criteria, unwarranted extra requirement, Friedman's further claim that his notion of equivalent frames connects with Anderson's symmetry group also fails to go through. The Anderson symmetry group of a theory is the largest group of transformations that will preserve the 'absolute objects' of the theory, the absolute objects being the fixed spacetime background of the theory (we consider these absolute objects further in the next section). Friedman's claim is that, given the principle of relativity, this symmetry group is identical to the indistinguishability group, which he defines as the group of all transformations connecting two equivalent frames. Since equivalent frames are defined in terms of transformations of affine connection and absolute velocity only, this ignores other possible absolute objects such as the metric and hence the need for the extra footnote apparently at odds with the development of the main text.

In addition to requiring the equivalence of Lorentz frames, Friedman adds a second requirement to his relativity principle. Equivalence of frames means that all boosts of allowed models (as described in the previous two paragraphs) are themselves dynamically allowed. This corresponds to the idea that one cannot distinguish experimentally between inertial frames. It is a formalised version of Galileo's thought experiment, in which no experiment performed in the closed cabin of a ship will reveal whether or not the ship is in motion. Friedman further requires that experimentally indistinguishable frames should also be theoretically identical. This requirement, referred to as 'well-behavedness', is added in order to disallow the type of situation which arises in Newtonian mechanics with absolute space, where inertial frames are experimentally equivalent but are theoretically distinguished by their differing absolute velocities. However, this additional requirement should arguably be seen not as part of a principle of relativity, but rather as a principle of metaphysical parsimony, or an application of Ockham's razor. Thus it is generally accepted that Newtonian physics, including the concept of absolute space, does in fact obey a principle of relativity. Newton's corollary V in the *Principia* is an explicit statement of classical

Newtonian relativity, and the principle is central in the collision theory of Huyghens. The wish to do away with absolute space springs from its redundancy rather than because the concept itself violates any principle of relativity, either phenomenological, or expressed in terms of form invariance of laws, or in terms of equivalence of frames and boosted models.

4. Further comments on 'spacetime' theories

An alternative formulation of the principle of relativity, which avoids Friedman's difficulty, can be formulated using what is essentially Earman's notion of the equivalence of frames (Earman 1974). Here the boosted model boosts all the geometric objects except the 'absolute' objects. 'Absolute' is here taken in Anderson's sense of those fixed background 'spacetime' objects which are unaffected by the dynamical interactions of the theory, for example absolute time in Newtonian physics or the Minkowski metric in special relativity. Although it is not easy to give a rigorous and completely general account of what constitutes an absolute object, i.e., of which objects are spacetime background and which are dynamical, (see, for example, the discussion in Earman 1990), in the case of special relativity and of Newtonian mechanics we may define them simply by listing them (spatial metric, temporal metric, absolute velocity and affine connection for the latter, and Minkowski metric and affine connection for the former). Thus we may define our models as $<M, \{A_i\}, O_1...O_n>$ where $\{A_i\}$ is the set of all absolute objects. Our corresponding boosted model becomes $<M, \{A_i\}, g^*O_1...g^*O_n>$. Frames are equivalent if, given an allowed model, the model obtained by boosting in this new way is also an allowed model. The principle of relativity requires inertial frames to be equivalent in this alternate sense.

Even using Earman's method for defining a boosted model, however, does not do full methodological justice to the principle of special relativity as envisaged by Einstein. For Einstein the principle was to be a heuristic guide in finding coordinate transformations and appropriate laws of nature. His route to finding the Lorentz transformations was *via* the use of the relativity principle. The spacetime approach presupposes the Lorentz transformations and the equations of the theory as given, and also assumes that the types of geometrical object which appear in the model are already known. The Earman/Friedman condition that boosted models be dynamically allowed by a theory is indeed a sufficient condition for that theory to satisfy the principle of relativity, but it is a *post hoc* condition which is of little help as a guide to the development of new relativistic theories. This is in complete contrast with the spirit in which the principle was used by Einstein himself. (For a fuller discussion of this point see Brown 1992). In this context it is interesting to note that in the seventeenth century Huyghens also used the principle of relativity to do work for him when he derived the laws of elastic collision by assuming that collisions taking place on a barge would be unaffected by any uniform motion of that barge. (See, for example, Barbour 1989, pp.464-467).

Einstein not only used the principle of relativity to derive the Lorentz transformations but went on to use it again in the derivation of the field transformations for the electromagnetic field. Here Friedman's account of the role of Maxwell's equations also needs further analysis. Friedman writes

> Einstein saw that the simplest way to implement a relativity principle for electrodynamics is to retain Maxwell's laws intact while changing the transformations connecting the inertial frames....[The Lorentz transformations] preserve the laws of Maxwell's electrodynamics. (Thus, for example, a simple calculation shows that the velocity c is preserved.) (p.15)

Again we read, "When we subject Maxwell's equations to a Galilean transformation, we obtain equations of a quite different form" (p.149).

In analyzing these statements, we note first that Einstein was anxious to derive the Lorentz transformations without appeal to the Maxwell equations. Because of his newly developed ideas on light quanta, he had some suspicion that in fact Maxwell's equations may only have statistical validity (see Brown and Maia 1992). Having established the Lorentz transformations, he then turned to the Maxwell-Hertz equations to discover what the field transformations would need to be in order for the equations to preserve their form. Even Lorentz himself had to employ a judicious mixture of intuition and juggling in order to simultaneously discover field transformations and coordinate transformations which together preserve the Maxwell-Lorentz equations. The idea that under a Galilean transformations "we obtain equations of a quite different form" is arguably misleading, since without some additional rules for transforming fields we do not obtain any dynamical equations at all. (Friedman does show (p.105) that Maxwell's equations cannot hold in a frame Galilean boosted from the 'ether' frame. But this demonstration does not and cannot provide any field equations in the boosted frame. Notice that in his subsequent discussion of "classical electrodynamics", Friedman does provide a theory in which field transformations accompany the Galilean coordinate transformations; a critique of this theory is found in Brown 1992.) The specification of transformation equations for dynamical quantities over and above coordinate transformations is crucial to satisfying the principle of relativity, but its role is masked in the spacetime theories approach because the dynamic geometrical objects are tensors whose transformation rules are in fact already given once the coordinate transformations are known. Much of the physics of the situation lies in the choice and construction of those tensors. Stating all our theories in the geometrical language of spacetime theories obscures the role of the principle of relativity in the discovery of the relevant geometrical objects.

A further limitation of this spacetime theory approach to the principle of relativity is that it only applies to theories which can be expressed in terms of geometric objects defined on a given manifold. This is not the case in quantum mechanics. The many-body wavefunction, or state vector, is not a geometric object in this spacetime sense. (For a discussion of this point and also of the status of quantum field theory, see Earman 1987, pp.470-473). Nor is it clear to us that classical thermodynamics can be expressed as a spacetime theory. The principle of relativity, as envisaged by Einstein, and before him by Galileo, Huyghens and Newton, is a more general principle thought to apply to all fundamental laws of physics and not just to theories of a particular limited mathematical form. The indistinguishability of inertial frames is not confined to spacetime theories alone and therefore should not be defined solely in terms of the language of such theories.

Notes

[1]We wish to thank Tim Budden for stimulating discussions related to the issues raised here.

References

Arntzenius, F. (1990), "Causal Paradoxes in Special Relativity", *British Journal for the Philosophy of Science* 41: 223-243.

Barbour, J. (1989), *Absolute or relative motion?* Volume 1. Cambridge: Cambridge University Press.

Brown, H.R. (forthcoming), "Correspondence, Invariance and Heuristics in the Emergence of Special Relativity" in *Correspondence, Invariance and Heuristics*, French, S. and Kamminga, K. (eds.).

Brown, H.R. and Maia, A. (forthcoming), "Light-speed constancy versus Light-speed Invariance in the Derivation of Relativistic Kinematics". *British Journal for the Philosophy of Science*.

Earman, J. (1974) "Covariance, Invariance, and the Equivalence of Frames", *Foundations of Physics* 4: 267-289.

______. (1987), "Locality, Nonlocality, and Action at a Distance: A Skeptical Review of some Dogmas", in *Kelvin's Baltimore Lectures and Modern Theoretical Physics*, Kargon, R. and Achinstein, P. (eds.). Cambridge: MIT Press, pp.449-490.

______. (1990), *World Enough and Spacetime*. Cambridge: MIT Press, appendix to Chapter 2.

Friedman, M. (1983), *Foundations of Space-Time Theories*. Princeton: Princeton University Press.

Wigner, E. (1956), "Relativistic Invariance in Quantum Mechanics". *Il Nuovo Cimento* 3: 517-532.

Space-Time and Isomorphism

Brent Mundy

Syracuse University

1. Introduction

Leibniz indiscernibility argument: If the points of space or space-time are real, then shifting everything by a geometric symmetry transformation f (e.g., translation or rotation) produces a physically different situation, since the same bodies or events now occupy different points: whatever occupied point p now occupies the distinct point fp. This may violate identity of indiscernibles, or other principles of verifiability or theoretical parsimony.

Leibniz's f was a *symmetry* of Euclidean space E, so the argument seems inapplicable to the inhomogeneous space-times of GTR. However, the underlying *manifold* M always possesses symmetries f, so one may take f also to move the geometric structure across M, along with the physical objects and events. The resulting variant models still represent physically distinct worlds, since the objects and geometric structure formerly at p are now at fp. Indeed there is more variability, since M has far *more* symmetries than E.

Hence, the *Einstein hole argument*: The symmetry group G_M of a manifold includes transformations f for which fp = p outside some region R of M (the "hole") while fp ≠ p inside R. Let <M,A> be a model of space-time theory T, 'A' representing both geometric and physical fields over M. We define variant fields A_f over M, satisfying $A_f(fp) = A(p)$. The models <M,A> and <M,A_f> assign different field values to points in R, but equal field values elsewhere. If <M,A_f> is a distinct model of T then T is *indeterministic,* since the totality of A-values outside R fails to determine the A-values inside R: these may be either A(r) or $A_f(r)$.

Einstein first thought this showed that no reasonable space-time theory T can be *generally covariant*, i.e., have <M,A_f> as a model whenever <M,A> is, for all f in G_M. Here he accepted the distinctness of <M,A> and <M,A_f>, but denied that both would be models of the same reasonable theory T. However, he later realized that general covariance follows from the formalism of differential geometry, and hence belongs to all reasonable theories. To avert indeterminism one must now *deny* that the variant models represent genuine alternatives: they must somehow be "physically

PSA 1992, Volume 1, pp. 515-527

equivalent". This principle of *Leibniz equivalence* (Earman and Norton 1987) is accepted in modern textbooks on GTR.

Earman and Norton now revive the hole argument with Einstein's original interpretation of the variant models: they argue that if the manifold M is physically real then <M,A> and $<M,A_f>$ do indeed represent physically distinct states of affairs. *M-realists* must then deny Leibniz equivalence and hence admit that generally covariant theories are radically indeterministic, as Einstein had first believed. Since we now accept general covariance, as Einstein then did not, this is now taken to undercut M-realism (cf. also Earman 1989, Ch. 9 and Norton 1988).

Against this I will argue that Leibniz equivalence is in fact fully compatible with M-realism when space-time theories are properly understood, thus undercutting both the revived hole argument and the original indiscernibility argument.

The positive argument uses a theorem on intrinsic axiomatizability of Riemannian geometry, given in the Appendix. This allows the models of GTR to be represented as first-order structures, like the models of classical axiomatic geometry. We can then apply formal semantics, as cannot meaningfully be done using coordinate representations. We find the variant models to be *structurally isomorphic*, and hence physically equivalent because theories determine their models only up to isomorphism; yet each model possesses full manifold structure, satisfying M-realism.

In addition, negatively, I challenge from this *structuralist* viewpoint certain assumptions concerning representation and identity of space-time points which support the inference from M-realism to Leibniz inequivalence. Coordinate representation creates an illusion of "rigid designation" of individual points, but by formal semantics all terms in a theoretical language are built from non-logical primitives, hence designate different objects under different interpretations. Modal metaphysics fosters the same illusion, *via* "transworld identity" between elements of different models. However, natural science requires identity only between elements of the same model.

I conclude that these structuralist considerations show the essential irrelevance of both the traditional indiscernibility and the revived hole argument to the real issue between relationism and M-realism.

2. Axiomatic and Coordinate Geometry

Geometric models are usually presented in *coordinate representation.* An n-dimensional manifold **M** is given by coordinate maps from regions of M into R^n. These determine a topological and differentiable structure on M, allowing definition of tangent vectors and other local geometric structure. All such geometric structures over a manifold **M** depend ultimately upon the coordinate-based characterization of **M** itself. We can prove *invariance* of these concepts under *change* to other compatible coordinates. However, this does not mean that we could define these quantities without reference to *any* coordinate system over M. Despite some claims in the physics literature, modern differential geometry is inherently coordinate-based because there is no way to define a manifold without reference to coordinate systems.

Euclidean geometry illustrates this contrast. Introduce Cartesian coordinates; define distance by the Pythagorean formula; then prove *invariance* under *change* of Cartesian coordinates. This does not prove Euclidean metric geometry *independent* of coordinates, since the Pythagorean formula still requires *some* Cartesian coordinate system in order to be meaningful, and attribution of Euclidean structure in this sense still requires initial postulation of one or more Cartesian coordinate systems for the space. Likewise, standard differential geometry still requires postulation of one or more manifold coordinate systems for the space, and hence is not "coordinate-free", despite its *invariance* under manifold recoordinatization.

This contrasts with *intrinsic* geometric axiomatizations, using truly coordinate-free primitives such as the affine betweenness relation B(p,q,r) and the segment congruence relation C(p,q,r,s) (Borsuk and Szmielew 1960). Euclidean geometry is expressible intrinsically through axioms T_E in a language L_E over these two primitives. Equivalence with the coordinate formulations is shown by a *representation theorem*: that each model of T_E has Cartesian coordinates, unique up to orthogonal transformations, representing B and C by the standard coordinate formulas.

Coordinate formulations are scientifically useful for their deductive power, but intrinsic formulations have semantic and philosophical advantages. Axiomatization yields a precise semantic characterization of a *model* of Euclidean geometry: it is a *first-order structure* of type <3,4> (a triple <S,B,C> with $B \subseteq S^3$ and $C \subseteq S^4$), satisfying the axioms of T_E. An *isomorphism* is a 1-1 onto function between structures of equal type preserving the basic relations in both directions. Formal semantics interprets languages in first-order structures, and formal theories determine their models only up to isomorphism: any statement or theory T true in a structure **S** is true in any structure **S'** isomorphic to **S**. Isomorphic models are semantically equivalent.

In contrast, coordinate formulations do not determine clear concepts of model or isomorphism. A *coordinate structure* may be defined as a set equipped with one or more coordinate systems, and there are various possible ways to relate such structures to one another and to numerically expressed theories, but there is no standard semantic theory giving natural definitions of 'model' and 'isomorphism' for coordinate theories and structures.

Philosophically, intrinsic formulations better describe their geometrical subject matter. *Coordinate* Euclidean geometry represents Euclidean space as a coordinate structure: a bare set together with one or more distinguished Cartesian coordinate systems. This is physically unreasonable: coordinate systems are mathematical constructs, not fundamental physical entities. This is why actual coordinate grids are defined using more basic geometrical facts about distances, angles and straight lines. It is physically unreasonable to interpret coordinate Euclidean geometry realistically: its theoretical structure cannot directly reflect the basic structure of physical space.

In contrast, *intrinsic* formulations of Euclidean geometry allow realist physical interpretation. Physical space is now a *first-order structure* **S** = <S,B,C> of type <3,4>; the points of space possess *intrinsic relations* B and C, satisfying the axioms of the theory T_E. This claim is supported by direct physical interpretation of the primitives B and C of L_E (e.g., by physical straight-edge and compass), and by direct empirical confirmation of the axioms of T_E so interpreted. Moreover it makes physical sense: intrinsic geometric relations among points of space could be fundamental physical facts, as coordinate systems cannot.

3. Axiomatizability of Riemannian Geometry

To make similarly explicit the underlying physical subject-matter of GTR, we need *intrinsic primitives* for Riemannian geometry, like the primitives B and C for Euclidean geometry. In fact the same relations suffice, under Riemannian reinterpretation. The Appendix shows B and C sufficient to axiomatize Riemannian geometry of any signature and any finite dimension. (Ehlers, Pirani and Schild 1972 state similar and deeper results, but their argument is sketchy and incomplete.) This means the following.

A *coordinate Riemannian structure* **R** is a differentiable manifold **M**, together with a non-singular differentiable rank-two symmetric tensor field g_{ij} over it. Using the coordinate structure of the underlying manifolds $\mathbf{M}_1$ and $\mathbf{M}_2$, we define an *R-diffeomorphism* as a 1-1 function f from M_1 onto M_2 preserving the differentiable structure and metric tensor. *Coordinate Riemannian geometry* studies the properties of coordinate Riemannian structures invariant under R-diffeomorphism.

Let '**<3,4>**' be the class of first-order structures of type <3,4>, and **Riem** the class of coordinate Riemannian structures. Our definitions of B and C associate with each **R** in **Riem** a structure K(**R**) = <M,B,C> in **<3,4>**, over the base-set M of **R**. Since the definitions of B and C are R-diffeomorphism invariant, R-diffeomorphic $\mathbf{R}_1$ and $\mathbf{R}_2$ have isomorphic K($\mathbf{R}_1$) and K($\mathbf{R}_2$). The theorem shows the converse: that whenever K($\mathbf{R}_1$) and K($\mathbf{R}_2$) are isomorphic then $\mathbf{R}_1$ and $\mathbf{R}_2$ are R-diffeomorphic. The functor K then maps the R-diffeomorphism classes of **Riem** 1-1 into the isomorphism classes of **<3,4>**. Any element S = <S,B,C> of <3,4> isomorphic to an element of K(**Riem**) therefore has a Riemannian structure uniquely up to R-diffeomorphism: namely, the R-diffeomorphism class of Riemannian structures over S whose common K-image is isomorphic to S. (Any S of continuum cardinality trivially carries all possible Riemannian structures; the <3,4>-structure of **S** singles out a *particular one* of these, uniquely up to R-diffeomorphism.)

This proves *axiomatizability* of coordinate Riemannian geometry using B and C, just as showing causal automorphisms to be Lorentz transformations proves causal axiomatizability of Minkowski geometry (Zeeman 1964). Since **Riem** is definable in higher-order logic, suitable theories T_R in higher-order languages L_R have as models <3,4>-structures isomorphic to elements of K(**Riem**). The axiomatizability result immediately yields the *representation theorem* for such T_R: that every model **S** = <S,B,C> of T_R determines a coordinate Riemannian structure over S, unique up to R-diffeomorphism, representing B and C by the standard Riemannian coordinate formulas. In practice one seeks a more constructive axiomatization, as had already been given for Minkowski geometry by Robb (1936). This can also be done for T_R in the manner of Ehlers, Pirani and Schild 1972, using intrinsic differentiability axioms.

This is fully analogous to the axiomatization T_E of Euclidean geometry (except that T_R is logically more complex). Any concepts expressible in coordinate Riemannian geometry (e.g., curvature concepts) are also expressible in the intrinsic language L_R over <3,4>-structures, since the criterion for meaningfulness in coordinate Riemannian geometry is R-diffeomorphism invariance, which corresponds under K to isomorphism of models of T_R. For example, the class of Riemannian geometries satisfying the vacuum equations of GTR is carried by K to a subclass of the models of T_R which is closed under <3,4>-isomorphism, and hence consists of the models of T_R satisfying some additional condition expressible in the higher-order language L_R.

This extends easily to space-time theories such as GTR containing non-geometrical physical primitives such as the energy-momentum tensor. These are representable in an extension L_R' of L_R containing additional primitive predicates of space-time points, using standard extensive measurement theory (Field 1980). In covariant theories these quantities are geometric objects, so the resulting equivalence-classes of coordinate structures remain diffeomorphism-closed, and hence still correspond to isomorphism-closed classes of L_R'-structures. Thus these theories continue to be intrinsically axiomatizable within L'_R so that every model of the intrinsic theory determines a coordinate model of the coordinate theory, uniquely up to R-diffeomorphism plus rescaling of the non-geometrical quantities.

4. Physical Interpretation of Riemannian Geometry

As before, this theorem has semantic and philosophical implications. We saw that Euclidean physical geometry is better interpreted semantically and physically as a theory of the intrinsic geometric relations B and C over the points of physical space, than as a theory of physical Cartesian coordinate structures. Similarly, Riemannian physical space-time geometry is better interpreted semantically and physically as a theory of the intrinsic geometric relations B and C over the points of physical space-time, than as a theory of physical manifold coordinate structures.

The argument is the same. Physical coordinate Riemannian geometry attributes to space-time a primitive family of *manifold coordinate systems*, which are quite unphysical entities. In contrast, our intrinsic formulation attributes to space-time points the intrinsic relations B and C, satisfying the theory T_R. As before, this formulation derives direct physical meaning and support from familiar physical interpretations of the primitives: C refers to metric behavior of clocks and rods, B refers to geodesic motion, and much of T_R thereby acquires direct observational support. In contrast, neither observation nor physical intuition favor manifold coordinate systems as fundamental physical entities. Physical Riemannian geometry is therefore better interpreted as a theory of such intrinsic relations over the bare points of space-time, than as a theory of a metric tensor field over a space-time coordinate manifold.

Note that this argument does not *assume* the physical equivalence of R-diffeomorphic structures. All concepts of standard coordinate Riemannian geometry and GTR *are* in fact R-diffeomorphism invariant, and hence, by our theorem, can be fully expressed as structural properties of <3,4>-structures. This holds whether R-diffeomorphic coordinate structures are physically equivalent or not: the theorem shows simply that the full content of the *standard theories*, which do treat them as equivalent, is captured by such axiomatizations.

However, if the standard theories are physically correct then it *follows* from this axiomatization that R-diffeomorphic Riemannian structures are physically equivalent, since they determine isomorphic <3,4>-structures. If space-time geometry is indeed simply the theory of these two relations over space-time points, then two coordinate structures giving the same instances of these relations describe the same geometric reality in different ways: they give two different coordinate representations of the same <3,4>-structure. Since theories determine their models only up to isomorphism, two such descriptions are equivalent in the strongest possible sense.

Our intrinsic formulations still support manifold realism, just as Euclidean axiomatic geometry does: the representation theorems imply that every model of T_E or T_R possesses a standard manifold structure. Intrinsic axiomatization therefore shows explicitly that manifold realism is fully compatible with Leibniz equivalence. I conclude that neither the revived hole argument nor the original indiscernibility argument bear against manifold realism. In both cases the variant M-realist models are *isomorphic* with respect to the physically meaningful geometrical relations, and therefore represent the same physical situation.

This completes the positive argument. I will now discuss two sources for the widespread contrary view that M-realism precludes Leibniz equivalence.

The standard direct argument was given earlier: that relocation to different points must define a different world. My response is given above: such variant models are *isomorphic* with respect to the physically meaningful theoretical predicates, and are therefore equivalent because theories determine their models only up to isomorphism. On this *structuralist* view, the "possible worlds" for a theory T are the *isomorphism classes* of its models, so isomorphic models always determine the same world.

Many writers on these topics consider forms of space-time structuralism, but none accept it. I see two main obstacles: over-reliance on coordinate representations, and acceptance of modal metaphysics.

Coordinates are names or labels for individual points, suggesting that one can *refer* to a particular element of a coordinate manifold **M** as, for example, "the point with coordinates x^i". We acknowledge that statements like "point p has coordinates x^i" are not diffeomorphism-invariant. However, the very occurrence of such statements within coordinate formulations suggests that individual points of a coordinate manifold must in *some* sense be distinguishable, if only for purposes of *assigning* coordinates. But this revives the indiscernibility and hole arguments: if one can indeed refer to individual points of the manifold **M**, then the variant models <M,A> and $<M,A_f>$ must indeed disagree, since they attach different field values to *the very same* point p of M.

Formal semantics reveals a confusion here between statements in the theoretical language L_R and statements in the semantic metalanguage. The semantic properties of a first-order structure are determined by the *structural pattern* of the basic relations over the elements of the domain: two structures bearing that same pattern are semantically equivalent. This is why theories determine their models only up to isomorphism. Under the functor K of section 3, the two coordinate Riemannian structures <M,A> and $<M,A_f>$ determine two <3,4>-structures S and S_f which are isomorphic, and therefore semantically indistinguishable in the language L_R.

However, these two <3,4>-structures also have the *same set* M as their domain. We may then look beyond their isomorphism as <3,4>-structures to compare also the respective positions in that common pattern occupied by given elements p of M. We then find differences expressible, for example, as '$A(p) \neq A_f(p)$'.

There is no contradiction here because the difference between the two models is not expressible in the theoretical language L_R: they are isomorphic, and therefore satisfy all of the same sentences. (Note that this is *theoretical*, not merely observational, equivalence.) This is because the language L_R does not and cannot contain any term which "rigidly designates" the point p itself, i.e., which refers to p when L_R is interpreted over S, and also refers to p when L_R is interpreted over S_f. This is impossible because the interpretation of all non-logical terms in L_R is determined by the interpretation of its primitives, as relations in the chosen <3,4>-structure. However, these relations may be assigned *arbitrarily* over the domain S. Even if L_R contains the symbol 'p', and p occurs in S, nothing requires 'p' to be interpreted as p. 'p' may refer to any distinguished element of S, just as 'B' may refer to any distinguished subset of S^3.

Formal theories refer to objects, not by rigid designators, but by *descriptions* D(x) formulated in the theoretical language L. Descriptions may always have different extensions in different interpretations, yet for *isomorphic* structures any statement of L true of the bearer of D in one is true of the bearer of D in the other. Descriptions pick out, not individuals *per se*, but particular *structural roles* in that structural pattern, and the same description must always pick out the occupant of the same structural role in any isomorphic structure.

Thus an element p of M satisfies a description D of L_R in S iff the corresponding element fp satisfies D in S_f: p and fp occupy the same structural roles in these two isomorphic models, so every statement true of p in S is true of fp in S_f. Therefore, since a theory identifies elements of its domain only by descriptions expressed in its language, two such models describe the same theoretical world. Only in our *semantic metalanguage* can we state a difference between them, but this difference does not reflect any fact expressible in the theoretical language.

This point can also be put in terms of representations. Although theories determine their models only up to isomorphism, we cannot study an abstract isomorphism-class. In semantic or mathematical analysis of theories it is therefore customary to construct *representations*. These are specific models constructed using familiar (often numerical) objects and relations. Study of a representation can yield conclusions valid for the whole isomorphism-class which it represents. However, we must use only properties which are *isomorphism-invariant*, hence independent of our choice of representation.

In these terms, the hole and indiscernibility arguments fail because $\langle M,A \rangle$ and $\langle M,A_f \rangle$ are merely *two different coordinate representations*, over the same coordinate manifold M, of the same abstract $\langle 3,4 \rangle$-structure. They are indeed *distinct* representations, since they assign distinct structural roles to the same elements of M. However, this distinction appears only in the semantic metalanguage in which we define these representations. It cannot express any *meaningful* difference between the worlds represented, because it is not $\langle 3,4 \rangle$-isomorphism invariant.

This distinction between a world and its representations requires formal semantics: our concepts of 'structure', 'structure-type', 'isomorphism', and 'truth' derive from the formal semantics of a standard logical language L, and apply only to theories T formulated in such an L. Only then do we have the basic metatheorem that isomorphic structures make true the same sentences of L, and hence are semantically equivalent for all such theories T.

In contrast, coordinate differential geometry is not formulated in a standard logical language. Rather, the models are defined directly as numerical objects: coordinate manifolds carrying tensor fields. There are no precise definitions of the basic semantic concepts, and hence no theorems giving necessary or sufficient conditions for semantic equivalence of models. The precise concepts and theorems of formal semantics give way to vague intuitions about which coordinate structures "should" count as physically equivalent, leading to intractable philosophical difficulties. Such questions require specification of the theory T and language L for which these coordinate structures are taken as models. Once that is done, through intrinsic axiomatization, these difficulties disappear. That is why the axiomatizability theorem is essential to the present structuralist interpretation of space-time geometry.

6. Trans-World Identity

A second philosophical obstacle to structuralism is modal metaphysics. We often consider possible situations involving actual objects: 'What would happen if I did X?' is a question about me, whether I do X or not. To express this question in a semantic framework of "possible worlds", we need some identity or "counterpart" relation between occupants of different worlds: a world or model in which I do X must somehow contain *me*. This creates a problem of modal metaphysics: to determine which occupants of different worlds are "the same" individual. (It is irrelevant here whether this counterpart relation is true identity or not; I will reject both for the same reason.)

This again revives the hole and indiscernibility arguments. If we can indeed identify "the same" space-time point p in a different model, then the difference between the variant coordinate models <M,A> and <M,A_f> again becomes meaningful: the models disagree because they assign different field values to *the same* point p.

Philosophers are welcome to construct modal extensions of physical theories. However, I claim that nothing in standard physical theory *supports* such extensions: no *scientific* problem requires introduction of any primitive relation extending across different models. The fact that transworld identity revives the hole argument is merely another objection to it.

This follows directly from the semantic form of standard physical theory. Physical laws, in either coordinate or intrinsic form, are statements constraining a *given model* of the theory: the quantifiers range over the domain of that model, and the laws assert that all elements *of that domain* have such-and-such properties. Whether a given structure is a *model* of the theory is determined entirely by information specified within that structure. For example, whether a given system of particle motions satisfies the laws of CPM is determined by the masses, motions and forces of the particles in that system or structure.

Physical theory is nonetheless counterfactually applicable, for this only requires each possible world *individually* to satisfy every true physical theory. We may then apply the laws of every true physical theory in each possible world. This requires no transworld identity, precisely *because* the theoretical sentences refer only to the relevant structural properties of the given world or model. Whether an orbital motion described in a given kinematic model satisfies CPM is determined entirely by the structure of that model, regardless of whether the orbiting object is identified as, say, Mars.

Transworld identity relations are thus entirely distinct from the scientific theories which support counterfactual reasoning. To determine what would happen if I did X, I consider a world in which someone *like* me does X, and apply in that world the theories I believe true. The transworld identification has no bearing whatever on the application of physical theory to that world, since that depends entirely upon the theoretical properties that individual possesses *in* that world. The theory will yield the same implications whether that individual is equated with me or not. Physical theory needs no transworld identity relation.

Transworld identity is often linked to the hole problem. For example, Earman (1989) considers some structuralist ideas, but concludes the book with a series of objections (196-207) based on transworld identity. He even mentions the central structuralist thesis that isomorphic worlds are identical (p. 198), but rejects it simply because it gives no criterion of transworld identity.

Two M-realist responses to the hole argument also reject structuralism in favor of modal metaphysics. (I consider only their rejection of structuralism, not their modal metaphysics, to which the foregoing would apply.) Most elaborate is that of Butterfield (1988 and elsewhere), who agrees that M-realism violates Leibniz equivalence, but hopes to justify this by finding *modal* differences between the inequivalent isomorphic models. To this end he argues that some objects, e.g., Hubert Humphrey, "could not be points", even if included in the domain of a structure satisfying the geometric axioms. Such a geometric model, "would not represent a world", though other models isomorphic to it might.

Following section 5 I respond that a possible world is simply an isomorphism-class of models, any element of which represents it equally well. Using Humphrey in a geometric representation does not imply that Humphrey *is* or *could be* a point, any more than using an abacus implies that numbers are or could be groups of beads on wires. Points, like numbers, are structural roles in isomorphism-classes of models of certain theories, which roles may be played by different objects in different models or representations. The occupant of a role is not *identical* to that role.

7. Realism

The most revealing response is Maudlin 1988 (p. 83-84, 89) who considers and rejects a representationalist reconciliation of M-realism with Leibniz equivalence, before turning to modal metaphysics. Remarkably, his objection (p. 84) is simply that this idea would also invalidate the original Leibniz indiscernibility argument, which he says that the M-realist *needs*. (In the Addendum, p. 90, he restricts this to "classical" M-realism, which he essentially *defines* as an M-realism which denies Leibniz equivalence.)

Maudlin here makes explicit a tendency implicit also in Earman and Norton, namely to incorporate denial of Leibniz equivalence into the *definition* of M-realism, rather than to *argue* substantively that M-realism in some more general sense precludes Leibniz equivalence. (Section 3 of Earman and Norton 1987 is revealingly entitled, "What is spacetime substantivalism?: denial of Leibniz equivalence".) But of course this way of strengthening the argument correspondingly diminishes its significance: it no longer applies to M-realism in general, but only to some special ("classical") form of it.

So: I take *manifold realism* ("M-realism") to be theoretical realism regarding certain theoretical entities in space-time theories, normally called "points", which are asserted by those theories to possess *manifold structure* in some standard mathematical sense. (Here "asserted by" means "has as a consequence".) Such a theory I call a *manifold theory*. The representation theorems for T_E and T_R show them to be manifold theories.

Theoretical realism here means, as usual, that the theoretical entities in question do really exist and do really have the properties attributed to them by the theory. Manifold realism is thus the view that some physical manifold theory, realistically interpreted, is *true*, and hence that the points of physical space do really exist and do really possess manifold structure. I agree with Earman, Norton and others that the underlying manifold of a modern space-time theory is a natural modern analog to *space* in older senses, so that the question whether the manifold is physically real or not is a natural modern descendant of the traditional issue between absolutism and relationism. Therefore, it is M-realism in *this* sense which a modern relationist such as myself wishes to deny.

Note that Leibniz equivalence does not figure at all in this *general* definition of M-realism. Now: the hole argument seems to have been viewed, both by its proponents (Earman and Norton) and by its respondents (Butterfield and the pre-Addendum Maudlin) as a significant argument against general M-realism in some such modern sense: not merely against some anachronistic "classical" M-realism whose denial of Leibniz equivalence is stipulated by definition. Therefore I take all of these authors (except the post-Addendum Maudlin) to have implicitly endorsed the *substantive inference* from general M-realism to Leibniz inequivalence. Only this inference gives the hole argument any force as an objection to general M-realism.

However, the preceding positive argument shows this inference unsound, by explicit specification of manifold theories satisfying Leibniz equivalence, realist acceptance of which would therefore constitute M-realism without Leibniz inequivalence. We also saw that GTR itself *is* such a manifold theory, so that realist acceptance of GTR is a form of M-realism satisfying Leibniz equivalence. The hole argument therefore has no force against *modern* M-realism based on realist acceptance of GTR or other manifold space-time theories.

Recall the steps in the argument: Like physical Euclidean geometry, physical Riemannian geometry is better interpreted as an intrinsic geometric theory of <3,4>-structures than as a theory of coordinate manifolds. Therefore the formal theory T_R stated in L_R provides a better account of the intrinsic physical content of GTR. However, L_R is a logical language carrying standard formal semantics, so Leibniz equivalence follows; yet the representation theorem for T_R shows it to be a manifold theory.

This is not just one possible response to the hole argument. Rather, I think that some such intrinsic reformulation of GTR *must* be accepted in preference to the standard coordinate formulations, since theories must be given physically meaningful intrinsic formulations in preference to extrinsic coordinate-based ones. (Similar views are expressed by Ehlers, Pirani and Schild 1972 and Woodhouse 1973, though their axiomatizations have technical flaws.) It then *follows* that the hole argument is empty, because intrinsic formulations yield Leibniz equivalence by formal semantics, just as coordinate formulations yield general covariance: both belong to any coherent theory.

What is the moral? Maudlin saw the conclusion, but stepped back from it. He is right that structuralism invalidates *both* the recent hole argument and the original indiscernibility argument. The correct conclusion is that, despite their great prominence both in the traditional and the recent literature on space and time, indiscernibility arguments in fact have no bearing at all on the real issue between M-realism and relationism.

Appendix: Axiomatizability of Riemannian Geometry

The relation B(p,q,r) of *geodesic betweenness* cannot be defined globally over geodesics which self-intersect or close smoothly. This does not affect the proof, which only uses local structure, but for the <3,4>-structure K(**R**) to be well-defined we must specify a suitable B for each **R** in **Riem**.

We define a *standard segment* as a geodesic segment along which the affine parameters are 1-1, and no two points of which lie on any other geodesic. We then define B(p,q,r) to hold iff there is some standard segment having q as *affine midpoint*, with p in one half and r in the other. This centering requirement limits the extension of B on closed or self-intersecting geodesics, making it globally non-standard, but still locally standard.

A *geodesic pair* of points and the associated *segment* pq are defined to satisfy: (a) There is exactly one geodesic G through p and q. (b) pq is a geodesic segment on G (not necessarily standard) bounded by p and q whose signed Riemannian length has smaller absolute value than that of any other geodesic segment in G bounded by p and q. The relation C(p,q,r,s) or C(pq,rs) of *unsigned Riemannian congruence* is defined to hold between any unordered geodesic pairs pq, rs whose segments have Riemannian length of equal absolute value.

From B and C we define two more intrinsic relations. The relation A(p,q,r) of *Riemannian orthogonality* holds just in case the tangents at p to the geodesic segments pq and pr are orthogonal. A(p,q,r) is definable from B and C using "infinitesimal isosceles triangles": Let p be the midpoint of sr, i.e., B(s,p,r) and C(ps,pr); and let q', r', s' uniformly approach p along their respective segments: B(s,s',p); B(r,r',p); B(q,q',p); C(ps',pr'); C(pr',pq'). Triangle qrs is isosceles iff C(qs,qr), which in flat space holds iff A(p,q,r). In Riemannian space A(p,q,r) holds iff q'r's' is "infinitesimally isosceles", i.e., length qs approaches length qr as q',r', and s' uniformly approach p. This is expressible using B and C.

Second, the relation $S(p,q_1,q_2,q_3,q_4)$ of *projective separation* holds iff the tangents to the geodesic segments pq_i bear the projective separation relation in the n-1 dimensional projective geometry of directions in the tangent space at p. This means: the four directions are co-planar in the tangent space, and continuous rotation of the first into the second within that plane passes through the third or the fourth. S suffices for axiomatization of projective geometry of any finite dimension (Borsuk and Szmielew 1960). $S(p,q_1,q_2,q_3,q_4)$ is definable from B and C by another "infinitesimal" construction, approximating the tangent plane through directions pq_i and pq_j using points q'_i, q'_j approaching p along pq_i, pq_j, and segments pr for r between q'_i and q'_j. $S(p,q_1,q_2,q_3,q_4)$ holds if, for example, the segments pq'_3 approach segments pr with $B(q_1',r,q_2')$, while the segments pq'_4 approach segments pr "on the other side of pq_2 from pq_1", i.e., $B(s,r,q'_2)$ and $B(s,p,q_1)$ for some s. These limit conditions are expressible using B and C.

Each R in Riem now determines a unique <3,4>-structure K(R). Following section 3, we must show coordinate structures R_1 and R_2 R-diffeomorphic in Riem iff $K(R_1)$ and $K(R_2)$ are isomorphic. The "only if" is immediate, because the coordinate definitions of B and C are R-diffeomorphism invariant. For example, while the sign and numerical value of the Riemannian length of a geodesic segment are coordinate-dependent and hence R-diffeomorphism variant, equality of Riemannian length (and hence the relation C) is invariant.

To prove the "if", take $\mathbf{R}_1$ and $\mathbf{R}_2$ in **Riem** and let f map $\mathbf{M}_1$ 1-1 onto $\mathbf{M}_2$ while preserving the relations B and C (and hence also A and S). We will prove f a diffeomorphism of M_1 and M_2 carrying the Riemannian metric of $\mathbf{R}_1$ onto that of $\mathbf{R}_2$. f has any degree of differentiability possessed by both spaces.

Take any point p_0 of $\mathbf{M}_1$. $\mathbf{M}_1$ and $\mathbf{M}_2$ give coordinate systems $x^i(p)$ around p_0 in $\mathbf{M}_1$ and $y^i(q)$ around $q_0 = f(p_0)$ in $\mathbf{M}_2$. Choose a neighborhood N_1 of p_0 small enough that the coordinates x^i cover N_1 and the coordinates y^i cover $f(N_1) = N_2$, and p_0,p and q_0,q are geodesic pairs for all p in N_1 and q in N_2. The metric coefficients of R_1 and R_2 in these coordinates are numerical functions $g^{[x]}_{ij}(p)$ and $g^{[x]}_{ij}(q)$ on N_1 and N_2 respectively. It suffices to show f diffeomorphic in this neighborhood N_1 of p_0.

We do this by defining differentiable changes to new coordinates $x'^i(p)$ and $y'^i(q)$ such that: (i) $x'^i(p) = y'^i(fp)$; and (ii) $g^{[x']}_{ij}(p) = g^{[y']}_{ij}(fp)$, for all p in N_1. (i) means that f carries each point p in the coordinate neighborhood $\langle N_1,x'\rangle$ to the point q in $\langle N_2,y'\rangle$ with the same coordinates as p. Such a map is always a local manifold diffeomorphism, since the coordinates are differentiable and the identity function is infinitely differentiable. (ii) means that f also preserves the local metric coefficients, and hence is an R-diffeomorphism. Inverting these R-diffeomorphic coordinate changes gives an explicit representation of f as an R-diffeomorphism with respect to the original coordinates x and y.

Every point p has some neighborhood N coverable by *Riemannian* coordinates x' (Eisenhart 1926, p. 53). These coordinates are unique up to a choice of orthonormal basis at p, and make all geodesics G through p satisfy linear equations throughout N: that is, the Riemannian coordinate ratios $x'^i(q)/x'^j(q)$ remain constant for all points q on G in N. These ratios are the *homogeneous projective coordinates* of the tangent direction to G at p, and their numerical values are determined by the chosen orthonormal basis at p. The coordinates $x'^i(q)$ are then fixed (up to Riemannian length scale) by the further stipulation that $x'^i(q)/x^i(q')$ along each G be proportional to the affine length ratio pq/pq'.

We now take a *Riemannian* coordinate neighborhood N_2' of q_0, and choose Riemannian coordinates x' in a neighborhood N_1' of p_0 which maps into N_2' under f. Because the coordinates x' are orthonormal, $g^{[x']}_{ij}(p_0)$ is a *unit* matrix (zero off-diagonal and 1 or -1 on-diagonal). The x' coordinate lines at p_0 are the geodesics in directions having homogeneous x' coordinates <0,0,.,0,1,0..> in the tangent space at p_0, and are non-null because the coordinates are orthonormal. Select points p_i in N_1' on each coordinate geodesic, with segments p_0p_i absolute Riemannian congruent in R_1. Rescale the coordinates x', taking as unsigned Riemannian length unit the common absolute length of the p_0p_i, keeping $g^{[x']}_{ij}(p_0)$ unit.

The segments p_0p_i are carried by f to *geodesic* segments $f(p_0)f(p_i) = q_0q_i$ through q_0 in N_2', because f preserves B. Only null segments satisfy C(pq,rs) identically, so the segments q_0q_i are non-null in R_2, because f preserves C. The basis directions at p_0 are orthogonal in $\mathbf{R}_1$, and hence the directions q_0q_i are also *orthogonal* in R_2, because f preserves A. Finally, because f preserves C the segments q_0q_i, q_0q_j are *absolute Riemannian congruent* in R_2. We can thus make a differentiable coördinate transformation with scale change to a Riemannian coordinate system y' for N_2', where the segments q_0q_i lie along the basis directions of y', have unit absolute Riemannian length, and $g^{[y']}_{ij}(q_0)$ is unit. Since the metrics of $\mathbf{R}_1$ and $\mathbf{R}_2$ have the same signature, these unit matrices can be made equal by signed rescaling of Riemannian length plus permutation of the basis directions.

The coordinates x', y' satisfy (i) and (ii). (i) holds because f preserves B, C and S. The Riemannian coordinates x'(p) for p in N_1' are determined by the constant homogenous projective ratios x'^i/x'^j along the geodesic segment p_0p at p_0, together with its absolute Riemannian length. By the coordinatization of projective geometry, these ratios are determined by the relations of the direction p_0p to the basis directions p_0p_i in the n-1 dimensional projective geometry of tangent directions at p_0. Since f preserves the projective separation relation S, f carries p_0p to a segment $q_0f(p)$ bearing those same projective relations to the basis directions q_0q_i of y'. Similarly, since f preserves C, $q_0f(p)$ bears the same ratio of absolute Riemannian length to the basis segments q_0q_i of y' that p_0p does to the basis segments p_0p_i of x'. Therefore the Riemannian coordinates y'(fp) have the same numerical values as the Riemannian coordinates x'(p), which is (i).

(ii) holds because f preserves C. We apply the standard limit definition of the metric coefficients $g_{ij}(p)$ using the Riemannian length ds of an infinitesimal coordinate displacement dx^i from p. By (i) f carries a coordinate displacement d(x') to the same numerical displacement d(y'), while the length ds in N_1' maps to the same length in N_2' by preservation of C. (The sign of ds is preserved because f preserves null lines, and the length ds of d(x') is the limiting length of the geodesic segments p,p+d(x').) We thus obtain the same numerical values for $g^{[x']}_{ij}(p)$ and $g^{[y']}_{ij}(fp)$.

References

Borsuk, K., and Szmielew, W. (1960), *Foundations of Geometry*. Amsterdam: North Holland.

Butterfield, J. (1988), "Albert Einstein Meets David Lewis", in *PSA 1988*, volume 2, A. Fine and J. Leplin (eds.). East Lansing: Philosophy of Science Association, pp.65-81.

Earman, J. (1989), *World Enough and Space-Time*. Cambridge: MIT Press.

______. and Norton, J (1987), "What Price Spacetime Substantivalism? The Hole Story", British Journal of Philosophy of Science 38, 515-525.

Ehlers, J., Pirani, F.A.E., and Schild, A. (1972), "The Geometry of Free Fall and Light Propagation", in *General Relativity; Papers in Honour of J. L. Synge*, O'Raifeartaigh, L., (ed.). Oxford: Clarendon Press, p. 63-84.

Eisenhart, L. (1926), *Riemannian Geometry*. Princeton: Princeton University Press.

Field, H. (1980), *Science Without Numbers*. Princeton: Princeton University Press.

Maudlin, T. (1988), "The Essence of Space-Time", in *PSA 1988*, volume 2, A. Fine and J. Leplin (eds.). East Lansing: Philosophy of Science Association, pp.82-91.

Norton, J. (1988), "The Hole Argument", in in *PSA 1988*, volume 2, A. Fine and J. Leplin (eds.). East Lansing: Philosophy of Science Association, pp.56-64.

Robb, A.A. (1936), *Geometry of Time and Space*. Cambridge, Cambridge U.P.

Woodhouse, N. (1973), "On the Differentiable and Causal Structure of Spacetime", *Journal of Mathematical Physics* 14, 495-501.

Zeeman, E.C. (1964), "Causality Implies the Lorentz Group", *Journal of Mathematical Physics* 5, 490-493.

The Relativity Principle and the Isotropy of Boosts[1]

Tim Budden

Oxford University

1. Introduction

In 1905 Einstein derived the Lorentz transformations linking inertial coordinate systems. He took as his central principles the relativity principle (RP) and the light postulate (LP), and assumed along the way a symmetry that has become known as 'spatial isotropy' (Berzi and Gorini 1969). This symmetry will be defined below but it basically requires that boosting has the same effect whether to the right or left or to the north or south etc. Since it really is a symmetry about boosts I shall call it 'boost isotropy'.

The purpose of this paper is to explore the relativistic kinematics (both Einsteinian and Galilean) resulting from implementing RP but breaking with the venerable tradition of implementing 'boost isotropy'. The inertial coordinate transformations shall be derived in the manner of Einstein as necessary conditions for RP (amongst other things) to hold and I shall go on to give examples of relativistic theories complete with dynamical fields in order to illuminate the dynamical origin of the more general kinematics. One example is a 1+1 version of special relativity with a massive Klein-Gordon field and the other is a 3+1 Galilean theory of gravitation.

After describing these theories I will go on to argue that they expose a deficiency in Friedman's (1983) practice of identifying inertial coordinate systems via the absolute objects of a modern spacetime theory, thereby creating the impression that kinematics is purely a function of the background spacetime. I will have shown that the dynamical fields from which 'rods' and 'clocks' are built also play a central role.

Despite my example theories being respectable and relativistic, their sets of indistinguishability transformations do not coincide with their sets of symmetry transformations when an attempt is made to cast them in the modern spacetime theory mould.

2. Boost Isotropy

As I have already mentioned the symmetry in question is more naturally termed 'boost isotropy'. This name also helps distance its flip-side, 'boost an isotropy' from the so-called 'spatial an isotropy' introduced by Edwards (1963) and Winnie (1970)

PSA 1992, Volume 1, pp. 528-541

by using a synchrony convention in special relativity (SR) which made the one-way velocity of light direction-dependent. When I consider 1+1 SR below I shall use the Einstein synchrony convention.

I shall cast my definition of boost isotropy in terms of a symmetry of the transformations between various pairs of coordinate systems, but it is equivalently considered a symmetry of the coordinate descriptions of variously boosted objects or processes. (See Berzi and Gorini 1969, pp.1520-1 for a demonstration of this.)

Coordinate systems are ways of labeling the events that occur in the world—in particular they are maps from the events to $\mathbb{R}^m$ (charts), m being the total number of spatial and temporal dimensions—and inertial coordinate systems (ICSs) are a special type of coordinate system. I shall be more specific about how they are special later. Given two ICSs r and s, $r \circ s^{-1}$ is a function from $\mathbb{R}^m$ to $\mathbb{R}^m$ and is called an 'inertial coordinate transformation' (or 'ICT' for short).

The ICTs I shall consider are linear and so if we assume, as I shall do from now on, standard configurations for origins (i.e., a unique event is mapped to (0, 0, 0, 0) in all ICSs) the ICTs can be represented as matrices. Consider an ICS r = (x, y, z, t) and an ICS s = (x', y', z', t') for which the y = constant plane of events is also a y' = constant plane, and similarly for the z and z' planes, and call any two such related ICSs 'x-related' (see Rindler 1982, pp.12-13). The sub-class of ICSs C_r = {s, s is x-related to r} is a one-parameter class and we can choose v, the velocity of s in r (i.e., dx/dt for fixed x') as our parameter. The ICTs $s \circ r^{-1}$ can then be represented as matrices whose elements are functions of v and when these elements are written with these functional dependencies explicit the matrix shall be said to have the 'canonical form' $M_r(v)$. The subscript on this matrix is a reflection of the fact that we are working independently of the relativity principle, so the canonical form might be different for different seeds, r, of the sub-class. Call r the 'resting frame' of C_r.

First I define 'boost directedness' for the one parameter class C_r :

BOOST DIRECTEDNESS. The one parameter sub-class of ICSs C_r = {s, s is x-related to r} is *boost directed about r* iff $\exists s \in C_r$ such that $\neg\exists u \in$ {velocities} such that $s^* \circ r^{*-1} = M_r(u)$, where r^*= (-x, y, z, t) and s^*= (-x', y', z', t').

In plainer language, boost *non*-directedness demands that x-inverted ICSs are related by a matrix of the canonical form, or that rigid bodies contract or stable processes dilate according only to the magnitude and not the direction of any boost they undergo in the sub-class (see Berzi and Gorini 1969, pp.1520-1). Notice that this does not imply Winnie's Principle of Equal Passage-Times (1970, p.230). In fact, $\neg$BD & RP $\Rightarrow$ EPT.

Now I can define 'boost isotropy' for the entire class of ICSs :

BOOST ISOTROPY. The class of ICSs is *boost isotropic* iff $\neg\exists r \in$ {ICSs} for which C_r is boost directed about r.

In general there will be many different one-parameter sub-classes of ICSs generated from r's which are not x-related, i.e., whose x-axes pick out different lines, and boost isotropy requires that none of these lines be boost directed.

The boost non-directedness of C_r about r also follows from $M_r(v)RM_r(v)R$ = identity & $\exists u \in \{velocities\}$, $M_r(v)^{-1} = M_r(u)$ where R is the matrix diag(-1, 1, 1, 1) and represents an x-inversion so that $s^* \circ r^{*-1} = RM_r(v)R$. Hitting the first conjunct with the second we get $RM_r(v)R = M_r(u)$ as required. So the first conjunct is a more efficient way of ensuring boost non-directedness *if* the second conjunct is guarantied, e.g., by the transformations forming a group and RP. Notice how the first conjunct disagrees with the commutator Torretti uses to define 'spatial isotropy' in (1983, p.79), which fails to capture this notion.

3. Inertial Coordinate Systems (ICSs)

Inertial coordinate systems are special ways of numbering the events which occur in the world—this much I take to be indisputable. I shall treat, however, their speciality as *consisting in* the way they coordinatise certain phenomena, such as electromagnetic and massive fields and this I presume to be contentious. It is certainly in marked contrast to Friedman, say, who takes as constitutive their form-giving nature to coordinate representations of absolute spacetime objects (Friedman 1983, p.60). One of the aims of this paper is to show that this latter attitude is unnecessarily restrictive.

I now turn to spelling out the essential special features which characterise Einsteinian and Galilean ICSs (in sections 4 and 6 respectively).

4. Boost Anisotropic 1+1 Special Relativity (SR).

In SR a coordinate system r = (x, y, z, t) is a member of the class of ICSs in virtue of it satisfying the following constraints[2]:

(A) Force-free particle motion[3] is coordinatised as coordinate straight lines, i.e., all the events in the history of such a particle lie on,

$$r^\mu = \lambda a^\mu + b^\mu,\ r^1 = \kappa,\ r^2 = y \text{ etc., } a^\mu, b^\mu \text{ constants.}$$

(B) Light emitted from a point source at (f, g, h, i) is coordinatised as coordinate null cones, i.e., all the events in the light history lie on,

$$(x - f)^2 + (y - g)^2 + (z - h)^2 - (t - i)^2 = 0$$

(C) Its scale is determinate, i.e., there is a way of comparing its coordinate intervals (e.g., the length of a given rod) to those of other members of the class.

(A) is well-known and (B) can be seen to contain what is now often called the light postulate (but is in fact Einstein's light postulate after the action of RP) along with Einstein's synchrony convention. This latter is easy to see if we focus on, say, the x-axis: (B) requires light from a source at the origin to satisfy,

$$x^2 - t^2 = (x - t)(x + t) = 0, \text{ i.e., 'left' and 'right' velocities are equal.}$$

The need for (C) is clear. If r is a chart which satisfies (A) and (B) then so does q, the chart which maps from event E to (kx(E), ky(E), kz(E), kt(E)) where k is a real constant (i.e., q is 'r scaled down by k'). Depending on whether we take r or q as belonging to the class of ICSs then, given an ICS, s, we will count $r \circ s^{-1}$ or $q \circ s^{-1}$ as ICTs and thus affect the structure of the set of ICTs.[4] Friedman runs a similar argument in (1983, pp.138-40).

The set of ICSs can carry physical information over and above the structural information about light cones and free particle trajectories that it already does. It can carry information about how rigid rods, stable clocks, massive fields or stable atoms (see Bell 1976, 69ff) contract and dilate, i.e., how their temporal and spatial aspects change as described from any fixed ICS as they are boosted. This dynamical information can be encoded by decreeing that the scaling of the ICSs must be such that the rest lengths or rest periods of rigid rods or stable clocks be invariant as they are boosted between ICSs and since exercising this rigid-dynamics-encoding capacity seems to me to be part and parcel of kinematics I shall construct my set of ICSs in line with the decree. This, then, is my way of satisfying (C). Notice that the link the decree forges between rigidity and the ICS set can function definitionally in both directions: given rigid behaviour we can define the ICS set, and given the ICS set we can define rigidity. Below I shall show that non-standard relativistic theories of rigid rods and stable clocks are allowed by special relativity by first constructing non-standardly scaled yet nonetheless relativistic ICS sets and taking them to encode rigidity and stability: there is more than one relativistic way to boost a rigid rod.

Although there is a well entrenched fear of talking about rods and clocks, for it risks the accusation of a certain dualism (see Einstein (1949, p.59) and Pauli (1981, final remarks of section I.5)), I shall refer to them fearlessly throughout but only as a shorthand for some scale-fixing phenomenon or process. I consider all such phenomena to be on a par.

Of course this rigid-dynamics-encoding feature can only function if suitable fields are amongst the ontology of the theory so that their dynamics can be encoded. Essential as I take this to be to kinematics, I must go further than denying the existence of a kinematics in empty spacetime: even with light and force-free particles the coordinate systems cannot latch on to the events securely enough for one to exist.

From now on in this section I shall restrict the discussion to one dimension of space and one of time ('1+1') and I do not intend any of my comments to be taken as generalisable to higher dimensions. (A) and (B) jointly constrain the ICTs to be linear (see Torretti 1983, p.75, bearing in mind Zeeman 1964, p.491) and in fact the Lorentz transformations or rotations only up to a velocity or angle dependent scale factor. The reader is referred to Appendix I for a proof that the 1+1 versions of (A) and (B) imply that for $r = (x, t)$ and $s = (x', t')$, the ICT $s \circ r^{-1}$ takes the form:

$$\begin{bmatrix} x' \\ t' \end{bmatrix} = k_r(v)\gamma(v) \begin{bmatrix} 1 & -v \\ -v & 1 \end{bmatrix} \begin{bmatrix} x \\ t \end{bmatrix} \qquad \text{where } k_r(v) > 0. \qquad (1)$$

The $\gamma(v)$ has been inserted, which is equivalent to defining k, in order to make the maths cleaner later. The subscript 'r' reminds us that RP has not yet been invoked.

As I have stressed, it is dynamical considerations that fix k(v), but we need not know all the precise details of the dynamics to deduce aspects of its character. Assuming the ICTs form a group, if the dynamics is relativistic we can locate k(v) within a small class of functions and if it is boost isotropic we can pin it down uniquely. (We do not need a *relativistic* dynamics to fix k(v). The Langevin-Bucherer theory of the deformable electron has ICSs which satisfy (A), (B) and (C) although there is a privileged resting frame (see Miller 1981, p.82 and Brown 1992, section 7)).

I shall take the relativity principle to be

RP. All laws of physics have the same form in all ICSs,

and, since the ICTs encode laws about how 'rods' and 'clocks' contract and dilate, I can apply RP to them. Thus RP requires that the canonical form $M_r(v)$ defined in section 2 be independent of r (i.e., independent of which frame we consider the resting frame), so that in our present context the subscript 'r' evaporates from $k_r(v)$ in (1). The group properties then restrict k(v) as follows:

(i) $\exists$ identity $\Rightarrow k(0) = 1$;
(ii) $\exists$ inverse $\Rightarrow k(-v) = 1/k(v)$

since,

$$\left(k(v)\gamma(v)\begin{bmatrix} 1 & -v \\ -v & 1 \end{bmatrix}\right)^{-1} = \gamma(v)/k(v)\begin{bmatrix} 1 & v \\ v & 1 \end{bmatrix}$$

(iii) closure $\Rightarrow k(u)\,k(v) = k((u + v)/(1 + uv))$
since,

$$k(u)\gamma(u)\begin{bmatrix} 1 & -u \\ -u & 1 \end{bmatrix} k(v)\gamma(v)\begin{bmatrix} 1 & -v \\ -v & 1 \end{bmatrix} = k(u)k(v)\gamma\begin{pmatrix} 1 & -v \\ -v & 1 \end{pmatrix}\begin{bmatrix} 1 & \frac{-u-v}{1+uv} \\ \frac{-u-v}{1+uv} & 1 \end{bmatrix}$$

(iv) associativity : trivially satisfied since the ICTs are matrices.

Notice that I only use the algebraic properties of the matrices of the form appearing in (1). From (iii) it can be seen that these already encode the special relativistic velocity addition rule.

These constraints are clear examples of the power of the form invariance reading of RP.

Normally boost isotropy would be played here. Since we are working in relativistic 1+1 we only need to enforce boost non-directedness. This requires that spatially reflected ICSs are related by a matrix of the canonical form. In fact they are related by

$$\begin{bmatrix} -1 & 0 \\ 0 & 1 \end{bmatrix} k(v)\gamma(v)\begin{bmatrix} 1 & -v \\ -v & 1 \end{bmatrix}\begin{bmatrix} -1 & 0 \\ 0 & 1 \end{bmatrix} = k(v)\gamma(v)\begin{bmatrix} 1 & v \\ v & 1 \end{bmatrix} \qquad (2)$$

and this is canonical, i.e., has the same form as (1) iff k(v) is even, $(k(v) = k(-v))$. Evenness along with (ii) forces k to be unity and so we home in on the Lorentz transformations. Einstein makes this move for the 3+1 case at (1905, p.47).[5] (Friedman requires equivalently in this context that the transformations be metric, i.e., that there exists a spacetime metric which is preserved under their action (1983, p.140)).

Since I am interested in the nature of special relativity here I prefer not to let my vision be clouded by boost isotropy or metricity, so I shall solve (i)-(iii) without further constraint. As the reader referring to Appendix II can see, the solutions are exhausted by,

$$k(v) = ((1 + v)/(1 - v))^n \qquad n \in \mathbb{R} \qquad (3)$$

For n non-zero boost isotropy fails although both RP and LP hold. Thus I claim that the boost anisotropic ICTs,

$$\begin{bmatrix} x' \\ t' \end{bmatrix} = \gamma(v)((1 + v)/(1 - v)^n\begin{bmatrix} 1 & -v \\ -v & 1 \end{bmatrix}\begin{bmatrix} x \\ t \end{bmatrix} \qquad (4)$$

are as special relativistic given their dynamical setting as the standard Lorentz transformations are given theirs and challenge the modern spacetime theorist to account for them.[6] They do not describe the symmetries of Minkowski spacetime.

5. An Example of a Special Relativistic Theory whose Covariance Transformations are not the Lorentz Transformations

Imagine a world populated by a light field and a force-free particle which 'emits' a massive scalar field obeying the following field equation in the ICS r.

$$[\partial^2/\partial t^2 - \partial^2/\partial x^2 + m^2(u)]\,\Phi = 0 \qquad (5)$$

The light field does not interact with the particle or its scalar field but, strangely, the mass parameter, m, which appears in (5) depends on the one-velocity, u, of the particle in which (5) is written and so is not of the usual scalar variety.

It is straightforward to show that such a world is special relativistic for certain one-velocity dependencies of the mass. That the light field and the force-free particle have descriptions of the same form in all ICSs is guarantied by (A) and (B) above, but it remains to be shown that the scalar field obeys a differential equation of the form (5) in all ICSs. This, in fact, is easily secured as it turns out to be equivalent to fixing the scale of the ICSs as required by (C). Thus the scale fixed ICSs will be physically equivalent and the ICTs the indistinguishability transformations.

Consider the ICSs r = (x, t) and s = (x', t') moving through r with velocity v. Define $f(u) \equiv ((1 + u)/(1 - u))$ and let

$$m(u) = M.f^p(u), \qquad M, p \text{ constants} \in \mathbb{R} \qquad (6)$$

By the considerations of the previous section the ICTs must be of the form (4) so that, as is easily verified,

$$(5) \,\&\, (6) \Rightarrow [\, f^{2n}(v)(\partial^2/\partial t'^2 - \partial^2/\partial x'^2) + M^2 f^{2p}(u)\,]\,\Phi' = 0$$

where the differential operators are now in terms of the coordinates of s, and I have used $\Phi = \Phi'$.

This implies $[\partial^2/\partial t'^2 - \partial^2/\partial x'^2 + M^2 f^{2p}(u) f^{-2n}(v)]\,\Phi' = 0$

which has the same form as (5) iff

$$M^2 f^{2p}(u) f^{-2n}(v) = m^2(u') = m^2((u - v)/(1 + uv)) \qquad (7)$$

where u' is the velocity of the particle in s and I have used the Lorentz velocity composition rule which carries over to the boost isotropic setting, as can be verified from (1).

(7) holds iff p = n for then

$$\begin{aligned} M^2 f^{2p}(u) f^{-2p}(v) &= M^2 f^{2p}(u) f^{2p}(-v) \\ &= M^2 f^{2p}((u - v)/(1 - uv)) = m^2(u') \end{aligned}$$

from (i) and (ii) of section 3 respectively.

So covariance of (5) under (4) given (6) forces n = p and so fixes the scale. Another way of seeing this is to consider a 'rod' constructed from the scalar field in one ICS and then boosting it into the others to effect a calibration. One such rod would be the spatial interval over which the non-oscillating field produced by a resting particle decays by a certain factor. The boosted rod resulting from boosting the particle from which it is being emitted will undergo a non-standard special relativistic contraction because of the velocity dependence of the mass parameter and this is the physical origin of the non-standard scaling.

Of course, a similar argument goes through when $p = 0$, i.e., when the mass is the scalar we are accustomed to and boost isotropy holds.

It might be objected that whereas I have used the notion of boosting the particle I have included in my theory only one force-free particle which by hypothesis cannot be accelerated. I could lean on the modal dimension here and bring in possible particles to underwrite calibration, or I could introduce some real dynamics by introducing more particles and interactions. Then I would have accelerating particles and this would introduce a coordinate-time-along-trajectory dependency in the mass parameter. Nevertheless the form invariance argument still goes through. This theory is strange but it is also special relativistic.

6. Boost Anisotropic 3+1 Galilean Physics

As told in Harvey Brown's fable (Brown 1993, section 4), Albert Keinstein in 1705, seeks a derivation of the Galilean coordinate transformations along the lines Einstein pursued in 1905. Just as Einstein invoked boost isotropy Keinstein invoked acceleration invariance, but in keeping with section 3 above I shall work without this restriction.

The class of Galilean ICSs is identified using the same (A) as used for the Einsteinian class, but light cannot here set the geometry of the ICSs and their synchrony simultaneously. Instead, the inertial time coordinates' speciality is to ensure that the one-way speed of a certain signal is infinite in all directions—Keinstein might have thought of casting the mysterious gravitational action-at-a-distance in this role—whilst (D) takes the place of the light postulate part of (B) above. As before, the scale of the charts defined by (A), (B) and (D) must be determined as (C) requires and as will be illustrated in the next section. The list then reads, for an ICS $r = (x, y, z, t,)$:

(A) Force-free particle motion is coordinatised as coordinate straight lines (see (A) in section (3)).

(B) The events in the history of the special signal all lie on $t = \text{constant}$ slices.

(D) All the events in the history of a signal of nature N emanating from a point source at (f, g, h, i) moving with velocity w along the x-axis lie on

$$(x - w(t - i))^2 + (y - g)^2 + (z - h)^2 = u^2(t - i)^2$$

where u depends only on N and not on the ICS, e.g., $u = c$ for light.

(C) Its scale is determined by features of the dynamical fields.

The linearity of the transformations follows from (A) and (D) since although (a) alone permits more general projective transformations, (D) rules out all but the linear ones (see Torretti 1983, p.75). Assuming that we are allowed the rotations so that we can arrange for the two ICSs r and $s = (x', y', z', t')$ to be x-related then, given standard origin configurations (see section 2 for both notions), $s \circ r^{-1}$ takes the following general matrix form:

$$\begin{bmatrix} x' \\ y' \\ z' \\ t' \end{bmatrix} = \begin{bmatrix} k & 0 & 0 & -kv \\ 0 & b & 0 & 0 \\ 0 & 0 & c & 0 \\ p & 0 & 0 & q \end{bmatrix} \begin{bmatrix} x \\ y \\ z \\ t \end{bmatrix} \tag{8}$$

where v is the velocity of s through r, and k, b, c and q are functions of v, now to be determined. Assuming that each primed coordinate increases as its unprimed counterpart increases and the other unprimeds remain fixed, all diagonal elements must be positive.

The standard spatial rotations and more generally their scaled counterparts also preserve (A), (B) and (D) and it is the dynamics which fixes the scaling. I shall assume this produces the standard rotation scaling, i.e., so that rods do not expand or contract when rotated. It might seem odd to retain this spatial isotropy whilst jettisoning boost isotropy, but I'm mostly interested in consistency here, not plausiblilty, so I proceed.[7]

(B) requires that the $t = \text{constant}$ events are $t' = \text{constant}$ events so that $p = 0$.

The implications of (D) can be seen by considering a source which emits a signal at (0, 0, 0, 0) and is stationary in r, and so moving with some velocity w in s. (D) requires,

$x^2 + y^2 + z^2 = (ut)^2$ (a) and $(x' - wt')^2 + y'^2 + z'^2 = (ut')^2$ (b).
(8) & (b) $\Rightarrow (k(x - vt) - wqt)^2 + (by)^2 + (cz)^2 = (uqt)^2$ (c)
and since $k, b, c, q > 0$, (a) & (b) $\Rightarrow$ $k = b = c = q$ & $w = -v$

Inserting these values into (8) we see that the Keinsteinian transformations for the one-parameter sub-class ICSs x-related to r take the form (x-boosts) :

$$\begin{bmatrix} x' \\ y' \\ z' \\ t' \end{bmatrix} = k_r(v) \begin{bmatrix} 1 & 0 & 0 & -v \\ 0 & 1 & 0 & 0 \\ 0 & 0 & 1 & 0 \\ 0 & 0 & 0 & 1 \end{bmatrix} \begin{bmatrix} x \\ y \\ z \\ t \end{bmatrix} \tag{9}$$

Clearly,

$$\begin{bmatrix} x' \\ y' \\ z' \\ t' \end{bmatrix} = k'_r(u) \begin{bmatrix} 1 & 0 & 0 & 0 \\ 0 & 1 & 0 & -u \\ 0 & 0 & 1 & 0 \\ 0 & 0 & 0 & 1 \end{bmatrix} \begin{bmatrix} x \\ y \\ z \\ t \end{bmatrix} \tag{10}$$

will be the canonical form for the sub-class y-related to r (y-boosts) and the form for the z-boost should be obvious. Notice that $k_r'(u)$ may be a different function to $k_r(u)$.

As for the 1+1 SR case, we now turn to the implications of RP. Again since they encode laws about rods and clocks, we can apply RP to the ICTs themselves, which means the subscripts to the resting frame on the canonical forms vanish. Assuming the ICTs form a group the group properties will restrict the class of k functions (to ones generable by Galilean relativistic dynamical fields). Focusing on just the x and t dimensions for simplicity and analogy with the 1+1 SR case,i.e., considering just the ICTs

$\begin{bmatrix} x' \\ t' \end{bmatrix} = k(v) \begin{bmatrix} 1 & -v \\ 0 & 1 \end{bmatrix} \begin{bmatrix} x \\ t \end{bmatrix}$ the restrictions are:

(i) $\exists$ identity $\Rightarrow k(0) = 1$;
(ii) $\exists$ inverse $\Rightarrow k(-v) = 1/k(v)$;
(iii) closure $\Rightarrow k(u)\,k(v) = k(u + v)$;

(iv) associativity : trivially satisfied since the ICTs are matrices.

Once again I refrain from implementing boost isotropy, which would result in k = 1 and admit the entire class of continuous solutions to (i) - (iii) which is, as is well-known,

$$k(v) = \exp nv \qquad n \text{ constant} \in \mathbb{R} \tag{11}$$

so that the 1+1 boost directed Galilean ICTs are

$$\begin{bmatrix} x' \\ t' \end{bmatrix} = \exp nv \begin{bmatrix} 1 & -v \\ 0 & 1 \end{bmatrix} \begin{bmatrix} x \\ t \end{bmatrix} \qquad n \text{ non-zero}. \tag{12}$$

The one-parameter family of boosts along any direction in full 3+1 must have this structure, where n can and, in fact must if is to be a continuous function of angle, be direction-dependent or zero.

If we wish the boosts in all directions to form a group,i.e., so that an x-boost followed by a y-boost is some other boost etc. this will place constraints on the direction-dependence of n.[8] The following three-parameter boost matrix is a solution to these constraints:

$$\begin{bmatrix} x' \\ y' \\ z' \\ t' \end{bmatrix} = \exp(n_0 b\cos\theta) \begin{bmatrix} 1 & 0 & 0 & -u \\ 0 & 1 & 0 & -v \\ 0 & 0 & 1 & -w \\ 0 & 0 & 0 & 1 \end{bmatrix} \begin{bmatrix} x \\ y \\ z \\ t \end{bmatrix} = \exp(n_0 u) \begin{bmatrix} 1 & 0 & 0 & -u \\ 0 & 1 & 0 & -v \\ 0 & 0 & 1 & -w \\ 0 & 0 & 0 & 1 \end{bmatrix} \begin{bmatrix} x \\ y \\ z \\ t \end{bmatrix} \tag{13}$$

where $b = (u^2 + v^2 + w^2)^{1/2}$, the speed of the boost, and θ is the angle between (u, v, w) and the x-axis.[9]

It represents boosts between ICSs which are (x,y,z)-related[10] to r for any (x,y,z) in a world where x-boosts from r are directed but y- and z- boosts are non-directed, yet it is a Galilean relativistic world nonetheless. The corresponding canonical form for a three parameter sub-class generated from an ICS *rotated* from r will be different,i.e., to maintain the directedness of the line picked out by the x-axis of r.

7. An example of a Galilean Relativistic Theory with Non-standard Covariance Transformations

In section 5, I showed how scale fixing in SR was equivalent to the covariance of a field equation governing a matter field. The same is true in Galilean relativistic worlds, and jettisoning boost isotropy allows us to construct a new theory of classical gravitation.

Consider a world inhabited by a scalar gravitational potential, a signal of nature N as mentioned in (D) in section 7 and various mass elements whose mass, strangely, depends on their three-velocity in any ICS. In r = (x,y,z,t), the masses generate the scalar field according to

$$[\partial^2/\partial x^2 + \partial^2/\partial y^2 + \partial^2/\partial z^2]\,\Phi = 4\pi g\rho \tag{14}$$

where g is a constant and ρ is the mass density field, and the field acts on the elements according to

$$\mathbf{acc} = -\,\mathbf{grad}\,\Phi. \tag{15}$$

(15) is covariant under (9), (10) and the general z-boost for any k as is easily verified, so we must look to (14) for the source of scale determination.

Suppose the mass of the ith element depends on its velocity (a,b,c) in such a way that the density it accounts for varies as,

$$\rho(a,b,c) = \rho_0 \exp pa \qquad p \text{ constant} \in \mathbb{R}. \tag{16}$$

The ICTs are of the form (13) so transforming the differential operators and the scalar field, (14) implies, for that part of the field linearly contributed by the ith element, Φ_i

$$\exp 2n_0u[\partial^2/\partial x'^2 + \partial^2/\partial y'^2 + \partial^2/\partial z'^2]\, \Phi_i' = 4\pi g\rho_0 \exp pa$$

$$[\partial^2/\partial x'^2 + \partial^2/\partial y'^2 + \partial^2/\partial z'^2]\, \Phi_i' = 4\pi g\rho_0 \exp(pa-2n_0u)$$

and this has the same form as (14) iff

$$\begin{aligned} \rho_0 \exp(pa-2n_0u) &= \rho(a',b',c') \\ &= \rho(a-u,b-v,c-w) = \rho_0 \exp p(a-u) \end{aligned} \tag{17}$$

where I have used the Galilean velocity composition rule which is valid as can be verified from (13).

Now (17) holds iff $2n_0 = p$, so covariance of (14) fixes the scale.

Alternatively, though equivalently, we can construct an argument in terms of rods: (14) gives $1/r$ solutions near isolated mass elements so scale calibration can proceed by arranging that the constant of proportionality take on an ICS independent value for a standard element at rest in the ICS being calibrated. On boosting the element to rest in other ICSs, the rod undergoes expansion or contraction because of the velocity dependence of the mass of the element and a non-standard kinematics emerges. That the rod undergoes a non-standard boost is quite independent of unit conventions: it is the physical fact underlying non-standard ICTs.

8. Modern Spacetime Theory

The Modern Spacetime Theorist adopts the mathematical architecture of the modern mathematical physicist's general relativity and extends it to other spacetime theories. Each theory is taken to consist of a set of models, each of the form:<M, {A}, {D}>. M is a differential manifold, the A's are 'absolute objects', so-called because they are the same in all models (in some notoriously elusive sense, see Friedman (1983, p.56-60)) and the D's are 'dynamical objects'. The absolute objects characterise the fixed structure of spacetime, if it has any, and the dynamical objects represent matter fields and any dynamic structure of spacetime. They are all taken to be tensor fields in the sense that they are all susceptible to recursive definitions as actions on lower order fields (see Wald 1984, Appendix C, C.1). The extent of the set of models is usually linked to the solution set of some intrinsically construed differential equations.

Can the theories I described in sections 5 and 7 be captured as modern spacetime theories? I have already mentioned that the mass in (5) and the density field in (17) are not the scalar and scalar field we are accustomed to and neither are they higher order fields, so it seems to me that the modern theorist must despair and can at best construct a spacetime theory which simply omits the velocity-dependent mass fields.

The other entities are perfectly at home, however, in the theory so the models might be structured like this:

in section 5 : <$\mathbb{R}^2$, g, D, light , particle history, Φ>

in section 7 : <$\mathbb{R}^4$, dt , h, D, signal N, element history, Φ>

The D's are connections, dt is the absolute time, h is a spatial metric and g a spacetime metric, all of which are absolute (see Friedman 1983 for details). Φ satisfies (5) and (14) respectively in the ICSs, so that we can recover the mass or density field *if* we know the ICSs, but it seems to me that the standard modern theoretic technique cannot provide them (not at least without making concessions to the status of the dynamical fields).

In his (1983), Friedman adopts the modern approach and identifies ICSs as coordinate systems in which a subset (not necessarily proper) of the absolute objects take on special coordinate representations (1983, p.60). For example in 3+1 SR the manifold is $\mathbb{R}^4$, and the absolute objects are the Minkowski metric and the metric connection. The inertial coordinate systems are then those in which the metric takes the unit diagonal form g = diag(-1, 1, 1, 1), and the Christoffel pseudo-tensor vanishes. Since I have adopted an alternative way of identifying ICSs I shall call those Friedman identifies, 'normal coordinate systems' (NCSs). But it is easily seen that the ICSs in the boost anisotropic relativistic theories developed in sections 4-7 are not NCSs. Take the 1+1 SR example. NCSs give the 1+1 Minkowski metric unit diagonal form so assuming that an ICS and an NCS coincide, the metric will have the unit diagonal form in the coordinates of the ICS. Now we can transform to another ICS using the boost anisotropic ICTs so that the metric will still have diagonal form but will pick up some k factors. In other words, since the scale of the ICSs is fixed by the dynamical fields permeating spacetime, they pick up feature quite independent of the absolute background. Another way of putting this is that clocks constructible out of relativistic matter are not thereby hodometers of spacetime (or absolute time in the case of Galilean clocks) as seems to be the folklore. (Brown's (1990, p.319) n = $\frac{1}{2}$ 1+1 SR presents a vivid illustration of this—the 'twin paradox', often explained by reference to the length of worldlines, is absent).[11]

Perhaps modern theorists will attempt to account for the boost anisotropic ICSs by altering the set of absolute objects the ICSs are supposed to be normal to. But they will search in vain to find *further* absolute objects to augment the current set to do the job as the set of NCSs does not contain the set of ICSs and so cannot be *restricted* to it. Nor should they diminish or radically change the normalising set, for the group theoretic structure of the ICT set is already isomorphic to that of the set of normal coordinate transformations (NCTs),[12] and it would be foolhardy to renounce this isomorphism in a search for the deeper coincidence between elements of the set.

The NCSs in Galilean and Minkowskian spacetimes are taken to be normal to the entire set of absolute objects (unlike in 'Newtonian spacetime') and thus are the symmetry transformations for the theory by definition (see Friedman 1983, p.56). Boost anisotropic theories are ones for which the coincidence between the set of symmetry transformations and the set of indistinguishability transformations (see Friedman 1983, p.154),i.e., the ICTs, fails in a subtler way than is seen in the modern theorist's Newtonian physics or 'classical electromagnetism'. Indeed it appears that coincidence fails even though the theories might well be classed well-behaved in Friedman's terms (1983, p.153-4), in the sense that the physically equivalent ICSs are all theoretically identical in that they are all NCSs *up to a scale factor*. The subtlety is that although at the group theoretic level the two sets of transformations are the same (this removing any motivation for tinkering with the background spacetime) the *sets* are distinct, and they are distinct because one is sensitive to the dynamical fields permeating the spacetime and the other is not. It seems to me that modern theorists must face up to this dynamics-related aspect if they are to be able to cope with boost anisotropic relativity and thereby relativity in general.

Appendix I

Consider two ICSs in 1+1 SR, $r = (x, t)$ and $s = (x', t')$ for which x' increase with x, for fixed t and t' increases with t, for fixed x. Since r and s satisfy (A) and (B) in section 4, $s \circ r^{-1}$ (Torretti 1983, p.75) and so if we assume $r^{-1}(0,0) = s^{-1}(0,0)$ we can write

$$\begin{bmatrix} x' \\ t' \end{bmatrix} = \begin{bmatrix} a & -av \\ c & d \end{bmatrix} \begin{bmatrix} x \\ t \end{bmatrix} \quad \text{where } a, d > 0 \text{ and } a, c, d \text{ are functions of } v.$$

and $G \equiv x'^2 - t'^2 = (a^2 - c^2)x^2 - (d^2 - a^2v)t^2 - 2(a^2v + cd)xt$.

Consider the events $r^{-1}(1, 1)$ and $r^{-1}(1,-1)$ on the history of light from a source at (0,0), so that (B) $\Rightarrow G = 0$ for both events,i.e., $a^2(1 - v)^2 = (c + d)^2$ & $a^2(1 + v)^2 = (c - d)^2$. Of the various ways of taking the square roots of these equations, $a(1 - v) = c + d$ & $a(1 + v) = d - c$ yield $c = -av$ and $d = a$ which is the only solution consistent with a, $d > 0$. So,

$$\begin{bmatrix} x' \\ t' \end{bmatrix} = a(v)\begin{bmatrix} 1 & -v \\ -v & 1 \end{bmatrix}\begin{bmatrix} x \\ t \end{bmatrix} = k(v)\gamma(v)\begin{bmatrix} 1 & -v \\ -v & 1 \end{bmatrix}\begin{bmatrix} x \\ t \end{bmatrix}$$

where $k(v) = a(v)/\gamma(v)$, $\gamma(v) \equiv (1 - v^2)^{-1/2} \Rightarrow k(v) > 0$.
I choose to define k(v) like this as it makes for cleaner maths in section 4.

Appendix II

The following proof was kindly provided by Andrew Hodges.

We wish to find the continuous solutions to

$$k(u)\, k(v) = k((u + v)/(1 + uv)) \qquad -1 < u, v < 1 \text{ and } k(u) > 0 \tag{1}$$

$$\text{Let } F(x) \equiv \log (k(\tanh x) \tag{2}$$

$$\begin{aligned} \text{so that } F(x) + F(y) &= \log (k(\tanh x).k(\tanh y)) \\ &= \log (k((\tanh x + \tanh y)/(1 + \tanh x \tanh y))) \\ &= \log (k(\tanh (x + y))) \end{aligned}$$

$$F(x) + F(y) = F(x + y) \tag{3}$$

The only continuous solutions to (3) are $F(x) = n\, x$, $n \in \mathbb{R}$ (4)
Inverting (2), recalling that $\tanh^{-1} v = 1/2 \log ((1 + v)/(1 - v))$,

$$k(v) = \exp(F(1/2 \log((1 + v)/(1 - v)))) \tag{5}$$

So substituting (4)

$$k(v) = ((1 + v)/(1 - v))^{n/2}$$

Notes

[1]I wish to thank Jeremy Butterfield for comments on a previous draft, Andrew Hodges for help with the maths and particularly Harvey Brown for lots of help and stimulating discussion. Any errors are my responsibility.

[2](A) and (B) are Torretti's in his (1983, p.75). The reader can check they imply that Torretti's constraints on inertial time (1983, pp.50-4) are satisfied.

[3](A) could be replaced by a constraint on the way free fields are coordinatised if the theory does not include particles amongst its ontology.

[4]Not at the group theoretic level, since any of the groups resulting from various scalings will all be isomorphic.

[5]In 3+1 SR the Lorentz transformations follow from what might plausibly be called a weaker assumption, namely that rods do not contract on rotation (see Budden et al. in preparation).

[6]The n = 1/2 case appeared in Brown (1990, p.319). Interestingly for this value of n the 'twin paradox' vanishes.

[7]For *special* relativity in worlds with more than one spatial dimension these two symmetries are bonded (see Budden et al. in preparation).

[8]i.e., k(u,v,w) k(a,b,c) = k(u+a, v+b, w+c).

[9]Suggested to me by Andrew Hodges.

[10]s is (x,y,z,)-related to r iff s moves along the line (x,y,z) in r and r moves along the line (x,y,z,) in s.

[11]The claim that there is more to kinematics than the absolute spacetime background in spacetime theories is also defended in Brown (1992).

[12]Two groups are isomorphic if there is a one-to-one mapping between their elements which preserves the group operation. If we have a group ($\{e_i\}$, o) and another group ($\{d_i = k_i e_i\}$, o), eg the group of NCTs and the group ICTs, where o is matrix multiplication, the isomorphism is $d_i \leftrightarrow e_i$.

References

Bell, J. (1976), "How to Teach Special Relativity" reprinted in *Speakable and Unspeakable in Quantum Mechanics*. Cambridge University Press 1987, pp.67-80.

Berzi, V. and Gorini, V. (1969), "Reciprocity Principle and the Lorentz Transformations", *Journal of Mathematical Physics* 10: 1518-24.

Brown, H.R. (1990), "Does the Principle of Relativity imply Winnie's (1970) Equal Passage Time Principle?", *Philosophy of Science* 57: 313-24.

_ _ _ _ _ _ _. (1992), "Correspondence, Invariance and Heuristics in the Emergence of Special Relativity" in *Correspondence, Invariance and Heuristics; a Festschrift for Heinz Post,* S. French and H. Kamminga (eds.). Dordrecht: Kluwer Academic Publishers.

Budden, T., Brown, H.R. and Hodges, A. (forthcoming), "The Role of Isotropy in the Derivation of Relativistic Kinematics".

Earman, J. (1974), "Covariance, Invariance and the Equivalence of Frames", *Foundations of Physics* 31: 267-89

Edwards, W. (1963), "Special Relativity in Anisotropic Space", *American Journal of Physics* 31: 482-89

Einstein, A. (1905), "Zur Elektrodynamik bewegter Körper", *Annalen der Physik* 17 : 891-921. English translation in *The Principle of Relativity,* W. Perrett and G. Jeffery (eds.). Dover 1952, pp.37-65

_ _ _ _ _ _ _. (1949), "Autobiographical Notes" in *Albert Einstein: Philosopher-Scientist,* P. Schilpp (ed.). Illinois: The Library of Living Philosophers.

Friedman, M. (1983), *Foundations of Space-time Theories.* Princeton: Princeton University Press.

Miller, A.I. (1981), *Albert Einstein's Theory of Relativity.* Reading: Addison-Wesley.

Pauli, W. (1981), *Theory of Relativity.* Dover. English translation of *Relativitätstheorie* (1921).

Rindler, W. (1982), *Introduction to Special Relativity.* Oxford: Clarendon Press.

Torretti, R. (1983), *Relativity and Geometry.* Oxford: Pergamon Press.

Wald, R. (1984), *General Relativity.* Chicago: University of Chicago Press.

Winnie, J. (1970), "Special Relativity Without One Way Velocity Assumptions", *Philosophy of Science* 37: 81-99, 223-38.

Zeeman, E.C. (1964), "Causality Implies te Lorentz Group", *Journal of Mathematical Physics 5,* 4: 490-3.

A New Look at Simultaneity[1]

Kent A. Peacock

University of Western Ontario

In recent years there has been renewed discussion of the problem of temporal becoming (Maxwell 1985, 1988; Dieks 1988; Stein 1991). The theme of this paper is that these discussions, timely and interesting as they are, do not go deeply enough into the question. I would like to suggest that a certain way of thinking about simultaneity, which as far as I know was first sketched in a science fiction novel by Robert A. Heinlein (1956), opens up a set of possibilities that deserve serious consideration.

More precisely, I will address the following question: is there anything in the formal structure of Minkowski spacetime which corresponds to our intuitive albeit imprecise notion of a global "present"? It is generally assumed that because of the relativity of optical simultaneity[2] the answer to this question is, resoundingly, *no*. I wish to suggest that the inference from the relativity of simultaneity in terms of equality of time coordinates to the non-availability in Minkowski spacetime of invariant structures that could represent an intuitively plausible "Now" may be a *non sequitur*, since equal-time hyperplanes may not be the most natural way to represent this intuitive notion. My suggestion for an alternative—which is due to the science fiction writer Robert Heinlein—is tentative; it has the drawback of a slight degree of arbitrariness, but the advantages of simplicity, mathematical elegance, and psychological naturalness. It may not be right, but if it is found to be wrong, it will be wrong, I believe, for an interesting reason. At any rate, Heinlein's ingenious hypothesis deserves to be better known, and it cannot be denied that some new ideas are desperately needed if there is to be any hope of moving this inquiry forward.

1. Statement of the Problem

To help put my present discussion in perspective, it will be useful to compare it with the approach taken in a recent paper by Howard Stein (1991). In this paper Stein restates and clarifies his objections (Stein 1968) to what he believes to be a serious and persistent misinterpretation of Minkowski geometry appearing in various guises in the works of N. Maxwell (1985, 1988), H. Putnam (1967) and C. W. Rietdijk (1966). Stein's major purpose is to evaluate Maxwell's claim that special relativity is incompatible with what Maxwell calls *probabilism*, the thesis that there is objective *becoming*. Stein says,

PSA 1992, Volume 1, pp. 542-552

> The issue is whether a notion of "real becoming" can be coherently formulated in terms of the structure of Einstein-Minkowski space-time. Now, such a notion requires that one distinguish "stages" of becoming, in such a way that, at each such stage, the entire history of the world is separated into a part that "has already become... and a part that "is not yet settled"... .
> In order to make a decisive attack upon this issue, it is necessary to agree on some general principles that the answers to these questions should be required to satisfy. I believe the following are uncontroversial:
> (i) The fundamental entity, relative to which the distinction of the "already definite" from the "still unsettled" is to be made, is...the *here and now* (Stein's emphasis); that is, the space-time point... . (Stein 1991, p.148)

The gist of Stein's reply to Maxwell is that an objective notion of becoming can indeed be defined in Minkowski spacetime but only relative to individual spacetime worldpoints. All and only the points in the past cone of a given point can be said to be definite with respect to that point, while (if I understand Stein correctly) all points in the future cone and sidecone of the given point are "unsettled" with respect to, or as of, the given point. Maxwell's fundamental error, in Stein's view, is to speak carelessly of an objective notion of a global present and to try to separate the unsettled from the definite in spacetime analogously to the way this would be done in a Newtonian universe, where the ontologically unsettled part of the world would be thought of as the locus of all points forward in time from some sort of hyperplane or hypersurface of absolute simultaneity. Since there can be no invariant concept of simultaneity in special relativity, Stein reminds us, there can be no global distinction of settled and unsettled points even though there is a perfectly sensible distinction relative to every given worldpoint. There is no invariant way to partition all of spacetime into two disjoint regions by means of a spacelike hypersurface or thin region bounded by two spacelike hypersurfaces such that one region would be agreed by all observers to be "still unsettled".

Stein states the orthodox view with vigour and clarity and indeed hints (1991, p.152) that suggestions that relativity theory allows a notion of "present (spatially distant) actualities" are so transparently fallacious that they should probably not "find continued publication". He suggests that our persistent intuition that there is a global present is merely an illusion or sheer mistake, akin to an untutored person's sense that it is paradoxical to think of the Earth as floating freely and unsupported in space (Stein 1991, p.162).

My object in this paper is to draw Stein's "uncontroversial" point (i) into question, and thus in effect to claim that there might, after all, be a sensible way within relativity theory to define present distant actualities. This is, obviously, a risky position to take; I entirely agree that if it is true that present actualities must necessarily be at the same *time* coordinate then making such a claim would be evidence of a simple failure to do necessary physics homework. If one could somehow succeed in doing this, however, then there might be room after all for the sort of global notion of becoming that Maxwell is interested in. However, I will not attempt to decide the question of becoming itself here, only suggest that there could be grounds upon which it could be decided other than those which Stein considers.

2. Heinlein on Time and Simultaneity

Robert A. Heinlein (1907—1988) was a best-selling American science fiction author whose novels and short stories contain wide-ranging speculations that have fascinated and delighted millions of readers—and at times infuriated not a few of them as well. He is best known for his novel, *Stranger in a Strange Land*, which, somewhat to Heinlein's own amazement, became a cult favourite of the "hippy" generation. Heinlein's stories are generally fast-paced adventure yarns, focussed on human interest. But Heinlein clearly had thought deeply and sometimes most originally about the scientific, political and philosophical concepts that he sketched in a misleadingly casual way as background to his stories.

In particular, he obviously did a great deal of thinking about the apparent and real contradictions between relativity theory and our intuitions of real global simultaneity and the real passage of time. Heinlein attempts to resolve or at least reconcile himself to these incongruities in many of his writings, but, I think, he comes closest to the heart of the problem in his novel, *Time for the Stars*. The premise of this novel is that there may be *another* relevant notion of simultaneity than the usual sense of equality of time coordinates, an *invariant* notion of simultaneity which is *not* expressed in terms of equality of time coordinates but in terms of some history-dependent relation between the *proper* times associated with spatially distant events.

I'll summarize Heinlein's story, leaving out the melodrama. We imagine a time in the not-too-distant future, in which the far-seeing Long Range Foundation has launched a program of interstellar exploration. The goal: *Lebensraum* for Earth's teeming billions of talking hominids. A very efficient rocket drive has been developed, based upon direct conversion of mass to energy. (Sadly, Heinlein omits to tell us the precise reaction sequences, containment methods, and other key details of the workings of his "torch".) It is now possible for a large spacecraft to be accelerated to nearly the speed of light. Several arcs are to be sent out in various directions, to search for habitable planets. There is a problem, however: an effective means of communication between the ships and home base must be found.

Fortunately, the LRF has also been sponsoring research aimed at enhancing the latent telepathic abilities possessed by identical twins. Heinlein was, of course, trading on the familiar folklore surrounding identicals; I don't know whether he actually believed that identical twins are or could be really telepathic. It turns out that with appropriate coaching some young identical twins can be trained to enhance their telepathic abilities to the point that they can literally converse mind-to-mind at will. What is most interesting and important is that the mind-to-mind interaction is *instantaneous* in a frame of reference in which the twins are mutually at rest; it is also completely undiminished and unaffected by distance, so that Heinlein's highly trained telepathic twins literally have a direct window into each other's minds regardless of where they are.

Heinlein's novel is a classic example of a "what-if" story: *what would happen if the twins in the infamous twin scenario could directly read each other's minds? Would the twin who had been accelerated actually perceive his brother's co-moving clock to run fast?* The fact that Heinlein may have been pulling our legs is not important; what makes his story so interesting is the brilliant and consistent way he plays out the implications of this fictional thought experiment.

Two teen-aged twin brothers, Tom and Pat, are recruited into the program; Tom embarks on an interstellar torch-ship while Pat remains at home on Earth. As the ship accelerates away from Earth they remain in mind-to-mind contact, and an amazing thing becomes apparent; Tom begins to perceive Pat's thoughts to be running *fast* in comparison with his own, while Pat, of course, perceives Tom's thoughts as correspondingly slow. They get so out of synchrony that Pat has a birthday days before Tom; then, as the ship approaches peak velocity (as close to c as they can push it) the disparity between mental rates becomes so severe that for weeks of Tom's time they cannot communicate at all. When the ship decelerates in its approach to a distant star enough that the boys can again contact each other, Tom discovers that *years* have passed for Pat, and when he finally returns home to Earth after many adventures, but having aged only two or three years, he finds brother Pat a cranky octogenarian in a wheel chair. Tom marries his grandniece and gets on with his life. (I have, of course, omitted many twists and turns of Heinlein's clever and well-written story; I highly recommend it.)

One might well ask what can be hoped to follow from a science fiction story which presumes and in fact crucially depends upon something—telepathy—which almost certainly does not exist. Let me make it crystal clear that I do not think that there is the slightest evidence for direct mind-to-mind contact beyond the kind of anecdotes and folklore that one hears around the campfire. However, there is no harm and sometimes much good in thinking hypothetically. The point is that Heinlein's story can give us some feel for what it would be like to talk about a *distant* event—a thought in my telepathic twin's mind—which was *present to* my mind in the way that my own immediate thoughts and sensations are. Howard Stein correctly points out that

> ... the fact that there is no experience of the presentness of remote events was one of Einstein's basic starting points. (1968, p.16, quoted in 1991, p.155)

However, Heinlein lets us see what might follow if there *could* be experience of the presentness of remote events. We have always automatically assumed that the Now as defined by sets of such events would be attached to common time slices; Heinlein gives us good reason to think that this unquestioned assumption deserves examination. The point of recounting the Heinlein story—and indeed the major point of this paper—is to make it plausible that the notion of the "Now" or specious present is not necessarily tied to equal time slices in spacetime, and that in fact there is a much more natural, and mathematically coherent, choice of spacetime structure to correspond to our psychological intuition of a "present".

3. The Mathematical Problem

Heinlein's story therefore leaves us with an interesting mathematical puzzle: given the initial conditions of the story, is there any way of identifying the particular proper time on Pat's worldline which is "present to" a given proper time on Tom's worldline? Heinlein does not give an explicit general definition of his notion of invariant simultaneity. Unlike most science fictional speculations, however, it is clearly enough sketched that we need do very little additional work to state it, and to some degree justify it, explicitly.

At one point in the story, Heinlein suggestively speaks of two spacelike separate events happening "at the same apparent instant by adjusted times" (p.89). Adjusted how? Earlier in the narrative (pp.77-83) he tells us enough to let us fill in the gaps. He recounts an experiment that Tom and Pat participate in, the purpose of which is to test the instantaneity of their mind-to-mind link.

Pat, on Earth, is asked to listen to a highly accurate metronome and "tick" in step with it *via* mindlink to his brother Tom on the ship (which by now has attained a high enough velocity with respect to Earth that relativistic effects should become apparent). Tom sounds off Pat's ticks as he perceives them; later, Tom sends metronome ticks back to his brother. (The shipboard metronome was synchronized with the Earthbound metronome before the voyage began.) Sure enough, we find that the metronome aboard the spaceship is running more slowly in direct comparison to the

Earthbound metronome. We can obtain a direct measurement of the difference in elapsed proper times between shipboard and Earthbound metronomes. How could we *calculate* this discrepancy, knowing the initial conditions and the acceleration schedule of the ship?

Another way of putting the problem is this: relativity as presently understood says that there is no invariant way of comparing the rate of time flow of the two twins while they are in flight and spatially distant from one another. In any given frame one can define a rate of change of Tom's proper time with respect to Pat's, but this quantity is frame dependent since the two proper times—i.e. clock readings—must be taken or measured at some particular times in some given frame of reference. One can only compare their proper times in an invariant way at the beginning and end of the journey, when they are coincident. However, if we grant for the sake of argument the science fictional hypothesis of telepathy, then there is a means of *directly* comparing clock readings at all points on the two worldlines; it would be as if Tom's clock and Pat's clock were effectively coincident even though they were far apart in space. If Tom and Pat really are telepathic in the sense that each can have direct experience of certain conscious thoughts and perceptions of the other, then Tom's local clock light years from Earth can be "present to" Pat back home; Pat can experience what Tom experiences. We can therefore again restate the problem: if Pat's proper time—the reading of his local co-moving clock—is τ, what particular reading τ' of Tom's clock will he (Pat) have direct experience of? That is, to use Heinlein's terms, how do we "adjust" their proper times to find their common "apparent instants"?

We can get an answer to this question if we make the assumption that each twin's mental processes keep in step with their local co-moving standard clocks, so that their thoughts and perceptions speed up or slow down in step with their clocks. This is not at all unreasonable if we grant that mental processes are linked to or dependent upon or even identical to certain kinds of physiological processes, since physiological processes would certainly keep step with the advance of proper time[3]. Then Pat has an unmediated "window" onto Tom's clock. What time will he read on it?

The suggestion implicit in Heinlein's story (although not stated exactly as I do here) is that we should answer this question in the simplest and most natural way:

$$\tau' = \sqrt{1-\beta^2}\,\tau. \tag{1}$$

(where β is the relative velocity of the twins, where we assume that their clocks were synchronized and initialized to zero at the beginning of the journey, and where we assume for simplicity that Tom was given only a very short burst of high acceleration and then allowed to coast freely. If you like, assume that Tom jumped from Earth onto a spaceship passing at relative velocity β.)

How do we justify this formula? It would be perhaps safest to regard it merely as an hypothesis or as a *definition* of what we shall mean by "same apparent instant", since there might be a danger of falling into circularity (and in effect falling into the very trap that Stein warns us against) if we try to justify this formula by appealing to some notion of global simultaneity. It would indeed be the simplest hypothesis and in fact the only possible non-arbitrary hypothesis) because the proper time τ' given by this formula is the only proper time along Tom's worldline that we have any particular *basis* for picking out, given the conditions of the problem. However, it *is* an hypothesis. Nevertheless, I think we can go some way toward making it plausible. We know that this formula would give us Tom's local clock reading if Tom has been somehow deflected elastically in mid-course and brought back to coincidence with Pat at Pat's local time τ. But that is *almost* what having telepathy amounts to; in effect, it is as if the two clock readings could be brought into coincidence at any point in their journeys even though they are far apart in space. Equation 1 is exactly the relationship we should expect to find between their readings whenever this could be done. (In a more realistic case Tom's elapsed proper time would be given by a line integral expressing the accelerations he was subjected to in his journey, but the principle is the same.)

If Pat—the Earth-bound twin—wants to know what proper time is "Now" for his brother, (i.e., what time his brother will tell him it is if they become telepathically linked at Pat's proper time τ), all he has to do is look at his own elapsed proper time and apply this formula. His brother, in turn, can apply the inverse of the above formula to find the corresponding time for the Earth-bound twin. Of course, the effect is not symmetrical, since only one twin underwent acceleration. This may seem suspicious to those of us who (perhaps like Herbert Dingle) believe that all velocity-dependent effects in relativity should be entirely symmetrical between the co-moving observers. But we are not dealing here with a purely velocity-dependent effect, but a path- or history-dependent effect, which certainly can be unsymmetrical between different observers depending upon what has happened to them in their pasts.[4]

At the risk of repetition, I will run through the argument once more in a slightly different way. Given the initial condition that the twins start out with synchronized standard clocks initialized to zero, if the twins had always remained relatively at rest their clocks would have continued to run at the same rate in all frames and their elapsed proper times would be equal whenever the clocks were brought into coincidence. In fact, their local clock readings would always agree with the global time coordinate in the inertial frame in which they were both at rest. Their mutually experienced present—if we allow them to be telepathic—would be comprised of events which would be simultaneous in this Lorentz frame. Now suppose that after initialization they undergo relative motion, and suppose—as it would seem most natural to do—that each twin's mental processes remains synchronous with his local co-moving clock. Then, to find the proper times which they would experience together in one apparent instant we would have to correct for their motion histories. That is, we would correct not for their relative motion but in terms of the line integral giving their elapsed proper times from their common initial point as a function of the accelerations they had experienced. Since the only relevant difference between the two twins is the difference in accelerations (or gravitational fields) they had experienced, it can only be this which would determine the difference in their proper times in their mutually experienced present.

In summary: the twins start off linked telepathically at equal proper times; their minds evolve in step with their local co-moving clocks; if we believe that they remain telepathically linked then each twin can directly compare his local clock rate and elapsed proper time with his twin's *as if the two clocks, although running at different rates, were side by side*; it is simplest to assume (although I am not sure that it follows with absolute necessity from the conditions of the problem) that the elapsed proper time that one twin "reads off" on the other's co-moving clock at his elapsed proper time τ is the elapsed proper time that he would find recorded on the other's clock if the two twins were made spatially coincident at τ; in terms of Heinlein's story, this is the same thing as the reading of Tom's clock that would be optically simultaneous with Pat's clock reading τ in the twin's initial inertial rest frame if Pat stayed at rest in that frame; this can be calculated according to the usual theory in terms of a line integral which will be a function of accelerations along the twin's route. That is, the events that are objectively simultaneous or belong to the same apparent instant are not those which are at the same time *coordinate* in some inertial frame or other but which (given suitable initial conditions) have *proper* times which differ *only* through adjustments due to acceleration history. Since Pat and Tom would experience the *same* proper times on their local clocks if it were not for the relative motions they had undergone, the simplest hypothesis is that the only difference in proper times they will note will be that due to their motion histories, calculable *as if* their two clocks could be brought into spatial coincidence.

It is important to emphasize that there is a degree of arbitrariness in this way of identifying the proper times which correspond to the supposed apparent present of the twins; I only claim that the way of doing it suggested here is very simple and natural. Since we obviously don't really know what telepathy would be like, we have to guess a little bit; I am trying to come as close as possible to work out what it would be like for one twin to simply be able to directly experience distant events experienced by the other. Even if I have in fact inadvertently imported some prior notion of global presentness or absolute time—and I do not feel entirely secure that I have not—I think that the model we have arrived at here is interesting enough to deserve study on its own merits.

A crucial and interesting point to emphasize is that two spacelike separate events which belong to the same apparent instant in the above sense will be simultaneous in the conventional sense in only one inertial frame. To put it another way, if they are linked telepathically as Heinlein imagines—that is, such that the link is instantaneous in some inertial frame when the two telepaths are relatively at rest in that frame—the link will not actually be instantaneous in all frames. Some frames will judge the reception to precede the transmission, others the other way around. (This is a point that Heinlein does not make very clear in his story. He speaks of the mindlink as instantaneous without being careful to state with respect to which inertial frame. I suspect that Heinlein understood the problem, but did not wish to overcomplicate his story.) Although this violates the presumed invariance of causal order, I do not think it is by itself a reason for rejecting the possibility of such a link. If there is any sort of spacelike causation then we would just have to accept that causal order is frame dependent. In fact, probably the most useful approach to this problem would be to distinguish between the *extrinsic* causal order of various events in terms of time coordinates as defined in various coordinate systems and *intrinsic* causal order in terms of proper time along the worldlines linking those events. The latter is, of course, invariant.

There is another interesting point we might note as well. It is reasonable to assume that a human's specious present is not infinitesimal in duration, but occupies a certain definite amount of proper time—some small fraction of a second—along that human's worldline. This means that the specious present of the spacefaring twin will be much longer than that of the Earthbound twin; a tick of Tom's clock might take a week in Pat's frame of reference, if their relative velocity is high enough. If this is so, then the hypersurfaces of intrinsic simultaneity as defined here in terms of psychologically apparent "instants" would not really be surfaces *per se* but flattish open regions bounded by two spacelike hypersurfaces. This is a point which would have to be clarified in a more detailed theory.

4. Intrinsic Simultaneity as Equality of Action

The spacelike separate event points that we have hypothesized would belong to the same apparent instant for the twins has a further interesting and elegant property—given certain initial conditions, they have the same *action*.

Return to the twin scenario, and again let Pat be the Earthbound twin and Tom the space voyager. Assume that just before the voyage began, both twins possessed the same initial energy E_0. Let τ be a given value of Pat's proper time after the voyage has begun. At this proper time τ Pat will possess an action $E_0\tau$. (Note the interesting fact that although action is an invariant, it is not conserved—i.e., assuming no net energy interchange with the environment, it increases as the time.) At Tom's corresponding proper time

$$\tau' = \sqrt{1-\beta^2}\,\tau.$$

(i.e., the reading of Tom's clock that I suggest is intrinsically simultaneous with Pat's clock reading τ), relativity tells us that Tom will have an energy

$$E = \frac{E_0}{\sqrt{1-\beta^2}}. \tag{2}$$

Hence Tom will have an action

$$E\tau' = \frac{E_0}{\sqrt{1-\beta^2}} \times \sqrt{1-\beta^2}\,\tau = E_0\tau \tag{3}$$

also, just the same as Pat's.

In other words, it may be that points in spacetime which (at least if we were telepathic) we would identify as belonging to the same apparent instant are just those which have equal *action*. If there is an invariant notion of global simultaneity, (which I suggest should be called *intrinsic* or *dynamic* simultaneity), perhaps it attaches to hypersurfaces of equal *action*, not coordinate-dependent hyperplanes of equal *time* (which would define sets of *extrinsically* or *kinematically* simultaneous events).

There are some obvious complications to this picture which should be mentioned. Real twins, in real spacecraft, will not always have precisely the same masses, and they will often exchange mass-energy with other parts of the universe in the course of their journeys. Therefore, intrinsically simultaneous events on different worldlines will not usually have precisely the same action; they will have a difference in action which depends only upon differences in initial conditions. This does not seem to be an essential complication, but it muddies the nice simplicity of the theory. We can salvage our original scheme, however, by noting that each macroscopic twin is really an assemblage of elementary particles. We do not know how many "elementary" particles there are (hundreds, perhaps an infinity) but we do know that the spectrum of particles is discrete. Each particle of each type (each electron, say) has precisely the same proper mass. Now, each particle of which each twin is made up will be affected by the overall time dilation that the twin experiences. Hence all the *particles* of which the twins are composed will have the same actions at intrinsically simultaneous events.

We can easily generalize this picture to one in which any number of spacefarers, or particles, radiate from a common origin. And we can go further, too: most contemporary cosmologists favour a model of the universe (the so-called Big Bang model) in which all matter and energy, and in fact all space itself, radiated outward from a highly compressed initial state. If this is correct, then we could identify sets of points on the worldlines of all particles which are globally or intrinsically simultaneous in the sense described here; we could then think of spacetime as being built up of "layers" of constant action which could be interpreted as invariant hypersurfaces of intrinsic or dynamic simultaneity. Let's call such layers the "action surfaces".

I should emphasize, if it is not already obvious, that nothing said here is an addition or correction to presently accepted spacetime physics. No one would dispute that given suitable initial conditions one can define something like action surfaces in the spacetime of general relativity. The question is whether these surfaces so defined have any special physical significance or interest; that is, whether they could (to borrow Dieks' phrase, 1988, p.456) play "a direct role as a determinant in physical processes." I do not claim that they do here (although I think that an argument can be made in favour of this claim); however, I do claim that they have a special psychological interest as the sets of hypersurfaces that *would* correspond to the most psychologically natural choice of sets of events belonging to one "apparent instant", if there could be such sets of events.

I think it is a very interesting fact that the spacelike separate points which on a natural psychological interpretation can be thought of as intrinsically simultaneous have this elegant property of having the same action. I do not profess to understand the significance of this odd fact, but I think that it should be better known.

I should note that C.W. Rietdijk (1985) has also introduced a notion of simultaneity in terms of equality of action, although from interestingly different considerations, having to do with quantum nonlocality.

5. Conclusions

I hope my readers have enjoyed this fictional excursion, even if they do not find the argument compelling. Indeed, much of the plausibility of Heinlein's view may come simply from his skill as a storyteller, which I can hardly convey here.

My conclusion is conditional: if there is a Now, it is associated not with equal time slices but with corrected proper time hypersurfaces, which, interestingly, given certain initial conditions are hypersurfaces of constant action. These hypersurfaces are invariant; so there is no contradiction between the psychological intuition (illusion, if you insist) of the universality of this sense of Nowness and the relativity or observer-dependence of simultaneity in the Einsteinian sense. It may, therefore, be just a mistake to insist that the common specious present of several observers must be tied to hyperplanes of constant time. If I am correct then it may, after all, be possible to define hypersurfaces in spacetime which invariantly represent what Stein (1991, p.162) calls a "cosmic present", in spite of the relativity of optical simultaneity. This would, of course, radically change our interpretation of the philosophical significance of relativity, although it would not make any difference to the mathematical form of the theory and might not make any difference to the way the theory is used and applied in physics. (I say only "might not" because the territory of overlap between relativity and quantum physics is still a conceptual demilitarized zone, and it is quite possible that a reconsideration of the meaning of "simultaneity" may help to resolve some of the enormous difficulties in constructing a truly satisfactory relativistic quantum theory.) However, nothing I have said in this paper directly resolves the question of probabilism that concerns Maxwell and Stein; it is thinkable that the corrected time hypersurfaces could represent a cosmic present which divides spacetime into two disjoint sets of actual and potential events, but it is also thinkable (and in fact I am sure that many would insist that this is the only thinkable alternative) that these hypersurfaces have no special physical significance beyond their psychological interest for human beings. Indeed, some authors (see, for instance, Rucker 1977) have suggested that the fact that we humans experience a only certain small interval of our worldline as "present" is only a peculiarity of the four-dimensional structure of the human brain/mind. My own guess is that the "corrected proper time" cosmic present indicated here does separate an actual unsettled future from a definite past; however, I could only defend this intuition by appealing to questions of quantum physics which are beyond the scope of this paper. This paper will have accomplished quite a lot if it has convinced anyone that there may be a new way of thinking about simultaneity (even though it is a way of thinking that is, after all, not all that new).

Notes

[1]I am very grateful to J.W. Crichton and C. Normore for encouragement in the early stages of development of these ideas, and G. Solomon for useful discussions. This work was supported by Social Sciences and Humanities Research Council Grant # 756-91-0068.

[2]By *optical simultaneity* I mean equality of coordinate times as defined by Einstein's synchronization procedure. This useful term is due to Brent Mundy (1986). For a very clear description of how spacetime coordinates are constructed in special relativity, see Taylor and Wheeler (1966).

[3]It is perhaps worth noting that there does not seem to be any *a priori* reason why a person's consciousness should keep right in step with their physiological clock, although anyone for whom this was not true would be very unusual indeed. For an idea of what might happen if there could be such a person, see Philip K. Dick's novel *The World Jones Made* (1956). Jones is a man whose specious present is about three days ahead of everyone else's. Jones has enormous difficulties adjusting to day-to-day life, since he has to teach himself to respond to three-day-old *memories* rather than direct perceptions. However, he possesses extraordinary powers of prediction—until three days before his own death. . . .

[4]There has been a long and wearisome controversy over whether the claim that the twins should be found to have aged differently when they are brought into coincidence marks an inconsistency within relativity. This debate is summarized engagingly in Marder (1971). It seems clear that Dingle and other like-minded authors have simply failed to grasp the basic distinction in relativity between relative-velocity dependent quantities such as apparent time rates and path-dependent quantities such as elapsed proper time. See Taylor and Wheeler (1966) for a very clear exposition of the relevant physics.

References

Dick, P.K. (1956), *The World Jones Made*. New York: A.A. Winn.

Dieks, D. (1988), "Discussion: Special Relativity and the Flow of Time", *Philosophy of Science* 55: 456-460.

Heinlein, R.A. (1956), *Time for the Stars*. New York, Scribners. Reprinted, 1978, New York: Ballantine Books.

Marder, L. (1971), *Time and the Space Traveller.* London: George Allen and Unwin. Reprinted, 1974, Philadelphia: University of Pennsylvania Press.

Maxwell, N. (1985), "Are Probabilism and Special Relativity Incompatible?", *Philosophy of Science* 52: 23-42.

_______. (1988), "Discussion: Are Probabilism and Special Relativity Compatible?", *Philosophy of Science* 55: 640-645.

Mundy, B. (1986), "The Physical Content of Minkowski Geometry", *British Journal for the Philosophy of Science* 37: 25-54.

Putnam, H. (1967), "Time and Physical Geometry", *The Journal of Philosophy* 64: 240-247.

Rietdijk, C.W. (1985), "On Nonlocal Influences", in *Open Questions in Quantum Physics*, G. Tarozzi and A. van der Merwe(eds.). Dordrecht: Reidel.

Rucker, R.v.B. (1977), *Geometry, Relativity and the Fourth Dimension.* New York: Dover Publications.

Stein, H. (1968), "On Einstein-Minkowski Space-Time", *The Journal of Philosophy* 65: 5-23.

_____. (1991), "On Relativity Theory and Openness of the Future", *Philosophy of Science* 58: 147-167.

Taylor, E.F. and Wheeler, J.A. (1966), *Spacetime Physics.* San Francisco: W.H. Freeman.